TESLA
SELECTED WRITINGS

DISCOVERY PUBLISHER

2014, Discovery Publisher

All rights reserved.
No part of this book may be reproduced in any form or by any electronic or mechanical means including information storage and retrieval systems, without permission in writing from the publisher.

Author : Nikola Tesla
Editor in Chief : Adriano Lucca

DISCOVERY PUBLISHER

616 Corporate Way
Valley Cottage, New York, 10989
www.discoverypublisher.com
books@discoverypublisher.com
facebook.com/DiscoveryPublisher
twitter.com/DiscoveryPB

New York • Tokyo • Paris • Hong Kong

TABLE OF CONTENTS

A New System of Alternating Current Motors and Transformers — 2
— Delivered before the American Institute of Electrical Engineers, May 1888

Phenomena of Alternating Currents of Very High Frequency — 17
— *Electrical World*, Feb. 21, 1891

The Tesla Effects with High Frequency and High Potential Currents — 30
   Introduction: The scope of the Tesla lectures — 30

Experiments with alternate currents of very high frequency and their application to methods of artificial illumination — 53
— Delivered before the American Institute of Electrical Engineers, Columbia College, N.Y., May 20, 1891

Experiments with Alternate Currents of High Potential and High Frequency — 98
— Delivered before the Institution of Electrical Engineers, London, February 1892

On Light and other High-Frequency Phenomena — 183
— Delivered before the Franklin Institute, Philadelphia, February 1893, and before the National Electric Light Association, St. Louis, March 1893

   Introductory — some thoughts on the eye — 183
   On the apparatus and method of conversion — 190
   On phenomena produced by electrostatic force — 203
   On current or dynamic electricity phenomena — 212
   Impedance phenomena — 222
   On electrical resonance — 223
   On the light phenomena produced by high-frequency currents of high potential and general remarks relating to the subject — 231

On the Dissipation of the Electrical Energy of the Hertz Resonator — 253
— *The Electrical Engineer* — December 21, 1892

Tesla's Oscillator and other Inventions — 258
— By Thomas Commerford Martin. *Century Magazine*. April 1895, pp. 916-933

   An Authoritative Account of Some of his Recent Electrical Work — 258
      The generation of current — 261
      Alternating currents — 262
      The tesla oscillator — 263
      The oscillator and the production of light — 268
      Lamps with buttons or bars in place of filaments — 269
      Light and photographs with tesla phosphorescent bulbs — 271
      Light from empty bulbs in free space — 272

Effects with attuned but widely separated circuits	273
Curious impedance phenomenon	276
Lamps lighted by currents passed through the human body	277
Transmission of intelligence by attuned or resonating currents	278
Disturbance and demonstration of the earth's electrical charge	279

Earth Electricity to Kill Monopoly — 283
— *The World Sunday Magazine*. March 8, 1896. A Way to Harness Free Electric Currents Discovered

On Electricity — 286
— The Address on the Occasion of the Commemoration of the Introduction of Niagara Falls Power in Buffalo at the Ellicot Club, January 12, 1897. *Electrical Review*, January 27, 1897

High Frequency Oscillators for Electro-Therapeutic and other Purposes — 298
— *The Electrical Engineer*. Vol. XXVI. November 17, 1898. No. 550.

Plans to Dispense with Artillery of the Present Type — 314
— *The Sun*, New York, November 21, 1898

Tesla Describes his Efforts in Various Fields of Work — 315
— *Electrical Review*. New York. Nov. 30, 1898

On Current Interrupters — 319
— *Electrical Review* — March 15, 1899

The Problem of Increasing Human Energy — 322
— With special references to the harnessing of the sun's energy *Century Illustrated Magazine*, June 1900

The onward movement of man — The energy of the movement — The three ways of increasing human energy.	322
The first problem: how to increase the human mass — The burning of atmospheric nitrogen	326
The second problem: how to reduce the force retarding the human mass — The art of telautomatics	332
The third problem: how to increase the force accelerating the human mass — The harnessing of the sun's energy	341
The source of human energy — The three ways of drawing energy from the sun	342
Great possibilities offered by iron for increasing human performance — Enormous waste in iron manufacture	344
Economical production of iron by a new process	345
The coming of age of aluminium — Doom of the copper industry — The great civilizing potency of the new metal	347
Efforts toward obtaining more energy from coal — The electric transmission — The gas-engine — The cold-coal battery	350
Energy from the medium — The windmill and the solar engine — Motive power from terrestrial heat — Electricity from natural sources	353
A departure from known methods — Possibility of a "self-acting" engine or machine, inanimate, yet capable, like a living being, of deriving energy from the medium — The ideal way of obtaining motive power	356
First efforts to produce the self-acting engine — The mechanical oscillator — Work of dewar and linde — Liquid air	359

Discovery of unexpected properties of the atmosphere — Strange experiments — Transmission of electrical energy through one wire without return — Transmission through the earth without any wire   362

"Wireless" telegraphy — The Secret of tuning — Errors in the hertzian investigations — A receiver of wonderful sensitiveness   366

Development of a new principle — The electrical oscillator — Production of immense electrical movements — The earth responds to man — Interplanetary communication now probable   370

Transmission of electrical energy to any distance without wires — Now practicable — The best means of increasing the force accelerating the human mass   377

Tesla's New Discovery   380
— *The Sun*, New York, January 30, 1901. Capacity of Electrical Conductors is Variable, Not Constant, and Formulas Will Have to Be Rewritten — Capacity Varies with Absolute Height Above Sea Level, Relative Height From Earth and Distance From the Sun

Talking with the Planets   385
— *Collier's Weekly*, February 19, 1901, page 4-5

  Signalling at 100,000,000 miles!   387
  Experiments in Colorado   388
  Terrified by success   390
  Communicating with the Martians   391

Inventor Tesla's Plant Nearing Completion   393
— *Brooklyn Eagle*, February 8, 1902. Buildings at wardencliff to be used in developing his electrical discoveries

  Sinking a Very Deep Well   393

The Transmission of Electrical Energy Without Wires   394
— *Electrical World* and Engineer, March 5, 1904

Electric Autos   400
— Special Correspondence, *Manufacturers' Record*. December 29, 1904

  Nicola Tesla's View of the Future in Motive Power   400

The Transmission of Electrical Energy Without Wires as a Means for Furthering Peace   402
— *Electrical World* and Engineer, January 7, 1905, pp. 21-24

Tuned Lightning   414
— *English Mechanic and World of Science*, March 8, 1907

Tesla's Wireless Torpedo   417
— *New York Times*, March 19, 1907

  Inventor Says He did Show that it Worked Perfectly   417

Possibilities of Wireless   418
— *New York Times*, October 22, 1907, p. 8, col. 6.

  Nikola Tesla says distance forms no obstacle to transmission of energy   418

The Future of the Wireless Art   419
  Wireless telegraphy & telephony   419

Mr. Tesla's Vision   422
— *New York Times*, April 21, 1908

How the electrician's lamp of Aladdin may construct New Worlds	422
Nikola Tesla's New Wireless — *The Electrical Engineer* — London, Dec. 24, 1909	424
Dr. Tesla Talks of Gas Turbines — *Motor World*, September 18, 1911	425
Tesla's New Monarch of Machines — *New York Herald*, Oct. 15, 1911	430
The Disturbing Influence of Solar Radiation on the Wireless Transmission of Energy — *Electrical Review* and *Western Electrician*, July 6, 1912	438
How Cosmic Forces Shape Our Destinies — *New York American*, February 7, 1915	444
Man as a machine	445
Natural forces influence us	446
How wars are started	448
War and the earthquake	449
Some Personal Recollections — *Scientific American*, June 5, 1915	451
The Wonder World to Be Created by Electricity — *Manufacturer's Record*, September 9, 1915	458
Electrical possibilities in coal and iron	460
Hydro-electric development	462
The next great achievement — Electrical control of atmospheric moisture	463
Economy in light and power — Electric propulsion	463
A few of the wonders to come	464
Telegraphic photography and other advances	465
Electric inventions in war	466
The power of the future	467
Nikola Tesla Sees a Wireless Vision	469
Correction by Mr. Tesla — Of Wireless Apparatus he meant to say "Ineffective," not "Inefficient"	472
Tesla's New Device Like Bolts of Thor	473
Wonders of the Future — *Collier's Weekly*, December 2, 1916	475
Electric Drive for Battle Ships — *New York Herald*, February 25, 1917	480
Controversial correspondence	480
The helical propeller	481
Advantages of types	482
Some fatal mistakes	482
Irreversible turbines	483

A mechanical tour de force	483
Superiority of electric drive	484
Saving of power	485
Question of weight	485
New cruisers' requirements	486
Disarmament impossible	487
To use war ships in peace	488

Presentation of the Edison Medal to Nikola Tesla — 489

Tesla's Views on Electricity and the War — 522
— By H. Winfield Secor, *Electrical Experimenter*, August, 1917

Famous Scientific Illusions — 527
— *Electrical Experimenter*, February 1919

    The singular misconception of the wireless — 527

The True Wireless — 531
— *Electrical Experimenter*, May 1919

Electrical Oscillators — 550
— *Electrical Experimenter* — July 1919

Rain Can Be Controlled and Hydraulic Force Provided, Says Nikola Tesla, if We Turn to Sun for Power — 566
— by Edward Marshall. *Syracuse Herald*, ca. February 29, 1920

    Famed inventor sees wide possibilities in new scientific theory — 566
        Water power ideal — 566
        On the eve of great feats — 567

When Woman Is Boss — 568
— *Colliers*, January 30, 1926

    Woman — free and regal — 569
    The queen is the center of life — 571
    We can only sit and wonder — 571

World System of Wireless Transmission of Energy — 573
— *Telegraph and Telegraph Age*, October 16, 1927

    High frequency dynamo and tesla coil — 573
    The "Magnifying Transmitter" and earth resonance — 574
    Short wave broadcasting and "beam" transmission — 575
    The "World System" — 577

Nikola Tesla Tells of New Radio Theories: an Interview with Nikola Tesla — 581
— *New York Herald Tribune*, September 22, 1929

    Disputes Hertz waves — 581

Our Future Motive Power — 584
— *Everyday Science and Mechanics*, December 1931

    Terrestrial energy — 589

Tesla Cosmic Ray Motor May Transmit Power Round Earth — 603
— *Brooklyn Eagle*. July 10, 1932

Not much power yet	603
Exceed velocity of light	604
Hopes to build large motor	604
Cited two principles	604

Pioneer Radio Engineer Gives Views on Power — 606
— New York Herald Tribune, September 11, 1932

Suggested short waves early	606
Waves go around world	607
Need radio channels	607

The Eternal Source of Energy of the Universe, Origin and Intensity of Cosmic Rays — 610
— October 13, 1932, New York

Tesla 'Harnesses' Cosmic Energy — 614
— Philadelphia Public Ledger, November 2, 1933

Tesla Invents Peace Ray — 615
— New York Sun, July 10, 1934

Tesla on Power Development and Future Marvels — 616
— New York World Telegram, July 24, 1934

Dr. Tesla Visions the End of Aircraft in War — 619
— Every Week Magazine, Oct. 21, 1934

The New Art of Projecting Concentrated Non-Dispersive Energy Through Natural Media — 623
— Circa. May 16, 1935.

Diagram indicating distribution of charges	628
Correspondence	633

A Machine to End War — 637

Tesla Predicts Ships Powered by Shore Beam — 643

Normandy uses U.S. cruiser system	643
Cites his force beam as one way	644
Started new idea in 1897	645
Pet scheme to light ocean	645

Tesla Tries to Prevent World War Two — 646
— The Unpublished Chapter 34 of Prodigal Genius, by John J. O'Neill

Mechanical Therapy — 651
— Undated

TESLA
SELECTED WRITINGS

A NEW SYSTEM OF ALTERNATING CURRENT MOTORS AND TRANSFORMERS

—Delivered before the American Institute of Electrical Engineers, May 1888.

I desire to express my thanks to Professor Anthony for the help he has given me in this matter. I would also like to express my thanks to Mr. Pope and Mr. Martin for their aid. The notice was rather short, and I have not been able to treat the subject so extensively as I could have desired, my health not being in the best condition at present. I ask your kind indulgence, and I shall be very much gratified if the little I have done meets your approval.

In the presence of the existing diversity of opinion regarding the relative merits of the alternate and continuous current systems, great importance is attached to the question whether alternate currents can be successfully utilized in the operation of motors. The transformers, with their numerous advantages, have afforded us a relatively perfect system of distribution, and although, as in all branches of the art, many improvements are desirable, comparatively little remains to be done in this direction. The transmission of power, on the contrary, has been almost entirely confined to the use of continuous currents, and notwithstanding that many efforts have been made to utilize alternate currents for this purpose, they have, up to the present, at least as far as known, failed to give the result desired. Of the various motors adapted to be used on alternate current circuits the following have been mentioned: 1. A series motor with subdivided field. 2. An alternate current generator having its field excited by continuous currents. 3. Elihu Thomson's motor. 4. A combined alternate and continuous current motor. Two more motors of this kind have suggested themselves to me. 1. A motor with one of its circuits in series with a transformer and the other in the secondary of the transformer. 2. A motor having its armature circuit connected to the generator and the field coils closed upon themselves. These, however, I mention only incidentally.

The subject which I now have the pleasure of bringing to your notice is a novel system of electric distribution and transmission of power by means of alternate currents, affording peculiar advantages, particularly in the way of motors, which I am confident will at once establish the superior adaptability of these currents to the transmission of power and will show that many results heretofore unattainable can be reached by their use; results which are very much desired in the practical operation of such systems and which cannot be accomplished by means of continuous currents.

Before going into a detailed description of this system, I think it necessary to make a few remarks with reference to certain conditions existing in continuous current generators and motors, which, although generally known, are frequently disregarded.

In our dynamo machines, it is well known, we generate alternate currents which we direct by means of a commutator, a complicated device and, it may be justly said, the source of most of the troubles experienced in the operation of the machines. Now, the currents so directed cannot be utilized in the motor, but they must—again by means of a similar unreliable device—be reconverted into their original state of alternate currents. The function of the commutator is entirely external, and in no way dues it affect the internal working of the machines. In reality, therefore, all machines are alternate current machines, the currents appearing as continuous only in the external circuit during their transit from generator to motor. In view simply of this fact, alternate currents would commend themselves as a more direct application of electrical energy, and the employment of continuous currents would only be justified if we had dynamos which would primarily generate, and motors which would be directly actuated by such currents.

But the operation of the commutator on a motor is twofold; firstly, it reverses the currents through the motor, and secondly, it effects, automatically, a progressive shifting of the poles of one of its magnetic constituents. Assuming, therefore, that both of the useless operations in the system, that is to say, the directing of the alternate currents on the generator and reversing the direct currents on the motor, be eliminated, it would still be necessary, in order to cause a rotation of the motor, to produce a progressive shifting of the poles of one of its elements, and the question presented itself,—How to perform this operation by the direct action of alternate currents? I will now proceed to show how this result was accomplished.

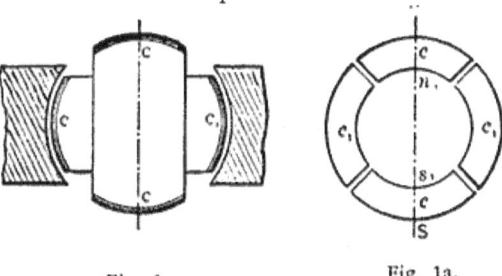

Fig. 1. Fig. 1a.

In the first experiment a drum-armature was provided with two coils at right angles to each other, and the ends of these coils were connected to two pairs of insulated contact-rings as usual. A ring was then made of thin insulated plates of sheet-iron and wound with four coils, each two opposite coils being connected together so as to produce free poles on diametrically opposite sides of the ring. The remaining free ends of the coils were then connected to the contact-rings of the generator armature so as to form two independent circuits, as indicated in figure 9. It may now be seen what results were secured in this combination, and with this view I would refer to the diagrams, figures 1 to 8a. The field of the generator being independently excited, the rotation of the armature sets up currents in the coils C C_1, varying in strength and direction in the well-known manner. In the position shown in figure 1 the current in coil C is nil while coil C_1 is traversed by its maximum current, and the connections my be such that the ring is magnetized by the coils c_1 c_1 as indicated by the letters N S in figure 1a, the magnetizing effect of the coils c c being nil, since these coils are included in the circuit of coil C.

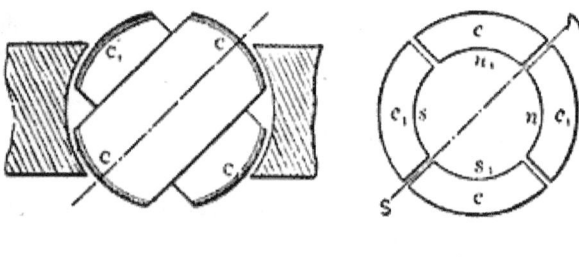

Fig. 2.   Fig. 2a.

In figure 2 the armature coils are shown in a more advanced position, one-eighth of one revolution being completed. Figure 2a illustrates the corresponding magnetic condition of the ring. At this moment the coil c_1 generates a current of the same direction as previously, but weaker, producing the poles n_1 s_1 upon the ring; the coil c also generates a current of the same direction, and the connections may be such that the coils c c produce the poles n s, as shown in figure 2a. The resulting polarity is indicated by the letters N S, and it will be observed that the poles of the ring have been shifted one-eighth of the periphery of the same.

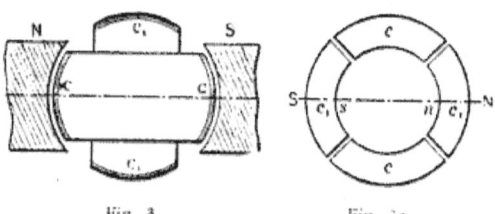

Fig. 3.   Fig. 3a.

In figure 3 the armature has completed one-quarter of one revolution. In this phase the current in coil C is maximum, and of such direction as to produce the poles N S in figure 3a, whereas the current in coil C_1 is nil, this coil being at its neutral position. The poles N S in figure 3a are thus shifted one-quarter of the circumference of the ring.

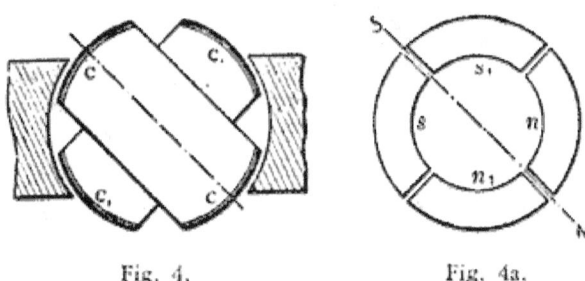

Fig. 4.   Fig. 4a.

Figure 4 shows the coils C C in a still more advanced position, the armature having completed three-eighths of one revolution. At that moment the coil C still generates a current of the same direction as before, but of less strength, producing the comparatively weaker poles n s in figure 4a, The current in the coil C_1 is of the same strength, but of opposite direction. Its effect is, therefore, to produce upon the ring the poles n_1 and s_1 as indicated, and a polarity, N S, results, the poles now being shifted three-eighths of the periphery of the ring.

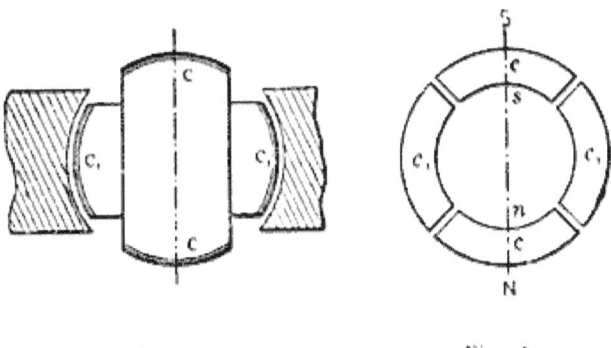

Fig. 5.   Fig. 5a.

In figure 5 one-half of one revolution of the armature is completed, and the resulting magnetic condition of the ring is indicated in figure 5a. Now, the current in coil C is nil, while the coil C_1 yields its maximum current, which is of the same direction as previously; the magnetizing effect is, therefore, due to the coils C_1 C_1 alone, and, referring to figure 5a, it will be observed that the poles N S are shifted one-half of the circumference of the ring. During the next half revolution the operations are repeated, as represented in the figures 6 to 8a.

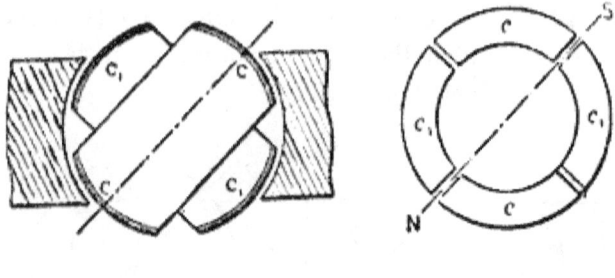

Fig. 6.   Fig. 6a.

A reference to the diagrams will make it clear that during one revolution of the armature the poles of the ring are shifted once around its periphery, and each revolution producing like effects, a rapid whirling of the poles in harmony with the rotation of the armature is the result. If the connections of either one of the circuits in the ring are reversed, the shifting of the poles is made to progress in the opposite direction, but the operation is identically the same. Instead of using four wires, with like result, three wires may be used, one forming a common return for both circuits.

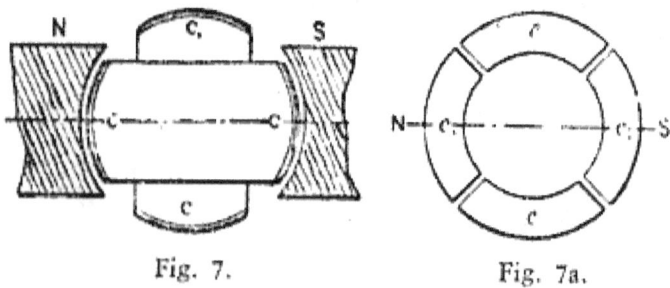

Fig. 7.   Fig. 7a.

This rotation or whirling of the poles manifests itself in a series of curious phenomena. If a delicately pivoted disc of steel or other magnetic metal is approached to the ring it is set in rapid rotation, the direction of rotation varying with the position of the disc. For instance, noting the direction outside of the ring it will be found that inside the ring it turns in an opposite direction, while it is unaffected if placed in a position symmetrical to the ring. This is easily explained. Each time that a pole approaches it induces an opposite pole in the nearest point on the disc, and an attraction is produced upon that point; owing to this, as the pole is shifted further away from the disc a tangential pull is exerted upon the same, and the action being constantly repeated, a more or less rapid rotation of the disc is the result. As the pull is exerted mainly upon that part which is nearest to the ring, the rotation outside and inside, or right and left, respectively, is in opposite directions, figure 9. When placed symmetrically to the ring, the pull on opposite sides of the disc being equal, no rotation results. The action is based on the magnetic inertia of the iron; for this reason a disc of hard steel is much more affected than a disc of soft iron, the latter being capable of very rapid variations of magnetism. Such a disc has proved to be a very useful instrument in all these investigations, as it has enabled me to detect any irregularity in the action. A curious effect is also produced upon iron filings. By placing some upon a paper and holding them externally quite close to the ring they are set in a vibrating motion, remaining in the same place, although the paper may be moved back and forth; but in lifting the paper to a certain height which seems to be dependent on the intensity of the poles and the speed of rotation, they are thrown away in a direction always opposite to the supposed movement of the poles. If a paper with filings is put flat upon the ring and the current turned on suddenly; the existence of a magnetic whirl may be easily observed.

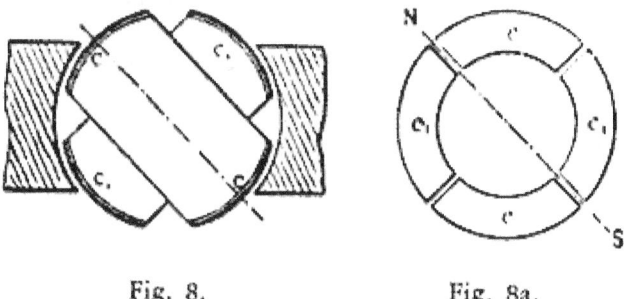

Fig. 8.   Fig. 8a.

To demonstrate the complete analogy between the ring and a revolving magnet, a strongly energized electro-magnet was rotated by mechanical power, and phenomena identical in every particular to those mentioned above were observed.

Obviously, the rotation of the poles produces corresponding inductive effects and may be utilized to generate currents in a closed conductor placed within the influence of the poles. For this purpose it is convenient to wind a ring with two sets of superimposed coils forming respectively the primary and secondary circuits, as shown in figure 10. In order to secure the most economical results the magnetic circuit should be completely closed, and with this object in view the construction may be modified at will.

The inductive effect exerted upon the secondary coils will be mainly due to the shifting or movement of the magnetic action; but there may also be currents set up in the circuits in consequence of the variations in the intensity of the poles. However, by properly designing the generator and determining the magnetizing effect of the primary coils the latter element may be made to disappear. The intensity of the poles being maintained constant, the action of the apparatus will be perfect, and the same result will be secured as though the shifting were effected by means of a commutator with an infinite number of bars. In such case the theoretical relation between the energizing effect of each set of primary coils and their resultant magnetizing effect may be expressed by the equation of a circle having its center coinciding with that of an orthogonal system of axes, and in which the radius represents the resultant and the co-ordinates both of its components. These are then respectively the sine and cosine of the angle U between the radius and one of the axes (O X). Referring to figure 11, we have $r_2 = x_2 + y_2$; where $x = r \cos a$, and $y = r \sin a$.

Assuming the magnetizing effect of each set of coils in the transformer to be proportional to the current—which may be admitted for weak degrees of magnetization—then $x = Kc$ and $y = Kc_1$, where K is a constant and c and cl the current in both sets of coils respectively. Supposing, further, the field of the generator to be uniform, we have for constant speed $c_1 = K_1 \sin a$ and $c = K_1 \sin(90o + a) = K_1 \cos a$, where K_1 is a constant. See figure 12.

Therefore,
```
x = Kc  = K K₁ cos a;
y = Kc₁ = K K₁ sin a; and
K K₁ = r.
```

That is, for a uniform field the disposition of the two coils at right angles will secure the theoretical result, and the intensity of the shifting poles will be constant. But from $r_2 = x_2 + y_2$ it follows that for $y = O, r = x$; it follows that the joint magnetizing effect of both sets of coils should be equal to the effect of one set when at its maximum action. In transformers and in a certain class of motors the fluctuation of the poles is not of great importance, but in another class of these motors it is desirable to obtain the theoretical result.

In applying this principle to the construction of motors, two typical forms of motor have been developed. First, a form having a comparatively small rotary effort at the start, but maintaining a perfectly uniform speed at all loads, which motor has been termed synchronous. Second, a form possessing a great rotary effort at the start, the speed being dependent on the load.

These motors may be operated in three different ways: 1. By the alternate currents of the source only. 2. By a combined action of these and of induced currents. 3. By the joint action of alternate and continuous currents.

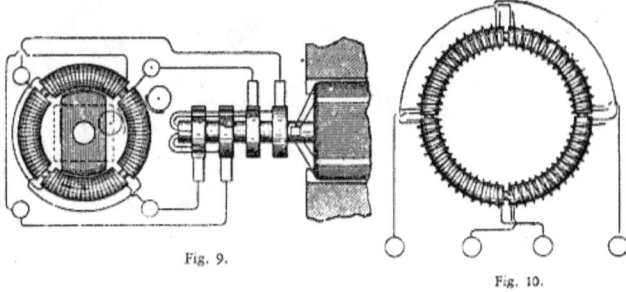

Fig. 9.

Fig. 10.

The simplest form of a synchronous motor is obtained by winding a laminated ring provided with pole projections with four coils, and connecting the same in the manner before indicated. An iron disc having a segment cut away on each side may be used as an armature. Such a motor is shown in figure 9. The disc being arranged to rotate freely within the ring in close proximity to the projections, it is evident that as the poles are shifted it will, owing to its tendency to place itself in such a position as to embrace the greatest number of the lines of force, closely follow the movement of the poles, and its motion will be synchronous with that of the armature of the generator; that is, in the peculiar disposition shown in figure 9, in which the armature produces by one revolution two current impulses in each of the circuits. It is evident that if, by one revolution of the armature, a greater number of impulses is produced, the speed of the motor will be correspondingly increased. Considering that the attraction exerted upon the disc is greatest when the same is in close proximity to the poles, it follows that such a motor will maintain exactly the same speed at all loads within the limits of its capacity.

To facilitate the starting, the disc may be provided with a coil closed upon itself. The advantage secured by such a coil is evident. On the start the currents set up in the coil strongly energize the disc and increase the attraction exerted upon the same by the ring, and currents being generated in the coil as long as the speed of the armature is inferior to that of the poles, considerable work may be performed by such a motor even if the speed be below normal. The intensity of the poles being constant, no currents will be generated in the coil when the motor is turning at its normal speed.

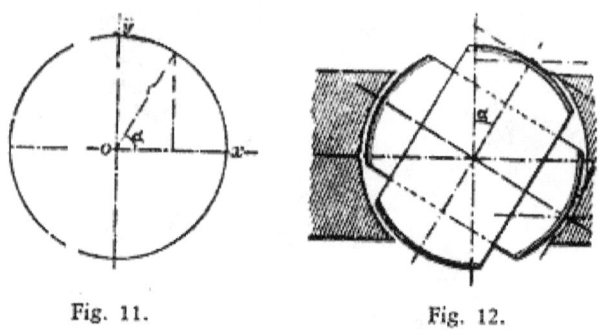

Fig. 11.   Fig. 12.

Instead of closing the coil upon itself, its ends may be connected to two insulated sliding rings, and a continuous current supplied to these from a suitable generator. The proper way to start such a motor is to close the coil upon itself until the normal speed is reached, or nearly so, and then turn on the continuous current. If the disc be very strongly energized by a continuous current the motor may not be able to start, but if it be weakly energized, or generally so that the magnetizing effect of the ring is preponderating it will start and reach the normal speed. Such a motor will maintain absolutely the same speed at all loads. It has also been found that if the motive power of the generator is not excessive, by checking the motor the speed of the generator is diminished in synchronism with that of the motor. It is characteristic of this form of motor that it cannot be reversed by reversing the continuous current through the coil.

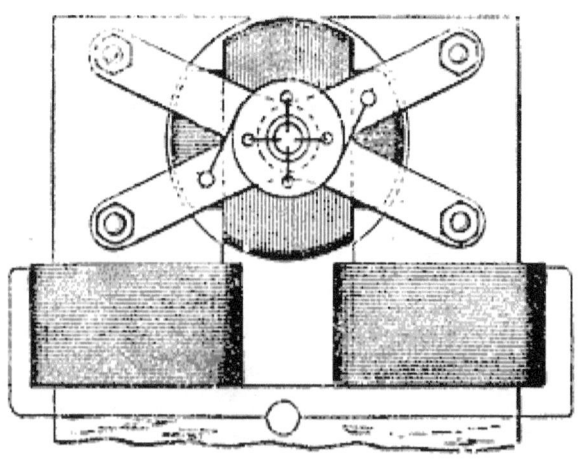

Fig. 13.

The synchronism of these motors may be demonstrated experimentally in a variety of ways. For this purpose it is best to employ a motor consisting of a stationary field magnet and an armature arranged to rotate within the same, as indicated in figure 13. In this case the shifting of the poles of the armature produces a rotation of the latter in the opposite direction. It results therefrom that when the normal speed is reached, the poles of the armature assume fixed positions relatively to the field magnet and the same is magnetized by induction, exhibiting a distinct pole on each of the pole-pieces. If a piece of soft iron is approached to the field magnet it will at the start be attracted with a rapid vibrating motion produced by the reversals of polarity of the magnet, but as the speed of the armature increases; the vibrations become less and less frequent and finally entirely cease. Then the iron is weakly but permanently attracted, showing that the synchronism is reached and the field magnet energized by induction.

The disc may also be used for the experiment. If held quite close to the armature it will turn as long as the speed of rotation of the poles exceeds that of the armature; but when the normal speed is reached, or very nearly so; it ceases to rotate and is permanently attracted.

A crude but illustrative experiment is made with an incandescent lamp. Placing the lamp in circuit with the continuous current generator, and in series with the magnet coil, rapid fluctuations are observed in the light in consequence of the induced current set up in the coil at the start; the speed increasing, the fluctuations occur at longer intervals, until they entirely disappear, showing that the motor has attained its normal speed. A telephone receiver affords a most sensitive instrument; when connected to any circuit in the motor the synchronism may be easily detected on the disappearance of the induced currents.

In motors of the synchronous type it is desirable to maintain the quantity of the shifting magnetism constant, especially if the magnets are not properly subdivided.

To obtain a rotary effort in these motors was the subject of long thought. In order to secure this result it was necessary to make such a disposition that while the poles of one element of the motor are shifted by the alternate currents of the source, the poles produced upon the other element should always be maintained in the proper relation to the former, irrespective of the speed of the motor. Such a condition exists in a continuous current motor; but in a synchronous motor, such as described, this condition is fulfilled only when the speed is normal.

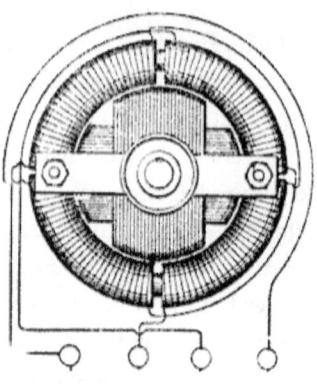

Fig. 14.

The object has been attained by placing within the ring a properly subdivided cylindrical iron core wound with several independent coils closed upon themselves. Two coils at right angles as in figure 14, are sufficient, but greater number may he advantageously employed. It results from this disposition that when the poles of the ring are shifted, currents are generated in the closed armature coils. These currents are the most intense at or near the points of the greatest density of the lines of force, and their effect is to produce poles upon the armature at right angles to those of the ring, at least theoretically so; and since action is entirely independent of the speed—that is, as far as the location of the poles is concerned—a continuous pull is exerted upon the periphery of the armature. In many respects these motors are similar to the continuous current motors. If load is put on, the speed, and also the resistance of the motor, is diminished and more current is made to pass through the energizing coils, thus increasing the effort. Upon the load being taken off, the counter-electromotive force increases and less current passes through the primary or energizing coils. Without any load the speed is very nearly equal to that of the shifting poles of the field magnet.

It will be found that the rotary effort in these motors fully equals that of the continuous current motors. The effort seems to be greatest when both armature and field magnet are without any projections; but as in such dispositions the field cannot be very concentrated, probably the best results will be obtained by leaving pole projections on one of the elements only. Generally, it may be stated that the projections diminish the torque and produce a tendency to synchronism.

A characteristic feature of motors of this kind is their capacity of being very rapidly reversed. This follows from the peculiar action of the motor. Suppose the armature to be rotating and the direction of rotation of the poles to be reversed. The apparatus then represents a dynamo machine, the power to drive this machine being the momentum stored up in the armature and its speed being the sum of the speeds of the armature and the poles.

If we now consider that the power to drive such a dynamo would be very nearly proportional to the third power of the speed, for this reason alone the armature should be quickly reversed. But simultaneously with the reversal another element is brought into action, namely, as the movement of the poles with respect to the armature is reversed, the motor acts like a transformer in which the resistance of the secondary circuit would be abnormally diminished by producing in this circuit an additional electromotive force. Owing to these causes the reversal is instantaneous.

If it is desirable to secure a constant speed, and at the same time a certain effort at the start, this result may be easily attained in a variety of ways. For instance, two armatures, one for torque and the other for synchronism, may be fastened on the same shaft, and any desired preponderance may be given to either one, or an armature may be wound for rotary effort, but a more or less pronounced tendency to synchronism may be given to it by properly constructing the iron core; and in many other ways.

As a means of obtaining the required phase of the currents in both the circuits, the disposition of the two coils at right angles is the simplest, securing the most uniform action; but the phase may be obtained in many other ways, varying with the machine employed. Any of the dynamos at present in use may be easily adapted for this purpose by making connections to proper points of the generating coils. In closed circuit armatures, such as used in the continuous current systems, it is best to make four derivations from equidistant points or bars of the commutator, and to connect the same to four insulated sliding rings on the shaft. In this case each of the motor circuits is connected to two diametrically opposite bars of the commutator. In such a disposition the motor may also be operated at half the potential and on the three-wire plan, by connecting the motor circuits in the proper order to three of the contact rings.

In multipolar dynamo machines, such as used in the converter systems, the phase is conveniently obtained by winding upon the armature two series of coils in such a manner that while the coils of one set or series are at their maximum production of current, the coils of the other will be at their neutral position, or nearly so, whereby both sets of coils may be subjected simultaneously or successively to the inducing action of the field magnets.

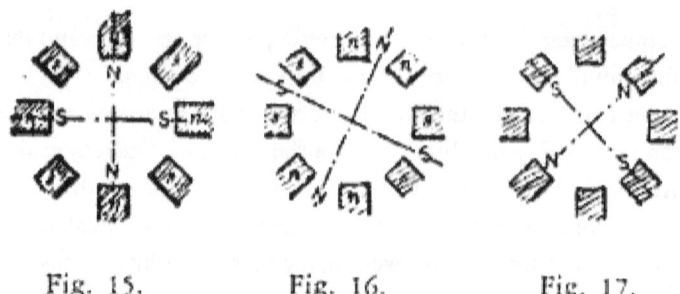

Fig. 15.   Fig. 16.   Fig. 17.

Generally the circuits in the motor will be similarly disposed, and various arrangements may be made to fulfill the requirements; but the simplest and most practicable is to arrange primary circuits on stationary parts of the motor, thereby obviating, at least in certain forms, the employment of sliding contacts. In such a case the magnet coils are connected alternately in both the circuits; that is 1, 3, 5... in one, and 2, 4, 6... in the other, and the coils of each set of series may be connected all in the same manner, or alternately in opposition; in the latter case a motor with half the number of poles will result, and its action will be correspondingly modified. The figures 15, 16 and 17, show three different phases, the magnet coils in each circuit being connected alternately in opposition. In this case there will be always four poles, as in figures 15 and 17, four pole projections will be neutral, and in figure 16 two adjacent pole projections will have the same polarity. If the coils are connected in the same manner there will be eight alternating poles, as indicated by the letters ns' in fig.15.

The employment of multipolar motors secures in this system an advantage much desired and unattainable in the continuous current system, and that is, that a motor may be made to run exactly at a predetermined speed irrespective of imperfections in construction, of the load, and, within certain limits, of electromotive force and current strength.

In a general distribution system of this kind the following plan should be adopted. At the central station of supply a generator should be provided having a considerable number of poles. The motors operated from this generator should be of the synchronous type, but possessing sufficient rotary effort to insure their starting. With the observance of proper rules of construction it may be admitted that the speed of each motor will be in some inverse proportion to its size, and the number of poles should be chosen accordingly. Still exceptional demands may modify this rule. In view of this, it will be advantageous to provide each motor with a greater number of pole projections or coils, the number being preferably a multiple of two and three. By this means, by simply changing the connections of the coils, the motor may be adapted to any probable demands.

If the number of the poles in the motor is even, the action will be harmonious and the proper result will be obtained; if this is not the case the best plan to be followed is to make a motor with a double number of poles and connect the same in the manner before indicated, so that half the number of poles result. Suppose, for instance, that the generator has twelve poles, and it would be desired to obtain a speed equal to 12/7 of the speed of the generator. This would require a motor with seven pole projections or magnets, and such a motor could not be properly connected in the circuits unless fourteen armature coils would be provided, which would necessitate the employment of sliding contacts. To avoid this the motor should be provided with fourteen magnets and seven connected in each circuit, the magnets in each circuit alternating among themselves. The armature should have fourteen closed coils. The action of the motor will not be quite as perfect as in the case of an even number of poles, but the drawback will not be of a serious nature.

However, the disadvantages resulting from this unsymmetrical form will be reduced in the same proportion as the number of the poles is augmented.

If the generator has, say, n, and the motor n_1 poles, the speed of the motor will be equal to that of the generator multiplied by n/n_1.

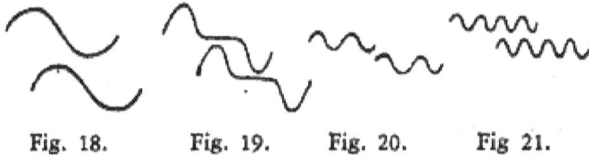

Fig. 18.   Fig. 19.   Fig. 20.   Fig 21.

The speed of the motor will generally be dependent on the number of the poles, but there may be exceptions to this rule. The speed may be modified by the phase of the currents in the circuits or by the character of the current impulses or by intervals between each or between groups of impulses. Some of the possible cases are indicated in the diagrams, figures 18, 19, 20 and 21, which are self-explanatory. Figure 18 represents the condition generally existing, and which secures the best result. In such a case, if the typical form of motor illustrated in figure 9 is employed, one complete wave in each circuit will produce one revolution of the motor. In figure 19 the same result will be effected by one wave in each circuit, the impulses being successive; in figure 20 by four, and in figure 21 by eight waves.

By such means any desired speed may be attained; that is, at least within the limits of practical demands. This system possesses this advantage besides others, resulting from simplicity. At full loads the motors show efficiency fully equal to that of the continuous current motors. The transformers present an additional advantage in their capability of operating motors. They are capable of similar modifications in construction, and will facilitate the introduction of motors and their adaptation to practical demands. Their efficiency should be higher than that of the present transformers, and I base my assertion on the following:

In a transformer as constructed at present we produce the currents in the secondary circuit by varying the strength of the primary or exciting currents. If we admit proportionality with respect to the iron core the inductive effect exerted upon the secondary coil will be proportional to the numerical sum of the variations in the strength of the exciting current per unit of time; whence it follows that for a given variation any prolongation of the primary current will result in a proportional loss. In order to obtain rapid variations in the strength of the current, essential to efficient induction, a great number of undulations are employed. From this practice various disadvantages result. These are, increased cost and diminished efficiency of the generator, more waste of energy in heating the cores, and also diminished output of the transformer, since the core is not properly utilized, the reversals being too rapid. The inductive effect is also very small in certain phases, as will be apparent from a graphic representation, and there may be periods of inaction, if there are intervals between the succeeding current impulses or waves. In producing a shifting of the poles in the transformer, and thereby inducing currents, the induction is of the ideal character, being always maintained at its maximum action. It is also reasonable to assume that by a shifting of the poles less energy will be wasted than by reversals.

May 1888

PHENOMENA OF ALTERNATING CURRENTS OF VERY HIGH FREQUENCY

—*Electrical World, Feb. 21, 1891*

Electrical journals are getting to be more and more interesting. New facts are observed and new problems spring up daily which command the attention of engineers. In the last few numbers of the English journals, principally in the Electrician there have been several new matters brought up which have attracted more than usual attention. The address of Professor Crookes has revived the interest in his beautiful and skillfully performed experiments, the effect observed on the Ferranti mains has elicited the expressions of opinion of some of the leading English electricians, and Mr. Swinburne has brought out some interesting points in connection with condensers and dynamo excitation.

The writer's own experiences have induced him to venture a few remarks in regard to these and other matters, hoping that they will afford some useful information or suggestion to the reader.

Among his many experiments Professor Crookes shows some performed with tubes devoid of internal electrodes, and from his remarks it must be inferred that the results obtained with these tubes are rather unusual. If this be so, then the writer must regret that Professor Crookes, whose admirable work has been the delight of every investigator, should not have availed himself in his experiments of a properly constructed alternate current machine—namely, one capable of giving, say 10,000 to 20,000 alternations per second. His researches on this difficult but fascinating subject would then have been even more complete. It is true that when using such a machine in connection with an induction coil the distinctive character of the electrodes—which is desirable, if not essential, in many experiments—is lost, in most cases both the electrodes behaving alike; but on the other hand, the advantage is gained that the effects may be exalted at will. When using a rotating switch or commutator the rate of change obtainable in the primary current is limited. When the commutator is more rapidly revolved the primary current diminishes, and if the current be increased, the sparking, which cannot be completely overcome by the condenser, impairs considerably the virtue of the apparatus. No such limitations exist when using an alternate current machine as any desired rate of change may be produced in the primary current. It is thus; possible to obtain excessively high electromotive forces in the secondary circuit with a comparatively small primary current; moreover, the perfect regularity in the working of the apparatus may be relied upon.

The writer will incidentally mention that any one who attempts for the first time to construct such a machine will have a tale of woe to tell. He will first start out, as a matter of course, by making an armature with the required number of polar projections. He will then get the satisfaction of having produced an apparatus which is fit to accompany a thoroughly Wagnerian opera. It may besides possess the virtue of converting mechanical energy into heat in a nearly perfect manner. If there is a reversal in the polarity of the projections, he will get heat out of the machine; if there is no reversal, the heating will be less, but the output will be next to nothing. He will then abandon the iron in the armature, and he will get from the Scylla to the Charybdis. He will look for one difficulty and will find another, but, after a few trials, he may get nearly what he wanted.

Among the many experiments which may be performed with such a machine, of not the least interest are those performed with a high-tension induction coil. The character of the discharge is completely changed. The arc is established at much greater distances, and it is so easily affected by the slightest current of air that it often wriggles around in the most singular manner. It usually emits the rhythmical sound peculiar to the alternate current arcs, but the curious point is that the sound may be heard with a number of alternations far above ten thousand per second, which by many is considered to be, about the limit of audition. In many respects the coil behaves like a static machine. Points impair considerably the sparking interval, electricity escaping from them freely, and from a wire attached to one of the terminals streams of light issue, as though it were connected to a pole of a powerful Toepler machine. All these phenomena are, of course, mostly due to the enormous differences of potential obtained. As a consequence of the self-induction of the coil and the high frequency, the current is minute while there is a corresponding rise of pressure. A current impulse of some strength started in such a coil should persist to flow no less than four ten-thousandths of a second. As this time is greater than half the period, it occurs that an opposing electromotive force begins to act while the current is still flowing. As a consequence, the pressure rises as in a tube filled with liquid and vibrated rapidly around its axis. The current is so small that, in the opinion and involuntary experience of the writer, the discharge of even a very large coil cannot produce seriously injurious effects, whereas, if the same coil were operated with a current of lower frequency, though the electromotive force would be much smaller, the discharge would be most certainly injurious. This result, however, is due in part to the high frequency. The writer's experiences tend to show that the higher the frequency the greater the amount of electrical energy which may be passed through the body without serious discomfort; whence it seems certain that human tissues act as condensers.

One is not quite prepared for the behavior of the coil when connected to a Leyden jar. One, of course, anticipates that since the frequency is high the capacity of the jar should be small. He therefore takes a very small jar, about the size of a small wine glass, but he finds that even with this jar the coil is practically short-circuited. He then reduces the capacity until he comes to about the capacity of two spheres, say, ten centimetres in diameter and two to four centimetres apart. The discharge then assumes; the form of a serrated band exactly like a succession of sparks viewed in a rapidly revolving mirror; the serrations, of course, corresponding to the condenser discharges. In this case one may observe a queer phenomenon. The discharge starts at the nearest points, works gradually up, breaks somewhere near the top of the spheres, begins again at the bottom; and so on. This goes on so fast that several serrated bands are seen at once. One may be puzzled for a few minutes, but the explanation is simple enough. The discharge begins at the nearest points; the air is heated and carries the arc upward until it breaks, when it is re-established at the nearest points, etc. Since the current passes easily through a condenser of even small capacity, it will be found quite natural that connecting only one terminal to a body of the same size, no matter how well insulated, impairs considerably the striking distance of the arc.

Experiments with Geissler tubes are of special interest. An exhausted tube, devoid of electrodes of any kind, will light up at some distance from the coil. If a tube from a vacuum pump is near the coil the whole of the pump is brilliantly lighted. An incandescent lamp approached to the coil lights up and gets perceptibly hot. If a lamp have the terminals connected to one of the binding posts of the coil and the hand is approached to the bulb, a very curious and rather unpleasant discharge from the glass to the hand takes place, and the filament may become incandescent. The discharge resembles to some extent the stream issuing from the plates of a powerful Toepler machine, but is of incomparably greater quantity. The lamp in this case acts as a condenser, the rarefied gas being one coating, the operator's hand the other. By taking the globe of a lamp in the hand, and by bringing the metallic terminals near td or in contact with a conductor connected to the coil, the carbon is brought to bright incandescence and the glass is rapidly heated. With a 100-volt 10 c.p. lamp one may without great discomfort stand as much current as will bring the lamp to a considerable brilliancy; but it can be held in the hand only for a few minutes, as the glass is heated in an incredibly short time. When a tube is lighted by bringing it near to the coil it may be made to go out by interposing a metal plate on the hand between the coil and tube; but if the metal plate be fastened to a glass rod or otherwise insulated, the tube may remain lighted if the plate be interposed, or may even increase in luminosity. The effect depends on the position of the plate and tube relatively to the coil, and may be always easily foretold by assuming that conduction takes place from one terminal of the coil to the other. According to the position of the plate, it may either divert from or direct the current to the tube.

In another line of work the writer has in frequent experiments maintained incandescent lamps of 50 or 100 volts burning at any desired candle power with both the terminals of each lamp connected to a stout copper wire of no more than a few feet in length. These experiments seem interesting enough, but they are not more so than the queer experiment of Faraday, which has been revived and made much of by recent investigators, and in which a discharge is made to jump between two points of a bent copper wire. An experiment may be cited here which may seem equally interesting.

If a Geissler tube, the terminals of which are joined by a copper wire, be approached to the coil, certainly no one would be prepared to see the tube light up. Curiously enough, it does light up, and, what is more, the wire does not seem to make much difference. Now one is apt to think in the first moment that the impedance of the wire might have something to do with the phenomenon. But this is of course immediately rejected, as for this an enormous frequency would be required. This result, however, seems puzzling only at first; for upon reflection it is quite clear that the wire can make but little difference. It may be explained in more than one way, but it agrees perhaps best with observation to assume that conduction takes place from the terminals of the coil through the space. On this assumption, if the tube with the wire be held in any position, the wire can divert little more than the current which passes through the space occupied by the wire and the metallic terminals of the tube; through the adjacent space the current passes practically undisturbed. For this reason, if the tube be held in any position at right angles to the line joining the binding posts of the coil, the wire makes hardly any difference, but in a position more or less parallel with that line it impairs to a certain extent the brilliancy of the tube and its facility to light up. Numerous other phenomena may be explained on the same assumption. For instance, if the ends of the tube be provided with washers of sufficient size and held in the line joining the terminals of the coil, it will not light up, and then nearly the whole of the current, which would otherwise pass uniformly through the space between the washers, is diverted through the wire. But if the tube be inclined sufficiently to that line, it will light up in spite of the washers. Also, if a metal plate be fastened upon a glass rod and held at right angles to the line joining the binding posts, and nearer to one of them, a tube held more or less parallel with the line will light up instantly when one of the terminals touches the plate, and will go out when separated from the plate. The greater the surface of the plate, up to a certain limit, the easier the tube will light up. When a tube is placed at right angles to the straight line joining the binding posts, and then rotated, its luminosity steadily increases until it is parallel with that line. The writer must state, however, that he does not favor the idea of a leakage or current through the space any more than as a suitable explanation, for he is convinced that all these experiments could not be performed with a static machine yielding a constant difference of potential, and that condenser action is largely concerned in these phenomena.

It is well to take certain precautions when operating a Ruhmkorff coil with very rapidly alternating currents. The primary current should not be turned on too long, else the core may get so hot as to melt the *guta-percha* or paraffin, or otherwise injure the insulation, and this may occur in a surprisingly short time, considering the current's strength. The primary current being turned on, the fine wire terminals may be joined without great risk, the impedance being so great that it is difficult to force enough current through the fine wire so as to injure it, and in fact the coil may be on the whole much safer when the terminals of the fine wire are connected than when they are insulated; but special care should be taken when the terminals are connected to the coatings of a Leyden jar, for with anywhere near the critical capacity, which just counteracts the self-induction at the existing frequency, the coil might meet the fate of St. Polycarpus. If an expensive vacuum pump is lighted up by being near to the coil or touched with a wire connected to one of the terminals, the current should be left on no more than a few moments, else the glass will be cracked by the heating of the rarefied gas in one of the narrow passages — in the writer's own experience *quod erat demonstrandum*.

There are a good many other points of interest which may be observed in connection with such a machine. Experiments with the telephone, a conductor in a strong field or with a condenser or arc, seem to afford certain proof that sounds far above the usual accepted limit of hearing would be perceived. A telephone will emit notes of twelve to thirteen thousand vibrations per second; then the inability of the core to follow such rapid alternations begins to tell. If, however, the magnet and core be replaced by a condenser and the terminals connected to the high-tension secondary of a transformer, higher notes may still be heard. If the current be sent around a finely laminated core and a small piece of thin sheet iron be held gently against the core, a sound may be still heard with thirteen to fourteen thousand alternations per second, provided the current is sufficiently strong. A small coil, however, tightly packed between the poles of a powerful magnet, will emit a sound with the above number of alternations, and arcs may be audible with a still higher frequency. The limit of audition is variously estimated. In Sir William Thomson's writings it is stated somewhere that ten thousand per second, or nearly so, is the limit. Other, but less reliable, sources give it as high as twenty-four thousand per second. The above experiments have convinced the writer that notes of an incomparably higher number of vibrations per second would be perceived provided they could be produced with sufficient power. There is no reason why it should not be so. The condensations and rarefactions of the air would necessarily set the diaphragm in a corresponding vibration and some sensation would be produced, whatever — within certain limits — the velocity, of transmission to their nerve centres, though it is probable that for want of exercise the ear would not be able to distinguish any such high note. With the eye it is different; if the sense of vision is based upon some resonance effect, as many believe, no amount of increase in the intensity of the ethereal vibration could extend our range of vision on either side of the visible spectrum.

The limit of audition of an arc depends on its size. The greater the surface by a given heating effect in the arc, the higher the limit of audition. The highest notes are emitted by the high-tension discharges of an induction coil in which the arc is, so to speak, all surface. If R be the resistance of an arc, and C the current, and the linear dimensions be n times increased, then the resistance is R/n, and with the same current density the current would be $n_2 C$; hence the heating effect is n_3 times greater, while the surface is only n_2 times as, great. For this reason very large arcs would not emit any rhythmical sound even with a very low frequency. It must be observed, however, that the sound emitted depends to some extent also on the composition of the carbon. If the carbon contain highly refractory material, this, when heated, tends to maintain the 'temperature' of the arc uniform and the sound is lessened; for this reason it would seem that an alternating arc requires such carbons:

With currents of such high frequencies it is possible to obtain noiseless arcs, but the regulation of the lamp is rendered extremely difficult on account of the excessively small attractions or repulsions between conductors conveying these currents:

An interesting feature of the arc produced by these rapidly alternating currents is its persistency. There are two causes for it, one of which is always present, the other sometimes only. One is due to the character of the current and the other to a property of the machine. The first cause is the more important one, and is due directly to the rapidity of the alternations. When an arc is formed by a periodically undulating current, there, is, a corresponding undulation in the temperature of the gaseous column, and, therefore, a corresponding undulation in the resistance of the arc. But the resistance of the arc varies enormously with the temperature of the gaseous column, being, practically infinite when the gas between the electrodes is cold. The persistence of the arc, therefore, depends on the inability of the column to cool. It is for this reason impossible to maintain an arc with the current alternating only a few times a second. On the other hand, with a practically continuous current, the arc is easily maintained, the column being constantly, kept at a high temperature and low resistance. The higher the frequency the smaller the time interval during which the arc may cool' and increase considerably in resistance. With a frequency of 10,000 per second or more in any arc of equal, size excessively small variations of temperature are superimposed upon a steady temperature, like ripples on the surface of a deep sea. The heating effect is practically continuous and the arc behaves like one produced, by a continuous current, with the exception, however, that it may not be quite as easily started, and that the electrodes are equally consumed; though the writer has observed 'some irregularities in this respect. The second cause alluded to, which possibly may not be present, is due to the tendency of a, machine of such high frequency td maintain a practically constant current. When the arc is lengthened, the electromotive force rises in proportion and the arc appears to be more persistent.

Such a machine is eminently adapted to maintain a constant current, but it is very unfit for a constant potential. As a matter of fact, in certain types of such machines a nearly constant current is an almost unavoidable result. As the number of poles or polar projections is greatly increased, the clearance becomes of great importance. One has really to do with' a great number of very small machines. Then there is the impedance in the armature, enormously augmented by the high frequency. Then, again, the magnetic leakage is facilitated. If there are' three or four hundred alternate poles, the leakage is so great that it is virtually the same as connecting, in a two-pole machine, the poles by a piece of iron. This disadvantage,, it is true, may be obviated more or less by using a field throughout of the same polarity, but then one encounters difficulties, of a different nature: All these things tend to maintain a constant' current in the armature circuit.

In this connection it is interesting to notice that even today engineers are astonished at the performance of a constant current machine, just as, some years ago, they used to consider it an extraordinary performance if a machine was capable of maintaining a constant, potential difference between the terminals. Yet one result is just as easily secured as the other. It must only be remembered that in an inductive apparatus of any kind, if constant potential is required, the inductive relation between the primary or exciting and secondary or armature circuit must be the closest possible; whereas, in an apparatus for constant current just the opposite is required. Furthermore, the opposition to the current's flow in the induced circuit must be as small as possible in the former and as great as possible in the latter case. But opposition to a current's flow may be caused in more than one way. It may be caused by ohmic resistance of self-induction. One may make the induced circuit of a dynamo machine or transformer of such high resistance that when operating devices of considerably smaller resistance within very wide limits a nearly constant current is maintained. But such high resistance involves a great loss in power, hence it is not practicable. Not so self-induction. Self-induction does not necessarily mean loss of power. The moral is, use self-induction instead of resistance. There is, however, a circumstance which favors the adoption of this plan, and this is, that a very high self-induction may be obtained cheaply by surrounding a comparatively small length of wire more or less completely with iron, and, furthermore, the effect may be exalted at will by causing a rapid undulation of the current. To sum up, the requirements for constant current are: Weak magnetic connection between the induced and inducing circuits, greatest possible self-induction with the least resistance, greatest practicable rate of change of the current. Constant potential, on the other hand, requires: Closest magnetic connection between the circuits, steady induced current, and, if possible, no reaction. If the latter conditions could be fully satisfied in a constant potential machine, its output would surpass many times that of a machine primarily designed to give constant current. Unfortunately, the type of machine in which these conditions may be satisfied is of little practical value, owing to the small electromotive force obtainable and the difficulties in taking off the current.

With their keen inventor's instinct, the now successful arc-light men have early recognized the desiderata of a constant current machine. Their arc light machines have weak fields, large armatures, with a great length of copper wire and few commutator segments to produce great variations in the current's strength and to bring self-induction into play. Such machines may maintain within considerable limits of variation in the resistance of the circuit a practically constant current. Their output is of course correspondingly diminished, and, perhaps with the object in view not to cut down the output too much, a simple device compensating exceptional variations is employed. The undulation of the current is almost essential to the commercial success of an arc-light system. It introduces in the circuit a steadying element taking the place of a large ohmic resistance, without involving a great loss in power, and, what is more important, it allows the use of simple clutch lamps, which with a current of a certain number of impulses per second, best suitable for each particular lamp, will, if properly attended to, regulate even better than the finest clock-work lamps. This discovery has been made by the writer—several years too late.

It has been asserted by competent English electricians that in a constant-current machine or transformer the regulation is effected by varying the phase of the secondary current. That this view is erroneous may be easily proved by using, instead of lamps, devices each possessing self-induction and capacity or self-induction and resistance—that is, retarding and accelerating components—in such proportions as to not affect materially the phase of the secondary current. Any number of such devices may be inserted or cut out, still it will be found that the regulation occurs, a constant current being maintained, while the electromotive force is varied with the number of the devices. The change of phase of the secondary current is simply a result following from the changes in resistance, and, though secondary reaction is always of more or less importance, yet the real cause of the regulation lies in the existence of the conditions above enumerated. It should be stated, however, that in the case of a machine the above remarks are to be restricted to the cases in which the machine is independently excited. If the excitation be effected by commutating the armature current, then the fixed position of the brushes makes any shifting of the neutral line of the utmost importance, and it may not be thought immodest of the writer to mention that, as far as records go, he seems to have been the first who has successfully regulated machines by providing a bridge connection between a point of the external circuit and the commutator by means of a third brush. The armature and field being properly proportioned, and the brushes placed in their determined positions, a constant current or constant potential resulted from the shifting of the diameter of commutation by the varying loads.

In connection with machines of such high frequencies, the condenser affords an especially interesting study. It is easy to raise the electromotive force of such a machine to four or five times the value by simply connecting the condenser to the circuit, and the writer has continually used the condenser for the purposes of regulation, as suggested by Blakesley in his book on alternate currents, in which he has treated the most frequently occurring condenser problems with exquisite simplicity and clearness. The high frequency allows the use of small capacities and renders investigation easy. But; although in most of the experiments the result may be foretold, some phenomena observed seem at first curious. One experiment performed three or four months ago with such a machine and a condenser may serve as an illustration. A machine was used giving about 20,000 alternations per second. Two bare wires about twenty feet long and two millimetres in diameter, in close proximity to each other, were connected to the terminals of the machine at the one end, and to a condenser at the other. A small transformer without an iron core, of course, was used to bring the reading within range of a Cardew voltmeter by connecting the voltmeter to the secondary. On the terminals of the condenser the electromotive force was about 120 volts, and from there inch by inch it gradually fell until at the terminals of the machine it was about 65 volts. It was virtually as though the condenser were a generator, and the line and armature circuit simply a resistance connected to it. The writer looked for a case of resonance, but he was unable to augment the effect by varying the capacity very carefully and gradually or by changing the speed of the machine. A case of pure resonance he was unable to obtain. When a condenser was connected to the terminals of the machine — the self-induction of the armature being first determined in the maximum and minimum position and the mean value taken — the capacity which gave the highest electromotive force corresponded most nearly to that which just counteracted the self-induction with the existing frequency. If the capacity was increased or diminished, the electromotive force fell as expected.

With frequencies as high as the above mentioned, the condenser effects are of enormous importance. The condenser becomes a highly efficient apparatus capable of transferring considerable energy.

The writer has thought machines of high frequencies may find use at least in cases when transmission at great distances is not contemplated. The increase of the resistance may be reduced in the conductors and exalted in the devices when heating effects are wanted, transformers may be made of higher efficiency and greater outputs and valuable results may be secured by means of condensers. In using machines of high frequency the writer has been able to observe condenser effects which would have otherwise escaped his notice. He has been very much interested in the phenomenon observed on the Ferranti main which has been so much spoken of. Opinions have been expressed by competent electricians, but up to the present all still seems : to be conjecture. Undoubtedly in the views expressed the truth must be contained, but as the opinions differ some must be erroneous. Upon seeing the diagram of M. Ferranti in the Electrician of Dec. 19 the writer has formed his opinion of the effect. In the absence of all the necessary data he must content himself to express in words the process which, in his opinion, must undoubtedly occur. The condenser brings about two effects: (1) It changes the phases of the currents in the branches; (2) it changes the strength of the currents. As regards the change in phase, the effect of the condenser is to accelerate the current in the secondary at Deptford and to retard it in the primary at London. The former has the effect diminishing the self-induction in the Deptford primary, and this means lower electromotive force on the dynamo. The retardation of the primary at London, as far as merely the phase is concerned, has little or no effect since the phase of the current in the secondary in London is not arbitrarily kept.

Now, the second effect of the condenser is to increase the current in both the branches. It is immaterial whether there is equality between the currents or not; but it is necessary to point out, in order . to see the importance of the Deptford step-up transformer, that an increase of the current in both the branches produces opposite effects. At Deptford it means further lowering of the electromotive force at the primary, and at London it means increase of the electromotive force. at the secondary., Therefore, all the things coact to bring about the phenomenon observed. Such actions, at least, have been formed to take place under similar conditions. When the dynamo is connected directly to the main, one can see that no such action can happen.

The writer has been particularly interested in, the suggestions and views expressed by Mr. Swinburne. Mr. Swinburne has frequently honored him by disagreeing with his views. Three years ago, when the writer, against the prevailing opinion of engineers, advanced' an open circuit transformer, Mr. Swinburne was the first to condemn it by stating in the Electrician: "The (Tesla) transformer must be inefficient; it has magnetic poles revolving, and has thus an open magnetic circuit." Two years later Mr. Swinburne becomes the champion of the open circuit transformer, and offers to convert him. But, *tempora mutantur, et nos mutamur in illis.*

The writer cannot believe in the armature reaction theory as expressed in Industries, though undoubtedly there is some truth in it. Mr. Swinburne's interpretation, however, is so broad that it may mean anything.

Mr. Swinburne seems to have been the first who has called attention to the heating of the condensers. The astonishment expressed at that by the ablest electrician is a striking illustration of 'the desirability to execute experiments on a large scale. To the scientific investigator, who deals with the minutest quantities, who observes the faintest effects, far more credit is due .than to one who experiments with apparatus on an industrial scale; and indeed history of science has recorded examples of marvelous skill, patience and keenness of observation. But however great the skill, and however keen the observer's perception, it can only be of advantage to magnify an effect and thus facilitate its study. Had Faraday carried out but one of his experiments on dynamic induction on a large scale it would have resulted in an incalculable benefit.

In the opinion of the writer, the heating of the condensers is due to three distinct causes: first, leakage or conduction; second, imperfect elasticity in the dielectric, and, third, surging of the charges in the conductor.

In many experiments he has been confronted with the problem of transferring the greatest possible amount of energy across a dielectric. For instance, he has made incandescent lamps the ends of the filaments being completely sealed in' glass, but attached to interior condenser coatings so that all the energy required had to be transferred across the glass with a condenser surface of no more than a few centimetres square. Such lamps would be a practical success with sufficiently high frequencies. With alternations as high as 15,000 per second it was easy to bring the filaments to incandescence. With lower frequencies this could also be effected, but the potential difference had, of course, to be increased. The writer has then found that the glass gets, after a while, perforated and the vacuum is impaired. The higher the frequency the longer the lamp can withstand. Such a deterioration of the dielectric always takes place when the amount of energy transferred across a dielectric of definite dimensions and by a given frequency is too great. Glass withstands best, but even glass is deteriorated. In this case the potential difference on the plates is of course too great and losses by conduction and imperfect elasticity result. If it is desirable to produce condensers capable to stand differences of potential, then the only dielectric which will involve no losses is a gas under pressure. The writer has worked with air under enormous pressures, but there are a great many practical difficulties in that direction. He thinks that in order to make the condensers of considerable practical utility, higher frequencies should be used: though such a plan has besides others the great disadvantage that the system would become very unfit for the operation of motors.

If the writer does not err Mr. Swinburne has suggested a way of exciting an alternator by means of a condenser. For a number of years past the writer has carried on experiments with the object in view of producing a practical self-exciting alternator: He has in a ,variety of ways succeeded in producing some excitation of the magnets by means of alternating currents, which were not commutated by mechanical devices. Nevertheless, his experiments have revealed a fact which stands as solid, as the rock of Gibraltar. No practical excitation can be obtained with a single periodically varying and not commutated current. The reason is that the changes in the strength of the exciting current produce corresponding changes in the field strength, with the result of inducing currents in the armature; and these currents interfere with these produced by the motion of the armature through the field, the former being a quarter phase in advance of the latter. If the field be laminated, no excitation can be produced; if it be not laminated, some excitation is produced, but .the magnets are heated. By combining two exciting currents—displaced by a quarter phase, excitation may be produced in both cases, and if the magnet be not laminated the heating effect is comparatively small, as a uniformity in the field strength is maintained, and, were it possible to produce a perfectly uniform field, excitation on this plan would give quite practical results. If such results are to be secured by the use of a condenser, as suggested by Mr. Swinburne, it is necessary to combine two circuits separated by a quarter phase; that is to say, the armature coils must be wound in two sets and connected to one or two independent condensers. The writer has done some work in that direction, but must defer the description of the devices for some future time.

Feb. 21, 1891

THE TESLA EFFECTS WITH HIGH FREQUENCY AND HIGH POTENTIAL CURRENTS (1/4)

Introduction: The scope of the Tesla lectures

Before proceeding to study the three Tesla lectures here presented, the reader may find it of some assistance to have his attention directed to the main points of interest and significance therein. The first of these lectures was delivered in New York, at *Columbia College*, before the *American Institute of Electrical Engineers*, May 20, 1891. The urgent desire expressed immediately from all parts of Europe for an opportunity to witness the brilliant and unusual experiments with which the lecture was accompanied, induced Mr. Tesla to go to England early in 1892, when he appeared before the Institution of Electrical Engineers, and a day later, by special request, before the Royal Institution. His reception was of the most enthusiastic and flattering nature on both occasions. He then went, by invitation, to France, and repeated his novel demonstrations before the *Societe Internationale des Electriciens*, and the *Societe Frangaise de Physique*. Mr. Tesla returned to America in the fall of 1892, and in February, 1893, delivered his third lecture before the Franklin Institute of Philadelphia, in fulfilment of a long standing promise to Prof. Houston. The following week, at the request of President James I. Ayer, of the *National Electric Light Association*, the same lecture was re-delivered in St. Louis. It had been intended to limit the invitations to members, but the appeals from residents in the city were so numerous and pressing that it became necessary to secure a very large hall. Hence it came about that the lecture was listened to by an audience of over 5,000 people, and was in some parts of a more popular nature than either of its predecessors. Despite this concession to the need of the hour and occasion, Mr. Tesla did not hesitate to show many new and brilliant experiments, and to advance the frontier of discovery far beyond any point he had theretofore marked publicly.

We may now proceed to a running review of the lectures themselves. The ground covered by them is so vast that only the leading ideas and experiments can here be touched upon; besides, it is preferable that the lectures should be carefully gone over for their own sake, it being more than likely that each student will discover a new beauty or stimulus in them. Taking up the course of reasoning followed by Mr. Tesla in his first lecture, it will be noted that he started out with the recognition of the fact, which he has now experimentally demonstrated, that for the production of light waves, primarily, electrostatic effects must be brought into play, and continued study has led him to the opinion that all electrical and magnetic effects may be referred to electrostatic molecular forces. This opinion finds a singular confirmation in one of the most striking experiments which he describes, namely, the production of a veritable flame by the agitation of electrostatically charged molecules. It is of the highest interest to observe that this result points out a way of obtaining a flame which consumes no material and in which no chemical action whatever takes place. It also throws a light on the nature of the ordinary flame, which Mr. Tesla believes to be due to electrostatic molecular actions, which, if true, would lead directly to the idea that even chemical affinities might be electrostatic in their nature and that, as has already been suggested, molecular forces in general may be referable to one and the same cause. This singular phenomenon accounts in a plausible manner for the unexplained fact that buildings are frequently set on fire during thunder storms without having been at all struck by any lightning. It may also explain the total disappearance of ships at sea.

One of the striking proofs of the correctness of the ideas advanced by Mr. Tesla is the fact that, notwithstanding the employment of the most powerful electromagnetic inductive effects, but feeble luminosity is obtainable, and this only in close proximity to the source of disturbance; whereas, when the electrostatic effects are intensified, the same initial energy suffices to excite luminosity at considerable distances from the source. That there are only electrostatic effects active seems to be clearly proved by Mr. Tesla's experiments with an induction coil operated with alternating currents of very high frequency. He shows how tubes may be made to glow brilliantly at considerable distances from any object when placed in a powerful, rapidly alternating, electrostatic field, and he describes many interesting phenomena observed in such a field. His experiments open up the possibility of lighting an apartment by simply creating in it such an electrostatic field, and this, in a certain way, would appear to be the ideal method of lighting a room, as it would allow the illuminating device to be freely moved about. The power with which these exhausted tubes, devoid of any electrodes, light up is certainly remarkable.

That the principle propounded by Mr. Tesla is a broad one is evident from the many ways in which it may be practically applied. We need only refer to the variety of the devices shown or described, all of which are novel in character and will, without doubt, lead to further important results at the hands of Mr. Tesla and other investigators. The experiment, for instance, of lighting up a single filament or block of refractory material with a single wire, is in itself sufficient to give Mr. Tesla's work the stamp of originality, and the numerous other experiments and effects which may be varied at will, are equally new and interesting. Thus, the incandescent filament spinning in an unexhausted globe, the well-known Crookes experiment on open circuit, and the many others suggested, will not fail to interest the reader. Mr. Tesla has made an exhaustive study of the various forms of the discharge presented by an induction coil when operated with these rapidly alternating currents, starting from the thread-like discharge and passing through various stages to the true electric flame.

A point of great importance in the introduction of high tension alternating current which Mr. Tesla brings out is the necessity of carefully avoiding all gaseous matter in the high tension apparatus. He shows that, at least with very rapidly alternating currents of high potential, the discharge may work through almost any practicable thickness of the best insulators, if air is present. In such cases the air included within the apparatus is violently agitated and by molecular bombardment the parts may be so greatly heated as to cause a rupture of the insulation. The practical outcome of this is, that, whereas with steady currents, any kind of insulation may be used, with rapidly alternating currents oils will probably be the best to employ, a fact which has been observed, but not until now satisfactorily explained. The recognition of the above fact is of special importance in the construction of the costly commercial induction coils which are often rendered useless in an unaccountable manner.

The truth of these views of Mr. Tesla is made evident by the interesting experiments illustrative of the behavior of the air between charged surfaces, the luminous streams formed by the charged molecules appearing even when great thicknesses of thinnest insulators are interposed between the charged surfaces. These luminous streams afford in themselves a very interesting study for the experimenter. With these rapidly alternating currents they become far more powerful and produce beautiful light effects when they issue from a wire, pinwheel or other object attached to a terminal of the coil; and it is interesting to note that they issue from a ball almost as freely as from a point, when the frequency is very high.

From these experiments we also obtain a better idea of the importance of taking into account the capacity and self-induction in the apparatus employed and the possibilities offered by the use of condensers in conjunction with alternate currents, the employment of currents of high frequency, among other things, making it possible to reduce the condenser to practicable dimensions. Another point of interest and practical bearing is the fact, proved by Mr. Tesla, that for alternate currents, especially those of high frequency, insulators are required possessing a small specific inductive capacity, which at the same time have a high insulating power.

Mr. Tesla also makes interesting and valuable suggestion in regard to the economical utilization of iron in machines and transformers. He shows how, by maintaining by continuous magnetization a flow of lines through the iron, the latter may be kept near its maximum permeability and a higher output and economy may be secured in such apparatus. This principle may prove of considerable commercial importance in the development of alternating systems. Mr. Tesla's suggestion that the same result can be secured by heating the iron by hysteresis and eddy currents, and increasing the permeability in this manner, while it may appear less practical, nevertheless opens another direction for investigation and improvement.

The demonstration of the fact that with alternating currents of high frequency, sufficient energy may be transmitted under practicable conditions through the glass of an incandescent lamp by electrostatic or electromagnetic induction may lead to a departure in the construction of such devices. Another important experimental result achieved is the operation of lamps, and even motors, with the discharges of condensers, this method affording a means of converting direct or alternating currents. In this connection Mr. Tesla advocates the perfecting of apparatus capable of generating electricity of high tension from heat energy, believing this to be a better way of obtaining electrical energy for practical purposes, particularly for the production of light.

While many were probably prepared to encounter curious phenomena of impedance in the use of a condenser discharged disruptively, the experiments shown were extremely interesting on account of their paradoxical character. The burning of an incandescent lamp at any candle power when connected across a heavy metal bar, the existence of nodes on the bar and the possibility of exploring the bar by means of an ordinary Garde voltmeter, are all peculiar developments, but perhaps the most interesting observation is the phenomenon of impedance observed in the lamp with a straight filament, which remains dark while the bulb glows.

Mr. Tesla's manner of operating an induction coil by means of the disruptive discharge, and thus obtaining enormous differences of potential from comparatively small and inexpensive coils, will be appreciated by experimenters and will find valuable application in laboratories. Indeed, his many suggestions and hints in regard to the construction and use of apparatus in these investigations will be highly valued and will aid materially in future research.

The London lecture was delivered twice. In its first form, before the

Institution of Electrical Engineers, it was in some respects an amplification of several points not specially enlarged upon in the New York lecture, but brought forward many additional discoveries and new investigations. Its repetition, in another form, at the Royal Institution, was due to Prof. Dewar, who with Lord Rayleigh, manifested a most lively interest in Mr. Tesla's work, and whose kindness illustrated once more the strong English love of scientific truth and appreciation of its votaries. As an indefatigable experimenter, Mr. Tesla was certainly nowhere more at home than in the haunts of Faraday, and as the guest of Faraday's successor. This Royal Institution lecture summed up the leading points of Mr. Tesla's work, in the high potential, high frequency field, and we may here avail ourselves of so valuable a summarization, in a simple form, of a subject by no means easy of comprehension until it has been thoroughly studied.

In these London lectures, among the many notable points made was first, the difficulty of constructing the alternators to obtain, the very high frequencies needed. To obtain the high frequencies it was necessary to provide several hundred polar projections, which were necessarily small and offered many drawbacks, and this the more as exceedingly high peripheral speeds had to be resorted to. In some of the first machines both armature and field had polar projections. These machines produced a curious noise, especially when the armature was started from the state of rest, the field being charged. The most efficient machine was found to be one with a drum armature, the iron body of which consisted of very thin wire annealed with special care. It was, of course, desirable to avoid the employment of iron in the armature, and several machines of this kind, with moving or stationary conductors were constructed, but the results obtained were not quite satisfactory, on account of the great mechanical and other difficulties encountered.

The study of the properties of the high frequency currents obtained from these machines is very interesting, as nearly every experiment discloses something new. Two coils traversed by such a current attract or repel each other with a force which, owing to the imperfection of our sense of touch, seems continuous. An interesting observation, already noted under another form, is that a piece of iron, surrounded by a coil through which the current is passing appears to be continuously magnetized.

This apparent continuity might be ascribed to the deficiency of the sense of touch, but there is evidence that in currents of such high frequencies one of the impulses preponderates over the other.

As might be expected, conductors traversed by such currents are rapidly heated, owing to the increase of the resistance, and the heating effects are relatively much greater in the iron. The hysteresis losses in iron are so great that an iron core, even if finely subdivided, is heated in an incredibly short time. To give an idea of this, an ordinary iron wire inch in diameter inserted within a coil having 250 turns, with a current estimated to be five amperes passing through the coil, becomes within two seconds' time so hot as to scorch wood. Beyond a certain frequency, an iron core, no matter how finely subdivided, exercises a dampening effect, and it was easy to find a point at which the impedance of a coil was not affected by the presence of a core consisting of a bundle of very thin well annealed and varnished iron wires.

Experiments with a telephone, a conductor in a strong magnetic field, or with a condenser or arc, seem to afford certain proof that sounds far above the usually accepted limit of hearing would be perceived if produced with sufficient power. The arc produced by these currents possesses several interesting features. Usually it emits a note the pitch of which corresponds to twice the frequency of the current, but if the frequency be sufficiently high it becomes noiseless, the limit of audition being determined principally by the linear dimensions of the arc. A curious feature of the arc is its persistency, which is due partly to the inability of the gaseous column to cool and increase considerably in resistance, as is the case with low frequencies, and partly to the tendency of such a high frequency machine to maintain a constant current.

In connection with these machines the condenser affords a particularly interesting study. Striking effects are produced by proper adjustments of capacity and self-induction. It is easy to raise the electromotive force of the machine to many times the original value by simply adjusting the capacity of a condenser connected in the induced circuit. If the condenser be at some distance from the machine, the difference of potential on the terminals of the latter may be only a small fraction of that on the condenser.

But the most interesting experiences are gained when the tension of the currents from the machine is raised by means of an induction coil. In consequence of the enormous rate of change obtainable in the primary current, much higher potential differences are obtained than with coils operated in the usual ways, and, owing to the high frequency, the secondary discharge possesses many striking peculiarities. Both the electrodes behave generally alike, though it appears from some observations that one current impulse preponderates over the other, as before mentioned.

The physiological effects of the high tension discharge are found to be so small that the shock of the coil can be supported without any inconvenience, except perhaps a small burn produced by the discharge upon approaching the hand to one of the terminals. The decidedly smaller physiological effects of these currents are thought to be due either to a different distribution through the body or to the tissues acting as condensers. But in the case of an induction coil with a great many turns the harmlessness is principally due to the fact that but little energy is available in the external circuit when the same is closed through the experimenter's body, on account of the great impedance of the coil.

In varying the frequency and strength of the currents through the primary of the coil, the character of the secondary discharge is greatly varied, and no less than five distinct forms are observed: A weak, sensitive thread discharge, a powerful naming discharge, and three forms of brush or streaming discharges. Each of these possesses certain noteworthy features, but the most interesting to study are the latter.

Under certain conditions the streams, which are presumably due to the violent agitation of the air molecules, issue freely from all points of the coil, even through a thick insulation. If there is the smallest air space between the primary and secondary, they will form there and surely injure the coil by slowly warming the insulation. As they form even with ordinary frequencies when the potential is excessive, the air-space must be most carefully avoided. These high frequency streamers differ in aspect and properties from those produced by a static machine. The wind produced by them is small and should altogether cease if still considerably higher frequencies could be obtained. A peculiarity is that they issue as freely from surfaces as from points. (having to this, a metallic vane, mounted in one of the terminals of the coil so as to rotate freely, and having one of its sides covered with insulation, is spun rapidly around. Such a vane would not rotate with a steady potential, but with a high frequency coil it will spin, even if it be entirely covered with insulation, provided the insulation on one side be either thicker or of a higher specific inductive capacity. A Crookes electric radiometer is also spun around when connected to one of the terminals of the coil, but only at very high exhaustion or at ordinary pressures.

There is still another and more striking peculiarity of such a high frequency streamer, namely, it is hot. The heat is easily perceptible with frequencies of about 10,000, even if the potential is not excessively high. The heating effect is, of course, due to the molecular impacts and collisions. Could the frequency and potential be pushed far enough, then a brush could be produced resembling in every particular a flame and giving light and heat, jet without a chemical process taking place.

The hot brush, when properly produced, resembles a jet of burning gas escaping under great pressure, and it emits an extraordinary strong smell of ozone. The great ozonizing action is ascribed to the fact that the agitation of the molecules of the air is more violent in such a brush than in the ordinary streamer of a static machine. But the most powerful brush discharges were produced by employing currents of much higher frequencies than it was possible to obtain by means of the alternators. These currents were obtained by disruptively discharging a condenser and setting up oscillations. In this manner currents of a frequency of several hundred thousand were obtained.

Currents of this kind, Mr. Tesla pointed out, produce striking effects. At these frequencies, the impedance of a copper bar is so great that a potential difference of several hundred volts can be maintained between two points of a short and thick bar, and it is possible to keep an ordinary incandescent lamp burning at full candle power by attaching the terminals of the lamp to two points of the bar no more than a few inches apart. When the frequency is extremely high, nodes are found to exist on such a bar, and it is easy to locate them by means of a lamp.

By converting the high tension discharges of a low frequency coil in this manner, it was found practicable to keep a few lamps burning on the ordinary circuit in the laboratory, and by bringing the undulation to a low pitch, it was possible to operate small motors.

This plan likewise allows of converting high tension discharges of one direction into low tension unidirectional currents, by adjusting the circuit so that there are no oscillations. In passing the oscillating discharges through the primary of a specially constructed coil, it is easy to obtain enormous potential differences with only few turns of the secondary.

Great difficulties were at first experienced in producing a successful coil on this plan. It was found necessary to keep all air, or gaseous matter in general, away from the charged surfaces, and oil immersion was resorted to. The wires used were heavily covered with gutta-percha and wound in oil, or the air was pumped out by means of a Sprengel pump. The general arrangement was the following: An ordinary induction coil, operated from a low frequency alternator, was used to charge Leyden jars. The jars were made to discharge over a single or multiple gap through the primary of the second coil. To insure the action of the gap, the arc was blown out by a magnet or air blast. To adjust the potential in the secondary a small oil condenser was used, or polished brass spheres of different sizes were screwed on the terminals and their distance adjusted.

When the conditions were carefully determined to suit each experiment, magnificent effects were obtained. Two wires, stretched through the room, each being connected to one of the terminals of the coil, emitted streams so powerful that the light from them allowed distinguishing the objects in the room; the wires became luminous even though covered with thick and most excellent insulation. When two straight wires, or two concentric circles of wire, are connected to the terminals, and set at the proper distance, a uniform luminous sheet is produced between them. It was possible in this way to cover an ana of more than one meter square completely with the streams. By attaching to one terminal a large circle of wire and to the other terminal a small sphere, the streams are focused upon the sphere, produce a strongly lighted spot upon the same, and present the appearance of a luminous cone. A very thin wire glued upon a plate of hard rubber of great thickness, on the opposite side of which is fastened a tinfoil coating, is rendered intensely luminous when the coating is connected to the other terminal of the coil. Such an experiment can be performed also with low frequency currents, but much less satisfactorily.

When the terminals of such a coil, even of a very small one, are separated by a rubber or glass plate, the discharge spreads over the plate in the form of streams, threads or brilliant sparks, and affords a magnificent display, which cannot be equaled by the largest coil operated in the usual ways. By a simple adjustment it is possible to produce with the coil a succession of brilliant sparks, exactly as with a Holtz machine.

Under certain conditions, when the frequency of the oscillation is very great, white, phantom-like streams are seen to break forth from the terminals of the coil. The chief interesting feature about them is, that they stream freely against the outstretched hand or other conducting object without producing any sensation, and the hand may be approached very near to the terminal without a spark being induced to jump. This is due presumably to the fact that a considerable portion of the energy is carried away or dissipated in the streamers, and the difference of potential between the terminal and the hand is diminished.

It is found in such experiments that the frequency of the vibration and the quickness of succession of the sparks between the knobs affect to a marked degree the appearance of the streams. When the frequency is very low, the air gives way in more or less the same manner as by a steady difference of potential, and the streams consist of distinct threads, generally mingled with thin sparks, which probably correspond to the successive discharges occurring between the knobs. But when the frequency is very high, and the arc of the discharge produces a sound which is loud and smooth (which indicates both that oscillation takes place and that the sparks succeed each other with great rapidity), then the luminous streams formed are perfectly uniform. They are generally of a purplish hue, but when the molecular vibration is increased by raising the potential, they assume a white color.

The luminous intensity of the streams increases rapidly when the potential is increased; and with frequencies of only a few hundred thousand, could the coil be made to withstand a sufficiently high potential difference, there is no doubt that the space around a wire could be made to emit a strong light, merely by the agitation of the molecules of the air at ordinary pressure.

Such discharges of very high frequency which render luminous the air at ordinary pressure we have very likely occasion to witness in the aurora borealis. From many of these experiments it seems reasonable to infer that sudden cosmic disturbances, such as eruptions on the sun, set the electrostatic charge of the earth in an extremely rapid vibration, and produce the glow by the violent agitation of the air in the upper and even in the lower strata. It is thought that if the frequency were low? or even more so if the charge were not at all vibrating, the lower dense strata would break down as in a lightning discharge. Indications of such breaking down have been repeatedly observed, but they can be attributed to the fundamental disturbances, which are few in number, for the superimposed vibration would be so rapid as not to allow a disruptive break.

The study of these discharge phenomena has led Mr. Tesla to the recognition of some important facts. It was found, as already stated, that uascous matter must be most carefully excluded from any dielectric which is subjected to great, rapidly changing electrostatic stresses. Since it is difficult to exclude the gas perfectly when solid insulators are used, it is necessary to resort to liquid dielectrics. When a solid dielectric is used, it matters little how thick and how good it is; if air be present, streamers form, which gradually heat the dielectric and impair its insulating power, and the discharge finally breaks through. Under ordinary conditions the best insulators are those which possess the highest specific inductive capacity, but such insulators are not the best to employ when working with these high frequency currents, for in most cases the higher specific inductive capacity is rather a disadvantage. The prime quality of the insulating medium for these currents is continuity. For this reason principally it is necessary to employ liquid insulators, such as oils. If two metal plates, connected to the terminals of the coil, are immersed in oil and set a distance apart, the coil may be kept working for any length of time without a break occurring, or without the oil being warmed, but if air bubbles are introduced, they become luminous; the air molecules, by their impact against the oil, heat it, and after some time cause the insulation to give way. If, instead of the oil, a solid plate of the best dielectric, even several times thicker than the oil intervening between the metal plates, is inserted between the latter, the air having free access to the charged surfaces, the dielectric i variably is warmed and breaks down.

The employment of oil is advisable or necessary even with low frequencies, if the potentials are such that streamers form, but only in such cases, as is evident from the theory of the action. If the potentials are so low that streamers do not form, then it is even disadvantageous to employ oil, for it may, principally by confining the heat, be the cause of the breaking down of the insulation.

The exclusion of gaseous matter is not only desirable on account of the safety of the apparatus, but also on account of economy, especially in a condenser, in which considerable waste of power may occur merely owing to the presence of air, if the electric density on the charged surfaces is great.

In the course of these investigations a phenomenon of special scientific interest was observed. It may be ranked among the brush phenomena, in fact it is a kind of brush which forms at, or near, a single terminal in high vacuum. In a bulb with a conducting electrode, even if the latter be of aluminum, the brush has only a very short existence, but it can be preserved for a considerable length of time in a bulb devoid of any conducting electrode. To observe the phenomenon it is found best to employ a large spherical bulb having in its centre a small bulb supported on a tube sealed to the neck of the former. The large bulb being exhausted to a high degree, and the inside of the small bulb being connected to one of the terminals of the coil, under certain conditions there appears a misty haze around the small bulb, which, after passing through some stages, assumes the form of a brush, generally at right angles to the tube supporting the small bulb. When the brush assumes this form it may be brought to a state of extreme sensitiveness to electrostatic and magnetic influence. The bulb hanging straight down, and all objects being remote from it, the approach of the observer within a few paces will cause the brush to fly to the opposite side, and if he walks around the bulb it will always keep on the opposite side. It may begin to spin around the terminal long before it reaches that sensitive stage. When it begins to turn around, principally, but also before, it is affected by a magnet, and at a certain stage it is susceptible to magnetic influence to an astonishing degree. A small permanent magnet, with its poles at a distance of no more than two centimetres will affect it visibly at a distance of two metres, slowing down or accelerating the rotation according to how it is held relatively to the brush.

When the bulb hangs with the globe down, the rotation is always clockwise. In the southern hemisphere it would occur in the opposite direction, and on the (magnetic) equator the brush should not turn at all. The rotation may be reversed by a magnet kept at some distance. The brush rotates best, seemingly, when it is at right angles to the lines of force of the earth. It, very likely rotates, when at its maximum speed, in synchronism with the alternations, say, 10,000 times a second. The rotation can be slowed down or accelerated by the approach or recession of the observer, or any conducting body, but it cannot be reversed by putting the bulb in any position. Very curious experiments may be performed with the brush when in its most sensitive state. For instance, the brush resting in one position, the experimenter may, by selecting a proper position, approach the hand at a certain considerable distance to the bulb, and he may cause the brush to pass off by merely stiffening the muscles of the arm, the mere change of configuration of the arm and the consequent imperceptible displacement being sufficient to disturb the delicate balance. When it begins to rotate slowly, and the hands are held at a proper distance, it is impossible to make even the slightest motion without producing a visible effect upon the brush. A metal plate connected to the other terminal of the coil affects it at a great distance, slowing down the rotation often to one turn a second.

Mr. Tesla hopes that this phenomenon will prove a valuable aid in the investigation of the nature of the forces acting in an electrostatic or magnetic field. If there is any motion which is measurable going on in the space, such a brush would be apt to reveal it. It is, so to speak, a beam of light, frictionless, devoid of inertia. On account of its marvellous sensitiveness to electrostatic or magnetic disturbances it may be the means of sending signals through submarine cables with any speed, and even of transmitting intelligence to a .distance without wires.

In operating an induction coil with these rapidly alternating currents, it is astonishing to note, for the first time, the great importance of the relation of capacity, self-induction, and frequency as bearing upon the general result. The combined effect of these elements produces many curious effects. For instance, two metal plates are connected to the terminals and set at a small distance, so that an arc is formed between them. This arc prevents a strong current from flowing through the coil. If the arc be interrupted by the interposition of a glass plate, the capacity of the condenser obtained counteracts the self-induction, and a stronger current is made to pass. The effects of capacity are the most striking, for in these experiments, since the self-induction and frequency both are high, the critical capacity is very small, and need be but slightly varied to produce a very considerable change. The experimenter brings his body in contact with the terminals of the secondary of the coil, or attaches to one or both terminals insulated bodies of very small bulk, such as exhausted bulbs, and he produces a considerable rise or fall of potential on the secondary, and greatly affects the flow of the current through the primary coil.

In many of the phenomena observed, the presence of the air, or, generally speaking, of a medium of a gaseous nature (using this term not to imply specific properties, but in contradistinction to homogeneity or perfect continuity) plays an important part, as it allows energy to be dissipated by molecular impact or bombardment. The action is thus explained: When an insulated body connected to a terminal of the coil is suddenly charged to high potential, it acts inductively upon the surrounding air, or whatever gaseous medium there might be. The molecules or atoms which are near it are, of course, more attracted, and move through a greater distance than the further ones. When the nearest molecules strike the body they are repelled, and collisions occur at all distances within the inductive distance. It is now clear that, if the potential be steady, bat little loss of energy can be caused in this way, for the molecules which are nearest to the body having had an additional charge imparted to them by contact, are not attracted until they have parted, if not with all, at least with most of the additional charge, which can be accomplished only after a great many collisions. This is inferred from the fact that with a steady potential there is but little loss in dry air. When the potential, instead of being steady, is alternating, the conditions are entirely different. In this case a rhythmical bombardment occurs, no matter whether the molecules after coming in contact with the body lose the imparted charge or not, and, what is more, if the charge is not lost, the impacts are all the more violent. Still, if the frequency of the impulses be very small, the loss caused by the impacts and collisions would not be serious unless the potential was excessive. But when extremely high frequencies and more or less high potentials are used, the loss may be very great, The total energy lost per unit of time is proportionate to the product of the number of impacts per second, or the frequency and the energy lost in each impact. But the energy of an impact must be proportionate to the square of the electric density of the body, on the assumption that the charge imparted to the molecule is proportionate to that density. It is concluded from this that the total energy lost must be proportionate to the product of the frequency and the square of the electric density; but this law needs experimental confirmation. Assuming the preceding considerations to be true, then, by rapidly alternating the potential of a body immersed in an insulating gaseous medium, any amount of energy may be dissipated into space. Most of that energy, then, is not dissipated in the form of long ether waves, propagated to considerable distance, as is thought most generally, but is consumed in impact and collisional losses that is, heat vibrations on the surface and in the vicinity of the body. To reduce the dissipation it is necessary to work with a small electric density the smaller, the higher the frequency.

The behavior of a gaseous medium to such rapid alternations of potential makes it appear plausible that electrostatic disturbances of the earth, produced by cosmic events, may have great influence upon the meteorological condition. When such disturbances occur both the frequency of the vibrations of the charge and the potential are in all probability excessive, and the energy converted into heat may be considerable. Since the density must be unevenly distributed, either in consequence of the irregularity of the earth's surface, or on account of the condition of the atmosphere in various places, the effect produced would accordingly vary from place to place. Considerable variations in the temperature and pressure of the atmosphere may in this manner be caused at any point of the surface of the earth. The variations may be gradual or very sudden, according to the nature of the original disturbance, and may produce rain and storms, or locally modify the weather in any way.

From many experiences gathered in the course of these investigations it appears certain that in lightning discharges the air is an element of importance. For instance, during a storm a stream may form on a nail or pointed projection of a building. If lightning strikes somewhere in the neighborhood* the harmless static discharge may, in consequence of the oscillations set up, assume the character of a high-frequency streamer, and the nail or projection may be brought to a high temperature by the violent impact of the air molecules. Thus, it is thought, a building may be set on fire without the lightning striking it. In like manner small metallic objects may be fused and volatilized as frequently occurs in lightning discharges merely because they are surrounded by air. Were they immersed in a practically continuous medium, such as oil, they would probably be safe, as the energy would have to spend itself elsewhere.

An instructive experience having a bearing on this subject is the following: A glass tube of an inch or so in diameter and several inches long is taken, and a platinum wire sealed into it, the wire running through the center of the tube from end to end. The tube is exhausted to a moderate degree. If a steady current is passed through the wire it is heated uniformly in all parts and the gas in the tube is of no consequence. But if high frequency discharges are directed through the wire, it is heated more on the ends than in the middle portion, and if the frequency, or rate of charge, is high enough, the wire might as well be cut in the middle as not, for most of the heating on the ends is due to the rarefied gas. Here the gas might only act as a conductor of no impedance, diverting the current from the wire as the impedance of the latter is enormously increased, and merely heating the ends of the wire by reason of their resistance to the passage of the discharge. But it is not at all necessary that the gas in the tube should he conducting; it might be at an extremely low pressure, still the ends of the wire would be heated; however, as is ascertained by experience, only the two ends would in such case not be electrically connected through the gaseous medium. Now, what with these frequencies and potentials occurs in an exhausted tube, occurs in the lightning discharge at ordinary pressure.

From the facility with which any amount of energy may be carried off through a gas, Mr. Tesla infers that the best way to render harmless a lightning discharge is to afford it in some way a passage through a volume of gas.

The recognition of some of the above facts has a bearing upon far-reaching scientific investigations in which extremely high frequencies and potentials are used. In such cases the air is an important factor to be considered. So, for instance, if two wires are attached to the terminals of the coil, and the streamers issue from' them, there is dissipation of energy in the form of heat and light, and the wires behave like a condenser of larger capacity. If the wires be immersed in oil, the dissipation of energy is prevented, or at least reduced, and the apparent capacity is diminished. The action of the air would seem to make it very difficult to tell, from the measured or computed capacity of a condenser in which the air is acted upon, its actual capacity or vibration period, especially if the condenser is of very small surface and is charged to a very high potential. As many important results are dependant upon the correctness of the estimation of the vibration period, this subject demands the most careful scrutiny of investigators.

In Leyden jars the loss due to the presence of air is comparatively small, principally on account of the great surface of the coatings and the small external action, but if there are streamers on the top, the loss may be considerable, and the period of vibration is affected. In a resonator, the density is small, but the frequency is extreme, and may introduce a considerable error. It appears certain, at any rate, that the periods of vibration of a charged body in a gaseous and in a continuous medium, such as oil, are different, on account of the action of the former, as explained.

Another fact recognized, which is of some consequence, is, that in similar investigations the general considerations of static screening are not applicable when a gaseous medium is present. This is evident from the following experiment : A short and wide glass tube is taken and covered with a substantial coating of bronze powder, barely allowing the light to shine a little through.

The tube is highly exhausted and suspended on a metallic clasp from the end of a wire. When the wire is connected with one of the terminals of the coil, the gas inside of the tube is lighted in spite of the metal coating. Here the metal evidently does not screen the gas inside as it ought to, even if it be very thin and poorly conducting. Yet, in a condition of rest the metal coating, however thin, screens the inside perfectly.

One of the most interesting results arrived at in pursuing these experiments, is the demonstration of the fact that a gaseous medium, upon which vibration is impressed by rapid changes of electrostatic potential, is rigid. In illustration of this result an experiment made by Mr. Tesla may by cited: A glass tube about one inch in diameter and three feet long, with outside condenser coatings on the ends, was exhausted to a certain point, when, the tube being suspended freely from a wire connecting the upper coating to one of the terminals of the coil, the discharge appeared in the form of a luminous thread passing through the axis of the tube. Usually the thread was sharply defined in the upper part of the tube and lost itself in the lower part. When a magnet or the finger was quickly passed near the upper part of the luminous thread, it was brought out of position by magnetic or electrostatic influence, and a transversal vibration like that of a suspended cord, with one or more distinct nodes, was set up, which lasted for a few minutes and gradually died out. By suspending from the lower condenser coating metal plates of different sizes, the speed of the vibration was varied. This vibration would seem to show beyond doubt that the thread possessed rigidity, at least to transversal displacements.

Many experiments were tried to demonstrate this property in air at ordinary pressure. Though no positive evidence has been obtained, it is thought, nevertheless, that a high frequency brush or streamer, if the frequency could be pushed far enough, would be decidedly rigid. A small sphere might then be moved within it quite freely, but if thrown against it the sphere would rebound. An ordinary flame cannot possess rigidity to a marked degree because the vibration is directionless; but an electric arc, it is believed, must possess that property more or less. A luminous band excited in a bulb by repeated discharges of a Leyden jar must also possess rigidity, and if deformed and suddenly released should vibrate.

From like considerations other conclusions of interest are readied. The most probable medium filling the space is one consisting of independent carriers immersed in an insulating fluid. If through' this medium enormous electrostatic stresses are assumed to act, which vary rapidly in intensity, it would allow the motion of a body through it, yet it would be rigid and elastic, although the fluid itself might be devoid of these properties. Furthermore, on the assumption that the independent carriers are of any configuration such that the fluid resistance to motion in one direction is greater than in another, a stress of that nature would cause the carriers to arrange themselves in groups, since they would turn to each other their sides of the greatest electric density, in which position the fluid resistance to approach would be smaller than to receding. If in a medium of the above characteristics a brush would be formed by a steady potential, an exchange of the carriers would go on continually, and there would be less carriers per unit of volume in the brush than in the space at some distance from the electrode, this corresponding to rarefaction. If the potential were rapidly changing, the result would be very different; the higher the frequency of the pulses, the slower would be the exchange of the carriers; finally, the motion of translation through measurable space would cease, and, with a sufficiently high frequency and intensity of the stress, the carriers would be drawn towards the electrode, and compression would result.

An interesting feature of these high frequency currents is that they allow of operating all kinds of devices by connecting the device with only one leading wire to the electric source. In fact, under certain conditions it may be more economical to supply the electrical energy with one lead than with two.

An experiment of special interest shown by Mr. Tesla, is the running, by the use of only one insulated line, of a motor operating on the principle of the rotating magnetic field enunciated by Mr. Tesla. A simple form of such a motor is obtained by winding upon a laminated iron core a primary and close to it a secondary coil, closing the ends of the latter and placing a freely movable metal disc within the influence of the moving field. The secondary coil may, however, be omitted. When one of the ends of the primary coil of the motor is connected to one of the terminals of the high frequency coil arid the other end to an insulated metal plate, which, it should be stated, is not absolutely necessary for the success of the experiment, the disc is set in rotation.

Experiments of this kind seem to bring it within possibility to operate a motor at any point of the earth's surface from a central source, without any connection to the same except through the earth. If, by means of powerful machinery, rapid variations of the earth's potential were produced, a grounded wire reaching up to some height would be traversed by a current which could be increased by connecting the free end of the wire to a body of some size. The current might be converted to low tension and used to operate a motor or other device. The experiment, which would be one of great scientific interest, would probably best succeed on a ship at sea. In this manner, even if it were not possible to operate machinery, intelligence might be transmitted quite certainly.

In the course of this experimental study special attention was devoted to the heating effects produced by these currents, which are not only striking, but open up the possibility of producing a more efficient illuminant. It is sufficient to attach to the coil terminal a thin wire or filament, to have the temperature of the latter perceptibly raised. If the wire or filament be enclosed in a bulb, the heating effect is increased by preventing the circulation of the air. If the air in the bulb be strongly compressed, the displacements are smaller, the impacts less violent, and the heating effect is diminished. On the contrary, if the air in the bulb be exhausted, an inclosed lamp filament is brought to incandescence, and any amount of light may thus be produced.

The heating of the inclosed lamp filament depends on so many things of a different nature, that it is difficult to give a generally applicable rule under which the maximum heating occurs. As regards the size of the bulb, it is ascertained that at ordinary or only slightly differing atmospheric pressures, when air is a good insulator, the filament is heated more in a small bulb, because of the better confinement of heat in this case. At lower pressures, when air becomes conducting, the heating effect is greater in a large bulb, but at excessively high degrees of exhaustion there seems to be, beyond a certain and rather small size of the vessel, no perceptible difference in the heating.

The shape of the vessel is also of some importance, and it has been found of advantage for reasons of economy to employ a spherical bulb with the electrode mounted in its centre, where the rebounding molecules collide.

It is desirable on account of economy that all the energy supplied to the bulb from the source should reach without loss the body to be heated. The loss in conveying the energy from the source to the body may be reduced by employing thin wires heavily coated with insulation, and by the use of electrostatic screens. It is to be remarked, that the screen, cannot be connected to the ground as under ordinary conditions.

In the bulb itself a large portion of the energy' supplied may be lost by molecular bombardment against the wire connecting the body to be heated with the source. Considerable improvement was effected by covering the glass stem containing the wire with a closely fitting conducting tube. This tube is made to project a little above the glass, and prevents the cracking of the latter near the heated body. The effectiveness of the conducting tube is limited to very high degrees of exhaustion. It diminishes the energy lost in bombardment for two reasons; first, the charge given up by the atoms spreads over a greater area, and hence the electric density at any point is small, and the atoms are repelled with less energy than if they would strike against a good insulator; secondly, as the tube is electrified by the atoms which first come in contact with it, the progress of the following atoms against the tube is more or less checked by the repulsion which the electrified tube must exert upon the similarly electrified atoms. This, it is thought, explains why the discharge through a bulb is established with much greater facility when an insulator, than when a conductor, is present.

During the investigations a great many bulbs of different construction, with electrodes of different material, were experimented upon, and a number of observations of interest were made. Mr. Tesla has found that the deterioration of the electrode is the less, the higher the frequency. This was to be expected, as then the heating is effected by many small impacts, instead by fewer and more violent ones, which quickly shatter the structure. The deterioration is also smaller when the vibration is harmonic. Thus an electrode, maintained at a certain degree of heat, lasts much longer with currents obtained from an alternator, than with those obtained by means of a disruptive discharge. One of the most durable electrodes was obtained from strongly compressed carborundum, which is a kind of carbon recently produced by Mr. E. G. Acheson, of Monongahela City, Pa. From experience, it is inferred, that to be most durable, the electrode should be in the form of a sphere with a highly polished surface.

In some bulbs refractory bodies were mounted in a carbon cup and put under the molecular impact. It was observed in such experiments that the carbon cup was heated at first, until a higher temperature was reached; then most of the bombardment was directed against the refractory body, and the carbon was relieved. In general, when different bodies were mounted in the bulb, the hardest fusible would be relieved, and would remain at a considerably lower temperature. This was necessitated by the fact that most of the energy supplied would find its way through the body which was more easily fused or "evaporated."

Curiously enough it appeared in some of the experiments made, that a body was fused in a bulb under the molecular impact by evolution" of less light than when fused by the application of heat in ordinary ways. This may be ascribed to a loosening of the structure of the body under the violent impacts and changing stresses.

Some experiments seem to indicate that under certain conditions a body, conducting or nonconducting, may, when bombarded, emit light, which to all appearances is due to phosphorescence, but may in reality be caused by the incandescence of an infinitesimal layer, the mean temperature of the body being comparatively small. Such might be the case if each single rythmical impact were capable of instantaneously exciting the retina, and the rhythm were just high enough to cause a continuous impression in the eye. According to this view, a coil operated by disruptive discharge would be eminently adapted to produce such a result, and it is found by experience that its power of exciting phosphorescence is extraordinarily great. It is capable of exciting phosphorescence at comparatively low degrees of exhaustion, and also projects shadows at pressures far greater than those at which the mean free path is comparable to the dimensions of the vessel. The latter observation is of some importance, inasmuch as it may modify the generally accepted views in regard to the "radiant state" phenomena.

A thought which early and naturally suggested itself to Mr. Tesla, was to utilize the great inductive effects of high frequency currents to produce light in a sealed glass vessel without the use of leading in wires. Accordingly, many bulbs were constructed in which the energy necessary to maintain a button or filament at high incandescence, was supplied through the glass by either electrostatic or electrodynamic induction. It was easy to regulate the intensity of the light emitted by means of an externally applied condenser coating connected to an insulated plate, or simply by means of a plate attached to the bulb which at the same time performed the function of a shade.

A subject of experiment, which has been exhaustively treated in England by Prof. J. J. Thomson, has been followed up independently by Mr. Tesla from the beginning of this study, namely, to excite by electrodynamic induction a luminous band in a closed tube or bulb. In observing the behavior of gases, and the luminous phenomena obtained, the importance of the electrostatic effects was noted and it appeared desirable to produce enormous potential differences, alternating with extreme rapidity. Experiments in this direction led to some of the most interesting results arrived at in the course of these investigations. It was found that by rapid alternations of a high electrostatic potential, exhausted tubes could be lighted at considerable distances from a conductor connected to a properly constructed coil, and that it was practicable to establish with the coil an alternating electrostatic field, acting through the whole room and lighting a tube wherever it was placed within the four walls. Phosphorescent bulbs may be excited in such a field, and it is easy to regulate the effect by connecting to the bulb a small insulated metal plate. It was likewise possible to maintain a filament or button mounted in a tube at bright incandescence, and, in one experiment, a mica vane was spun by the incandescence of a platinum wire.

Coming now to the lecture delivered in Philadelphia and St. Louis, it may be remarked that to the superficial reader, Mr. Tesla's introduction, dealing with the importance of the eye, might appear as a digression, but the thoughtful reader will find therein much food for meditation and speculation. Throughout his discourse one can trace Mr. Tesla's effort to present in a popular way thoughts and views on the electrical phenomena which have in recent years captivated the scientific world, but of which the general public has even yet merely received an inkling. Mr. Tesla also dwells rather extensively on his well-known method of high-frequency conversion; and the large amount of detail information will be gratefully received by students and experimenters in this virgin field. The employment of apt analogies in explaining the fundamental principles involved makes it easy for all to gain a clear idea of their nature. Again, the ease with which, thanks to Mr. Tesla's efforts, these high-frequency currents may now be obtained from circuits carrying almost any kind of current, cannot fail to result in an extensive broadening of this field of research, which offers so many possibilities. Mr. Tesla, true philosopher as he is, does not hesitate to point out defects in some of his methods, and indicates the lines which to him seem the most promising. Particular stress is laid by him upon the employment of a medium in which the discharge electrodes should be immersed in order that this method of conversion may be brought to the highest perfection. He has evidently taken pains to give as much useful information as possible to those who wish to follow in his path, as he shows in detail the circuit arrangements to be adopted in all ordinary cases met with in practice, and although some of these methods were described by him two years before, the additional information is still timely and welcome.

In his experiments he dwells first on some phenomena produced by electrostatic force, which he considers in the light of modern theories to be the most important force in nature for us to investigate. At the very outset he shows a strikingly novel experiment illustrating the effect of a rapidly varying electrostatic force in a gaseous medium, by touching with one hand one of the terminals of a 200,000 volt transformer and bringing another hand to the opposite terminal. The powerful streamers which issued from his hand and astonished his audiences formed a capital illustration of some of the views advanced, and afforded Mr. Tesla an opportunity of pointing out the true reasons why, with these currents, such an amount of energy can be passed through the body with impunity. He then showed by experiment the difference between a steady and a rapidly varying force upon the dielectric. This difference is most strikingly illustrated in the experiment in which a bulb attached to the end of a wire in connection with one of the terminals of the transformer is ruptured, although all extraneous bodies are remote from the bulb. He next illustrates how mechanical motions are produced by a varying electrostatic force acting through a gaseous medium. The importance of the action of the air is particularly illustrated by an interesting experiment.

Taking up another class of phenomena, namely, those of dynamic electricity, Mr. Tesla produced in a number of experiments a variety of effects by the employment of only a single wire with the evident intent of impressing upon his audience the idea that electric vibration or current can be transmitted with ease, without any return circuit; also how currents so transmitted can be converted and used for many practical purposes. A number of experiments are then shown, illustrating the effects of frequency, self-induction and capacity; then a number of ways of operating motive and other devices by the use of a single lead. A number of novel impedance phenomena are also shown which cannot fail to arouse interest.

Mr. Tesla next dwelt upon a subject which he thinks of great importance, that is, electrical resonance, which he explained in a popular way. He expressed his firm conviction that by observing proper conditions, intelligence, and possibly even power, can be transmitted through the medium or through the earth; and he considers this problem worthy of serious and immediate consideration.

Coming now to the light phenomena in particular, lie illustrated the four distinct kinds of these phenomena in an original way, which to many must have been a revelation. Mr. Tesla attributes these light effects to molecular or atomic impacts produced by a varying electrostatic stress in a gaseous medium. Fie illustrated in a series of novel experiments the effect of the gas surrounding the conductor and shows beyond a doubt that with high frequency and high potential currents, the surrounding gas is of paramount importance in the heating of the conductor. He attributes the heating partially to a conduction current and partially to bombardment, and demonstrates that in many cases the heating may be practically due to the bombardment alone. He pointed out also that the skin effect is largely modified by the presence of the gas or of an atomic medium in general. He showed also some interesting experiments in which the effect of convection is illustrated. Probably one of the most curious experiments in this connection is that in which a thin platinum wire stretched along the axis of an exhausted tube is brought to incandescence at certain points corresponding to the position of the striae, while at others it remains dark. This experiment throws an interesting light upon the nature of the strife and may lead to important revelations.

Mr. Tesla also demonstrated the dissipation of energy through an atomic medium and dwelt upon the behavior of vacuous space in conveying heat, and in this connection showed the curious behavior of an electrode stream, from which he concludes that the molecules of a gas probably cannot be acted upon directly at measurable distances.

Mr. Tesla summarized the chief results arrived at in pursuing his investigations in a manner which will serve as a valuable guide to all who may engage in this work. Perhaps most interest will centre on his general statements regarding the phenomena of phosphorescence, the most important fact revealed in this direction being that when exciting a phosphorescent bulb a certain definite potential gives the most economical result.

The lectures will now be presented in the order of their date of delivery.

EXPERIMENTS WITH ALTERNATE CURRENTS OF VERY HIGH FREQUENCY AND THEIR APPLICATION TO METHODS OF ARTIFICIAL ILLUMINATION (2/4)

—Delivered before the American Institute of Electrical Engineers, Columbia College, N.Y., May 20, 1891.

There is no subject more captivating, more worthy of study, than nature. To understand this great mechanism, to discover the forces which are active, and the laws which govern them, is the highest aim of the intellect of man.

Nature has stored up in the universe infinite energy. The eternal recipient and transmitter of this infinite energy is the ether. The recognition of the existence of ether, and of the functions it performs, is one of the most important results of modern scientific research. The mere abandoning of the idea of action at a distance, the assumption of a medium pervading all space and connecting all gross matter, has freed the minds of thinkers of an ever present doubt, and, by opening a new horizon—new and unforeseen possibilities—has given fresh interest to phenomena with which we are familiar of old. It has been a great step towards the understanding of the forces of nature and their multifold manifestations to our senses. It has been for the enlightened student of physics what the understanding of the mechanism of the firearm or of the steam engine is for the barbarian. Phenomena upon which we used to look as wonders baffling explanation, we now see in a different light. The spark of an induction coil, the glow of an incandescent lamp, the manifestations of the mechanical forces of currents and magnets are no longer beyond our grasp; instead of the incomprehensible, as before, their observation suggests now in our minds a simple mechanism, and although as to its precise nature all is still conjecture, yet we know that the truth cannot be much longer hidden, and instinctively we feel that the understanding is dawning upon us. We still admire these beautiful phenomena, these strange forces, but we are helpless no longer; we can in a certain measure explain them, account for them, and we are hopeful of finally succeeding in unraveling the mystery which surrounds them.

In how far we can understand the world around us is the ultimate thought of every student of nature. The coarseness of our senses prevents us from recognizing the ulterior construction of matter, and astronomy, this grandest and most positive of natural sciences, can only teach us something that happens, as it were, in our immediate neighborhood; of the remoter portions of the boundless universe, with its numberless stars and suns, we know nothing, But far beyond the limit of perception of our senses the spirit still can guide us, and so we may hope that even these unknown worlds — infinitely small and great — may in a measure became known to us. Still, even if this knowledge should reach us, the searching mind will find a barrier, perhaps forever unsurpassable, to the true recognition of that which seems to be, the mere appearance of which is the only and slender basis of all our philosophy.

Of all the forms of nature's immeasurable, all-pervading energy, which ever and ever changing and moving; like a soul animates the inert universe, electricity and magnetism are perhaps the most fascinating. The effects of gravitation, of heat and light we observe daily, and soon we get accustomed to them, and soon they lose for us the character of the marvelous and wonderful; but electricity and magnetism, with their singular relationship, with their seemingly dual character, unique among the forces in nature, with their phenomena of attractions, repulsions and rotations, strange manifestations of mysterious agents; stimulate and excite the mind to thought and research. What is electricity, and what is magnetism? These questions have been asked again and again. The most able intellects have ceaselessly wrestled with the problem; still the question has not as yet been fully answered. But while we cannot even today state what these singular forces are, we have made good headway towards the solution of the problem. We are now confident that electric and magnetic phenomena are attributable to ether, and we are perhaps justified in saying that the effects of static electricity are effects of ether under strain, and those of dynamic electricity and electro-magnetism effects of ether in motion. But this still leaves the question, as to what electricity and magnetism are, unanswered.

First, we naturally inquire, What is electricity, and is there such a thing as electricity? In interpreting electric phenomena: we may speak of electricity or of an electric condition, state or effect. If we speak of electric effects we must distinguish two such effects, opposite in character and neutralizing each other, as observation shows that two such opposite effects exist. This is unavoidable, for in a medium of the properties of ether, we cannot possibly exert a strain, or produce a displacement or motion of any kind, without causing in the surrounding medium an equivalent and opposite effect. But if we speak of electricity, meaning a thing, we must, I think, abandon the idea of two electricities, as tie existence of two such things is highly improbable. For how can we imagine that there should be two things, equivalent in amount, alike in their properties, but of opposite character, both clinging to matter, both attracting and completely neutralizing each other? Such an assumption, though suggested by many phenomena, though most convenient for explaining them, has little to commend it. If there is such a thing as electricity, there can be only one such thing, and; excess and want of that one thin, possibly; but more probably its condition determines the positive and negative character. The old theory of Franklin, though falling short in some respects; is, from a certain point of view, after all, the most plausible one. Still, in spite of this, the theory of the two electricities is generally accepted, as it apparently explains electric phenomena in a more satisfactory manner. But a theory which better explains the facts is not necessarily true. Ingenious minds will invent theories to suit observation, and almost every independent thinker has his own views on the subject.

It is not with the, object of advancing an opinion; but with the desire of acquainting you better with some of the results, which I will describe, to show you the reasoning I have followed, the departures I have made — that I venture to express, in a few words, the views and convictions which have led me to these results.

I adhere to the idea that there is a thing which we have been in the habit of calling electricity. The question is, What is that thing? or, What, of all things, the existence of which we know, have we the best reason to call electricity? We know that it acts like an incompressible fluid; that there must be a constant quantity of it in nature; that it can be neither produced nor destroyed; and, what is more important, the electro-magnetic theory of light and all facts observed teach us that electric and ether phenomena are identical. The idea at once suggests itself, therefore, that electricity might be called ether. In fact, this view has in a certain sense been advanced by Dr. Lodge. His interesting work has been read by everyone and many have been convinced by his arguments. Isis great ability and the interesting nature of the subject, keep the reader spellbound; but when the impressions fade, one realizes that he has to deal only with ingenious explanations. I must confess, that I cannot believe in two electricities, much less in a doubly-constituted ether. The puzzling behavior of tile ether as a solid waves of light anti heat, and as a fluid to the motion of bodies through it, is certainly explained in the most natural and satisfactory manner by assuming it to be in motion, as Sir William Thomson has suggested; but regardless of this, there is nothing which would enable us to conclude with certainty that, while a fluid is not capable of transmitting transverse vibrations of a few hundred or thousand per second, it might not be capable of transmitting such vibrations when they range into hundreds of million millions per second. Nor can anyone prove that there are transverse ether waves emitted from an alternate current machine, giving a small number of alternations per second; to such slow disturbances, the ether, if at rest, may behave as a true fluid.

Returning to the subject, and bearing in mind that the existence of two electricities is, to say the least, highly improbable, we must remember, that we have no evidence of electricity, nor can we hope to get it, unless gross matter is present. Electricity, therefore, cannot be called ether in the broad sense of the term; but nothing would seem to stand in the way of calling electricity ether associated with matter, or bound ether; or, in other words, that the so-called static charge of the molecule is ether associated in some way with the molecule. Looking at it in that light, we would be justified in saying, that electricity is concerned in all molecular actions.

Now, precisely what the ether surrounding the molecules is, wherein it differs from ether in general, can only be conjectured. It cannot differ in density, ether being incompressible; it must, therefore, be under some strain or is motion, and the latter is the most probable. To understand its functions, it would be necessary to have an exact idea of the physical construction of matter, of which, of course, we can only form a mental picture.

But of all the views on nature, the one which assumes one matter and one force, and a perfect uniformity throughout, is the most scientific and most likely to be true. An infinitesimal world, with the molecules and their atoms spinning and moving in orbits, in much the same manner as celestial bodies, carrying with them and probably spinning with them ether, or in other words; carrying with them static charges, seems to my mind the most probable view, and one which, in a plausible manner, accounts for most of the phenomena observed. The spinning of the molecules and their ether sets up the ether tensions or electrostatic strains; the equalization of ether tensions sets up ether motions or electric currents, and the orbital movements produce the effects of electro- and permanent magnetism.

About fifteen, years ago, Prof. Rowland demonstrated a most interesting and important fact; namely, that a static charge carried around produces the effects of an electric current. Leaving out of consideration the precise nature of the mechanism, which produces the attraction and repulsion of currents, and conceiving the electrostatically charged molecules in motion, this experimental fact gives us a fair idea of magnetism. We can conceive lines or tubes of force which physically exist, being formed of rows of directed moving molecules; we can see that these lines must be closed, that they must tend to shorten and expand, etc. It likewise explains in a reasonable way, the most puzzling phenomenon of all, permanent magnetism, and, in general, has all the beauties of the Ampere theory without possessing the vital defect of the same, namely, the assumption of molecular currents. Without enlarging further upon the subject, I would say, that I look upon all electrostatic, current and magnetic phenomena as being due to electrostatic molecular forces.

The preceding remarks I have deemed necessary to a full understanding; of the subject as it presents itself to my mind.

Of all these phenomena the most important to study are the current phenomena, on account of the already extensive and ever-growing use of currents for industrial purposes. It is now a century since the first practical source of current was produced, and, ever since, the phenomena which accompany the flow of currents have been diligently studied, and through the untiring efforts of scientific men the simple laws which govern them have been discovered. But these laws are found to hold good only when the currents are of a steady character. When the currents are rapidly varying in strength, quite different phenomena, often unexpected, present themselves, and quite different laws hold good, which even now have not been determined as fully as is desirable, though through the work, principally, of English scientists, enough knowledge has been gained on the subject to enable us to treat simple cases which now present themselves in daily practice.

The phenomena which are peculiar to the changing character of the currents are greatly exalted when the rate of change is increased, hence the study of these currents is considerably facilitated by the employment of properly constructed apparatus. It was with this and other objects in view that I constructed alternate current machines capable of giving more than two million reversals of current per minute, and to this circumstance it is principally due, that I am able to bring to your attention some of the results thus far reached; which I hope will prove to be a step in advance on account of their direct bearing upon one of the most important problems, namely, the production of a practical and efficient source of light.

The study of such rapidly alternating currents is very interesting. Nearly every experiment discloses something new. Many results may, of course, be predicted, but many more are unforeseen. The experimenter makes many interesting observations. For instance, we take a piece of iron and hold it against a magnet. Starting from low alternations and running up higher and higher we feel the impulses succeed each other faster and faster, get weaker and weaker, and finally disappear. We then observe a continuous pull; the pull, of course, is not continuous; it only appears so to us; our sense of touch is imperfect.

We may next establish an arc between the electrodes and observe, as the alternations rise, that the note which accompanies alternating arcs gets shriller and shriller, gradually weakens, and finally ceases. The air vibrations, of course, continue, but they are too weak to be perceived; our sense of hearing fails us.

We observe the small physiological effects, the rapid heating of the iron cores and conductors, curious inductive effects, interesting condenser phenomena, and still more interesting light phenomena with a high tension induction coil. All these experiments and observations would be of the greatest interest to the student, but their description would lead me too far from the principal subject. Partly for this reason, and partly on account of their vastly greater importance, I will confine myself to the description of the light effects produced by these currents.

In the experiments to this end a high tension induction coil or equivalent apparatus for converting currents of comparatively low into currents of high tension is used.

If you will be sufficiently interested in the results I shall describe as to enter into an experimental study of this subject; if you will be convinced of the truth of the arguments I shall advance—your aim will be to produce high frequencies and high potentials; in other words, powerful electrostatic effects. You will then encounter many difficulties, which, if completely overcome, would allow us to produce truly wonderful results.

First will be met the difficulty of obtaining the required frequencies by means of mechanical apparatus, and, if they be obtained otherwise, obstacles of a different nature will present themselves. Next it will be found difficult to provide the requisite insulation without considerably increasing the size of the apparatus, for the potentials required are high, and, owing to the rapidity of the alternations, the insulation presents peculiar difficulties. So, for instance, when a gas is present, the discharge may work, by the molecular bombardment of the gas and consequent heating, through as much as an inch of the best solid insulating material, such as glass, hard rubber, porcelain, sealing wax, etc.; in fact, through any known insulating substance. The chief requisite in the insulation of the apparatus is, therefore, the exclusion of any gaseous matter.

In general my experience tends to show that bodies which possess the highest specific inductive capacity, such as glass, afford a rather inferior insulation to others, which, while they are good insulators, have a much smaller specific inductive capacity, such as oils, for instance, the dielectric losses being no doubt greater in the former. The difficulty of insulating, of course, only exists when the potentials are excessively high, for with potentials such as a few thousand volts there is no particular difficulty encountered in conveying currents from a machine giving, say, 20,000 alternations per second, to quite a distance. This number of alternations, however, is by far too small for many purposes, though quite sufficient for some practical applications. This difficulty of insulating is fortunately not a vital drawback; it affects mostly the size of the apparatus, for, when excessively high potentials would be used, the light-giving devices would be located not far from the apparatus, and often they would be quite close to it. As the air-bombardment of the insulated wire is dependent on condenser action, the loss may be reduced to a trifle by using excessively thin wires heavily insulated.

Another difficulty will be encountered in the capacity and self-induction necessarily possessed by the coil. If the coil be large, that is, if it contain a great length of wire, it will be generally unsuited for excessively high frequencies; if it be small, it may be well adapted for such frequencies, but the potential might then not be as high as desired. A good insulator, and preferably one possessing a small specific inductive capacity, would afford a two-fold advantage. First, it would enable us to construct a very small coil capable of withstanding enormous differences of potential; and secondly, such a small coil, by reason of its smaller capacity and self-induction, would be capable of a quicker and more vigorous vibration. The problem then of constructing a coil or induction apparatus of any kind possessing the requisite qualities I regard as one of no small importance, and it has occupied me for a considerable time.

The investigator who desires to repeat the experiments which I will describe, with an alternate current machine, capable of supplying currents of the desired frequency, and an induction coil, will do well to take the primary coil out and mount the secondary in such a manner as to be able to look through the tube upon which the secondary is wound. He will then be able to observe the streams which pass from the primary to the insulating tube, and from their intensity he will know how far he can strain the coil. Without this precaution he is sure to injure the insulation. This arrangement permits, however, an easy exchange of the primaries, which is desirable in these experiments.

The selection of the type of machine best suited for the purpose must be left to the judgment of the experimenter. There are here illustrated three distinct types of machines, which, besides others, I have used in my experiments.

Fig. 97.

Fig. 97 represents the machine used in my experiments before this Institute. The field magnet consists of a ring of wrought iron with 384 pole projections. The armature comprises a steel disc to which is fastened a thin, carefully welded rim of wrought iron. Upon the rim are wound several layers of fine, well annealed iron wire, which, when wound, is passed through shellac. The armature wires are wound around brass pins, wrapped with silk thread. The diameter of the armature wire in this type of machine should not be more than 1/6 of the thickness of the pole projections, else the local action will be considerable.

Fig. 98.

Fig. 98 represents a larger machine of a different type. The field magnet of this machine consists of two like parts which either enclose an exciting coil, or else are independently wound. Each part has 480 pole projections, the projections of one facing those of the other. The armature consists of a wheel of hard bronze, carrying the conductors which revolve between the projections of the field magnet. To wind the armature conductors, I have found it most convenient to proceed in the following manner. I construct a ring of hard bronze of the required size. This ring and the rim a the wheel are provided with the proper number of pins, and both fastened upon a plate. The armature conductors being wound, the pins are cut off and the ends of the conductors fastened by two rings which screw to the bronze ring and the rim of the wheel, respectively. The whole may then be taken off and forms a solid structure. The conductors in such a type of machine should consist of sheet copper, the thickness of which, of course, depends on the thickness of the pale projections; or else twisted thin wires should be employed.

Fig. 99.

Fig. 99 is a smaller machine, in many respects similar to the former, only here the armature conductors and the exciting coil are kept stationary, while only a block of wrought iron is revolved.

It would be uselessly lengthening this description were I to dwell more on the details of construction of these machines. Besides, they have been described somewhat more elaborately in The Electrical Engineer, of March 18, 1891. I deem it well, however, to call the attention of the investigator to two things, the importance of which, though self evident, he is nevertheless apt to underestimate; namely, to the local action in the conductors which must be carefully avoided, and to the clearance, which must be small. I may add, that since it is desirable to use very high peripheral speeds, the armature should he of very large diameter in order to avoid impracticable belt speeds. Of the several types of these machines which have been constructed by me,

I have found that the type illustrated in Fig. 97 caused me the least trouble in construction, as well as in maintenance, and on the whole, it has been a good experimental machine.

In operating an induction coil with very rapidly alternating currents, among the first luminous phenomena noticed are naturally those, presented by the high-tension discharge. As the number of alternations per second is increased, or as — the number being high — the current through the primary is varied, the discharge gradually changes in appearance. It would be difficult to describe the minor changes which occur, and the conditions which bring them about, but one may note five distinct forms of the discharge.

Fig. 100a.

Fig. 100b.

First, one may observe a weak, sensitive discharge in the form of a thin, feeble-colored thread (Fig. 100a). It always occurs when, the number of alternations per second being high, the current through the primary is very small. In spite of the excessively small current, the rate of change is great, and the difference of potential at the terminals of the secondary is therefore considerable, so that the arc is established at great distances; but the quantity of "electricity" set in motion is insignificant, barely sufficient to maintain a thin, threadlike arc. It is excessively, sensitive and may be made so to such a degree that the mere act of breathing near the coil will affect it, and unless it is perfectly well protected from currents of air, it wriggles around constantly. Nevertheless, it is in this form excessively persistent, and when the terminals are approached to, say, one-third of the striking distance, it can be blown out only with difficulty. This exceptional persistency, when short, is largely due to the arc being excessively thin; presenting, therefore, a very small surface to the blast. Its great sensitiveness, when very long, is probably due to the motion of the particles of dust suspended in the air.

When the current through the primary is increased, the discharge gets broader and stronger, and the effect of the capacity of the coil becomes visible until, finally, under proper conditions, a white flaming arc, Fig. 100b, often as thick as one's finger, and striking across the whole coil, is produce. It develops remarkable heat, and may be further characterized by the absence of the high note which accompanies the less powerful discharges. To take a shock from .the coil under these conditions would not be advisable, although under different conditions the potential being much higher; a shock from the coil may be taken with impunity. To produce this kind of discharge the number of alternations per second must not be too great for the coil used; and, generally speaking, certain relations between capacity, self-induction and frequency must be observed.

The importance of these elements in an alternate current circuit is now well-known, and under ordinary conditions, the general rules are applicable. But in an induction coil exceptional conditions prevail. First, the self-induction is of little importance before the arc is established, when it asserts itself, but perhaps never as prominently as in ordinary alternate current circuits, because the capacity is distributed all along the coil, and by reason of the fact that the coil usually discharges through very great resistances; hence the currents are exceptionally small. Secondly, the capacity goes on increasing continually as the potential rises, in consequence of absorption which takes place to a considerable extent. Owing to this there exists no critical relationship between these quantities, and ordinary rules would not seem: to be applicable: As the potential is increased either in consequence of the increased frequency or of the increased current through the primary, the amount of the energy stored becomes greater and greater, and the capacity gains more and more in importance. Up to a certain point the capacity is beneficial, but after that it begins to be an enormous drawback. It follows from this that each coil gives the best result with a given frequency and primary current. A very large coil, when operated with currents of very high frequency, may not give as much as 1/8 inch spark. By adding capacity to the terminals, the condition may be improved, but what the coil really wants is a lower frequency.

When the flaming discharge occurs, the conditions are evidently such that the greatest current is made to flow through the circuit. These conditions may be attained by varying the frequency within wide limits, but the highest frequency at which the flaming arc can still be produced, determines, for a given primary current, the maximum striking distance of the coil. In the flaming discharge the eclat effect of the capacity is not perceptible; the rate at which the energy is being stored then just equals the rate at which it can be disposed of through the circuit. This kind of discharge is the severest test for a coil; the break, when it occurs, is of the nature of that in an overcharged Leyden jar. To give a rough approximation I would state that, with an ordinary coil of, say, 10,000 ohms resistance, the most powerful arc would be produced with about 12,000 alternations per second.

When the frequency is increased beyond that rate, the potential, of course, rises, but the striking distance may, nevertheless, diminish, paradoxical as it may seem. As the potential rises the coil attains more and more the properties of a static machine until, finally, one may observe the beautiful phenomenon of the streaming discharge, Fig. 101, which may be produced across the whole length of the coil. At that stage streams begin to issue freely from all points and projections. These streams will also be seen to pass in abundance in the space between the primary and the insulating tube. When the potential is excessively high they will always appear; even if the frequency be low, and even if the primary be surrounded by as much as an inch of wax, hard rubber, glass, or any other insulating substance. This limits greatly the output of the coil, but I will later show how I have been able to overcome to a considerable extent this disadvantage in the ordinary coil.

FIG. 101.   FIG. 102.

Besides the potential, the intensity of the streams depends on the frequency; but if the coil be very large they show themselves, no matter how low the frequencies used. For instance, in a very large coil of a resistance of 67,000 ohms, constructed by me some time ago, they appear with as low as 100 alternations per second and less, the insulation of the secondary being 3/4 inch of ebonite. When very intense they produce a noise similar to that produced by the charging of a Holtz machine, but much more powerful, and they emit a strong smell of ozone. The lower the frequency, the more apt they are to suddenly injure the coil. With excessively high frequencies they may pass freely without producing any other effect than to heat the insulation slowly and uniformly.

The existence of these streams shows the importance of constructing an expensive coil so as to permit of one's seeing through the tube surrounding the primary, and the latter should be easily exchangeable; or else the space between the primary and secondary should be completely filled up with insulating material so as to exclude all air. The non-observance of this simple rule in the construction of commercial coils is responsible for the destruction of many an expensive coil.

At the stage when the streaming discharge occurs, or with somewhat higher frequencies, one may, by approaching the terminals quite nearly, and regulating properly the effect of capacity, produce a veritable spray of small silver-white sparks, or a bunch of excessively thin silvery threads (Fig. 102) amidst a powerful brush — each spark or thread possibly corresponding to one alternation. This, when produced under proper conditions, is probably the most beautiful discharge, and when an air blast is directed against it, it presents a singular appearance. The spray of sparks, when received through the body, causes some inconvenience, whereas, when the discharge simply streams, nothing at all is likely to be felt if large conducting objects are held in the hands to protect them from receiving small burns.

Fig. 103.   Fig. 104.

If the frequency is still more increased, then the coil refuses to give any spark unless at comparatively small distances, and the fifth typical form of discharge may be observed (Fig. 103). The tendency to stream out and dissipate is then so great that when the brush is produced at one terminal no sparking occurs; even if, as I have repeatedly tried, the hand, or any conducting object, is held within the stream; and, what is mere singular, the luminous stream is not at all easily deflected by the approach of a conducting body.

At this stage the streams seemingly pass with the greatest freedom through considerable thicknesses of insulators, and it is particularly interesting to study their behavior. For this purpose it is convenient to connect to the terminals of the coil two metallic spheres which may be placed at any desired distance, Fig. 104. Spheres arc preferable to plates, as the discharge can be better observed. By inserting dielectric bodies between the spheres, beautiful discharge phenomena tray be observed. If the spheres be quite close and the spark be playing between them, by interposing a thin plate of ebonite between the spheres the span: instantly ceases and the discharge spread;

into an intensely luminous circle several inches in diameter, provided the spheres are sufficiently large. The passage of the streams heats, and; after a while, softens, the rubber so much that two plates may be made to stick together in this manner. If the spheres are so far apart that no spark occurs, even if they are far beyond the striking distance, by inserting a thick plate of mass the discharge is instantly induced to pass from the spheres to the glass is the form of luminous streams. It appears almost as though these streams pass through the dielectric. In reality this is not the case, as the streams are due to the molecules of the air which are violently agitated in the space between the oppositely charged surfaces of the spheres. When no dielectric other than air is present, the bombardment goes on, but is too weak to be visible; by inserting, a dielectric the inductive effect is much increased, and besides, the projected air molecules find an obstacle and the bombardment becomes so intense that the streams become luminous. If by any mechanical means we could effect such a violent agitation of the molecules we could produce the same phenomenon. A jet of air escaping through a small hole under enormous pressure and striking against an insulating substance, such as glass, may be luminous in the dark, and it might be possible to produce a phosphorescence of the gloss or other insulators in this manner.

The greater the specific inductive capacity of the interposed dielectric, the more powerful the effect produced. Owing to this, the streams show themselves with excessively high potentials even if the glass be as much as one and one-half to two inches thick. But besides the heating due to bombardment, some heating goes on undoubtedly in the dielectric, being apparently greater in glass than in ebonite. I attribute this to the greater specific inductive capacity of the glass; in consequence of which, with the same potential difference, a greater amount of energy is taken up in it than in rubber. It is like connecting to a battery a copper and a brass wire of the same dimensions. The copper wire, though a more perfect conductor, would heat more by reason of its taking more current. Thus what is otherwise considered a virtue of the glass is here a defect. Glass usually gives way much quicker than ebonite; when it is heated to a certain degree, the discharge suddenly breaks through at one point, assuming then the ordinary form of an arc.

The heating effect produced by molecular bombardment of the dielectric would, of course, diminish as the pressure of tile air is increased, and at enormous pressure it would be negligible, unless the frequency would increase correspondingly.

It will be often observed in these experiments that when the spheres are beyond the striking distance, the approach of a glass plate, for instance, may induce the spark to jump between the spheres. This occurs when the capacity of the spheres is somewhat below the critical value which gives the greatest difference of potential at the terminals of the coil. By approaching a dielectric, the specific inductive capacity of the space between the spheres is increased, producing the same effect as if the capacity of the spheres were increased. The potential at the terminals may then rise so high that the air space is cracked. The experiment is best performed with dense glass or mica.

Another interesting observation is that a plate of insulating material, when the discharge is passing through it, is strongly attracted by either of the spheres, that is by the nearer one, this being obviously due to the smaller mechanical effect of the bombardment on that side, and perhaps also to the greater electrification.

From the behavior of the dielectrics in these experiments; we may conclude that the best insulator for these rapidly alternating currents would be the one possessing the smallest specific inductive capacity and at the same time one capable of withstanding the greatest differences of potential; and thus two diametrically opposite ways of securing the required insulation are indicated, namely, to use either a perfect vacuum or a gas under great pressure; but the former would be preferable. Unfortunately neither of these two ways is easily carried out in practice.

Fig. 105.

Fig. 106.

It is especially interesting to note the behavior of an excessively high vacuum in these experiments. If a test tube, provided with external electrodes and exhausted to the highest possible degree, be connected to the terminals of the coil, Fig. 105, the electrodes of the tube are instantly brought to a high temperature and the glass at each end of the tube is rendered intensely phosphorescent, but the middle appears comparatively dark, and for a while remains cool.

When the frequency is so high that the discharge shown in Fig. 103 is, observed, considerable dissipation no doubt occurs in the coil. Nevertheless the coil may be worked for a long time, as the heating is gradual.

In spite of the fact that the difference of potential may be enormous, little is felt when the discharge is passed through the body, provided the hands are armed. This is to some extent due to the higher frequency, but principally to the fact that less energy is available externally, when the difference of potential reaches an enormous value, owing to the circumstance that, with the rise of potential, the energy absorbed in the coil increases as the square of the potential. Up to a certain point the energy available externally increases with the rise of potential, then it begins to fall off rapidly. Thus, with the ordinary high tension induction coil, the curious paradox exists, that, while with a given current through the primary the shock might be fatal, with many times that current it might be perfectly harmless, even if the frequency be the same. With high frequencies and excessively high potentials when the terminals are not connected to bodies of some size, practically all the energy supplied to the primary is taken up by the coil. There is no breaking through, no local injury, but all the material, insulating and conducting, is uniformly heated.

To avoid misunderstanding in regard to the physiological effect of alternating currents of very high frequency, I think it necessary to state that, while it is an undeniable fact that they are incomparably less dangerous than currents of low frequencies; it should not be thought that they are altogether harmless. What has just been said refers only to currents from an ordinary high tension induction coil, which currents are necessarily very small; if received directly from a machine or from a secondary of low resistance, they produce more or less powerful effects, and may cause serious injury, especially when used in conjunction with condensers.

The streaming discharge of a high tension induction coil differs in many respects from that of a powerful static machine. In color it has neither the violet of the positive, nor the brightness of the negative, static discharge, but lies somewhere between, being, of course, alternatively positive and negative. But since the streaming is more powerful when the point or terminal is electrified positively, than when electrified negatively, it follows that the point of the brush is more like the positive, and the root more like the negative, static discharge. In the dark, when the brush is very powerful, the root may appear almost white. The wind produced by the escaping streams, though it may be very strong—often indeed to such a degree that it may be felt quite a distance from the coil—is, nevertheless, considering the quantity of the discharge, smaller than that produced by the positive brush of a static machine, and it affects the flame much less powerfully: From the nature of the phenomenon we can conclude that the higher the frequency, the smaller must, of course, be the wind produced by the streams, and with sufficiently high frequencies no wind at all would be produced at the ordinary atmospheric pressures. With frequencies obtainable by means of a machine, the mechanical effect is sufficiently great to revolve, with considerable speed, large pin-wheels, which in the dark present beautiful appearance owing to the abundance of the streams (Fig. 106).

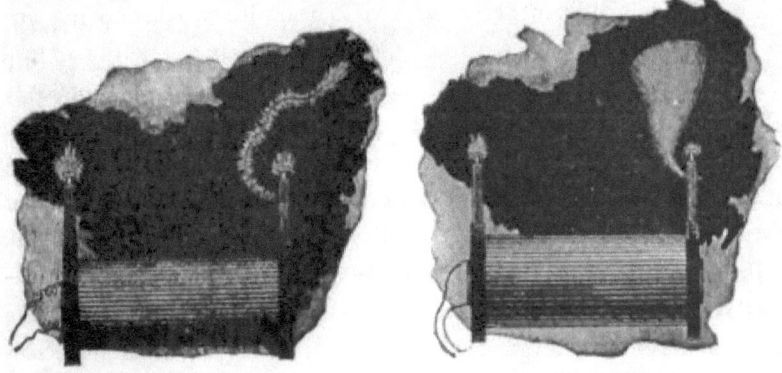

FIG. 107.   FIG. 108.

In general, most of the experiments usually performed with a static machine can be performed with an induction coil when operated with very rapidly alternating currents. The effects produced, however, are much more striking; being of incomparably greater power. When a small length of ordinary cotton covered wire, Fig. 107, is attached to one terminal of the coil, the streams issuing from all points of the wire may be so intense as to produce a considerable light effect. When the potentials and frequencies are very high, a wire insulated with *gutta percha* or rubber and attached to one of the terminals, appears to be covered with a luminous film A very thin bare wire when attached to a terminal emits powerful streams and vibrates continually to and fro or spins in a circle, producing a singular effect (Fig. 108). Some of these experiments have been described by me in *The Electrical World*, of February 21, 1891.

Another peculiarity of the rapidly alternating discharge of the induction coil is its radically different behavior with respect to points and rounded surfaces.

If a thick wire, provided with a ball at one end and with a point at the other, be attached to the positive terminal of a static machine, practically all the charge will be lost through the point, on account of the enormously greater tension, dependent on the radius of curvature. But if such a wire is attached to one of the terminals of the induction coil, it, will be observed that with very high frequencies streams issue from the ball almost as copiously as from the point (Fig. 109).

It is hardly conceivable that we could produce such a condition to an equal degree in a static machine, for the simple reason, that the tension increases as the square of the density, which in turn is proportional to the radius of curvature; hence, with a steady potential an enormous charge would be required to make streams issue from a polished ball while it is connected with a point. But with an induction coil the discharge of which alternates with great rapidity it is different: Here we have to deal with two distinct tendencies. First, there is the tendency to escape which exists in a condition of rest, and which depends on the radius of curvature; second, there is the tendency to dissipate into the surrounding air by condenser action, which depends on the surface. When one of these tendencies is at a maximum, the other is at a minimum. At the point the luminous stream is principally due to the air molecules coming bodily in contact with the point; they are attracted and repelled, charged and discharged, and, their atomic charges being thus disturbed; vibrate and emit light waves. At the ball, on the contrary, there is no doubt that the effect is to a great extent produced inductively, the air molecules not necessarily coming in contact with the ball, though they undoubtedly do so. To convince ourselves of this we only need to exalt the condenser action, for instance, by enveloping the ball, at some distance, by a better conductor than the surrounding medium, the conductor being, of course, insulated; or else by surrounding it with a better dielectric and approaching an insulated conductor; in both cases the streams will break forth more copiously. Also, the larger the ball with a given frequency, or the higher the frequency, the more will the ball have the advantage over the point. But, since a certain intensity of action is required to render the streams visible, it is obvious that in the experiment described the ball should not be taken too large.

In consequence of this two-fold tendency, it is possible to produce by means of points, effects identical to those produced by capacity. Thus, for instance, by attaching to one terminal of the coil a small length of soiled wire, presenting many points and offering great facility to escape, the potential of the coil may be raised to the same value as by attaching to the terminal a polished ball of a surface many times greater than that of the wire.

An interesting experiment, showing the effect of the points, may be performed in the following manner: Attach to one of the terminals of the coil a cotton covered wire about two feet in length, and adjust the conditions so that streams issue from the wire. In this experiment the primary coil should be preferably placed so that it extends only about half way into the secondary coil. Now touch the free terminal of the secondary with a conducting object held in the hand, or else connect it to an insulated body of some size. In this manner the potential on the wire may be enormously raised. The effect of this will be either to increase, or to diminish, the streams: if they increase, the wire is too short; if they diminish, it is too long. By adjusting the length of the wire, a point is found where the touching of the other terminal does not at all affect the streams. In this case the rise of potential is exactly counteracted by the drop through the coil. It will be observed that small lengths of wire produce considerable difference in the magnitude and luminosity of the streams. The primary coil is placed sidewise for two reasons: first, to increase the potential at the wire: and, second, to,increase the drop through the coil. The sensitiveness is thus augmented.

Fig. 109.   Fig. 110.

There is still another and far more striking peculiarity of the brush discharge produced by very rapidly alternating currents. To observe this it is best to replace the usual terminals of the coil by two metal columns insulated with a good thickness of ebonite. It is also well to close all fissures and cracks with wax so that the brushes cannot form anywhere except at the tops of the columns. If the conditions are carefully adjusted — which, of course, must be left to the skill of the experimenter — so that the potential rises to an enormous value, one may produce two powerful brushes several inches long, nearly white at their roots, which in the dart: bear a striking resemblance two flames of a gas escaping under pressure (Fig. 110). But they do not only resemble, they are veritable flames, for they are hot. Certainly they are not as hot as a gas burner, *but they would be so if the frequency and the potential would be sufficiently high.* Produced with, say, twenty thousand alternations per second, the heat is easily perceptible even if the potential is not excessively high. The heat developed is, of course, due to the impact of the air molecules against the terminals and against each other. As, at the ordinary pressures, the mean free path is excessively small, it is possible that in spite of the enormous initial speed imparted to each molecule upon coming in contact with the terminal, its progress — by collision with other molecules — is retarded to such an extent, that it does not get away far from the terminal, but may strike the same many times in succession. The higher the frequency, the less the molecule is able to get away, and this the more so, as for a given effect the potential required is smaller; and a frequency is conceivable — perhaps even obtainable — at which practically the same molecules would strike the terminal. Under such conditions the exchange of the molecules would be very slow, and the heat produced at, and very near, the terminal would be excessive. But if the frequency would go on increasing constantly, the heat produced would begin to diminish for obvious reasons. In the positive brush of a static machine the exchange of the molecules is very rapid, the stream is constantly of one direction, and there are fewer collisions; hence the heating effect must be very small. Anything that impairs the facility of exchange tends to increase the local heat produced. Thus, if a bulb be held over the terminal of the coil so as to enclose the brush, the air contained in the bulb is very quickly brought to a high temperature. If a glass tube be held over the brush so as to allow the draught to carry the brush upwards, scorching hot air escapes at the top of the tube. Anything held within the brush is, of course, rapidly heated, and the possibility of using such heating effects for some purpose or other suggests itself.

When contemplating this singular phenomenon of the hot brush, we cannot help being convinced that a similar process must take place in the ordinary flame, and it seems strange that after all these centuries past of familiarity with the flame, now, in this era of electric lighting and heating; we are finally led to recognize, that since time immemorial we have, after all, always had "electric light and : heat" at our disposal. It is also of no little interest to contemplate, that we have a possible way of producing — by other than chemical means — a veritable flame; which would give light and heat without any material being consumed, without any chemical process taking place, and to accomplish this, we only need to perfect methods of producing enormous frequencies and potentials. I have no doubt that if the potential could be made to alternate with sufficient rapidity and power, the brush formed at the end of a wire would lose its electrical characteristics and would become flame-like. The flame must be due to electrostatic molecular action.

This phenomenon now explains in a manner which can hardly be doubted the frequent accidents occurring in storms. It is well known that objects are often set on fire without the lightning striking them. We shall presently see how this can happen. On a nail in a roof, for instance, or on a projection of any kind, more or less conducting, or rendered so by dampness, a powerful brush may appear. If the lightning strikes somewhere in .the neighborhood the enormous potential may be made to alternate or fluctuate perhaps many million times a second. The air molecules are violently attracted and repelled, and by their impact produce such a powerful heating effect that a fire is started. It is conceivable that a ship at sea may, in this manner, catch fire at many points at once. When we consider, that even with the comparatively low frequencies obtained from a dynamo machine, and with potentials of no more than one or two hundred thousand volts, the heating effects are considerable, we may imagine how much more powerful they must be with frequencies and potentials many times greater: and the above explanation seems, to say the least, very probable. Similar explanations may have been suggested, but I am not aware that, up to the present; the heating effects of a brush produced by a rapidly alternating potential have been experimentally demonstrated, at least not to such a remarkable degree.

Fig. 111.

By preventing completely the exchange of the air molecules, the local heating effect may be so exalted as to bring a body to incandescence. Thus, for instance, if a small button, or preferably a very thin wire or filament be enclosed in an unexhausted globe and connected with the terminal of the coil, it may be rendered incandescent. The phenomenon is made much more interesting by the rapid spinning round in a circle of the top of the filament, thus presenting the appearance of a luminous funnel, Fig. 111, which widens when the potential is increased. When the potential is small the end of the filament may perform irregular motions, suddenly changing from one to the other, or it may describe an ellipse; but when the potential is very high it always spins in a circle; and so does generally a thin straight wire attached freely to the terminal of the coil. These motions are, of course, due to the impact of the molecules, and the irregularity in the distribution of the potential, owing to the roughness and dissymmetry of the wire or filament. With a perfectly symmetrical and polished wire such motions would probably not occur. That the motion is not likely to be due to other causes is evident from the fact that it is not of a definite direction, and that in a very highly exhausted globe it ceases altogether. The possibility of bringing a body to incandescence in an exhausted globe, or even when not at all enclosed, would seem to afford a possible way of obtaining light effects, which, in perfecting methods of producing rapidly alternating potentials, might be rendered available for useful purposes,

In employing a commercial coil; the production of very powerful brush effects is attended with considerable difficulties, for when these high frequencies and enormous potentials are used, the best insulation is apt to give way. Usually the coil is insulated well enough to stand the strain from convolution to convolution, since two double silk covered paraffined wires will withstand a pressure of several thousand volts; the difficulty lies principally in preventing the breaking through from the secondary to the primary, which is greatly facilitated by the streams issuing from the latter. In the coil, of course, the strain is greatest from section to section — but usually in a larger coil there are so many sections that the danger of a sudden giving way is not very great. No difficulty will generally be encountered in that direction, and besides, the liability of injuring the coil internally is very much reduced by the fact that the effect most likely to be produced is simply a gradual heating, which, when far enough advanced, could not fail to be observed. The principal necessity is then to prevent the streams between he primary and the tube, not only on account of the heating and possible injury, but also because the streams may diminish very considerably the potential difference available at the terminals. A few hints as to how this may be accomplished will probably be found useful in most of these experiments with the ordinary induction coil.

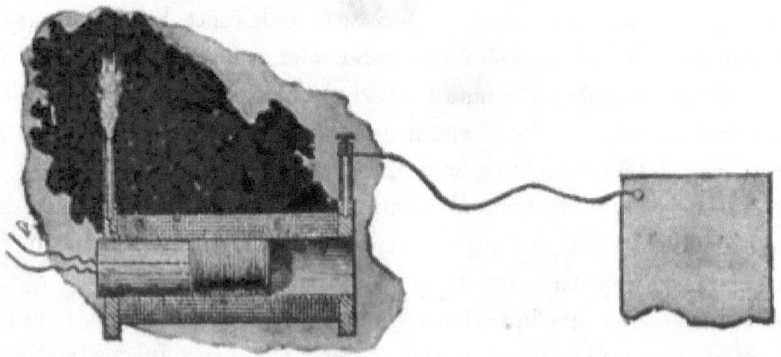

FIG. 112a.

One of the ways is to wind a short primary, Fig. 112a, so that the difference of potential is not at that length great enough to cause the breaking forth of the streams through the insulating tube. The length of the primary should be determined by experiment. Both the ends of the coil should be brought out on one end through a plug of insulating material fitting in the tube as illustrated. In such a disposition one terminal of the secondary is attached to a body, the surface of which is determined with the greatest care so as to produce the greatest rise in the potential. At the other terminal a powerful brush appears, which may be experimented upon.

Fig. 112b.

The above plan necessitates the employment of a primary of comparatively small size, and it is apt to heat when powerful effects are desirable for a certain length of time. In such a case it is better to employ a larger coil, Fig. 112b, and introduce it from one side of the tube, until the streams begin to appear. In this case the nearest terminal of the secondary may be connected to the primary or to the ground, which is practically the same thing, if the primary is connected directly to the machine. In the case of ground connections it is well to determine experimentally the frequency which is best suited under the conditions of the test. Another way of obviating the streams, more or less, is to make the primary in sections and supply it from separate, well insulated sources.

In many of these experiments, when powerful effects are wanted for a short time, it is advantageous to use iron cores with the primaries. In such case a very large primary coil may be wound and placed side by side with the secondary, and, the nearest terminal of the latter being connected to the primary, a laminated iron core is introduced through the primary into the secondary as far as the streams will permit. Under these conditions an excessively powerful brush, several inches long, which may be appropriately called "St. Elmo's hot fire", may be caused to appear at the other terminal of the secondary, producing striking effects. It is a most powerful ozonizer, so powerful indeed, that only a few minutes are sufficient to fill the whole room with the smell of ozone, and it undoubtedly possesses the quality of exciting chemical affinities.

For the production of ozone, alternating currents of very high frequency are eminently suited, not only on account of the advantages they offer in the way of conversion but also because of the fact, that the ozonizing action of a discharge is dependent on the frequency as well as on the potential, this being undoubtedly confirmed by observation.

In these experiments if an iron core is used it should be carefully watched, as it is apt to get excessively hot in an incredibly short time. To give an idea of the rapidity of the heating, I will state, that by passing a powerful current through a coil with many turns, the inserting within the same of a thin iron wire for no more than one seconds time is sufficient to heat the wire to something like 100° C.

But this rapid heating need not discourage us in the use of iron cores in connection with rapidly alternating currents. I have for a long time been convinced that in tile industrial distribution by means of transformers, some such plan as the following might be practicable. We may use a comparatively small iron core, subdivided, or perhaps not even subdivided. We may surround this core with a considerable thickness of material which is fire-proof and conducts the heat poorly, and on top of that we may place the primary and secondary windings. By using either higher frequencies or greater magnetizing forces, we may by hysteresis and eddy currents heat the iron core so far as to bring it nearly to its maximum permeability, which, as Hopkinson has shown, may be as much as sixteen times greater than that at ordinary temperatures. If the iron core were perfectly enclosed, it would not be deteriorated by the heat, and, if the enclosure of fire-proof material would be sufficiently thick, only a limited amount of energy could be radiated in spite of the high temperature. Transformers have been constructed by me on that plan, but for lack of time, no thorough tests have as yet been made.

Another way of adapting the iron core to rapid alternations, or, generally speaking, reducing the frictional losses, is to produce by continuous magnetization a flow of something like seven thousand or eight thousand lines per square centimetre through the core, and then work with weak magnetizing forces and preferably high frequencies around the point of greatest permeability. A higher efficiency of conversion and greater output are obtainable in this manner. I have also employed this principle in connection .with machines in which there is no reversal of polarity. In these types of machines, as long as there are only few pole projections, there is no great gain; as the maxima and minima of magnetization are far from the point of maximum permeability; but when the number of the pole projections is very great, the required rate of change may be obtained, without the magnetization varying so far as to depart greatly from the point of maximum permeability, and the gain is considerable.

The above described arrangements refer only to the use of commercial coils as ordinarily constructed. If it is desired to construct a coil for the express purpose of performing with it such experiments as I have described, or, generally, rendering it capable of withstanding the greatest possible difference of potential, then a construction as indicated in Fig. 113 will be found of advantage. The coil in this case is formed of two independent parts which are wound oppositely, the connection between both being made near the primary. The potential in the middle being zero, there is not much tendency to jump to the primary and not much insulation is required. In some cases the middle point may, however, be connected to the primary or to the ground. In such a coil the places of greatest difference of potential are far apart and the coil is capable of withstanding an enormous strain. The two parts may be movable so as to allow a slight adjustment of the capacity effect.

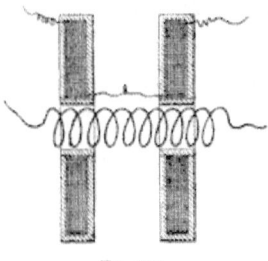

Fig. 113.

As to the manner of insulating the coil, it will be found convenient to proceed in the following way: First, the wire should be boiled in paraffine until all the air is out; then the coil is wound by running the wire through melted paraffine, merely for the purpose of fixing the wire. The coil is then taken off from the spool, immersed in a cylindrical vessel filled with pure melted wax and boiled for a long time until the bubbles cease to appear. The whole is then left to cool down thoroughly, and then the mass is taken out of the vessel and turned up in a lathe. A coil made in this manner and with care is capable of withstanding enormous potential differences.

It may be found convenient to immerse the coil in paraffine oil or some other hind of oil; it is a most effective way of insulating, principally on account of the perfect exclusion of air, but it may be found that, after all, a vessel filled with oil is not a very convenient thing to handle in a laboratory.

If an ordinary coil can be dismounted, the primary may be taken out of the tube and the latter plugged up at one end, filled with oil, and the primary reinserted. This affords an excellent insulation and prevents the formation of the streams.

Of all the experiments which may be performed with rapidly alternating currents the most interesting are those which concern the production of a practical illuminant. It cannot be denied that the present methods, though they were brilliant advances, are very wasteful. Some better methods must be invented, some more perfect apparatus devised. Modern research has opened new possibilities for the production of an efficient source of light, and the attention of all has been turned in the direction indicated by able pioneers. Many have been carried away by the enthusiasm and passion to discover, but in their zeal to reach results, some have been misled. Starting with the idea of producing electro-magnetic waves, they turned their attention, perhaps, too much to the study of electro-magnetic effects, and neglected the study of electrostatic phenomena. Naturally, nearly every investigator availed himself of an apparatus similar to that used in earlier experiments. But in those forms of apparatus, while the electro-magnetic inductive effects are enormous, the electrostatic effects are excessively small.

In the Hertz experiments, for instance, a high tension induction coil is short circuited by an arc, the resistance of which is very small, the smaller, the more capacity is attached to the terminals; and the difference of potential at these is enormously diminished: On the other hand, when the discharge is not passing between the terminals, the static effects may be considerable, but only qualitatively so, not quantitatively, since their rise and fall is very sudden, and since their frequency is small. In neither case, therefore, are powerful electrostatic effects perceivable. Similar conditions exist when, as in some interesting experiments of Dr. Lodge, Leyden jars are discharged disruptively. It has been thought—and I believe asserted—that in such cases most of the energy is radiated into space. In the light of the experiments which I have described above, it will now not be thought so. I feel safe in asserting that in such cases most of the energy is partly taken up and converted into heat. in the arc of the discharge and in the conducting and insulating material of the jar, some energy being, of course, given off by electrification of the air; but the amount of the directly radiated energy is very small.

When a high tension induction coil, operated by currents alternating only 20,000 times a second, has its terminals closed through even a very small jar, practically all the energy passes through the dielectric of the jar, which is heated, and the electrostatic effects manifest themselves outwardly only to a very weak degree. Now the external circuit of a Leyden jar, that is, the arc and the connections of the coatings, may be looked upon as a circuit generating alternating currents of excessively high frequency and fairly high potential, which is closed through the coatings and the dielectric between them, and from the above it is evident that the external electrostatic effects must be very small, even if a recoil circuit be used. These conditions make it appear that with the apparatus usually at hand, the observation of powerful electrostatic effects was impossible, and what experience has been gained in that direction is only due to the great ability of the investigators.

But powerful electrostatic effects are a sine qua non of light production on the lines indicated by theory. Electro-magnetic effects are primarily unavailable, for the reason that to produce the required effects we would have to pass current impulses through a conductor; which, long before the required frequency of the impulses could be reached, would cease to transmit them. On the other hand, electro-magnetic waves many times longer than those of light, and producible by sudden discharge of a condenser, could not be utilized, it would seem, except we avail ourselves of their effect upon conductors as in the present methods, which are wasteful. We could not affect by means of such waves the static molecular or atomic charges of a gas, cause them to vibrate and to emit light. Long transverse waves cannot, apparently, produce such effects, since excessively small electro-magnetic disturbances may pass readily through miles of air. Such dark waves, unless they are of the length of true light waves, cannot, it would seem, excite luminous radiation in a Geissler tube; and the luminous effects, which are producible by induction in a tube devoid of electrodes, I am inclined to consider as being of an electrostatic nature.

To produce such luminous effects, straight electrostatic thrusts are required; these, whatever be their frequency, may disturb the molecular charges and produce light. Since current impulses of the required frequency cannot pass through a conductor of measurable dimensions, we must work with a gas, and then the production of powerful electrostatic effects becomes an imperative necessity.

It has occurred to me, however, that electrostatic effects are in many ways available for the production of light. For instance, we may place a body of some refractory material in a closed; and preferably more or less exhausted, globe, connect it to a source of high, rapidly alternating potential, causing the molecules of the gas to strike it many times a second at enormous speeds, and in this manner, with trillions of invisible hammers, pound it until it, gets incandescent: or we may place a body in a very highly exhausted globe, in a non-striking vacuum, and, by employing very high frequencies and potentials, transfer sufficient energy from it to other bodies in the vicinity, or in general to the surroundings, to maintain it at any degree of incandescence; or we may, by means of such rapidly alternating high potentials, disturb the ether carried by the molecules of a gas or their static charges, causing them to vibrate and to emit light.

But, electrostatic effects being dependent upon the potential and frequency, to produce the most powerful action it is desirable to increase both as far as practicable. It may be possible to obtain quite fair results by keeping either of these factors small, provided the other is sufficiently great; but we are limited in both directions. My experience demonstrates that we cannot go below a certain frequency, for, first, the potential then becomes so great that it is dangerous; and, secondly, the light production is less efficient.

I have found that, by using the ordinary low frequencies, the physiological effect of the current required to maintain at a certain degree of brightness a tube four feet long, provided at the ends with outside and inside condenser coatings, is so powerful that, I think, it might produce serious injury to those not accustomed to such shocks: whereas, with twenty thousand alternations per second, the tube may be maintained at the same degree of brightness without any effect being felt. This is due principally to the fact that a much smaller potential is required to produce the same light effect, and also to the higher efficiency in the light production. It is evident that the efficiency in such cases is the greater, the higher the frequency, for the quicker the process of charging and discharging the molecules, the less energy will be lost in the form of dark radiation. But, unfortunately, we cannot go beyond a certain frequency on account of the difficulty of producing and conveying the effects.

I have stated above that a body inclosed in an unexhausted bulb may be intensely heated by simply connecting it with a source of rapidly alternating potential. The heating in such a case is, in all probability, due mostly to the bombardment of the molecules of the gas contained in the bulb. When the bulb is exhausted, the heating of the body is much more rapid, and there is no difficulty whatever in bringing a wire or filament to any degree of incandescence by simply connecting it to one terminal of a coil of the proper dimensions. Thus, if the well-known apparatus of Prof. Crookes, consisting of a bent platinum wire with vanes mounted over it (Fig. 114), be connected to one terminal of the coil—either one or both ends of the platinum wire being connected—the wire is rendered almost instantly incandescent, and the mica vanes are rotated as though a current from a battery were used: A thin carbon filament, or, preferably, a button of some refractory material (Fig. 115), even if it be a comparatively poor conductor, inclosed in an exhausted globe, may be rendered highly incandescent; and in this manner a simple lamp capable of giving any desired candle power is provided.

The success of lamps of this kind would depend largely on the selection of the light-giving bodies contained within the bulb. Since, under the conditions described, refractory bodies—which are very poor conductors and capable of withstanding for a long time excessively high degrees of temperature—may be used, such illuminating devices may be rendered successful.

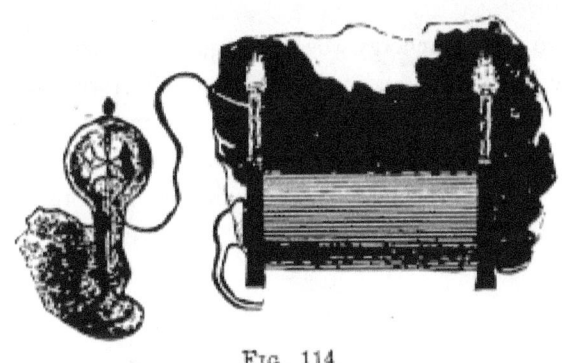
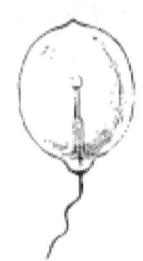

Fig. 114. Fig. 115.

It might be thought at first that if the bulb, containing the filament or button of refractory material, be perfectly well exhausted — that is, as far as it can be done by the use of the best apparatus — the heating would be much less intense, and that in a perfect vacuum it could not occur at all. This is not confirmed by my experience; quite the contrary, the better the vacuum the more easily the bodies are brought to incandescence. This result is interesting for many reasons.

At the outset of this work the idea presented itself to me, whether two bodies of refractory material enclosed in a bulb exhausted to such a degree that the discharge of a large induction coil, operated in the usual manner, cannot pass through, could be rendered incandescent by mere condenser action. Obviously, to reach this result enormous potential differences and very high frequencies are required, as is evident from a simple calculation.

But such a lamp would possess a vast advantage over an ordinary incandescent lamp in regard to efficiency. It is well-known that the efficiency of a lamp is to some extent a function of the degree of incandescence, and that, could we but work a filament at many times higher degrees of incandescence, the efficiency would be much greater. In an ordinary lamp this is impracticable on account of the destruction of the filament, and it has been determined by experience how far it is advisable to push the incandescence. It is impossible to tell how much higher efficiency could be obtained if the filament could withstand indefinitely, as the investigation to this end obviously cannot be carried beyond a certain stage; but there are reasons for believing that it would be very considerably higher. An improvement might be made in the ordinary lamp by employing a short and thick carbon; but then the leading-in wires would have to be thick, and, besides, there are many other considerations which render such a modification entirely impracticable. But in a lamp as above described, the leading-in wires may be very small, the incandescent refractory material may be in the shape of blocks offering a very small radiating surface, so that less energy would be required to keep them at the desired incandescence; and in addition to this, the refractory material need not be carbon, but may be manufactured from mixtures of oxides, for instance, with carbon or other material, or may be selected from bodies which are practically non-conductors, and capable of withstanding enormous degrees of temperature.

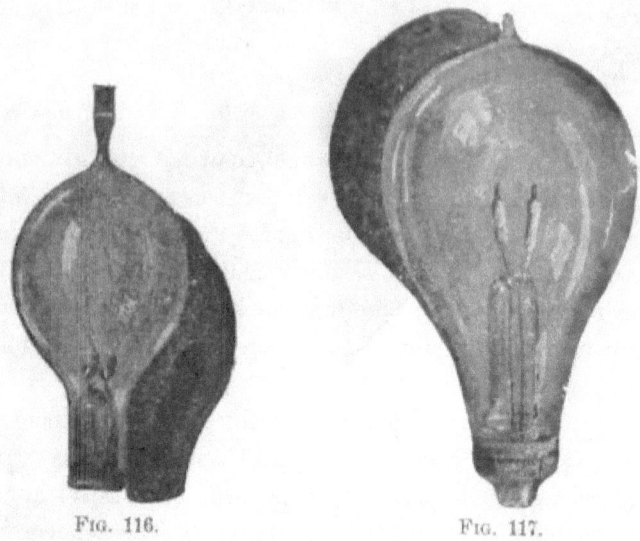

Fig. 116.   Fig. 117.

All this would point to the possibility of obtaining a much higher efficiency with such a lamp than is obtainable in ordinary lamps. In my experience it has been demonstrated that the blocks are brought to high degrees of incandescence with much lower potentials than those determined by calculation, and the blocks may be set at greater distances from each other. We may freely assume, and it is probable, that the molecular bombardment is an important element in the heating, even if the globe be exhausted with the utmost care, as I have done; for although the number of the molecules is, comparatively speaking, insignificant, yet on account of the mean free path being very great, there are fewer collisions, and the molecules may reach much higher speeds, so that the heating effect due to this cause may be considerable, as in the Crookes experiments with radiant matter.

But it is likewise possible that we have to deal here with an increased facility of losing the charge in very high vacuum, when the potential is rapidly alternating, in which case most of the heating would be directly due to the surging of the charges in the heated bodies. Or else the observed fact may be largely attributable to the effect of the points which I have mentioned above, in consequence of which the blocks or filaments contained in the vacuum are equivalent to condensers of many times greater surface than that calculated from their geometrical dimensions. Scientific men still differ in opinion as to whether a charge should, or should not, be lost in a perfect vacuum, or. in other words, whether ether is, or is not, a conductor. If the former were the case, then a thin filament enclosed in a perfectly exhausted globe, and connected to a source of enormous, steady potential, would be brought to incandescence.

Various forms of lamps on the above described principle, with the refractory bodies in the form of filaments, Fig. 116, or blocks, Fig. 117, have been constructed and operated by me, and investigations are being carried on in this line. There is no difficulty in reaching such high degrees of incandescence that ordinary carbon is to all appearance melted and volatilized. If the vacuum could be made absolutely perfect, such a lamp, although inoperative with apparatus ordinarily used, would, if operated with currents of the required character, afford an illuminant which would never be destroyed, and which would be far more efficient than an ordinary incandescent lamp. This perfection can, of course, never be reached; and a very slow destruction and gradual diminution in size always occurs, as in incandescent filaments; but there is no possibility of a sudden and premature disabling which occurs in the latter by the breaking of the filament, especially when the incandescent bodies are in the shape of blocks.

With these rapidly alternating potentials there is, however, no necessity of enclosing two blocks in a globe, but a single block, as in Fig. 115, or filament, Fig. 118, may be used. The potential in this case must of course be higher, but is easily obtainable, and besides it is not necessarily dangerous.

Fig. 118.

The facility with which the button or filament in such a lamp is brought to incandescence, other things being equal, depends on the size of the globe. If a perfect ,vacuum could be obtained, the size of the globe would not be of importance, for then the heating would be wholly due to the surging of the charges, and all the energy would be given off to the surroundings by radiation. But this can never occur in practice. There is always some gas left in the globe, and although the exhaustion may be carried to the highest degree, still the space inside of the bulb must be considered as conducting when such high potentials are used, and I assume that, in estimating the energy that may be given off from the filament to the surroundings, we may consider the inside surface of the bulb as one coating of a condenser, the air and other objects surrounding the bulb forming the other coating. When the alternations are very low there is no doubt that a considerable portion of the energy is given off by the electrification of the surrounding air.

In order to study this subject better, I carried on some experiments with excessively high potentials and low frequencies. I then observed that when the hand is approached to the bulb,—the filament being connected with one terminal of the coil,—a powerful vibration is felt, being due to the attraction and repulsion of the molecules of the air which are electrified by induction through the glass. In some cases when the action is very intense I have been able to hear a sound, which must be due to the same cause.

When the alternations are low, one is apt to get an excessively powerful shock from the bulb. In general, when one attaches bulbs or objects of some size to the terminals of the coil, one should look out for the rise of potential, for it may happen that by merely connecting a bulb or plate to the terminal, the potential may rise to many times its original value. When lamps are attached to the terminals, as illustrated in Fig. 119, then the capacity od the bulbs should be such as to give the maximum rise of potential under the existing conditions. In this manner one may obtain the required potential with fewer turns of wire.

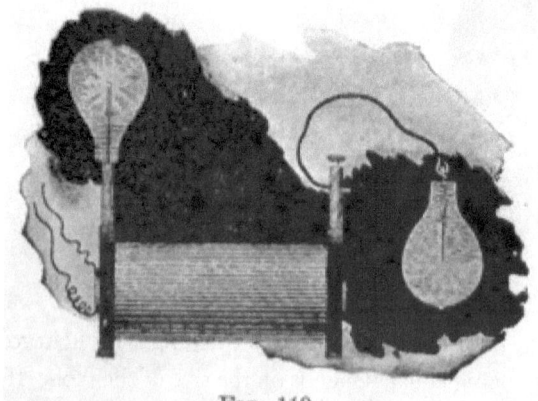
Fig. 119.

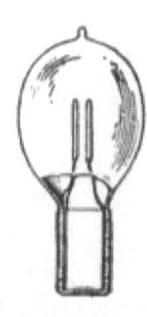
Fig. 120.

The life of such lamps as described above depends, of course, largely on the degree of exhaustion, but to some extent also on the shape of the block of refractory material. Theoretically it would seem that a small sphere of carbon enclosed in a sphere of glass would not suffer deterioration from molecular bombardment, for, the matter in the globe being radiant, the molecules would move in straight lines, and would seldom strike the sphere obliquely. An interesting thought in connection with such a lamp is, that in it "electricity" and electrical energy apparently must move in the same lines.

The use of alternating currents of very high frequency makes it possible to transfer, by electrostatic or electromagnetic induction through the glass of a lamp, sufficient energy to keep a filament at incandescence and so do away with the leading-in wires. Such lamps have been proposed, but for want of proper apparatus they have not been successfully operated. Many forms of lamps on this principle with continuous and broken filaments have been constructed by me and experimented upon. When using a secondary enclosed within the lamp, a condenser is advantageously combined with the secondary. When the transference is effected by electrostatic induction, the potentials used are, of course, very high with frequencies obtainable from a machine. For instance, with a condenser surface of forty square centimetres, which is not impracticably large, and with glass of good quality 1 mm. thick, using currents alternating twenty thousand times a second, the potential required is approximately 9,000 volts. This may seem large, but since each lamp may be included in the secondary of a transformer of very small dimensions, it would not be inconvenient, and, moreover, it would not produce fatal injury. The transformers would all be preferably in series. The regulation would offer no difficulties, as with currents of such frequencies it is very easy to maintain a constant current.

In the accompanying engravings some of the types of lamps of this kind are shown. Fig. 120 is such a lamp with a broken filament, and Figs. 121a and 121b one with a single outside and inside coating and a single filament. I have also made lamps with two outside and inside coatings and a continuous loop connecting the latter. Such lamps have been operated by me with current impulses of the enormous frequencies obtainable by the disruptive discharge of condensers.

The disruptive discharge of a condenser is especially suited for operating such lamps—with no outward electrical connections—by means of electromagnetic induction, the electromagnetic inductive effects being excessively high; and I have been able to produce the desired incandescence with only a few short turns of wire. Incandescence may also be produced in this manner in a simple closed filament.

Fig. 121a. Fig. 121b.

Leaving now out of consideration the practicability of such lamps, I would only say that they possess a beautiful and desirable feature, namely, that they can be rendered, at will, more or less brilliant simply by altering the relative position of the outside and inside condenser coatings, or inducing and induced circuits.

Fig. 122. Fig. 123.

When a lamp is lighted by connecting it to one terminal only of the source, this may be facilitated by providing the globe with an outside condenser coating, which serves at the same time as a reflector, and connecting this to an insulated body of some size. Lamps of this kind are illustrated in Fig. 122 and Fig. 123. Fig. 124 shows the plan of connection. The brilliancy of the lamp may, in this case, be regulated within wide limits by varying the size of the insulated metal plate to which the coating is connected.

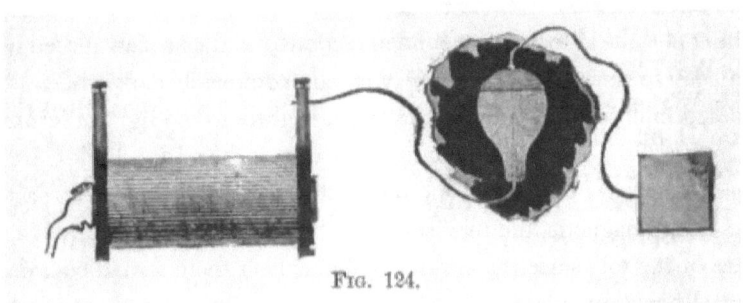

Fig. 124.

It is likewise practicable to light with one leading wire lamps such as illustrated in Fig. 116 and Fig. 117, by connecting one terminal of the lamp to one terminal of the source, and the other to an insulated body of the required size. In all cases the insulated body serves to give off the energy into the surrounding space, and is equivalent to a return wire. Obviously, in the two last-named cases, instead of connecting the wires to an insulated body, connections may be made to the ground.

The experiments which will prove most suggestive and of most interest to the investigator are probably those performed with exhausted tubes. As might be anticipated, a source of such rapidly alternating potentials is capable of exciting the tubes at a considerable distance, and the light effects produced are remarkable.

During my investigations in this line I endeavored to excite tubes, devoid of any electrodes, by electromagnetic induction, snaking the tube the secondary of the induction device, and passing through the primary the discharges of a Leyden jar. These tubes were made of many shapes, and I was able to obtain luminous effects which I then thought were due wholly to electromagnetic induction. But on carefully investigating the phenomena I found that the effects produced were more of an electrostatic nature. It may be attributed to this circumstance that this mode of exciting tubes is very wasteful, namely, the primary circuit being closed, the potential, and consequently the electrostatic inductive effect, is much diminished.

When an induction coil, operated as above described, is used, there is no doubt that the tubes are excited by electrostatic induction, and that electromagnetic induction has little, if anything, to do with the phenomena.

This is evident from many experiments. For instance, if a tube be taken in one hand, the observer being near the coil, it is brilliantly lighted and remains so no matter in what position it is held relatively to the observer's body. Were the action electromagnetic, the tube could not be lighted when the observer's body is interposed between it and the coil, or at least its luminosity should be considerably diminished. When the tube is held exactly over the centre of the coil—the latter being wound in sections and the primary placed symmetrically to the secondary—it may remain completely dark, whereas it is rendered intensely luminous by moving it slightly to the right or left from the centre of the coil. It does not light because in the middle both halves of the coil neutralize each other, and the electric potential is zero. If the action were electromagnetic, the tube should light best in the plane through the centre of the toil, since the electromagnetic effect there should be a maximum. When an arc is established between the terminals, the tubes and lamps in the vicinity of the coil go out, out light up again when the arc is broken, on account of the rise of potential. Yet the electromagnetic effect should be practically the same in both cases.

By placing a tube at some distance from the coil, and nearer to one terminal—preferably at a point on the axis of the coil—one may light it by touching the remote terminal with an insulated body of some size or with the hand, thereby raising the potential at that terminal nearer to, the tube. If the tube is shifted nearer to the coil so that it is lighted by the action of the nearer terminal, it may be made to go out by holding, on an insulated support, the end of a wire connected to the remote terminal, in the vicinity of the nearer terminal, by this means counteracting the action of the latter upon the tube. These effects are evidently electrostatic. Likewise, when a tube is placed it a considerable distance from the coil, the observer may, standing upon an insulated support between coil and tube, light the latter by approaching the hand to it; or he may even render it luminous by simply stepping between it and the coil. This would be impossible with electro-magnetic induction, for the body of the observer would act as a screen.

When the coil is energized by excessively weak currents, the experimenter may, by touching one terminal of the coil with the tube, extinguish the latter, and may again light it by bringing it out of contact with the terminal and allowing a small arc to form. This is clearly due to the respective lowering and raising of the potential at that terminal. In the above experiment, when the tube is lighted through a small arc, it may go out when the arc is broken, because the electrostatic inductive effect alone is too weak, though the potential may be much higher; but when the arc is established, the electrification of the end of the tube is much greater, and it consequently lights.

If a tube is lighted by holding it near to the coil, and in the hand which is remote, by grasping the tube anywhere with the other hand, the part between the hands is rendered dark, and the singular effect of wiping out the light of the tube may he produced by passing the hand quickly along the tube and at the same time withdrawing it gently from the coil, judging properly tile distance so that the tube remains dark afterwards.

If the primary coil is placed sidewise, as in Fig. 112b for instance, and an exhausted tube be introduced from the other side in the hollow space, the tube is lighted most intensely because of the increased condenser action, and in this position the striae are most sharply defined. In all these experiments described, and in many others, the action is clearly electrostatic.

The effects of screening also indicate the electrostatic nature of the phenomena and show something of the nature of electrification through the air. For instance, if a tube is placed in the direction of the axis of the coil, and an insulated metal plate be interposed, the tube will generally increase in brilliancy, or if it be too far from the coil to light, it may even be rendered luminous by interposing an insulated metal plate. The magnitude of the effects depends to some extent on the size of the plate. But if the metal plate be connected by a wire to the ground, its interposition will always make the tube go put even if it be very near the coil. In general, the interposition of a body between the coil and tube, increases or diminishes the brilliancy of the tube, or its facility to light up, according to whether it increases or diminishes the electrification. When experimenting with an insulated plate, the plate should not be taken too large, else it will generally produce a weakening effect by reason of its great facility for giving off energy to the surroundings.

If a tube be lighted at some distance from the coil, and a plate of hard rubber or other insulating substance be interposed, the tube may be made to go .out. The interposition of the dielectric in this case only slightly increases the inductive effect, but diminishes considerably the electrification through the air.

In all cases, then, when we excite luminosity in exhausted tubes by means of such a coil, the effect is due to the rapidly alternating electrostatic' potential; and, furthermore, it must be attributed to the harmonic alternation produced directly by the machine, and not to any superimposed vibration which might be thought to exist. Such superimposed vibrations are impossible when we work with an alternate current machine. If a spring be gradually tightened and released, it does not perform independent vibrations; for this a sudden release is necessary. So with the alternate currents from a dynamo machine; the medium is harmonically strained and released, this giving rise to only one kind of waves; a sudden contact or break, or a sudden giving way of the dielectric, as in the disruptive discharge of a Leyden jar, are essential for the production of superimposed waves.

In all the last described experiments, tubes devoid of any electrodes may be used, and there is no difficulty in producing by their means sufficient light to read by. The light effect is, however, considerably increased by the use of phosphorescent bodies such as yttrium, uranium glass, etc. A difficulty will be found when the phosphorescent material is used, for with these powerful effects, it is carried gradually away, and it is preferable to use material in the form of a solid.

Instead of depending on induction at a distance to light the tube, the same may be provided with an external—and, if desired, also with an internal—condenser coating, and it may then be suspended anywhere in the room from a conductor connected to one terminal of the coil, and in this manner a soft illumination may be provided.

The ideal way of lighting a hall or room would, however, be to produce such a condition in it that an illuminating device could be moved and put anywhere, and that it is lighted, no matter where it is put and without being electrically connected to anything. I have been able to produce such a condition by creating in the room a powerful, rapidly alternating electrostatic field. For this purpose I suspend a sheet of metal a distance from the ceiling on insulating cords and connect it to one terminal of the induction coil, the other terminal being preferably connected to the ground. Or else I suspend two sheets as illustrated in Fig. 125, each sheet being connected with one of the terminals of the coil, and their size being carefully determined. An exhausted tube may then be carried in the hand anywhere between the sheets or placed anywhere, even a certain distance beyond them; it remains always luminous.

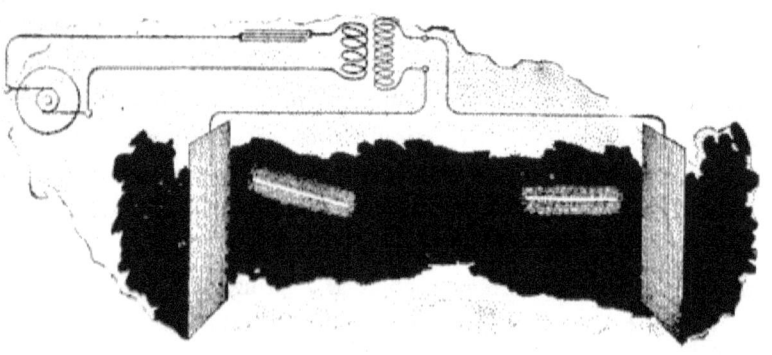

Fig. 125

In such an electrostatic field interesting phenomena may be observed, especially if the alternations are kept low and the potentials excessively high. In addition to the luminous phenomena mentioned, one may observe that any insulated conductor gives sparks when the hand or another object is approached to it, and the sparks may often be powerful. When a large conducting object is fastened on an insulating support, and the hand approached to it, a vibration, due to the rhythmical motion of the air molecules is felt, and luminous streams may be perceived when the hand is held rear a pointed projection. When a telephone receiver is made to touch with one or both of its terminals art insulated conductor of some size, the telephone emits a loud sound; it also emits a sound when a length of wire is attached to one or both terminals, and with very powerful fields a sound may be perceived even without any wire.

How far this principle is capable of practical application, the future will tell. It might be thought that electrostatic effects are unsuited for such action at a distance. Electromagnetic inductive effects, if available for the production of light, might be thought better suited. It is true the electrostatic effects diminish nearly with the cube of the distance from the coil, whereas the electromagnetic inductive effects diminish simply with the distance. But when we establish an electrostatic field of force, the condition is very different, for then, instead of the differential effect of both the terminals, we get their conjoint effect. Besides, I would call attention to the effect that in an alternating electrostatic field, a conductor, such as an exhausted tube, for instance, tends to take up most of the energy, whereas in an electromagnetic alternating field the conductor tends to take up the least energy, the waves being reflected with but little loss. This is one reason why it is difficult to excite an exhausted tube, at a distance, by electromagnetic induction. I have wound coils of very large diameter and of many turns of wire, and connected a Geissler tube to the ends of the coil with the object of exciting the tube at a distance; but even with the powerful inductive effects producible by Leyden jar discharges, the tube could not be excited unless at a very small distance, although some judgment was used as to the dimensions of the coil. I have also found that even the most powerful Leyden jar discharges are capable of exciting only feeble luminous effects in a closed exhausted tube, and even these effects upon thorough examination I have been forced to consider of an electrostatic nature.

How then can we hope to produce the required effects at a distance by means of electromagnetic action, when even in the closest proximity to the source of disturbance, under the most advantageous conditions, we can excite but faint luminosity? It is true that when acting at a distance we have the resonance to help us out. We can connect an exhausted tube, or whatever the illuminating device may be, with an insulated system of the proper capacity, and so it may be possible to increase the effect qualitatively, and only qualitatively, for we would not get snore energy through the device. So we may, by resonance effect, obtain the required electromotive force in an exhausted tube, and excite faint luminous effects, but we cannot get enough energy to render the light practically available, and a simple calculation, based on experimental results, shows that even if all the energy which a tube would receive at a certain distance from the source should be wholly converted into light, it would hardly satisfy the practical requirements. Hence the necessity of directing, by means of a conducting circuit, the energy to the place of transformation. But in so doing we cannot very sensibly depart from present methods, and all we could do would be to improve the apparatus.

From these considerations it would seem that if this ideal way of lighting is to rendered practicable it will be only by the use of electrostatic effects. In such a case the most powerful electrostatic inductive effects are needed; the apparatus employed must, therefore, be capable of producing high electrostatic potentials changing in value with extreme rapidity. High frequencies are especially wanted, for practical considerations make it desirable to keep down the potential. By the employment of machines, or, generally speaking, of any mechanical apparatus, but low frequencies can be reached; recourse must, therefore, be had to some other means. The discharge of a condenser affords us a means of obtaining frequencies by far higher than are obtainable mechanically, and I have accordingly employed condensers in the experiments to the above end.

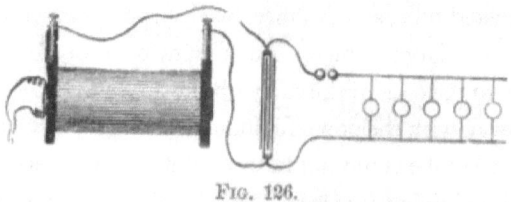

FIG. 126.

When the terminals of a high tension induction coil, Fig. 126, are connected to a Leyden jar, and the latter is discharging disruptively into a circuit, we may look upon the arc playing between the knobs as being a source of alternating, or generally speaking, undulating currents, and then we have to deal with the familiar system of a generator of such currents, a circuit connected to it, and a condenser bridging the circuit. The condenser in such case is a veritable transformer, and since the frequency is excessive, almost any ratio in the strength of the currents in both the branches may be obtained. In reality the analogy is not quite complete, for in the disruptive discharge we have most generally a fundamental instantaneous variation of comparatively low frequency, and a superimposed harmonic vibration, and the laws governing the flow of currents are not the: same for both.

In converting in this manner, the ratio of conversion should not be too great, for the loss in the arc between the knobs increases with the square of the current, and if the jar be discharged through very thick and short conductors, with the view of obtaining a very rapid oscillation, a very considerable portion of the energy stored is lost. On the other hand, too small ratios are not practicable for many obvious reasons.

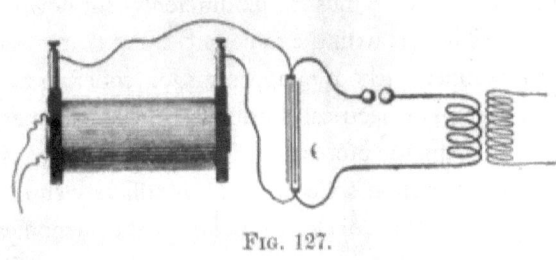

FIG. 127.

As the converted currents flow in a practically closed circuit, the electrostatic effects are necessarily small, and I therefore convert them into currents or effects of the required character. I have effected such conversions in several ways. The preferred plan of connections is illustrated in Fig. 127. The manner of operating renders it easy to obtain by means of a small and inexpensive apparatus enormous differences of potential which have been usually obtained by means of large and expensive coils. For this it is only necessary to take an ordinary small coil, adjust to it a condenser and discharging circuit, forming the primary of an auxiliary small coil, and convert upward. As the inductive effect of the primary currents is excessively great, the second coil need have comparatively but very few turns. By properly adjusting the elements, remarkable results may be secured.

In endeavoring to obtain the required electrostatic effects in this manner, I have, as might be expected, encountered many difficulties which I have been gradually overcoming, but I am not as yet prepared to dwell upon my experiences in this direction.

I believe that the disruptive discharge of a condenser will play an important part in the future, for it offers vast possibilities, not only in the way of producing light in a more efficient manner and in the line indicated by theory, but also in many other respects.

For years the efforts of inventors have been directed towards obtaining electrical energy from heat by means of the thermopile. It might seem invidious to remark that but few know what is the real trouble with the thermopile. It is not the inefficiency or small output—though these are great drawbacks—but the fact that the thermopile has its phylloxera, that is, that by constant use it is deteriorated, which has thus far prevented its introduction on an industrial scale. Now that all modern research seems to point with certainty to the use of electricity of excessively high tension, the question must present itself to many whether it is not possible to obtain in a practicable manner this form of energy from heat. We have been used to look upon an electrostatic machine as a plaything, and somehow we couple with it the idea of the inefficient and impractical. But now we must think differently, for now we know that everywhere we have to deal with the same forces, and that it is a mere question of inventing proper methods or apparatus for rendering them available.

In the present systems of electrical distribution, the employment of the iron with its wonderful magnetic properties allows us to reduce considerably the size of the apparatus; but, in spite of this, it is still very cumbersome. The more we progress in the study of electric and magnetic phenomena, the more we become convinced that the present methods will be short-lived. For the production of light, at least, such heavy machinery would seem to be unnecessary. The energy required is very small, and if light can be obtained as efficiently as, theoretically, it appears possible, the apparatus need have but a very small output. There being a strong probability that the illuminating methods of the future will involve the use of very high potentials, it seems very desirable to perfect a contrivance capable of converting the energy of heat into energy of the requisite form. Nothing to speak of has been done towards this end, for the thought that electricity of some 50,000 or 100,000 volts pressure or more, even if obtained, would be unavailable for practical purposes, has deterred inventors from working in this direction.

In Fig. 126 a plan of connections is shown for converting currents of high, into currents of low, tension by means of the disruptive discharge of a condenser. This plan has been used by me frequently for operating a few incandescent lamps required in the laboratory. Some difficulties have been encountered in the arc of the discharge which I have been able to overcome to a great extent; besides this, and the adjustment necessary for the proper working, no other difficulties have been met with, and it was easy to operate ordinary lamps; and even motors, in this manner. The line being connected to the ground, all the wires could be handled with perfect impunity, no matter how high the potential at the terminals of the condenser. In these experiments a high tension induction coil, operated from a battery or from an alternate current machine, was employed to charge the condenser; but the induction coil might be replaced by an apparatus of a different kind, capable of giving electricity of such high tension. In this manner, direct or alternating currents may be converted, and in both cases the current-impulses may be of any desired frequency. When the currents charging the condenser are of the same direction, and it is desired that the converted currents should also be of one direction, the resistance of the discharging circuit should, of course, be so chosen that there are no oscillations.

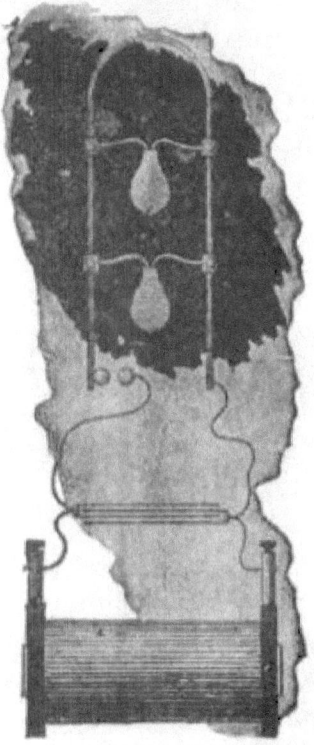

FIG. 128.

In operating devices on the above plan I have observed curious phenomena of impedance which are of interest. For instance if a thick copper bar be bent, as indicated in Fig. 128, and shunted by ordinary incandescent lamps, then, by passing the discharge between the knobs, the lamps may be brought to incandescence although they are short-circuited. When a large induction coil is employed it is easy to obtain nodes on the bar, which are rendered evident by the different degree of brilliancy of the lamps, as shown roughly in Fig. 128. The nodes are never clearly defined, but they are simply maxima and minima of potentials along the bar. This is probably due to the irregularity of the arc between the knobs. In general when the above-described plan of conversion from high to low tension is used, the behavior of the disruptive discharge may be closely studied. The nodes may also be investigated by means of an ordinary Cardew voltmeter which should be well insulated. Geissler tubes may also be lighted across the points of the bent bar; in this case, of course, it is better to employ smaller capacities. I have found it practicable to light up in this manner a lamp, and even a Geissler tube, shunted by a short, heavy block of metal, and this result seems at first very curious. In fact, the thicker the copper bar in Fig. 128; the better it is for the success of the experiments, as they appear more striking. When lamps with long slender filaments are used it will be often noted that the filaments are from time to time violently vibrated, the vibration being smallest at the nodal points. This vibration seems to be due to an electrostatic action between the filament and the glass of the bulb.

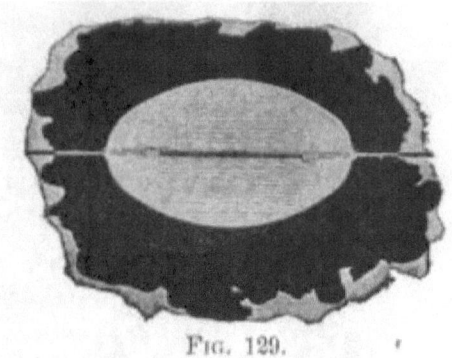

Fig. 129.

In some of the above experiments it is preferable to use special lamps having a straight filament as shown in Fig. 129. When such a lamp is used a still more curious phenomenon than those described may be observed. The lamp may be placed across the copper bar and lighted, and by using somewhat larger capacities, or, in other words, smaller frequencies or smaller impulsive impedances, the filament may be brought to any desired degree of incandescence. But when the impedance is increased, a point is reached when comparatively little current passes through the carbon, and most of it through the rarefied gas; or perhaps it may be more correct to state that the current divides nearly evenly through both, its spite of the enormous difference in the resistance, and this would be true unless the Las and the filament behave differently. It is then noted that the whole bulb is brilliantly illuminated, and the ends of the leading-in wires become incandescent and often throw off sparks in consequence of the violent bombardment, but the carbon filament remains dark. This is illustrated in Fig. 129. Instead of the filament a single wire extending through the whole bulb may be used, and in this case the phenomenon would seen to be still more interesting.

From the above experiment it will be evident, that when ordinary lamps are operated by the converted currents, those should be preferably taken in which the platinum wires are far apart, and the frequencies used should not be too great, else the discharge will occur at the ends of the filament or in the base of the lamp between the leading-in wires, and the lamp might then be damaged.

In presenting to you these results of my investigation on the subject under consideration, I have paid only a passing notice to facts upon which I could have dwelt at length, and among many observations I have selected only those which I thought most likely to interest you. The field is wide and completely unexplored, and at every step a new truth is gleaned, a novel fact observed.

How far the results here borne out are capable of practical applications will be decided in the future. As regards the production of light, some results already reached are encouraging and make me confident in asserting that the practical solution of the problem lies in the direction I have endeavored to indicate. Still, whatever may be the immediate outcome of these experiments I am hopeful that they will only prove a step in further developments towards the ideal and final perfection. The possibilities which are opened by modern research are so vast that even the most reserved must feel sanguine of the future. Eminent scientists consider the problem of utilizing one kind of radiation without the others a rational one. In an apparatus designed for the production of light by conversion from any form of energy into that of light, such a result can never be reached, for no matter what the process of producing the required vibrations, be it electrical, chemical or any other, it will not be possible to obtain the higher light vibrations without going through the lower heat vibrations. It is the problem of imparting to a body a certain velocity without passing through all lower velocities. But there is a possibility of obtaining energy not only in the form of light, but motive power, and energy of any other form, in some more direct way from the medium. The time will be when this will be accomplished, and the time has come when one may utter such words before an enlightened audience without being considered a visionary. We are whirling through endless space with an inconceivable speed, all around us everything is spinning, everything is moving, everywhere is energy. There mart be some way of availing ourselves of this energy more directly. Then; with the light obtained from the medium, with the power derived from it, with every form of energy obtained without effort, from the store forever inexhaustible, humanity will advance with giant strides. The mere contemplation of these magnificent possibilities expands our minds, strengthen our hopes and fills our hearts with supreme delight.

May 20, 1891

EXPERIMENTS WITH ALTERNATE CURRENTS OF HIGH POTENTIAL AND HIGH FREQUENCY (3/4)

—Delivered before the Institution of Electrical Engineers, London, February 1892.

I cannot find words to express how deeply I feel the honor of addressing some of the foremost thinkers of the present time, and so many able scientific men, engineers and electricians, of the country greatest in scientific achievements.

The results which I have the honor to present before such a gathering I cannot call my own. There are among you not a few who can lay better claim than myself on any feature of merit which this work may contain. I need not mention many names which are world-known — names of those among you who are recognized as the leaders in this enchanting science; but one, at least, I must mention — a name which could not be omitted in a demonstration of this kind. It is a name associated with the most beautiful invention ever made: it is Crookes!

When I was at college, a good time ago; I read, in a translation (for then I was not familiar with you magnificent language), the description of his experiments on radiant matter. I read it only once in my life — that time — yet every detail about that charming work I can remember this day. Few are the books, let me say, which can make such an impression upon the mind of a student.

But if, on the present occasion, I mention this name as one of many your institution can boast of, it is because I have more than one reason to do so. For what I have to tell you and to show you this evening concerns, in a large measure, that same vague world which Professor Crookes has so ably explored; and, more than this, when I trace back the mental process which led me to these advances — which even by myself cannot be considered trifling, since they are so appreciated by you — I believe that their real origin, that which started me to work in this direction, and brought me to them, after a long period of constant thought, was that fascinating little book which I read many years ago.

And now that I have made a feeble effort to express my homage and acknowledge my indebtedness to him and others among you, I will make a second effort, which I hope you will not find so feeble as the first, to entertain you.

Give me leave to introduce the subject in a few words.

A short time ago I had the honor to bring before our *American Institute of Electrical Engineers* some results then arrived at by me in a novel line of work. I need not assure you that the many evidences which I have received that English scientific men and engineers were interested in this work have been for me a great reward and encouragement. I will not dwell upon the experiments already described, except with the view of completing, or more clearly expressing, some ideas advanced by me before, and also with the view of rendering the study here presented self-contained, and my remarks on the subject of this evening's lecture consistent.

This investigation, then, it goes without saying, deals with alternating currents, and, to be more precise, with alternating currents of high potential and high frequency. Just in how much a very high frequency is essential for the production of the results presented is a question which, even with my present experience, would embarrass me to answer. Some of the experiments may be performed with low frequencies; but very high frequencies are desirable, not only on account of the many effects secured by their use, but also as a convenient means of obtaining, in the induction apparatus employed, the high potentials, which in their turn are necessary to the demonstration of most of the experiments here contemplated.

Of the various branches of electrical investigation, perhaps the most interesting and immediately the most promising is that dealing with alternating currents. The progress in this branch of applied science has been so great in recent years that it justifies the most sanguine hopes. Hardly have we become familiar with one fact, when novel experiences are met with and new avenues of research are opened. Even at this hour possibilities not dreamed of before are, by the use of these currents, partly realized. As In nature all is ebb and tide, all is wave motion, so it seems that in all branches of industry alternating currents—electric wave motion—will have the sway.

One reason, perhaps, why this brand of science is being so rapidly developed is to be found in the interest which is attached to its experimental study. We wind a simple ring of iron with coils; we establish the connections to the generator, and with wonder and delight we note the effects of strange forces which we bring into play, which allow us to transform, to transmit and direct energy at will. We arrange the circuits properly, and we see the mass of iron and wires behave as though it were endowed with life, spinning a heavy armature, through invisible connections, with great speed and power with the energy possibly conveyed from a great distance. We observe how the energy of an alternating current traversing the wire manifests itself—not so much in the wire as in the surrounding space—in the most surprising manner, taking the forms of heat, light, mechanical energy, and, most surprising of all, even chemical affinity. All these observations fascinate us, and fill us with an intense desire to know more about the nature of these phenomena. Each day we go to our work in the hope of discovering—in the hope that some one, no matter who, may find a solution of one of the pending great problems,—and each succeeding day we return to our task with renewed ardor; and even if we are unsuccessful, our work has not been in vain, for in these strivings, in these efforts, we have hours of untold pleasure, and we have directed our energies to the benefit of mankind.

We may take—at random, if you choose—any of the many experiments which may be performed with alternating currents; a few of which only, and by no means the mast striking, form the subject of this evening's demonstration; they are all equally interesting, equally inciting to thought.

Here is a simple glass tube from which the air has been partially exhausted. I take hold of it; I bring my body in contact with a wire conveying alternating currents of high potential, and the tube in my hand is brilliantly lighted. In whatever position I may put it, wherever I may move it in space, as far as I can reach, its soft, pleasing light persists with undiminished brightness.

Here is an exhausted bulb suspended from a single wire. Standing on an insulated support, I grasp it, and a platinum button mounted in it is brought to vivid incandescence.

Here, attached to a leading wire is another bulb, which, as I touch its metallic socket, is filled with magnificent colors of phosphorescent light.

Here still another, which by my fingers' touch casts a shadow—the Crookes shadow, of the stem inside of it.

Here, again, insulated as I stand on this platform, I bring my body in contact with one of the terminals of the secondary of this induction coil—with the end of a wire many miles long—and you see streams of light break forth from its distant end, which is set in violent vibration.

Here, once more, attach these two plates of wire gauze to the terminals of the coil, I set them a distance apart, and I set the coil to work. You may see a small spark pass between the plates. I insert a thick plate of one of the best dielectrics between them, and instead of rendering altogether impossible, as we are used to expect, I aid the passage of the discharge, which, as I insert the plate, merely changes in appearance and assumes the form of luminous streams.

Is there, I ask, can there be, a more interesting study than that of alternating currents?

In all these investigations, in all these experiments, which are so very, very interesting, for many years past—ever since the greatest experimenter who lectured in this hall discovered its principle—we have had a steady companion, an appliance familiar to every one, a plaything once, a thing of momentous importance now—the induction coil. There is no dearer appliance to the electrician. From the ablest among you, I dare say, down to the inexperienced student, to your lecturer, we all have passed many delightful hours in experimenting with the induction coil. We have watched its play, and thought and pondered over the beautiful phenomena which it disclosed to our ravished eyes. So well known is this apparatus, so familiar are these phenomena to every one, that my courage nearly fails me when I think that I have ventured to address so able an audience, that I have ventured to entertain you with that same old subject. Here in reality is the same apparatus, and here are the same phenomena, only the apparatus is operated somewhat differently, the phenomena are presented in a different aspect. Some of the results we find as expected, others surprise us, but all captivate our attention, for in scientific investigation each novel result achieved may be the centre of a new departure, each novel fact learned may lead to important developments.

Usually in operating an induction foil we have set up a vibration of moderate frequency in the primary, either by means of an interrupter or break, or by the use of an alternator. Earlier English investigators, to mention only Spottiswoode and J. E. H. Gordon, have used a rapid break in connection with the coil. Our knowledge and experience of today enables us to see clearly why these coils under the conditions of the tests did not disclose any remarkable phenomena, and why able experimenters failed to perceive many of the curious effects which have since been observed.

In the experiments such as performed this evening, we operate the coil either from a specially constructed alternator capable of giving many thousands of reversals of current per second, or, by disruptively discharging a condenser through the primary, we set up a vibration in the secondary circuit of a frequency of many hundred thousand or millions per second, if we so desire; and in using either of these means we enter a field as yet unexplored.

It is impossible to pursue an investigation in any novel line without finally making some interesting observation or learning some useful fact. That this statement is applicable to the subject of this lecture the many curious and unexpected phenomena which we observe afford a convincing proof. By way of illustration, take for instance the most obvious phenomena, those of the discharge of the induction coil.

Here is a coil which is operated by currents vibrating with extreme rapidity, obtained by disruptively discharging a Leyden jar. It would not surprise a student were the lecturer to say that the secondary of this coil consists of a small length of comparatively stout wire; it would not surprise him were the lecturer to state that, in spite of this, the coil is capable of giving any potential which the best insulation of the turns is able to withstand; but although he may be prepared, and even be indifferent as to the anticipated result, yet the aspect of the discharge of the coil will surprise and interest him. Every one is familiar with the discharge of an ordinary coil; it need not be reproduced here. But, by way of contrast, here is a form of discharge of a coil, the primary current of which is vibrating several hundred thousand times per second. The discharge of an ordinary coil appears as a simple line or band of light. The discharge of this coil appears in the form of powerful brushes and luminous streams issuing from all points of the two straight wires attached to the terminals of the secondary (Fig. 1.)

Now compare this phenomenon which you have just witnessed with the discharge of a Holtz or Wimshurst machine — that other interesting appliance, so dear to the experimenter. What a difference there is between these phenomena! And yet, had I made the necessary arrangements — which could have been made easily, were it not that they would interfere with other experiments — I could have produced with this coil sparks which, had I the coil hidden from your view and only two knobs exposed, even the keenest observer among you would find it difficult, if not impossible, to distinguish from those of an influence or friction machine. This may be done in many ways — for instance, by operating the induction coil which charges the condenser from an alternating-current machine of very low frequency, and preferably adjusting the discharge circuit so that there are no oscillations set up in it. We then obtain in the secondary circuit, if the knobs are of the required size and properly set, a more or less rapid succession of sparks of great intensity and small quantity, which possess the same brilliancy, and are accompanied by the same sharp crackling sound, as those obtained from a friction or influence machine.

Another way is to pass through two primary circuits, having a common secondary, two currents of a slightly different period, which produce in the secondary circuit sparks occurring at comparatively long intervals. But, even with the means at hand this evening, I may succeed in imitating the spark of a Holtz machine. For this purpose I establish between the terminals of the coil which charges the condenser a long, unsteady arc, which is periodically interrupted by the upward current of air produced by it. To increase the current of air I place on each side of the arc, and close to it, a large plate of mica. The condenser charged from this coil discharge into the primary circuit of a second coil through a small air gap, which is necessary to produce a sudden rush of current through the primary. The scheme of connections in the present experiment is indicated in Fig. 2.

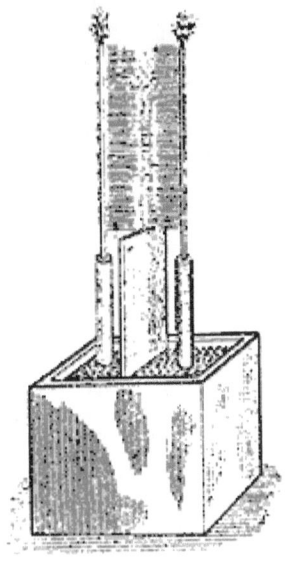

Fig. 1.

G is an ordinarily constructed alternator, supplying the primary P of an induction coil, the secondary S of which charges the condensers or jars C C. The terminals of the secondary are connected to the inside coatings of the jars, the outer coatings being connected to the ends of the primary p p of a second induction coil. This primary p p has a small air gap a b.

The secondary s of this coil is provided with knobs or spheres K K of the proper size and set at a distance suitable for the experiment.

A long arc is established between the terminals A B of the first induction coil. M M are the mica plates.

Each time the arc is broken between A and B the jars are quickly charged and discharged through the primary p p, producing a snapping spark between the knobs K K. Upon the arc forming between A and B the potential falls, and the jars cannot be charged to such high potential as to break through the air gap a b until the arc is again broken by the draught.

In this manner sudden impulses, at long intervals, are produced in the primary P P, which in the secondary s give a corresponding number of impulses of great intensity. If the secondary knobs or spheres K K are of the proper size, the sparks show much resemblance to those of a Holtz machine. But these two effects, which to the eye appear so very different, are only two of the many discharge phenomena. We only need to change the conditions of the test, and again we make other observations of interest.

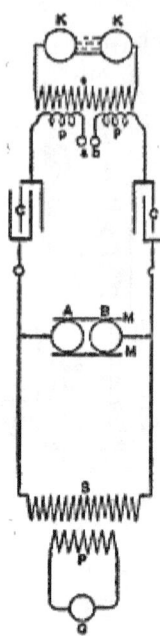

Fig. 2.

When, instead of operating the induction coil as in the last two experiments, we operate it from a high frequency alternator, as in the next experiment, a systematic study of the phenomena is rendered much more easy. In such case, in varying the strength and frequency of the currents through the primary, we may observe five distinct forms of discharge, which I have described in my former paper on the subject before the *American Institute of Electrical Engineers*, May 20, 1891 [Experiments With Alternate Currents of Very High Frequency and Their Application to Methods of Artificial Illumination] (see *The Electrical World*, July 11, 1891).

It would take too much time, and it would lead us too far from the subject presented this evening, to reproduce all these forms, but it seems to me desirable to show you one of them. It is a brush discharge, which is interesting in more than one respect. Viewed from a near position it resembles much a jet of gas escaping under great pressure. We know that the phenomenon is due to the agitation of the molecules near the terminal, and we anticipate that some heat must be developed by the impact of the molecules against the terminal or against each other. Indeed, we find that the brush is hot, and only a little thought leads us to the conclusion that, could we but reach sufficiently high frequencies, we could produce a brush which would give intense light and heat, and which would resemble in every particular an ordinary flame, save, perhaps, that both phenomena might not be due to the same agent—save, perhaps, that chemical affinity might not be electrical in its nature.

As the production of heat and light is here due to the impact of the molecules, or atoms of air, or something else besides, and, as we can augment the energy simply by raising the potential, we might, even with frequencies obtained from a dynamo machine, intensify the action to such a degree as to bring the terminal to melting heat. But with such low frequencies we would have to deal always with something of the nature of an electric current. If I approach a conducting object to the brush, a thin little spark passes, yet, even with the frequencies used this evening, the tendency to spark is not very great. So, for instance, if I hold a metallic sphere at some distance above the terminal you may see the whole space between the terminal and sphere illuminated by the streams without the spark passing; and with the much higher frequencies obtainable by the disruptive discharge of a condenser, were it not for the sudden impulses, which are comparatively few in number, sparking would not occur even at very small distances. However, with incomparably higher frequencies, which we may yet find means to produce efficiently, and provided that electric impulses of such high frequencies could be transmitted through a conductor, the electrical characteristics of the brush discharge would completely vanish — no spark would pass, no shock would be felt — yet we would still have to deal with an electric phenomenon, but in the broad, modern interpretation of the word. In my first paper before referred to I have pointed out the curious properties of the brush, and described the best manner of producing it, but I have thought it worth while to endeavor to express myself more clearly in regard to this phenomenon, because of its absorbing interest.

When a coil is operated with currents of very high frequency, beautiful brush effects may be produced, even if the coil be of comparatively small dimensions. The experimenter may vary them in many ways, and, if it were nothing else, they afford a pleasing sight. What adds to their interest is that they may be produced with one single terminal as well as with two — in fact, often better with one than with two.

But of all the discharge phenomena observed, the most pleasing to the eye, and the most instructive, are those observed with a coil which is operated by means of the disruptive discharge of a condenser. The power of the brushes, the abundance of the sparks, when the conditions are patiently adjusted, is often amazing. With even a very small coil, if it be so well insulated as to stand a difference of potential of several thousand volts per turn, the sparks may be so abundant that the whole coil may appear a complete mass of fire.

Curiously enough the sparks, when the terminals of the coil are set at a considerable distance, seem to dart in every possible direction as though the terminals were perfectly independent of each other. As the sparks would soon destroy the insulation it is necessary to prevent them. This is best done by immersing the coil in a good liquid insulator, such as boiled-out oil. Immersion in a liquid may be considered almost an absolute necessity for the continued and successful working of such a coil.

It is, of course, out of the question, in an experimental lecture, with only a few minutes at disposal for the performance of each experiment, to show these discharge phenomena to advantage, as to produce each phenomenon at its best a very careful adjustment is required. But even if imperfectly produced, as they are likely to be this evening, they are sufficiently striking to interest an intelligent audience.

Before showing some of these curious effects I must, for the sake of completeness, give a short description of the coil and other apparatus used in the experiments with the disruptive discharge this evening.

It is contained in a box B (Fig. 3) of thick boards of hard wood, covered on the outside with zinc sheet Z, which is carefully soldered all around. It might be advisable, in a strictly scientific investigation, when accuracy is of great importance, to do away with the metal cover, as it might introduce many errors, principally on account of its complex action upon the coil, as a condenser of very small capacity and as an electrostatic and electromagnetic screen. When the coil is used for such experiments as are here contemplated, the employment of the metal cover offers some practical advantages, but these are not of sufficient importance to be dwelt upon.

The coil should be placed symmetrically to the metal cover, and the space between should, of course, not be too small, certainly not less than, say, five centimeters, but much more if possible; especially the two sides of the zinc box, which are at right angles to the axis of the coil, should be sufficiently remote from the latter, as otherwise they might impair its action and be a source of loss.

The coil consists of two spools of hard rubber R R held apart at a distance of 10 centimetres by bolts c and nuts n, likewise of hard rubber. Each spool comprises a tube T of approximately 8 centimetres inside diameter, and 3 millimetres thick, upon which are screwed two flanges F F, 24 centimetres square, the space between the flanges being about 3 centimetres. The secondary, S S, of the best gutta percha-covered wire, has 26 layers, 10 turns in each, giving for each half a total of 260 turns. The two halves are wound oppositely and connected in series, the connection between both being made over the primary. This disposition besides being convenient, has the advantage that when the coil is well balanced—that is, when both of its terminals $T_1 T_1$ are connected to bodies or devices of equal capacity—there is not much danger of breaking through to the primary, and the insulation between the primary and the secondary need not be thick. In using the coil it is advisable to attach to both terminals devices of nearly equal capacity, as, when the capacity of the terminals is not equal, sparks will be apt to pass to the primary. To avoid this, the middle point of the secondary may be connected to the primary, but this is not always practicable.

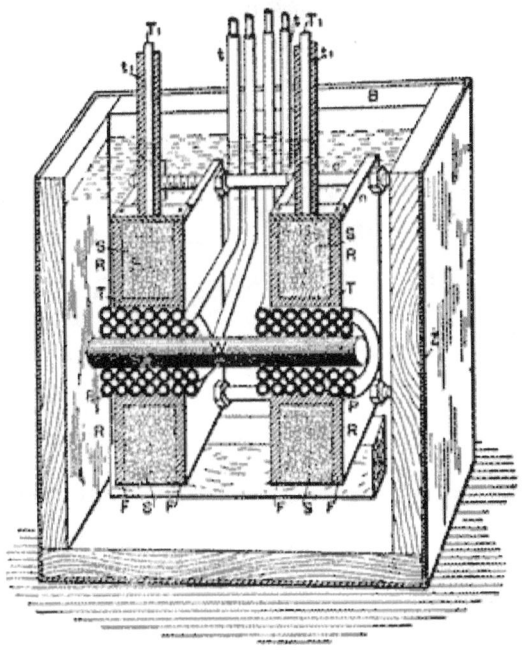

Fig. 3.

The primary P P is wound in two parts, and oppositely, upon a wooden spool W, and the four ends are led out of the oil through hard rubber tubes t t. The ends of the secondary $T_1 T_1$ are also led out of the oil through rubber tubes $t_1 t_1$ of great thickness. The primary and secondary layers are insulated by cotton cloth, the thickness of the insulation, of course, bearing some proportion to the difference of potential between the turns of the different layers. Each half of the primary has four layers, 24 turns in each, this giving a total of 96 turns. When both the parts are connected in series, this gives a ratio of conversion of about 1:2.7, and with the primaries in multiple, 1:5.4 but in operating with very rapidly alternating currents this ratio does not convey even an approximate idea of the ratio of the E.M.Fs. in the primary and secondary circuits. The coil is held in position in the oil on wooden supports, there being about 5 centimetres thickness of oil all round. Where the oil is not specially needed, the space is filled with pieces of wood, and for this purpose principally the wooden box B surrounding the whole is used.

The construction here shown is, of course, not the best on general principles, but I believe it is a good and convenient one for the production of effects in which an excessive potential and a very small current are needed.

In connection with the coil I use either the ordinary form of discharger or a modified form. In the former I have introduced two changes which secure some advantages, and which are obvious. If they are mentioned, it is only in the hope that some experimenter may find them of use.

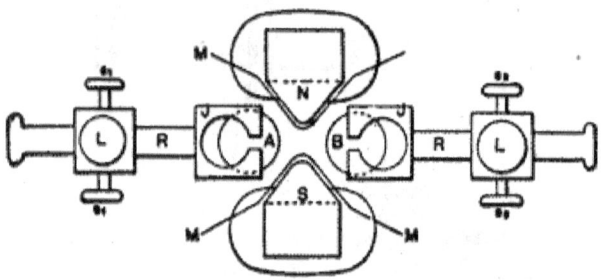

Fig. 4.

One of the changes is that the adjustable knobs A and B (Fig. 4), of the discharger are held in jaws of brass, J J, by spring pressure, this allowing of turning them successively into different positions, and so doing away with the tedious process of frequent polishing up.

The other change consists in the employment of a strong electromagnet N S, which is placed with its axis at right angles to the line joining the knobs A and B, and produces a strong magnetic field between them. The pole pieces of the magnet are movable and properly formed so as to protrude between the brass knobs, in order to make the field as intense as possible; but to prevent the discharge from jumping to the magnet the pole pieces are protected by a layer of mica, M M, of sufficient thickness, $s_1 s_1$ and $s_2 s_2$ are screws for fastening the wires. On each side one of the screws is for large and the other for small wires. L L are screws for fixing in position the rods R R, which support the knobs.

In another arrangement with the magnet I take the discharge between the rounded pole pieces themselves, which in such case are insulated and preferably provided with polished brass caps.

The employment of an intense magnetic field is of advantage principally when the induction coil or transformer which charges the condenser is operated by currents of very low frequency. In such a case the number of the fundamental discharges between the knobs may be so small as to render the currents produced in the secondary unsuitable for many experiments. The intense magnetic field then serves to blow out the arc between the knobs as soon as it is formed, and the fundamental discharges occur in quicker succession.

Instead of the magnet, a draught or blast of air may be employed with some advantage. In this case the arc is preferably established between the knobs A B, in Fig. 2 (the knobs a b being generally joined, or entirely done away with), as in this disposition the arc is long and unsteady, and is easily affected by the draught.

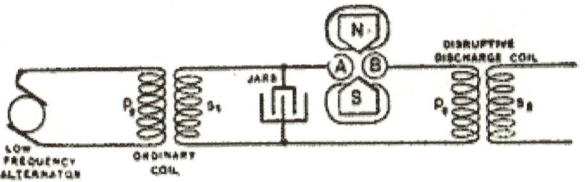

Fig. 5.

When a magnet is employed to break the arc, it is better to choose the connection indicated diagrammatically in Fig 5, as in this case the currents forming the arc are much more powerful, and the magnetic field exercises a greater influence. The use of the magnet permits, however, of the arc being replaced by a vacuum tube, but I have encountered great difficulties in working with an exhausted tube.

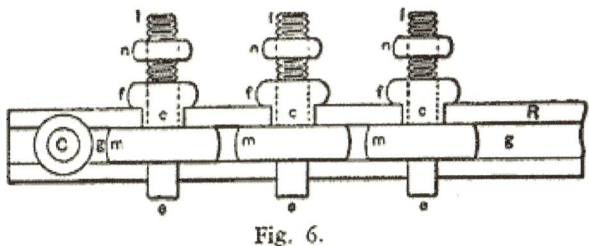

Fig. 6.

The other form of discharger used in these and similar experiments is indicated in Figs. 6 and 7. It consists of a number of brass pieces c c (Fig. 6), each of which comprises a spherical middle portion m with an extension e below — which is merely used to fasten the piece in a lathe when polishing up the discharging surface — and a column above, which consists of a knurled flange f surmounted by a threaded stem I carrying a nut n, by means of which a wire is fastened to the column. The flange f conveniently serves for holding the brass piece when fastening the wire, and also for turning it in any position when it becomes necessary to present a fresh discharging surface. Two stout strips of hard rubber R R, with planed grooves g g (Fig. 7) to fit the middle portion of the pieces c c, serve to clamp the latter and hold them firmly in position by means of two bolts C C (of which only one is shown) passing through the ends of the strips.

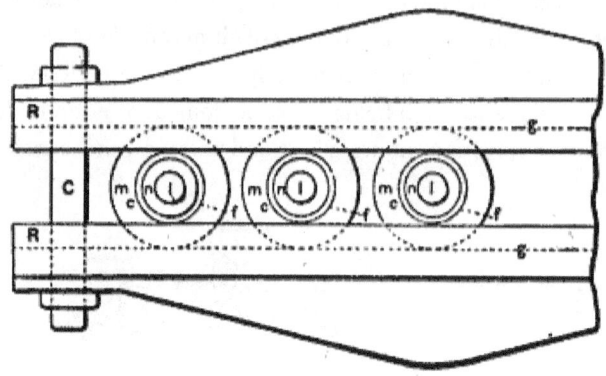

Fig. 7.

In the use of this kind of discharger I have found three principal advantages over the ordinary form. First, the dielectric strength of a given total width of air space is greater when a great many small air gaps are used instead of one, which permits of working with a smaller length of air gap, and that means smaller loss and less deterioration of the metal; secondly by reason of splitting the arc up into smaller arcs, the polished surfaces are made to last much longer; and, thirdly, the apparatus affords some gauge in the experiments. I usually set the pieces by putting between them sheets of uniform thickness at a certain very small distance which is known from the experiments of Sir William Thomson to require a certain electromotive force to be bridged by the spark.

It should, of course, be remembered that the sparking distance is much diminished as the frequency is increased. By taking any number of spaces the experimenter has a rough idea of the electromotive force, and he finds it easier to repeat an experiment, as he has not the trouble of setting the knobs again and again. With this kind of discharger I have been able to maintain an oscillating motion without any spark being visible with the naked eye between the knobs, and they would not show a very appreciable rise in temperature. This form of discharger also lends itself to many arrangements of condensers and circuits which are often very convenient and timesaving. I have used it preferably in a disposition similar to that indicated in Fig. 2, when the currents forming the arcs are small.

I may here mention that I have also used dischargers with single or multiple air gaps, in which the discharge surfaces were rotated with great speed. No particular advantage was, however, gained by this method, except in cases where the currents from the condenser were large and the keeping cool of the surfaces was necessary, and in cases when, the discharge not being oscillating of itself, the arc as soon as established was broken by the air current, thus starting the vibration at intervals in rapid succession. I have also used mechanical interrupters in many ways. To avoid the difficulties with frictional contacts, the preferred plan adopted was to establish the arc and rotate through it at great speed a rim of mica provided with many holes and fastened to a steel plate.

It is understood, of course, that the employment of a magnet, air current, or other interrupter, produces an effect worth noticing, unless the self-induction, capacity and resistance are so related that there are oscillations set up upon each interruption.

I will now endeavor to show you some of the most noteworthy of these discharge phenomena.

I have stretched across the room two ordinary cotton covered wires, each about 7 metres in length. They are supported on insulating cords at a distance of about 30 centimetres. I attach now to each of the terminals of the coil one of the wires and set the coil in action. Upon turning the lights off in the room you see the wires strongly illuminated by the streams issuing abundantly from their whole surface in spite of the cotton covering, which may even be very thick. When the experiment is performed under good conditions, the light from the wires is sufficiently intense to allow distinguishing the objects in a room. To produce the best result it is, of course, necessary to adjust carefully the capacity of the jars, the arc between the knobs and the length of the wires. My experience is that calculation of the length of the wires leads, in such case, to no result whatever. The experimenter will do best to take the wires at the start very long, and then adjust by cutting off first long pieces, and then smaller and smaller ones as he approaches the right length.

A convenient way is to use an oil condenser of very small capacity, consisting of two small adjustable metal plates, in connection with this and similar experiments. In such case I take wires rather short and set at the beginning the condenser plates at maximum distance. If the streams for the wires increase by approach of the plates, the length of the wires is about right; if they diminish the wires are too long for that frequency and potential. When a condenser is used in connection with experiments with such a coil, it should be an oil condenser by all means, as in using an air condenser considerable energy might be wasted. The wires leading to the plates in the oil should be very thin, heavily coated with some insulating compound, and provided with a conducting covering—this preferably extending under the surface of the oil. The conducting cover should not be too near the terminals, or ends, of the wire, as a spark would be apt to jump from the wire to it. The conducting coating is used to diminish the air losses, in virtue of its action as an electrostatic screen. As to the size of the vessel containing the oil and the site of the plates, the experimenter gains at once an idea from a rough trial. The size of the plates in oil is, however, calculable, as the dielectric losses are very small.

In the preceding experiment it is of considerable interest to know what relation the quantity of the light emitted bears to the frequency and potential of the electric impulses. My opinion is that the heat as well as light effects produced should be proportionate, under otherwise equal conditions of test, to the product of frequency and square of potential, but the experimental verification of the law, whatever it may be, would be exceedingly difficult. One thing is certain, at any rate, and that is, that in augmenting the potential and frequency we rapidly intensify the streams; and, though it may be very sanguine, it is surely not altogether hopeless to expect that we may succeed in producing a practical illuminant on these lines. We would then be simply using burners or flames, in which there would be no chemical process, no consumption of material, but merely a transfer of energy, and which would, in all probability emit more light and less heat than ordinary flames.

The luminous intensity of the streams is, of course, considerably increased when they are focused upon a small surface. This may be shown by the following experiment:

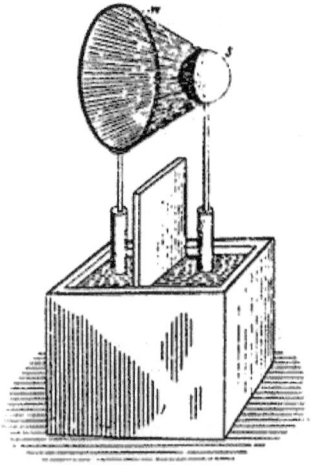

Fig. 8.

I attach to one of the terminals of the coil a wire w (Fig. 8), bent in a circle of about 30 centimetres in diameter, and to the other terminal I fasten a small brass sphere s, the surface of the wire being preferably equal to the surface of the sphere, and the centre of the latter being in a line at right angles to the plane of the wire circle and passing through its centre. When the discharge is established under proper conditions, a luminous hollow cone is formed, and in the dark one-half of the brass sphere is strongly illuminated, as shown in the cut.

By some artifice or other, it is easy to concentrate the streams upon small surfaces and to produce very strong light effects. Two thin wires may thus be rendered intensely luminous. In order to intensify the streams, the wires should be very thin and short; but as in this case their capacity would be generally too small for the coil—at least, for such a one as the present—it is necessary to augment the capacity to the required value, while, al the same time, the surface of the wires remains very small. This may be done in many ways.

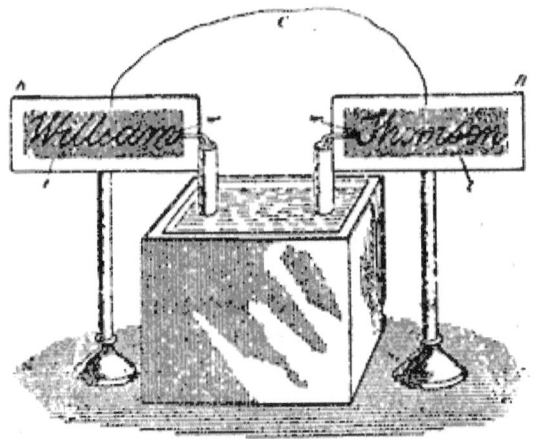

Fig. 9.

Here, for instance, I have two plates R R, of hard rubber (Fig. 9), upon which I have glued two very thin wires w w, so as to form a name. The wires may be bare or covered with the best insulation—it is immaterial for the success of the experiment. Well-insulated wires, if anything, are preferable. On the back of each plate, indicated by the shaded portion, is a tinfoil coating t t. The plates are placed in line at a sufficient distance to prevent a spark passing from one to the other wire. The two tinfoil coatings I have joined by a conductor C, and the two wires I presently connect to the terminals of the coil. It is now easy, by varying the strength and frequency of the currents through the primary, to find a point at which the capacity of the system is best suited to the conditions, and the wires become so strongly luminous that, when the light in the room is turned off the name formed by them appears in brilliant letters.

It is perhaps preferable to perform this experiment with a coil operated from an alternator of high frequency, as then, owing to the harmonic rise and fall, the streams are very uniform, though they are less abundant than when produced with such a coil as the present. This experiment, however, may be performed with low frequencies, but much less satisfactorily.

When two wires, attached to the terminals of the coil, are set at the proper distance, the streams between them may be so intense as to produce a continuous luminous sheet. To show this phenomenon I have here two circles, C and c (Fig. 10), of rather stout wire, one being about 80 centimetres and the other 30 centimetres in diameter. To each of the terminals of the coil I attach one of the circles. The supporting wires are so bent that the circles may be placed in the same plane, coinciding as nearly as possible. When the light in the room is turned off and the coil set to work, you see the whole space between the wires uniformly filled with streams, forming a luminous disc, which could be seen from a considerable distance, such is the intensity of the streams. The outer circle could have been much larger than the present one; in fact, with this coil I have used much larger circles, and I have been able to produce a strongly luminous sheet, covering an area of more than one square metre, which is a remarkable effect with this very small coil. To avoid uncertainty, the circle has been taken smaller, and the area is now about 0.43 square metre.

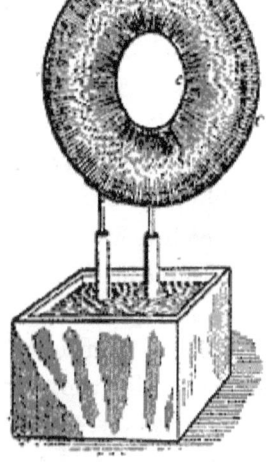

Fig. 10.

The frequency of the vibration, and the quickness of succession of the sparks between the knobs, affect to a marked degree the appearance of the streams. When the frequency is very low, the air gives way in more or less the same manner, as by a steady difference of potential, and the streams consist of distinct threads, generally mingled with thin sparks, which probably correspond to the successive discharges occurring between the knobs. But when the frequency is extremely high, and the arc of the discharge produces a very loud but smooth sound—showing both that oscillation takes place and that the sparks succeed each other with great rapidity—then the luminous streams formed are perfectly uniform. To reach this result very small coils and jars of small capacity should be used. I take two tubes of thick Bohemian glass, about 5 centimetres in diameter and 20 centimetres long. In each of the tubes I slip a primary of very thick copper wire. On the top of each tube I wind a secondary of much thinner gutta-percha covered wire. The two secondaries I connect in series, the primaries preferably in multiple arc. The tubes are then placed in a large glass vessel, at a distance of 10 to 15 centimetres from each other, on insulating supports, and the vessel is filled with boiled out oil, the oil reaching about an inch above the tubes. The free ends of the secondary are lifted out of the oil and placed parallel to each other at a distance of about 10 centimetres. The ends which are scraped should be dipped in the oil. Two four-pint jars joined in series may be used to discharge through the primary. When the necessary adjustments in the length and distance of the wires above the oil and in the arc of discharge are made, a luminous sheet is produced between the wires, which is perfectly smooth and textureless, like the ordinary discharge through a moderately exhausted tube.

I have purposely dwelt upon this apparently insignificant experiment. In trials of this kind the experimenter arrives at the startling conclusion that, to pass ordinary luminous discharges through gases, no particular degree of exhaustion is needed, but that the gas may be at ordinary or even greater pressure. To accomplish this, a very high frequency is essential; a high potential is likewise required, but this is a merely incidental necessity. These experiments teach us that, in endeavoring to discover novel methods of producing light by the agitation of atoms, or molecules, of a gas, we need not limit our research to the vacuum tube, but may look forward quite seriously to the possibility of obtaining the light effects without the use of any vessel whatever, with air at ordinary pressure.

Such discharges of very high frequency, which render luminous the air at ordinary pressures, we have probably often occasion to witness in Nature. I have no doubt that if, as many believe, the aurora borealis is produced by sudden cosmic disturbances, such as eruptions at the sun's surface, which set the electrostatic charge of the earth in an extremely rapid vibration the red glow observed is not confined to the upper rarefied strata of the air, but the discharge traverses, by reason of its very high frequency, also the dense atmosphere in the form of a glow, such as we ordinarily produce in a slightly exhausted tube. If the frequency were very low or even more so, if the charge were not at all vibrating, the dense air would break down as in a lightning discharge. Indications of such breaking down of the lower dense strata of the air have been repeatedly observed at the occurrence of this marvelous phenomenon; but if it does occur; it can only be attributed to the fundamental disturbances, which are few in number, for the vibration produced by them would be far too rapid to allow a disruptive break. It is the original and irregular impulses which affect the instruments; the superimposed vibrations probably pass unnoticed.

When an ordinary low frequency discharge is passed through moderately rarefied air, the air assumes a purplish hue. If by some means or other we increase the intensity of the molecular, or atomic, vibration, the gas changes to a white color. A similar change occurs at ordinary pressures with electric impulses of very high frequency. If the molecules of the air around a wire are moderately agitated, the brush formed is reddish or violet; if the vibration is rendered sufficiently intense, the streams become white. We may accomplish this in various ways. In the experiment before shown with the two wires across the room, I have endeavored to secure the result by pushing to a high value both the frequency and potential; in the experiment with the thin wires glued on the rubber plate I have concentrated the action upon a very small surface—in other words, I have worked with a great electric density.

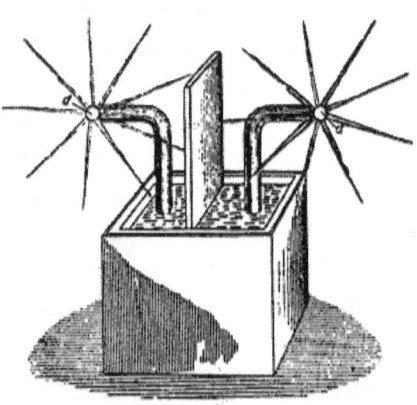

Fig. 11.

A most curious form of discharge is observed with such a coil when the frequency and potential are pushed to the extreme limit. To perform the experiment, every part of the coil should be heavily insulated, and only two small spheres — or, better still, two sharp-edged metal discs (d d, Fig. 11) of no more than a few centimetres in diameter — should be exposed to the air. The coil here used immersed in oil, and the ends of the secondary reaching out of the oil are covered with an airtight cover of hard rubber of great thickness. All cracks, if there are any, should be carefully stopped up, so that the brush discharge cannot form anywhere except on the small spheres or plates which are exposed to the air. In this case, since there are no large plates or other bodies of capacity attached to the terminals, the coil is capable of an extremely rapid vibration. The potential may be raised by increasing, as far as the experimenter judges proper, the rate of change of the primary current. With a coil not widely differing from the present, it is best to connect the two primaries in multiple arc; but if the secondary should have a much greater number of turns the primaries should preferably be used in series, as otherwise the vibration might be too fast for the secondary. It occurs under these conditions that misty white streams break forth from the edges of the discs and spread out phantom-like into space. With this coil, when fairly well produced, they are about 25 to 30 centimetres long. When the hand is held against them no sensation is produced, and a spark, causing a shock, jumps from the terminal only upon the hand being brought much nearer. If the oscillation of the primary current is rendered intermittent by some means or other, there is a corresponding throbbing of the streams, and now the hand or other conducting object may be brought in still greater proximity to the terminal without a spark being caused to jump.

Among the many beautiful phenomena which may be produced with such a coil I have here selected only those which appear to possess some features of novelty, and lead us to some conclusions of interest. One will not find it at all difficult to produce in the laboratory, by means of it, many other phenomena which appeal to the eye even more than these here shown, but present no particular feature of novelty.

Early experimenters describe the display of sparks produced by an ordinary large induction coil upon an insulating plate separating the terminals. Quite recently Siemens performed some experiments in which fine effects were obtained, which were seen by many with interest. No doubt large coils, even if operated with currents of low frequencies, are capable of producing beautiful effects. But the largest coil ever made could not, by far, equal the magnificent display of streams and sparks obtained from such a disruptive discharge coil when properly adjusted. To give an idea, a coil such as the present one will cover easily a plate of 1 metre in diameter completely with the streams. The best way to perform such experiments is to take a very thin rubber or a glass plate and glue on one side of it a narrow ring of tinfoil of very large diameter, and on the other a circular washer, the centre of the latter coinciding with that of the ring, and the surfaces of both being preferably equal, so as to keep the coil well balanced. The washer and ring should be connected to the terminals by heavily insulated thin wires. It is easy in observing the effect of the capacity to produce a sheet of uniform streams, or a fine network of thin silvery threads, or a mass of loud brilliant sparks, which completely cover the plate.

Since I have advanced the idea of the conversion by means of the disruptive discharge, in my paper before the *American Institute of Electrical Engineers* at the beginning of the past year, the interest excited in it has been considerable. It affords us a means for producing any potentials by the aid of inexpensive coils operated from ordinary systems of distribution, and—what is perhaps more appreciated—it enables us to convert currents of any frequency into currents of any other lower or higher frequency. But its chief value will perhaps be found in the help which it will afford us in the investigations of the phenomena of phosphorescence, which a disruptive discharge coil is capable of exciting in innumerable cases where ordinary coils, even the largest, would utterly fail.

Considering its probable uses for many practical purposes, and its possible introduction into laboratories for scientific research, a few additional remarks as to the construction of such a coil will perhaps not be found superfluous.

It is, of course, absolutely necessary to employ in such a coil wires provided with the best insulation.

Good coils may be produced by employing wires covered with several layers of cotton, boiling the coil a long time in pure wax, and cooling under moderate pressure. The advantage of such a coil is that it can be easily handled, but it cannot probably give as satisfactory results as a coil immersed in pure oil. Besides, it seems that the presence of a large body of wax affects the coil disadvantageously, whereas this does not seem to be the case with oil. Perhaps it is because the dielectric losses in the liquid are smaller.

I have tried at first silk and cotton covered wires with oil immersion; but I have been gradually led to use gutta-percha covered wires, which proved most satisfactory. Gutta-percha insulation adds, of course, to the capacity of the coil, and this, especially if the coil be large, is a great disadvantage when extreme frequencies are desired; but, on the other hand, gutta-percha will withstand much more than an equal thickness of oil, and this advantage should be secured at any price. Once the coil has been immersed, it should never be taken out of the oil for more than a few hours, else the gutta-percha will crack up and the coil will not be worth half as much as before. Gutta-percha is probably slowly attacked by the oil, but after an immersion of eight to nine months I have found no ill effects.

I have obtained in commerce two kinds of gutta-percha wire: in one the insulation sticks tightly to the metal, in the other it does not. Unless a special method is followed to expel all air, it is much safer to use the first kind. I wind the coil within an oil tank so that all interstices are filled up with the oil. Between the layers I use cloth boiled out thoroughly in oil, calculating the thickness according to the difference of potential between the turns. There seems not to be a very great difference whatever kind of oil is used; I use paraffin or linseed oil.

To exclude more perfectly the air, an excellent way to proceed, and easily practicable with small coils, is the following: Construct a box of hard wood of very thick boards which have been for a long time boiled in oil. The boards should be so joined as to safely withstand the external air pressure. The coil being placed and fastened in position within the box, the latter is closed with a strong lid, and covered with closely fitting metal sheets, the joints of which are soldered very carefully. On the top two small holes are drilled, passing through the metal sheet and the wood, and in these holes two small glass tubes are inserted and the joints made air-tight. One of the tubes is connected to a vacuum pump and the other with a vessel containing a sufficient quantity of boiled-out oil. The latter tube has a very small hole at the bottom, and is provided with a stopcock. When a fairly good vacuum has been obtained, the stopcock is opened and the oil slowly fed in. Proceeding in this manner, it is impossible that any big bubbles, which are the principal danger, should remain between the turns. The air is most completely excluded, probably better than by boiling out, which, however, when gutta-percha coated wires are used, is not practicable.

For the primaries I use ordinary line wire with thick cotton coating. Strands of very thin insulated wires properly interlaced would, of course, be the best to employ for the primaries, but they are not to be had.

In an experimental coil the size of the wires is not of great importance. In the coil here used the primary is No, 12 and the secondary No. 24 Brown & Sharpe gauge wire; but the sections maybe varied considerably. I would only imply different adjustments; the results aimed at would not be materially affected.

I have dwelt at some length upon the various forms of brush discharge because, in studying them, we not only observe phenomena which please our eye, but also afford us food for thought, and lead us to conclusions of practical importance. In the use of alternating currents of very high tension, too much precaution cannot be taken to prevent the brush discharge. In a main conveying such currents, in an induction coil or transformer, or in a condenser, the brush discharge is a source of great danger to the insulation. In a condenser especially the gaseous matter must be most carefully expelled, for in it the charged surfaces are near each other, and if the potentials are high, just as sure as a weight will fall if let go, so the insulation will give way if a single gaseous bubble of some site be present, whereas, if all gaseous matter were carefully excluded, the condenser would safely withstand a much higher difference of potential. A main conveying alternating currents of very high tension may be injured merely by a blowhole or small crack in the insulation, the more so as a blowhole is apt to contain gas at low pressure; and as it appears almost impossible to completely obviate such little imperfections, I am led to believe that in our future distribution of electrical energy by currents of very high tension liquid insulation will be used. The cost is a great drawback, but if we employ an oil as an insulator the distribution of electrical energy with something like 100,000 volts, and even more, become, at least with higher frequencies, so easy that they could be hardly called engineering feats. With oil insulation and alternate current motors transmissions of power can be effected with safety and upon an industrial basis at distances of as much as a thousand miles.

A peculiar property of oils, and liquid insulation in general, when subjected to rapidly changing electric stresses, is to disperse any gaseous bubbles which may be present, and diffuse them through its mass, generally long before any injurious break can occur. This feature may be easily observed with an ordinary induction coil by taking the primary out, plugging up the end of the tube upon which the secondary is wound, and fining it with some fairly transparent insulator, such as paraffin oil. A primary of s diameter something like six millimetres smaller than the inside of the tube may be inserted in the oil. When the coil is set to work one may see, looking from the top through the oil, many luminous points — air bubbles which are caught by inserting the primary, and which ate rendered luminous in consequence of the violent bombardment. The occluded air, by its impact against the oil, beats it; the oil begins to circulate, carrying some of the air along with it, until the bubbles are dispersed and the luminous points disappear. In this manner, unless large bubbles are occluded in such way that circulation is rendered impossible, a damaging break is averted, the only effect being a moderate warming up of the oil. If, instead of the liquid, a solid insulation, no matter how thick, were used, a breaking through and injury of the apparatus would be inevitable.

The exclusion of gaseous matter from any apparatus in which the dielectric is subjected to more or less rapidly changing electric forces is, however, not only desirable in order to avoid a possible injury of the apparatus, but also on account of economy. In a condenser, for instance, as long as only a solid or only a liquid dielectric is used, the loss is small; but if a gas under ordinary or small pressure be present the loss may be very great. Whatever the nature of the force acting in the dielectric may be, it seems that in a solid or liquid the molecular displacement produced by the force is small: hence the product of force and displacement is insignificant, unless the force be very great; but in a gas the displacement, and, therefore, this product is considerable; the molecules are free to move, they reach high speeds, and the energy of their impact is lost in heat or otherwise. If the gas be strongly compressed, the displacement due to the force is made smaller, and the losses are reduced.

In most of the succeeding experiments I prefer, chiefly on account of the regular and positive action, to employ the alternator before referred to. This is one of the several machines constructed by me for the purposes of these investigations. It has 384 pole projections, and is capable of giving currents of a frequency of about 10,000 per second. This machine has been illustrated and briefly described in my first paper before the *American Institute of Electrical Engineers*, May 20, 1831, to which I have already referred. A more detailed description, sufficient to enable any engineer to build a similar machine, will be found in several electrical journals of that period.

The induction coils operated from the machine are rather small, containing from 5,000 to 15,000 turns in the secondary. They are immersed in boiled-out linseed oil, contained in wooden boxes covered with zinc sheet.

I have found it advantageous to reverse the usual position of the wires, and to wind, in these coils, the primaries on the top; this allowing the use of a much bigger primary, which, of course, reduces the danger of overheating and increases the output of the coil. I make the primary on each side at least one centimetre shorter than the secondary, to prevent the breaking through on the ends, which would surely occur unless the insulation on the top of the secondary be very thick, and this, of course, would be disadvantageous.

When the primary is made movable, which is necessary in some experiments, and many times convenient for the purposes of adjustment, I cover the secondary with wax, and turn it off in a lathe to a diameter slightly smaller than the inside of the primary coil. The latter I provide with a handle reaching out of the oil, which serves to shift it in any position along the secondary.

I will now venture to make, in regard to the general manipulation of induction coils, a few observations bearing upon points which have not been fully appreciated in earlier experiments with such coils, and are even now often overlooked.

The secondary of the coil possesses usually such a high self-induction that the current through the wire is inappreciable, and may be so even when the terminals ate joined by a conductor of small resistance. If capacity is added to the terminals, the self-induction is counteracted, and a stronger current is made to flow through the secondary, though its terminals are insulated from each other. To one entirely unacquainted with the properties of alternating currents nothing will look more puzzling. This feature was illustrated in the experiment performed at the beginning with the top plates of wire gauze attached to the terminals and the rubber plate. When the plates of wire gauze were close together, and a small arc passed between them, the arc prevented a strong current from passing through the secondary, because it did away with the capacity on the terminals; when the rubber plate was inserted between, the capacity of the condenser formed counteracted the self-induction of the secondary, a stronger current passed now, the coil performed more work, and the discharge was by far more powerful.

The first thing, then, in operating the induction coil is to combine capacity with the secondary to overcome the self-induction. If the frequencies and potentials are very high gaseous matter should be carefully kept away from the charged surfaces. If Leyden jars are used, they should be immersed in oil, as otherwise considerable dissipation may occur if the jars are greatly strained. When high frequencies are used, it is of equal importance to combine a condenser with the primary. One may use a condenser connected to the ends of the primary or to the terminals of the alternator, but the latter is not to be recommended, as the machine might be injured. The best way is undoubtedly to use the condenser in series with the primary and with the alternator, and to adjust its capacity so as to annul the self-induction of both the latter. The condenser should be adjustable by very small steps, and for a finer adjustment a small oil condenser with movable plates may be used conveniently.

I think it best at this juncture to bring before you a phenomenon, observed by me some time ago, which to the purely scientific investigator may perhaps appear more interesting than any of the results which I have the privilege to present to you this evening.

It may be quite properly ranked among the brush phenomena—in fact, it is a brush, formed at, or near, a single terminal in high vacuum.

In bulbs provided with a conducting terminal, though it be of aluminium, the brush has but an ephemeral existence, and cannot, unfortunately, be indefinitely preserved in its most sensitive state, even in a bulb devoid of any conducting electrode. In studying one phenomenon, by all means a bulb having no leading-in wire should be used. I have found it best to use bulbs constructed as indicated in Figs. 12 and 13.

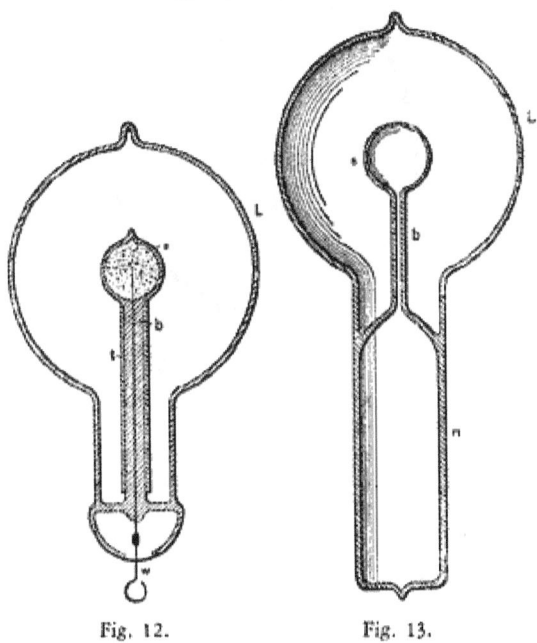

Fig. 12.   Fig. 13.

In Fig. 12 the bulb comprises an incandescent lamp globe L, in the neck of which is sealed a barometer tube b, the end of which is blown out to form a small sphere s. This sphere should be sealed as closely as possible in the centre of the large globe. Before sealing, a thin tube t, of aluminium sheet, may be slipped in the barometer tube, but it is not important to employ it.

The small hollow sphere s is filled with some conducting powder, and a wire w is cemented in the neck for the purpose of connecting the conducting powder with the generator.

The construction shown in Fig. 13 was chosen in order to remove from the brush any conducting body which might possibly affect it. The bulb consists in this case of a lamp globe L, which has a neck n, provided with a tube b and small sphere s, sealed to it, so that two entirely independent compartments are formed, as indicated in the drawing. When the bulb is in use, the neck n is provided with a tinfoil coating, which is connected to the generator and acts inductively upon the moderately rarefied and highly conducting gas enclosed in the neck. From there the current passes through the tube b into the small sphere s, to act by induction upon the gas contained in the globe L.

It is of advantage to make the tube t very thick, the hole through it very small, and to blow the sphere s very thin. It is of the greatest importance that the sphere s be placed in the centre of the globe L.

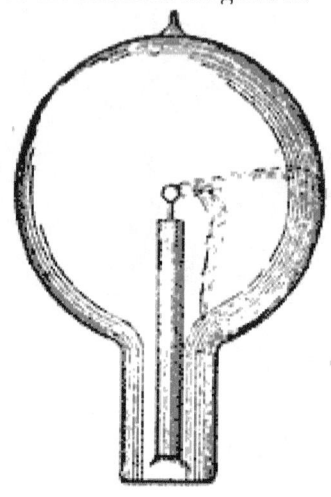

Fig. 14.

Figs. 14, 15 and 16 indicate different forms, or stages, of the brush. Fig. 14 shows the brush as it first appears in a bulb provided with a conducting terminal: but, as in such a bulb it very soon disappears — often after a few minutes — I will confine myself to the description of the phenomenon as seen in a bulb without conducting electrode. It is observed under the following conditions:

When the globe L (Figs. 12 and 13) is exhausted to a very high degree, generally the bulb is not excited upon connecting the wire w (Fig. 12) or the tinfoil coating of the bulb (Fig. 13) to the terminal of the induction coil. To excite it, it is usually sufficient to grasp the globe L with the hand. An intense phosphorescence then spreads at first over the globe, but soon gives place to a white, misty light. Shortly afterward one may notice that the luminosity is unevenly distributed in the globe, and after passing the current for some time the bulb appears as in Fig. 15. From this stage the phenomenon will gradually pass to that indicated in Fig. 16, after some minutes, hours, days or weeks, according as the bulb is worked. Warming the bulb or increasing the potential hastens the transit.

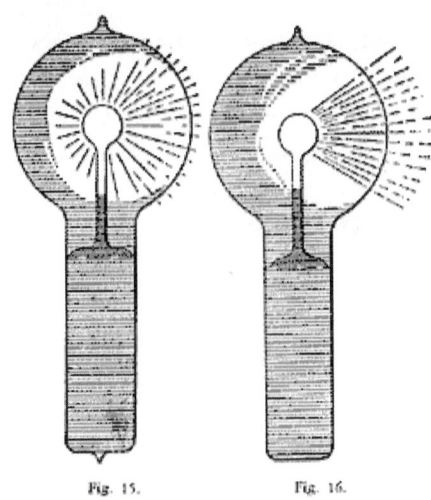

Fig. 15.   Fig. 16.

When the brush assumes the form indicated in Fig. 16, it may be brought to a state of extreme sensitiveness to electrostatic and magnetic influence. The bulb hanging straight down from a wire, and all objects being remote from it, the approach of the observer at a few paces from the bulb will cause the brush to fly to the opposite side, and if he walks around the bulb it will always keep on the opposite side. It may begin to spin around the terminal long before it reaches that sensitive stage. When it begins to turn around principally, but also before, it is affected by a magnet and at a certain stage it is susceptible to magnetic influence to an astonishing degree. A small permanent magnet, with its poles at a distance of no more than two centimetres, will affect it visibly at a distance of two metres, slowing down or accelerating the rotation according to how it is held relatively to the brush. I think I have observed that at the stage when it is most sensitive to magnetic, it is not most sensitive to electrostatic, influence. My explanation is, that the electrostatic attraction between the brush and the glass of the bulb, which retards the rotation, grows much quicker than the magnetic influence when the intensity of the stream is increased.

When the bulb hangs with the globe L down, the rotation is always clockwise. In the southern hemisphere it would occur in the opposite direction and on the equator the brush should not turn at all. The rotation may be reversed by a magnet kept at some distance. The brush rotates best, seemingly, when it is at right angles to the lines of force of the earth. It very likely rotates, when at its maximum speed, in synchronism with the alternations, say 10,000 times a second. The rotation can be slowed down or accelerated by the approach or receding of the observer or any conducting body, but it cannot be reversed by putting the bulb in any position. When it is in the state of the highest sensitiveness and the potential or frequency be varied the sensitiveness is rapidly diminished. Changing either of these but little will generally stop the rotation. The sensitiveness is likewise affected by the variations of temperature. To attain great sensitiveness it is necessary to have the small sphere s in the centre of the globe L, as otherwise the electrostatic action of the glass of the globe will tend to stop the rotation. The sphere s should be small and of uniform thickness; any dissymmetry of course has the effect to diminish the sensitiveness.

The fact that the brush rotates in a definite direction in a permanent magnetic field seems to show that in alternating currents of very high frequency the positive and negative impulses are not equal, but that one always preponderates over the other.

Of course, this rotation in one direction may be due to the action of two elements of the same current upon each other, or to the action of the field produced by one of the elements upon the other, as in a series motor, without necessarily one impulse being stronger than the other. The fact that the brush turns, as far as I could observe, in any position, would speak for this view. In such case it would turn at any point of the earth's surface. But, on the other hand, it is then hard to explain why a permanent magnet should reverse the rotation, and one must assume the preponderance of impulses of one kind.

As to the causes of the formation of the brush or stream, I think it is due to the electrostatic action of the globe and the dissymmetry of the parts. If the small bulb s and the globe L were perfect concentric spheres, and the glass throughout of the same thickness and quality, I think the brush would not form, as the tendency to pass would be equal on all sides. That the formation of the stream is due to an irregularity is apparent from the fact that it has the tendency to remain in one position, and rotation occurs most generally only when it is brought out of this position by electrostatic or magnetic influence. When in an extremely sensitive state it rests in one position, most curious experiments may be performed with it. For instance, the experimenter may, try selecting a proper position, approach the hand at a certain considerable distance to the bulb, and he may cause the brush to pass off by merely stiffening the muscles of the arm. When it begins to rotate slowly, and the hands are held at a proper distance, it is impossible to make even the slightest motion without producing a visible effect upon the brush. A metal plate connected to the other terminal of the coil affects it at a great distance, slowing down the rotation often to one turn a second.

I am firmly convinced that such a brush, when we learn how to produce it properly, will prove a valuable aid in the investigation' of the nature of the forces acting in an electrostatic or magnetic field. If there is any motion which is measurable going on in the space, such a brush ought to reveal it. It is, so to speak, a beam of light, frictionless, devoid of inertia.

I think that it may find practical applications in telegraphy. With such a brush it would be possible to send dispatches across the Atlantic, for instance, with any speed, since its sensitiveness may be so great that the slightest changes will affect it. If it were possible to make the stream more intense and very narrow, its deflections could be easily photographed.

I have been interested to find whether there is a rotation of the stream itself, or whether there is simply a stress traveling around in the bulb. For this purpose I mounted a light mica fan so that its vanes were in the path of the brush. If the stream itself was rotating the fan would be spun around. I could produce no distinct rotation of the fan, although I tried the experiment repeatedly; but as the fan exerted a noticeable influence on the stream, and the apparent rotation of the latter was, in this case, never quite satisfactory, the experiment did not appear to be conclusive.

I have been unable to produce the phenomenon with the disruptive discharge coil, although every other of these phenomena can be tell produced by it—many, in fact, much better than with coils operated from an alternator.

It may be possible to produce the brush by impulses of one direction, or even by a steady potential, in which case it would be still more sensitive to magnetic influence.

In operating an induction coil with rapidly alternating currents, we realize with astonishment, for the first time, the great importance of the relation of capacity, self-induction and frequency as regards the general result. The effects of capacity are the most striking, for in these experiments, since the self-induction and frequency both are high, the critical capacity is very small, and need be but slightly varied to produce a very considerable change. The experimenter may bring his body in contact with the terminals of the secondary of the coil, or attach to one or both terminals insulated bodies of very small bulk, such as bulbs, and he may produce a considerable rise or fall of potential, and greatly affect the flow of the current through the primary. In the experiment before shown, in which a brush appears at a wire attached to one terminal, and the wire is vibrated when the experimenter brings his insulated body in contact with the other terminal of the coil, the sudden rise of potential was made evident.

I may show you the behavior of the coil in another manner which possesses a feature of some interest. I have here a little light fan of aluminium sheet, fastened to a needle and arranged to rotate freely in a metal piece screwed to one of the terminals of the coil. When the coil is set to work, the molecules of the air are rhythmically attracted and repelled. As the force with which they are repelled is greater than that with which they are attracted, it results that there is repulsion exerted on the surfaces of the fan. If the fan were made simply of a metal sheet, the repulsion would be equal on the opposite sides, and would produce no effect. But if one of the opposite surfaces is screened, or if, generally speaking, the bombardment on this side is weakened in some wag or other, there remains the repulsion exerted upon the other, and the fan is set in rotation. The screening is best effected by fastening upon one of the opposing sides of the fan insulated conducting coatings, or, if the fan is made in the shape of an ordinary propeller screw, by fastening on one side, and close to it, an insulated metal plate. The static screen may however, be omitted and simply a thickness of insulating material fastened to one of the sides of the fan.

To show the behavior of the coil, the fan may be placed upon the terminal and it will readily rotate when the coil is operated by currents of very high frequency. With a steady potential, of course, and even with alternating currents of very low frequency, it would not turn, because of the very slow exchange of air and, consequently, smaller bombardment; but in the latter case it might turn if the potential were excessive. With a pin wheel, quite the opposite rule holds good; it rotates best with a steady potential, and the effort is the smaller the higher the frequency. Now, it is very easy to adjust the conditions so that the potential is normally not sufficient to turn the fan, but that by connecting the other terminal of the coil with an insulated body it rises to a much greater value, so as to rotate the fan, and it is likewise possible to stop the rotation by connecting to the terminal a body of different size, thereby diminishing the potential.

Instead of using the fan in this experiment, we may use the "electric" radiometer with similar effect. But in this case it will be found that the vanes will rotate only at high exhaustion or at ordinary pressures; they will not rotate at moderate pressures, when the air is highly conducting. This curious observation was made conjointly by Professor Crookes and myself. I attribute the result to the high conductivity of the air, the molecules of which then do not act as independent carriers of electric charges, but act all together as a single conducting body. In such case, of course, if there is any repulsion at all of the molecules from the vanes, it must be very small. It is possible, however, that the result is in part due to the fact that the greater part of the discharge passes from the leading-in wire through the highly conducting gas, instead of passing off from the conducting vanes.

In trying the preceding experiment with the electric radiometer the potential should not exceed a certain limit, as then the electrostatic attraction between the vanes and the glass of the bulb may be so great as to stop the rotation.

A most curious feature of alternate currents of high frequencies and potentials is that they enable us to perform many experiments by the use of one wire only. In many respects this feature is of great interest.

In a type of alternate current motor invented by me some years ago I produced rotation by inducing, by means of a single alternating current passed through a motor circuit, in the mass or other circuits of the motor, secondary currents, which, jointly with the primary or inducing current, created n moving field of force. A simple but crude form of such a motor is obtained by winding upon an iron core a primary, and close to it a secondary coil, joining the ends of the latter and placing a freely movable metal disc within the influence of the field produced by both. The iron core is employed for obvious reasons, but it is not essential to the operation. To improve the motor, the iron core is made to encircle the armature. Again to improve, the secondary coil is made to overlap partly the primary, so that it cannot free itself from a strong inductive action of the latter, repel its lines as it may. Once more to improve, the proper difference of phase is obtained between the primary and secondary currents by a condenser, self-induction, resistance or equivalent windings.

I had discovered, however, that rotation is produced by means of a single coil and core; my explanation of the phenomenon, and leading thought in trying the experiment, being that there must be a true time lag in the magnetization of the core. I remember the pleasure I had when, in the writings of Professor Ayrton, which came later to my hand, I found the idea of the time lag advocated. Whether there is a true time lag, whether the retardation is due to eddy currents circulating in minute paths, must remain an open question, but the fact is that a coil wound upon an iron core and traversed by an alternating current creates a moving field of force, capable of setting an armature in rotation. It is of some interest, in conjunction with the historical Arago experiment, to mention that in lag or phase motors I have produced rotation in the opposite direction to the moving field, which means that in that experiment the magnet may not rotate, or may even rotate in the opposite direction to the moving disc. Here, then, is a motor (diagrammatically illustrated in Fig. 17), comprising a coil and iron core, and a freely movable copper disc in proximity to the latter.

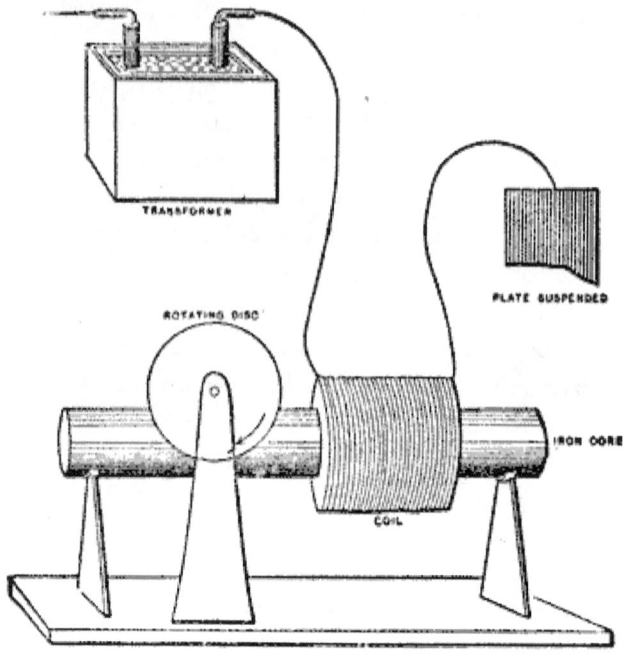

Fig. 17.

To demonstrate a novel and interesting feature, I have, for a reason which I will explain, selected this type of motor. When the ends of the coil are connected to the terminals of an alternator the disc is set in rotation. But it is not this experiment, now well known, which I desire to perform. What I wish to show you is that this motor rotates with one single connection between it and the generator; that is to say, one terminal of the motor is connected to one terminal of the generator—in this case the secondary of a high-tension induction coil—the other terminals of motor and generator being insulated in space. To produce rotation it is generally (but not absolutely) necessary to connect the free end of the motor coil to an insulated body of some size. The experimenter's body is more than sufficient. If he touches the free terminal with an object held in the hand, a current passes through the coil and the copper disc is set in rotation. If an exhausted tube is put in series with the coil, the tube lights brilliantly, showing the passage of a strong current. Instead of the experimenter's body, a small metal sheet suspended on a cord may be used with the same result. In this case the plate acts as a condenser in series with the coil. It counteracts the self-induction of the latter and allows a strong current to pass. In such a combination, the greater the self-induction of the coil the smaller need be the plate, and this means that a lower frequency, or eventually a lower potential, is required to operate the motor. A single coil wound upon a core has a high self-induction; for this reason principally, this type of motor was chosen to perform the experiment. Were a secondary closed coil wound upon the core, it would tend to diminish the self-induction, and then it would be necessary to employ a much higher frequency and potential. Neither would be advisable, for a higher potential would endanger the insulation of the small primary coil, and a higher frequency would result in a materially diminished torque.

It should be remarked that when such a motor with a closed secondary is used, it is not at all easy to obtain rotation with excessive frequencies, as the secondary cuts off almost completely the lines of the primary—and this, of course, the more, the higher the frequency—and allows the passage of but a minute current. In such a case, unless the secondary is closed through a condenser, it is almost essential, in order to produce rotation, to make the primary and secondary coils overlap each other more or less.

But there is an additional feature of interest about this motor, namely, it is not necessary to have even a single connection between the motor and generator, except, perhaps, through the ground; for not only is an insulated plate capable of giving off energy into space, but it likewise capable of deriving it from an alternating electrostatic field, though in the latter case the available energy is much smaller. In this instance one of the motor terminals is connected to the insulated plate or body located within the alternating electrostatic field, and the other terminal preferably to the ground.

It is quite possible, however, that such "no-wire" motors, as they might be called, could be operated by conduction through the rarefied air at considerable distances. Alternate currents, especially of high frequencies, pass with astonishing freedom through even slightly rarefied gases. The upper strata of the air are rarefied. To reach a number of miles out into space requires the overcoming of difficulties of a merely mechanical nature. There is no doubt that with the enormous potentials obtainable by the use of high frequencies and oil insulation, luminous discharges might be passed through many miles of rarefied air, and that, by thus directing the energy of many hundreds or thousands of horse-power, motors or lamps might be operated at considerable distances from stationary sources. But such schemes are mentioned merely as possibilities. We shall have no need to transmit power at all. Ere many generations pass, our machinery will be driven by a power obtainable at any point of the universe. This idea is not novel. Men have been led to it long ago by instinct or reason; it has been expressed in many ways, and in many places, in the history of old and new. We find it in the delightful myth of Antheus [Antaeus], who derives power from the earth; we find it among the subtle speculations of one of your splendid mathematicians and in many hints and statements of thinkers of the present time. Throughout space there is energy. Is this energy static or kinetic? If static our hopes are in vain; if kinetic — and this we know it is, for certain — then it is a mere question of time when men will succeed in attaching their machinery to the very wheelwork of nature. Of all, living or dead, Crookes came nearest to doing it. His radiometer will turn in the light of day and in the darkness of the night; it will turn everywhere where there is heat, and heat is everywhere. But, unfortunately, this beautiful little machine, while it goes down to posterity as the most interesting, must likewise be put on record as the most inefficient machine ever invented!

The preceding experiment is only one of many equally interesting experiments which may be performed by the use of only one wire with alternate currents of high potential and frequency. We may connect an insulated line to a source of such currents, we may pass an inappreciable current over the line, and on any point of the same we are able to obtain a heavy current, capable of fusing a thick copper wire. Or we may, by the help of some artifice, decompose a solution in any electrolytic cell by connecting only one pole of the cell to the line or source of energy. Or we may, by attaching to the line, or only bringing into its vicinity, light up an incandescent lamp, an exhausted tube, or a phosphorescent bulb.

However impracticable this plan of working may appear in many cases, it certainly seems practicable, and even recommendable, in the production of light. A perfected lamp would require but little energy, and if wires were used at all we ought to be able to supply that energy without a return wire.

It is now a fact that a body may be rendered incandescent or phosphorescent b) bringing it either in single contact or merely in the vicinity of a source of electric impulses of the proper character, and that in this manner a quantity of light sufficient to afford a practical illuminant may be produced. It is, therefore, to say the least, worth while to attempt to determine the best conditions and to invent the best appliances for attaining this object.

Some experiences have already been gained in this direction, and I will dwell on them briefly, in the hope that they might prove useful.

The heating of a conducting body inclosed in a bulb, and connected to a source of rapidly alternating electric impulses, is dependent on so many things of a different nature, that it would be difficult to give a generally applicable rule under which this maximum heating occurs. As regards the size of the vessel, I have lately found that at ordinary or only slightly differing atmospheric pressures, when air is a good insulator, and hence practically the same amount of energy by a certain potential and frequency is given off from the body, whether the bulb be small or large, the body is brought to a higher temperature if inclosed in a small bulb, because of the better confinement of heat in this case.

At lower pressures, when air becomes more or less conducting, or if the air be sufficiently warmed as to become conducting, the body is rendered more intensely incandescent in a large bulb, obviously because, under otherwise equal conditions of test, more energy may be given off from the body when the bulb is large.

At very high degrees of exhaustion, when the matter in the bulb becomes "radiant," a large bulb has still an advantage, but a comparatively slight one, over the small bulb. Finally, at excessively high degrees of exhaustion, which cannot be reached except by the employment of special means, there seems to be, beyond a certain and rather small size of vessel, no perceptible difference in the heating.

These observations were the result of a number of experiments, of which one, showing the effect of the size of the bulb at a high degree of exhaustion may be described and shown here, as it presents a feature of interest. Three spherical bulbs of 2 inches, 3 inches and 4 inches diameter were taken, and in the centre of each was mounted an equal length of an ordinary incandescent lamp filament of uniform thickness. In each bulb the piece of filament was fastened to the leading-in wire of platinum, contained in a glass stem sealed in the bulb; care being taken, of course, to make everything as nearly alike as possible. On each glass stem in the inside of the bulb was slipped a highly polished tube made of aluminium sheet, which fitted the stem and was held on it by spring pressure. The function of this aluminium tube will be explained subsequently. In each bulb an equal length of filament protruded above the metal tube. It is sufficient to say now that under these conditions equal lengths of filament of the same thickness—in other words, bodies of equal bulk—were brought to incandescence. The three bulbs were sealed to a glass tube, which was connected to a Sprengel pump. When a high vacuum had been reached, the glass tube carrying the bulbs was sealed off. A current was then turned on successively on each bulb, and it was found that the filaments came to about the same brightness, and, if anything, the smallest bulb, which was placed midway between the two larger ones, may have been slightly brighter. This result was expected, for when either of the bulbs was connected to the coil the luminosity spread through the other two, hence the three bulbs constituted really one vessel. When all the three bulbs were connected in multiple arc to the coil, in the largest of them the filament glowed brightest, in the next smaller it was a little less bright, and in the smallest it only came to redness. The bulbs were then sealed off and separately tried.

The brightness of the filaments was now such as mould have been expected on the supposition that the energy given off was proportionate to the surface of the bulb, this surface in each case representing one of the coatings of a condenser. Accordingly, there was less difference between the largest and the middle sited than between the latter and the smallest bulb.

An interesting observation was made in this experiment. The three bulbs were suspended from a straight bare wire connected to a terminal of the coil, the largest bulb being placed at the end of the wire, at some distance from it the smallest bulb, and an equal distance from the latter the middle-sized one. The carbons glowed then to both the larger bulbs about as expected, but the smallest did not get its share by far. This observation led me to exchange the position of the bulbs, and I then observed that whichever of the bulbs was in the middle it was by far less bright than it was in any other position. This mystifying result was, of course, found to be due to the electrostatic action between the bulbs. When they were placed at a considerable distance, or when they were attached to the corners of an equilateral triangle of copper wire, they glowed about in the order determined by their surfaces.

As to the shape of the vessel, it is also of some importance, especially at high degrees of exhaustion. Of all the possible constructions, it seems that a spherical globe with the refractory body mounted in its centre is the best to employ. In experience it has been demonstrated that in such a globe a refractory body of a given bulk is more easily brought to incandescence than when otherwise shaped bulbs are used. There is also an advantage in giving to the incandescent body the shape of a sphere, for self-evident reasons. In any case the body should be mounted in the centre, where the atoms rebounding from the glass collide. This object is best attained in the spherical bulb; but it is also attained in a cylindrical vessel with one or two straight filaments coinciding with its axis, and possibly also in parabolical or spherical bulbs with the refractory body or bodies placed in the focus or foci of the same; though the latter is not probable, as the electrified atoms should in all cases rebound normally from the surface they strike, unless the speed were excessive, in which case they would probably follow the general law of reflection. No matter what shape the vessel may have, if the exhaustion be low, a filament mounted in the globe is brought to the same degree of incandescence in all parts; but if the exhaustion be high and the bulb be spherical or pear-shaped, as usual, focal points form and the filament is heated to a higher degree at or near such points.

To illustrate the effect, I have here two small bulbs which are alike, only one is exhausted to a low and the other to a very high degree. When connected to the coil, the filament in the former glows uniformly throughout all its length; whereas in the latter, that portion of the filament which is in the centre of the bulb glows far more intensely than the rest. A curious point is that the phenomenon occurs even if two filament: are mounted in a bulb, each being connected to one terminal of the coil, and, what is still more curious, if they be very near together, provided the vacuum be very high. I noted in experiments with such bulbs that the filaments would give way usually at a certain point, and in the first trials I attributed it to a defect in the carbon. But when that phenomenon occurred many times in succession I recognized its real cause.

In order to bring a refractory body inclosed in a bulb to incandescence, it is desirable, on account of economy, that all the energy supplied to the bulb from the source should reach without lass the body to be heated; from there, and from nowhere else, it should be radiated. It is, of course, out of the question to reach this theoretical result, but it is possible by a proper construction of the illuminating device to approximate it more or less.

For many reasons, the refractory body is placed in the centre of the bulb and it is usually supported on a glass stem containing the leading-in wire. As the potential of this wire is alternated, the rarefied gas surrounding the stem is acted upon inductively, and the glass stem is violently bombarded and heated. In this manner by far the greater portion of the energy supplied to the bulb — especially when exceedingly high frequencies are used — may be lost for the purpose contemplated. To obviate this loss, or at least to reduce it to a minimum, I usually screen the rarefied gas surrounding the stem from the inductive action of the leading-in wire by providing; the stem with a tube or coating of conducting material. It seems beyond doubt that the best among metals to employ for this purpose is aluminium, on account of its many remarkable properties. Its only fault is that it is easily fusible and, therefore, its distance from the incandescing: body should be properly estimated. Usually, a thin tube, of a diameter somewhat smaller than that of the glass stem, is made of the finest aluminium sheet, and slipped on the stem. The tube is conveniently prepared by wrapping around a rod fastened in a lathe a piece of aluminium sheet of the proper size, grasping the sheet firmly with clean chamois leather or blotting paper, and spinning the rod very fast. The sheet is wound tightly around the rod, and a highly polished tube of one or three layers of the sheet is obtained. When slipped on the stem, the pressure is generally sufficient to prevent it from slipping off, but, for safety, the lower edge of the sheet may be turned inside. The upper inside corner of the sheet — that is, the one which is nearest to the refractory incandescent body — should be cut out diagonally, as it often happens that, in consequence of the intense heat, this corner turns toward the inside and comes very near to, or in contact with, the wire, or filament, supporting the refractory body. The greater part of the energy supplied to the bulb is then used up in heating the metal tube, and the bulb is rendered useless for the purpose. The aluminium sheet should project above the glass stem more or less—one inch or so—or else, if the glass be too close to the incandescing body, it may be strongly heated and become more or less conducting, whereupon it may be ruptured, or may, by its conductivity, establish a good electrical connection between the metal tube and the leading-in wire, in which case, again, most of the energy will be lost in heating the former. Perhaps the best way is to make the top of the glass tube for about an inch, of a much smaller diameter. To still further reduce the danger arising from the heating of the glass stem, and also with the view of preventing an electrical connection between the metal tube and the electrode, I preferably wrap; the stem with several layers of thin mica which extends at least as far as the metal tube. In some bulbs I have also used an outside insulating cover.

The preceding remarks are only made to aid the experimenter in the first trials, for the difficulties which he encounters he may soon find means to overcome in his own way.

To illustrate the effect of the screen, and the advantage of using it, I have here two bulbs of the same size, with their stems, leading-in wires and incandescent lamp filaments tied to the latter, as nearly alike as possible. The stem of one bulb is provided with an aluminium tube, the stem of the other has none. Originally the two bulbs were joined by a tube which was connected to a Sprengel pump. When a high vacuum had been reached, first the connecting tube, and then the bulbs, were sealed off; they are therefore of the same degree of exhaustion. When they are separately connected to the coil giving a certain potential, the carbon filament in the bulb provided with the aluminium screen in rendered highly incandescent, while the filament in the other bulb may, with the same potential, not even come to redness, although in reality the latter bulb takes generally more energy than the former. When they are both connected together to the terminal, the difference is even more apparent, showing the importance of the screening. The metal tube placed in the stem containing the leading-in wire performs really two distinct functions: First, it acts more or less as an electrostatic screen, thus economizing the energy supplied to the bulb; and, second, to whatever extent it may fail to act electrostatically, it acts mechanically, preventing the bombardment, and consequently intense heating and possible deterioration of the slender support of the refractory incandescent body, or of the glass stem containing the leading-in wire. I say slender support, for it is evident that in order to confine the heat more completely to the incandescing body its support should be very thin, so as to carry away the smallest possible amount of heat by conduction. Of all the supports used I have found an ordinary incandescent lamp filament to be the best, principally because among conductors it can withstand the highest degrees of heat.

The effectiveness of the metal tube as an electrostatic screen depends largely on the degree of exhaustion.

At excessively high degrees of exhaustion—which are reached by using great care and special means in connection with the Sprengel pump—when the matter in the globe is in the ultra-radiant state, it acts most perfectly. The shadow of the upper edge of the tube is then sharply defined upon the bulb.

At a somewhat lower degree of exhaustion, which is about the ordinary "non-striking" vacuum, and generally as long as the matter moves predominantly in straight lines, the screen still does well. In elucidation of the preceding remark it is necessary to state that what is a "non-striking" vacuum for a coil operated, as ordinarily, by impulses, or currents, of low frequency, is not, by far, so when the coil is operated by currents of very high frequency. In such case the discharge may pass with great freedom through the rarefied gas through which a low-frequency discharge may not pass, even though the potential be much higher. At ordinary atmospheric pressures just the reverse rule holds good: the higher the frequency, the less the spark discharge is able to jump between the terminals, especially if they are knobs or spheres of some site. Finally, at very low degrees of exhaustion, when the gas is well conducting, the metal tube not only does not act as an electrostatic screen, but even is a drawback, aiding to a considerable extent the dissipation of the energy laterally from the leading-in wire. This, of course, is to be expected. In this case, namely, the metal tube is in good electrical connection with -the leading-in wire, and most of the bombardment is directed upon the tube. As long as the electrical connection is not good, the conducting tube is always of some advantage for although it may not greatly economize energy, still it protects the support of the refractory button, and is a means for concentrating more energy upon the same.

To whatever extent the aluminium tube performs the function of a screen, its usefulness is therefore limited to very high degrees of exhaustion when it is insulated from the electrode—that is, when the gas as a whole is non-conducting, and the molecules, or atoms, act as independent carriers of electric charges.

In addition to acting as a more or less effective screen, in the true meaning of the word, the conducting tube or coating may also act, by reason of its conductivity, as a sort of equalizer or dampener of the bombardment against the stem. To be explicit, I assume the action as follows: Suppose a rhythmical bombardment to occur against the conducting tube by reason of its imperfect action as a screen, it certainly must happen that some molecules, or atoms, strike the tube sooner than others. Those which come first in contact with it give up their superfluous charge, and the tube is electrified, the electrification instantly spreading over its surface. But this must diminish, the energy lost in the bombardment for two reasons: first, the charge given up by the atoms spreads over a great area, and hence the electric density at any point is small, and the atoms are rebelled with less energy than they would be if they would strike against a good insulator; secondly, as the tube is electrified by the atoms which first come in contact with it, the progress of the following atoms against the tube is more or less checked by, the repulsion which the electrified tube must exert upon the similarly electrified atoms. This repulsion may perhaps be sufficient to prevent a large portion of the atoms from striking the tube, but at any rate it must diminish the energy of their impact. It is clear that when the exhaustion is very low, and the rarefied gas well conducting, neither of the above effects can occur, and, on the other hand, the fewer the atoms, with the greater freedom they move; in other words, the higher the degree of exhaustion, up to a limit, the more telling will be both the effects:

What I have just said may afford an explanation of the phenomenon observed by Prof. Crookes, namely, that a discharge through a bulb is established with much greater facility when an insulator than when a conductor is present in the same. In my opinion, the conductor acts as a dampener of the motion of the atoms in the two ways pointed out; hence, to cause a visible discharge to pass through the bulb, a much higher potential is needed if a conductor, especially of many surfaces, be present.

For the sake of clearness of some of the remarks before made, I must now refer to Figs. 18, 19, and 20, which illustrate various arrangements with a type of bulb most generally used.

Fig. 18 is a section though a spherical bulb L, with the glass stem s, containing the leading-in wire w, which has a lamp filament l fastened to it, serving to support the refractory button m in the centre. M is a sheet of thin mica wound in several layers around the stem s, and a is the aluminium tube.

Fig. 19 illustrates such a bulb in a somewhat more advanced stage of perfection. A metallic tube S is fastened by means of some cement to the neck of the tube. In the tube is screwed a plug P, of insulating material, in the centre of which is fastened a metallic terminal t, for the connection to the lead-in wire w. This terminal must be well insulated from the metal tube S, therefore, if the cement used is conducting and most generally it is sufficiently so — the space between the plug P and the neck of the bulb should be filled with some good insulating material, as mica powder.

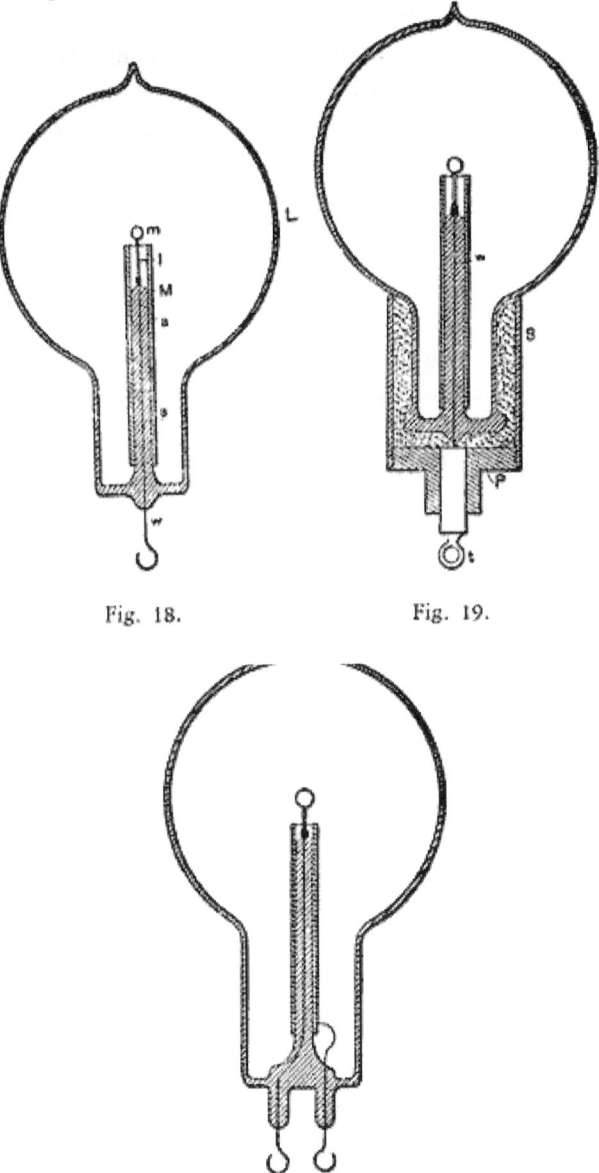

Fig. 18.   Fig. 19.

Fig. 20.

Fig. 20 shows a bulb made for experimental purposes. In this bulb the aluminium tube is provided with an external connection, which serves to investigate the effect of the tube under various conditions. It is referred to chiefly to suggest a line of experiment followed.

Since the bombardment against the stem containing the leading-in wire is due to the inductive action of the latter upon the rarefied gas, it is of advantage to reduce this action as far as practicable by employing a very thin wire, surrounded by a very thick insulation of glass or other material, and by making the wire passing through the rarefied gas as short as practicable. To combine these features I employ a large tube T (Fig. 21), which protrudes into the bulb to some distance, and carries on the top a very short glass stem s, into which is sealed the leading-in wire w, and I protect the top of the glass stem against the heat by a small, aluminium tube a and a layer of mica underneath the same, as usual. The wire w, passing through the large tube to the outside of the bulb, should be well insulated—with a glass tube, for instance—and the space between ought to be filled out with some excellent insulator. Among many insulating powders I have tried, I have found that mica powder is the best to employ. If this precaution is not taken, the tube T, protruding into the bulb, will surely be cracked in consequence of the heating by the brushes which are apt to form in the upper part of the tube, near the exhausted globe, especially if the vacuum be excellent, and therefore the potential necessary to operate the lamp be very high.

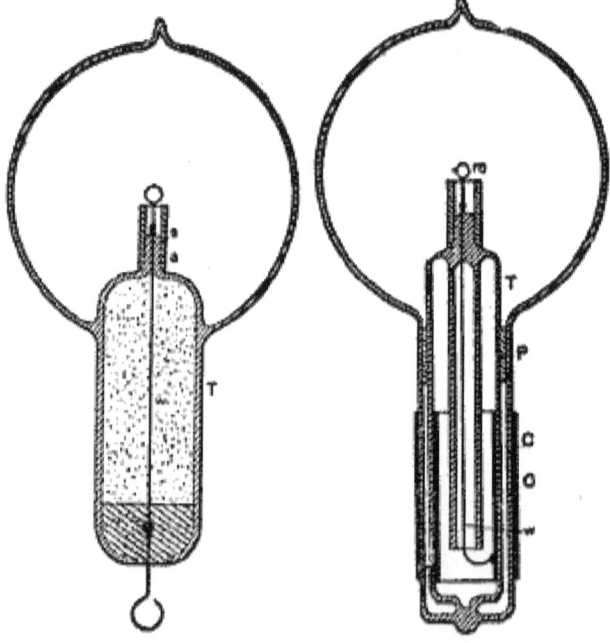

Fig. 21.   Fig. 22.

Fig. 22 illustrates a similar arrangement, with a large tube T protruding into the part of the bulb containing the refractory button m. In this case the wire leading from the outside into the bulb is omitted, the energy required being supplied through condenser coatings C C. The insulating packing P should in this construction be tightly fitting to the glass, and rather wide, or otherwise the discharge might avoid passing through the wire w, which connects the inside condenser coating to the incandescent button m.

The molecular bombardment against the glass stem in the bulb is a source of great trouble. As illustration I will cite a phenomenon only too frequently and unwillingly observed. A bulb, preferably a large one, may be taken, and a good conducting body, such as a piece of carbon, may be mounted in it upon a platinum wire sealed in the glass stem. The bulb may be exhausted to a fairly high degree, nearly to the point when phosphorescence begins to appear. When the bulb is connected with the coil, the piece of carbon, if small, may become highly incandescent at first, but its brightness immediately diminishes, and then the discharge may break through the glass somewhere in the middle of the stem, in the form of bright sparks, in spite of the fact that the platinum wire is in good electrical connection with the rarefied gas through the piece of carbon or metal at the top. The first sparks are singularly bright, recalling those drawn from a clear surface of mercury. But, as they heat the glass rapidly, they, of course, lose their brightness, and cease when the glass at the ruptured place becomes incandescent, or generally sufficiently hot to conduct. When observed for the first time the phenomenon must appear very curious, and shows in a striking manner how radically different alternate currents, or impulses, of high frequency behave, as compared with steady currents, or currents of low frequency. With such currents — namely, the latter — the phenomenon would of course not occur. When frequencies such as are obtained by mechanical means are used, I think that the rupture of the glass is more or less the consequence of the bombardment, which warms it up and impairs its insulating power; but with frequencies obtainable with condensers I have no doubt that the glass may give way without previous heating. Although this appears most singular at first, it is in reality what we might expect to occur. The energy supplied to the wire leading into the bulb is given off partly by direct action through the carbon button, and party by inductive action through the glass surrounding the wire. The case is thus analogous to that in which a condenser shunted by a conductor of low resistance is connected to a source of alternating currents. As long as the frequencies are low, the conductor gets the most, and the condenser is perfectly safe; but when the frequency becomes excessive, the role of the conductor may become quite insignificant. In the latter case the difference of potential at the terminals of the condenser may become so great as to rupture the dielectric, notwithstanding the fact that the terminals are joined by a conductor of low resistance.

It is, of course, not necessary, when it is desired to produce the incandescence of a body inclosed in a bulb by means of these currents, that the body should be a conductor, for even a perfect non-conductor may be quite as readily heated. For this purpose it is sufficient to surround a conducting electrode with a non-conducting material, as, for instance, in the bulb described before in Fig. 21, in which a thin incandescent lamp filament is coated with a non-conductor, and supports a button of the same material on the top. At the start the bombardment goes on by inductive action through the non-conductor, until the same is sufficiently heated to become conducting, then the bombardment continues in the ordinary way.

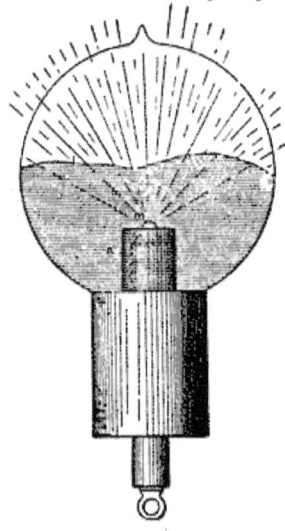

Fig. 23.

A different arrangement used in some of the bulbs constructed is illustrated in Fig. 23. In this instance a non-conductor m is mounted in a piece of common arc light carbon so as to project some small distance above the latter. The carbon piece is connected to the leading-in wire passing through a glass stem, which is wrapped with several layers of mica. An aluminium tube a is employed as usual for screening. It is so arranged that it reaches very nearly as high as the carbon and only the non-conductor m projects a little above it. The bombardment goes at first against the upper surface of carbon, the lower parts being protected by the aluminium tube. As soon, however, as the non-conductor m is heated it is rendered good conducting, and then it becomes the centre of the bombardment, being most exposed to the same.

I have also constructed during these experiments many such single-wire bulbs with or without internal electrode, in which the radiant matter was projected against, or focused upon, the body to be rendered incandescent. Fig. 24 illustrates one of the bulbs used. It consists of a spherical globe L, provided with a long neck n, on the top, for increasing the action in some cases by the application of an external conducting coating. The globe L is blown out on the bottom into a very small bulb b, which serves to hold it firmly in a socket S of insulating material into which it is cemented. A fine lamp filament f, supported on a wire w, passes through the centre of filament is rendered incandescent In the middle portion, where the bombardment proceeding from the lower inside surface of the globe is most intense. The lower portion of the globe, as far as the socket S reaches, is rendered conducting, either by g tinfoil coating or otherwise, and the external electrode is connected to a terminal of the coil.

The arrangement diagrammatically indicated in Fig. 24 was found to be an inferior one when it was desired to render incandescent a filament or button supported in the centre of the globe, but it was convenient when the object was to excite phosphorescence.

In many experiments in which bodies of a different kind were mounted in the bulb as, for instance, indicated in Fig. 23, some observations of interest were made.

It was found, among other things, that in such cases, no matter where the bombardment began, just as soon as a high temperature was reached there was generally one of the bodies which seemed to take most of the bombardment upon itself, the other, or others, being thereby relieved. This quality appeared to depend principally on the point of fusion, and on the facility with which the body was "evaporated," or, generally speaking, disintegrated—meaning by the latter term not only the throwing off of atoms, but likewise of larger lumps. The observation made was in accordance with generally accepted notions. In a highly exhausted bulb electricity is carried off from the electrode by independent carriers, which are partly the atoms, or molecules, of the residual atmosphere, and partly the atoms, molecules, or lumps thrown off from the electrode. If the electrode is composed of bodies of different character, and if one of these is more easily disintegrated than the others, most of the electricity supplied is carried off from that body, which is then brought to a higher temperature than the others, and this the more, as upon an increase of the temperature the body is still more easily disintegrated.

It seems to me quite probable that a similar process takes place in the bulb even with a homogenous electrode, and I think it to be the principal cause of the disintegration. There is bound to be some irregularity, even if the surface is highly polished, which, of course, is impossible with most of the refractory bodies employed as electrodes. Assume that a point of the electrode gets hotter, instantly most of the discharge passes through that point, and a minute patch is probably fused and evaporated. It is now possible that in consequence of the violent disintegration the spot attacked sinks in temperature, or that a counter force is created, as in an arc; at any rate, the local tearing off meets with the limitations incident to the experiment, where upon the same process occurs on another place. To the eye the electrode appears uniformly brilliant, but there are upon it points constantly shifting and wandering around, of a temperature far above the mean, and this materially hastens the process of deterioration. That some such thing occurs, at least when the electrode is at a lower temperature, sufficient experimental evidence can be obtained in the following manner: Exhaust a bulb to a very high degree, so that with a fairly high potential the discharge cannot pass — that is, not a luminous one, for a weak invisible discharge occurs always, in all probability. Now raise slowly and carefully the potential, leaving the primary current on no mote than for an instant. At a certain point, two, three, or half a dozen phosphorescent spots mill appear on the globe. These places of the glass are evidently more violently bombarded than others, this being due to the unevenly distributed electric density, necessitated, of course, by sharp projections, or, generally speaking, irregularities of the electrode. But the luminous patches are constantly changing in position, which is especially well observable if one manages to produce very few, and this indicates that the configuration of the electrode is rapidly changing.

From experiences of this kind I am led to infer that, in order to be most durable, the refractory button in the bulb should be in the form of a sphere with a highly polished surface. Such a small sphere could be manufactured from a diamond or some other crystal, but a better way would be to fuse, by the employment of extreme degrees of temperature, some oxide — as, for instance, zirconia — into a small drop, and then keep it in the bulb at a temperature somewhat below its point of fusion.

Interesting and useful results can no doubt be reached in the direction of extreme degrees of heat. How can such high temperatures be arrived at? How are the highest degrees of heat reached in nature? By the impact of stars, by high speeds and collisions. In a collision any rate of heat generation may be attained. In a chemical process we are limited. When oxygen and hydrogen combine, they fall, metaphorically speaking, from a definite height. We cannot go very far with a blast, nor by confining heat in a furnace, but in an exhausted bulb we can concentrate any amount of energy upon a minute button.

Leaving practicability out of consideration, this, then, would be the means which, in my opinion, would enable us to reach the highest temperature. But a great difficulty when proceeding in this way is encountered, namely, in most cases the body is carried off before it can fuse and form a drop. This difficulty exists principally with an oxide such as zirconia, because it cannot be compressed in so hard a cake that it would not be carried off quickly. I endeavored repeatedly to fuse zirconia, placing it in a cup or arc light carbon as indicated in Fig. 23. It glowed with a most intense light, and the stream of the particles projected out of the carbon cup was of a vivid white; but whether it was compressed in a cake or made into a paste with carbon, it was carried off before it could be fused. The carbon cup containing the zirconia had to be mounted very low in the neck of a large bulb, as the heating of the glass by the projected particles of the oxide was so rapid that in the first trial the bulb was cracked almost in an instant when the current was turned on. The heating of the glass by the projected particles was found to be always greater when the carbon cup contained a body which was rapidly carried off — I presume because in such cases, with the same potential, higher speeds were reached, and also because, per unit of time, more matter was projected — that is, more particles would strike the glass.

The before mentioned difficulty did not exist, however, when the body mounted in the carbon cup offered great resistance to deterioration. For instance, when an oxide was first fused in an oxygen blast and then mounted in the bulb, it melted very readily into a drop.

Generally during the process of fusion magnificent light effects were noted, of which it would be difficult to give an adequate idea. Fig. 23 is intended to illustrate the effect observed with a ruby drop. At first one may see a narrow funnel of white light projected against the top of the globe, where it produces an irregularly outlined phosphorescent patch. When the point of the ruby fuses the phosphorescence becomes very powerful; but as the atoms are projected with much greater speed from the surface of the drop, soon the glass gets hot and "tired," and now only the outer edge of the patch glows. In this manner an intensely phosphorescent, sharply defined line, corresponding to the outline of the drop, is produced, which spreads slowly: over the globe as the drop gets larger. When the mass begins to boil, small bubbles and cavities are formed, which cause dark colored spots to sweep across the globe. The bulb may be turned downward without fear of the drop falling off, as the mass possesses considerable viscosity.

I may mention here another feature of some interest, which I believe to have noted in the course of these experiments, though the observations do not amount to a certitude. It appeared that under the molecular impact caused by the rapidly alternating potential the body was fused and maintained in that state at a lower temperature in a highly exhausted bulb than was the case at normal pressure and application of heat in the ordinary way—that is, at least, judging from the quantity of the light emitted. One of the experiments performed may be mentioned here by way of illustration. A small piece of pumice stone was stuck on a platinum wire, and first melted to it in a gas burner. The wire was next placed between two pieces of charcoal and a burner applied so as to produce an intense heat, sufficient to melt down the pumice stone into a small glass-like button. The platinum wire had to be taken of sufficient thickness to prevent its melting in the fire. While in the charcoal fire, or when held in a burner to get a better idea of the degree of heat, the button glowed with great brilliancy. The wire with the button was then mounted in a bulb, and upon exhausting the same to a high degree, the current was turned on slowly so as to prevent the cracking of the button. The button was heated to the point of fusion, and when it melted it did not, apparently, glow with the same brilliancy as before, and this would indicate a lower temperature. Leaving out of consideration the observer's possible, and even probable, error, the question is, can a body under these conditions be brought from a solid to a liquid state with evolution of less light?

When the potential of a body is rapidly alternated it is certain that the structure is jarred. When the potential is very high, although the vibrations may be few—say 20,000 per second—the effect upon the structure may be considerable. Suppose, for example, that a ruby is melted into a drop by a steady application of energy. When it forms a drop it will emit visible and invisible waves, which will be in a definite ratio, and to the eye the drop will appear to be of a certain brilliancy. Next, suppose we diminish to any degree we choose the energy steadily supplied, and, instead, supply energy which rises and falls according to a certain law. Now, when the drop is formed, there will be emitted from it three different kinds of vibrations—the ordinary visible, and two kinds of invisible waves: that is, the ordinary dark waves of all lengths, and, in addition, waves of a well-defined character. The latter would not exist by a steady supply of the energy; still they help to jar and loosen the structure. If this really be the case, then the ruby drop will emit relatively less visible and more invisible waves than before. Thus it would seem that when a platinum wire, for instance, is fused by currents alternating with extreme rapidity, it emits at the point of fusion less light and more invisible radiation than it does when melted by a steady current, though the total energy used up in the process of fusion is the same in both cases, Or, to cite another example, a lamp filament is not capable of withstanding as long with currents of extreme frequency as it does with steady currents, assuming that it be worked at the same luminous intensity. This means that for rapidly alternating currents the filament should be shorter and thicker. The higher the frequency—that is, the greater the departure from the steady flow—the worse it would be for the filament. But if the truth of this remark were demonstrated, it would be erroneous to conclude that such a refractory button as used in these bulbs would be deteriorated quicker by currents of extremely high frequency than by steady or low frequency currents. From experience I may say that just the opposite holds good: the button withstands the bombardment better with currents of very high frequency. But this is due to the fact that a high frequency discharge passes through a rarefied gas with much greater freedom than a steady or low frequency discharge, and this will say that with the former we can work with a lower potential or with a less violent impact. As long, then, as the gas is of no consequence, a steady or low frequency current is better; but as soon as the action of the gas is desired and important, high frequencies are preferable.

In the course of these experiments a great many trials were made with all kinds of carbon buttons. Electrodes made of ordinary carbon buttons were decidedly more durable when the buttons were obtained by the application of enormous pressure. Electrodes prepared by depositing carbon in well known ways did not show up well; they blackened the globe very quickly. From many experiences I conclude that lamp filaments obtained in this manner can be advantageously used only with low potentials and low frequency currents. Some kinds of carbon withstand so well that, in order to bring them to the point of fusion, it is necessary to employ very small buttons. In this case the observation is rendered very difficult on account of the intense heat produced. Nevertheless there can be no doubt that all kinds of carbon are fused under the molecular bombardment, but the liquid state must be one of great instability. Of all the bodies tried there were two which withstood best—diamond and carborundum. These two showed up about equally, but the latter was preferable, for many reasons. As it is more than likely that this body is not yet generally known, I will venture to call your attention to it.

It has been recently produced by Mr. E. G. Acheson, of Monongahela City, Pa., U. S. A. It is intended to replace ordinary diamond powder for polishing precious stones, etc., and I have been informed that it accomplishes this object quite successfully. I do not know why the name "carborundum" has been given to it, unless there is something in the process of its manufacture which justifies this selection. Through the kindness of the inventor, I obtained a short while ago some samples which I desired to test in regard to their qualities of phosphorescence and capability of withstanding high degrees of heat.

Carborundum can be obtained in two forms — in the form of "crystals" and of powder. The former appear to the naked eye dark colored, but are very brilliant; the latter is of nearly the same color as ordinary diamond powder, but very much finer. When viewed under a microscope the samples of crystals given to me did not appear to have any definite form, but rather resembled pieces of broken up egg coal of fine quality. The majority were opaque, but there were some which were transparent and colored. The crystals are a kind of carbon containing some impurities; they are extremely hard, and withstand for a long time even an oxygen blast. When the blast is directed against them they at first form a cake of some compactness, probably in consequence of the fusion of impurities they contain. The mass withstands for a very long time the blast without further fusion; but a slow carrying off, or burning, occurs, and, finally, a small quantity of a glass-like residue is left, which, I suppose, is melted alumina. When compressed strongly they conduct very well, but not as well as ordinary carbon. The powder, which is obtained from the crystals in some way, is practically non-conducting. It affords a magnificent polishing material for stones.

The time has been too short to make a satisfactory study of the properties of this product, but enough experience has been gained in a few weeks I have experimented upon it to say that it does possess some remarkable properties in many respects. It withstands excessively high degrees of heat, it is little deteriorated by molecular bombardment, and it does not blacken the globe as ordinary carbon does. The only difficulty which I have found in its use in connection with these experiments was to find some binding material which would resist the heat and the effect of the bombardment as successfully as carborundum itself does.

I have here a number of bulbs which I have provided with buttons of carborundum. To make such a button of carborundum crystals I proceed in the following manner: I take an ordinary lamp filament and dip its point in tar, or some other thick substance or paint which may be readily carbonized. I next pass the point of the filament through the crystals, and then hold it vertically over a hot plate. The tar softens and forms a drop on the point of the filament, the crystals adhering to the surface of the drop. By regulating the distance from the plate the tar is slowly dried out and the button becomes solid. I then once more dip the button in tar and hold it again over a plate until the tar is evaporated, leaving only a hard mass which firmly binds the crystals. When a larger button is required I repeat the process several times, and I generally also cover the filament a certain distance below the button with crystals. The button being mounted in a bulb, when a good vacuum has been reached, first a weak and then a strong discharge is passed through the bulb to carbonize the tar and expel all gases, and later it is brought to a very intense incandescence.

When the powder is used I have found it best to proceed as follows: I make a thick paint of carborundum and tar, and pass a lamp filament through the paint. Taking then most of the paint off by rubbing the filament against a piece of chamois leather, I hold it over a hot plate until the tar evaporates and the coating becomes firm. I repeat this process as many times as it is necessary to obtain a certain thickness of coating. On the point of the coated filament I form a button in the same manner.

There is no doubt that such a button—properly prepared under great pressure—of carborundum, especially of powder of the best quality, will withstand the effect of the bombardment fully as well as anything we know. The difficulty is that the binding material gives way, and the carborundum is slowly thrown off after some time. As it does not seem to blacken the globe in the least, it might be found useful for coating the filaments of ordinary Incandescent lamps, and I think that it is even possible to produce thin threads or sticks of carborundum which will replace the ordinary filaments in an incandescent lamp. A carborundum coating seems to be more durable than other coatings, not only because the carborundum can withstand high degrees of heat, but also because it seems to unite with the carbon better than any other material I have tried. A coating of zirconia or any other oxide, for instance, is far more quickly destroyed. I prepared buttons of diamond dust in the same manner as of carborundum, and these came in durability nearest to those prepared of carborundum, but the binding paste gave way much more quickly in the diamond buttons: this, however, I attributed to the site and irregularity of the grains of the diamond.

It was of interest to find whether carborundum possesses the quality of phosphorescence. One is, of course, prepared to encounter two difficulties: first, as regards the rough product, the "crystals," they are good conducting, and it is a fact that conductors do not phosphoresce; second, the powder, being exceedingly fine, would not be apt to exhibit very prominently this quality, since we know that when crystals, even such as diamond or ruby, are finely powdered, they lose the property of phosphorescence to a considerable degree.

The question presents itself here, can a conductor phosphoresce? What is there in such a body as a metal, for instance, that would deprive it of the quality of phosphorescence, unless it is that property which characterizes it as a conductor? For it is a fact that most of the phosphorescent bodies lose that quality when they are sufficiently heated to become more or less conducting. Then, if a metal be in a large measure, or perhaps entirely deprived of that property, it should be capable of phosphorescence. Therefore it is quite possible that at some extremely high frequency, when behaving practically as a non-conductor, a metal of any other conductor might exhibit the quality of phosphorescence, even though it be entirely incapable of phosphorescing under the impact of a low-frequency discharge. There is, however, another possible way how a conductor might at least appear to phosphoresce.

Considerable doubt still exists as to what really is phosphorescence, and as to whether the various phenomena comprised under this head are due to the same causes. Suppose that in an exhausted bulb, under the molecular impact, the surface of a piece of metal or other conductor is rendered strongly luminous, but at the same time it is found that it remains comparatively cool, would not this luminosity be called phosphorescence? Now such a result, theoretically at least, is possible, for it is a mere question of potential of speed. Assume the potential of the electrode, and consequently the speed of the projected atoms, to be sufficiently high, the surface of the metal piece against which the atoms are projected would be rendered highly incandescent, since the process of heat generation would be incompatibly faster than that of radiating or conducting away from the surface of the collision. In the eye of the observer a single impact of the atoms would cause an instantaneous flash, but if the impact were repeated with sufficient rapidity they would produce a continuous impression upon his retina. To him then the surface of the metal would appear continuously incandescent and of constant luminous intensity, while in reality the light would be either intermittent or at least changing periodically in intensity. The metal piece would rise in temperature until equilibrium was attained — that is, until the energy continuously radiated would equal that intermittently supplied. But the supplied energy might under such conditions not be sufficient to bring the body to any more than a very moderate mean temperature, especially if the frequency of the atomic impacts be very low — just enough that the fluctuation of the intensity of the light emitted could not be detected by the eye. The body would now, owing to the manner in which the energy is supplied, emit a strong light, and yet be at a comparatively very low mean temperature. How could the observer call the luminosity thus produced? Even if the analysis of the light would teach him something definite, still he would probably rank it under the phenomena of phosphorescence. It is conceivable that in such a way both conducting and nonconducting bodies may be maintained at a certain-luminous intensity, but the energy required would very greatly vary with the nature and properties of the bodies.

These and some foregoing remarks of a speculative nature were made merely to bring out curious features of alternate currents or electric impulses. By their help we may cause a body to emit more light, while at a certain mean temperature, than it would emit if brought to that temperature by a steady supply; and, again, we may bring a body to the point of fusion, and cause it to emit less light than when fused by the application of energy in ordinary ways. It all depends on how we supply the energy, and what kind of vibrations we set up: in one case the vibrations are more, in the other less, adapted to affect our sense of vision.

Some effects, which I had not observed before, obtained with carborundum in the first trials, I attributed to phosphorescence, but in subsequent experiments it appeared that it was devoid of that quality. The crystals possess a noteworthy feature. In a bulb provided with a single electrode in the shape of a small circular metal disc, for instance, at a certain degree of exhaustion the electrode is covered with a milky film, which is separated by a dark space from the glow filling the bulb. When the metal disc is covered with carborundum crystals, the film is far more intense, and snow-white. This I found later to be merely an effect of the bright surface of the crystals, for when an aluminium electrode was highly polished it exhibited more or less the same phenomenon. I made a number of experiments with the samples of crystals obtained, principally because it would have been of special interest to find that they are capable of phosphorescence, on account of their being conducting. I could not produce phosphorescence distinctly, but I must remark that a decisive opinion cannot be formed until other experimenters have gone over the same ground.

The powder behaved in some experiments as though it contained alumina, but it did not exhibit with sufficient distinctness the red of the latter. Its dead color brightens considerably under the molecular impact, but I am now convinced it does not phosphoresce. Still, the tests with the powder are not conducive, because powdered carborundum probably does not behave like a phosphorescent sulphide, for example, which could be finely powdered without impairing the phosphorescence, but rather like powdered ruby or diamond, and therefore it would be necessary, in order to make a decisive test, to obtain it in a large lump and polish up the surface.

If the carborundum proves useful in connection with these and similar experiments, its chief value will be found in the production of coatings, thin conductors, buttons, or other electrodes capable of withstanding extremely high degrees of heat.

The production of a small electrode capable of withstanding enormous temperatures I regard as of the greatest importance in the manufacture of light. It would enable us to obtain, by means of currents of very high frequencies, certainly 20 times, if not more, the quantity of light which is obtained in the present incandescent lamp by the same expenditure of energy. This estimate may appeal — to many exaggerated, but in reality I think it is far from being so. As this statement might be misunderstood I think it necessary to expose clearly the problem with which in this line of work we are confronted, and the manner in which, in my opinion, a solution will be arrived at.

Any one who begins a study of the problem will be apt to think that what is wanted in a lamp with an electrode is a very high degree of incandescence of the electrode. There he will be mistaken. The high incandescence of the button is a necessary evil, but what is really wanted is the high incandescence of the gas surrounding thee button. In other words, the problem in such a lamp is to bring a mass of gas to the highest possible incandescence. The higher the incandescence, the quicker the mean vibration, the greater is the economy of the light production. But to maintain a mass of gas at a high degree of incandescence in a glass vessel, it will always be necessary to keep the incandescent mass away from the glass; that is, to confine it as much as possible to the central portion of the globe.

In one of the experiments this evening a brush was produced at the end of a wire. This brush was a flame, a source of heat and light. It did not emit much perceptible heat, nor did it glow with an intense light; but is it the less a flame because it does not scorch my hand? Is it the less a flame because it does not hurt my eye by its brilliancy? The problem is precisely to produce in the bulb such a flame, much smaller in site, but incomparably more powerful. Were there means at hand for producing electric impulses of a sufficiently high frequency, and for transmitting them, the bulb could be done away with, unless it were used to protect the electrode, or to economize the energy by confining the heat. But as such means are not at disposal, it becomes necessary to place The terminal in a bulb and rarefy the air in the same. This is done merely to enable the apparatus to perform the work which it is not capable of performing at ordinary air pressure. In the bulb we are able to intensify the action to any degree — so far that the brush emits a powerful light.

The intensity of the light emitted depends principally on the frequency and potential of the impulses, and on the electric density on the surface of the electrode. It is of the greatest importance to employ the smallest possible button, in order to push the density very far. Under the violent impact of the molecules of the gas surrounding it, the small electrode is of course brought to an extremely high temperature, but around it is a mass of highly incandescent gas, a flame photosphere, many hundred times the volume of the electrode. With a diamond, carborundum or zircon button the photosphere can be as much as one thousand times the volume of the button. Without much reflecting one would think that in pushing so far the incandescence of the electrode it would be instantly volatilized. But after a careful consideration he would find that, theoretically, it should not occur, and in this fact—which, however, is experimentally demonstrated—lies principally the future value of such a lamp.

At first, when the bombardment begins, most of the work is performed on the surface of the button, but when a highly conducting photosphere is formed the button is comparatively relieved. The higher the incandescence of the photosphere the more it approaches in conductivity to that of the electrode, and the more, therefore, the solid and the gas form one conducting body. The consequence is that the further is forced the incandescence the more work, comparatively, is performed on the gas, and the less on the electrode. The formation of a powerful photosphere is consequently the very means for protecting the electrode. This protection, of course, is a relative one, and it should not be thought that by pushing the incandescence higher the electrode is actually less deteriorated. Still, theoretically, with extreme frequencies, this result must be reached, but probably at a temperature too high for most of the refractory bodies known. Given, then, an electrode which can withstand to a very high limit the effect of the bombardment and outward strain, it would be safe no matter how much it is forced beyond that limit. In an incandescent lamp quite different considerations apply. There the gas is not at all concerned: the whole of the work is performed on the filament; and the life of the lamp diminishes so rapidly with the increase of the degree of incandescence the economical reasons compel us to work it at a low incandescence. But if an incandescent lamp is operated with currents of very high frequency, the action of the gas cannot be neglected, and the rules for the most economical working must be considerably modified.

In order to bring such a lamp with one or two electrodes to a great perfection, it is necessary to employ impulses of very high frequency. The high frequency secures, among others, two chief advantages, which have a most important bearing upon the economy of the light production. First, the deterioration of the electrode is reduced by reason of the fact that we employ a great many small impacts, instead of a few violent ones, which shatter quickly the structure; secondly, the formation of a large photosphere is facilitated.

In order to reduce the deterioration of the electrode to the minimum, it is desirable that the vibration be harmonic, for any suddenness hastens the process of destruction. An electrode lasts much longer when kept at incandescence by currents, or impulses, obtained from a high-frequency alternator, which rise and fall more or less harmonically, than by impulses obtained from a disruptive discharge coil. In the latter case there is no doubt that most of the damage is done by the fundamental sudden discharges.

One of the elements of loss in such a lamp is the bombardment of the globe. As the potential is very high, the molecules are projected with great speed; they strike the glass, and usually excite a strong phosphorescence. The effect produced is very pretty but for economical reasons it would be perhaps preferable to prevent, or at least reduce to the minimum, the bombardment against the globe, as in such case it is, as a result, not the object to excite phosphorescence, and as some loss of energy results from the bombardment. This loss in the bulb is principally dependent on the potential of the impulses and on the electric density on the surface of the electrode. In employing very high frequencies the loss of energy by the bombardment is greatly reduced, for, first, the potential needed to perform a given amount of work is much smaller; and, secondly, by producing a highly conducting photosphere around the electrode, the same result is obtained as though the electrode were much larger, which is equivalent to a smaller electric density. But be it by the diminution of the maximum potential or of the density, the gain is effected in the same manner, namely, by avoiding violent shocks, which strain the glass much beyond its limit of elasticity. If the frequency could be brought high enough, the loss due to the imperfect elasticity of the glass would be entirely negligible. The loss due to bombardment of the globe may, however, be reduced by using two electrodes instead of one. In such case each of the electrodes may be connected to one of the terminals; or else, if it is preferable to use only one wire, one electrode may be connected to one terminal and the other to the ground or to an insulated body of some surface, as, for instance, a shade on the lamp. In the latter case, unless some judgment is used, one of the electrodes might glow more intensely than the other.

But on the whole I find it preferable when using such high frequencies to employ only one electrode and one connecting wire. I am convinced that the illuminating device of the near future will not require for its operation more than one lead, and, at any rate, it will have no leading-in wire, since the energy required can be as well transmitted through the glass. In experimental bulbs the leading-in wire is most generally used on account of convenience, as in employing condenser coatings in the manner indicated in Fig. 22, for example, there is some difficulty in fitting the parts, but these difficulties would not exist if a great many bulbs were manufactured; otherwise the energy can be conveyed through the glass as well as through a wire, and with these high frequencies the losses are very small. Such illuminating deices will necessarily involve the use of very high potentials, and this, in the eyes of practical men, might be an objectionable feature. Yet, in reality, high potentials are not objectionable — certainly not in the least as far as the safety of the devices is concerned.

There are two ways of rendering an electric appliance safe. One is to use low potentials, the other is to determine the dimensions of the apparatus so that it is safe no matter how high a potential is used. Of the two the latter seems to me the better way, for then the safety is absolute, unaffected by any possible combination of circumstances which might render even a low-potential appliance dangerous to life and property. But the practical conditions require not only the judicious determination of the dimensions of the apparatus; they likewise necessitate the employment of energy of the proper kind. It is easy, for instance, to construct a transformer capable of giving, when operated from an ordinary alternate current machine of low tension, say 50,000 volts, which might be required to light a highly exhausted phosphorescent tube, so that, in spite of the high potential, it is perfectly safe, the shock from it producing no inconvenience. Still, such a transformer would be expensive, and in itself inefficient; and, besides, what energy was obtained from it would not be economically used for the production of light. The economy demands the employment of energy in the form of extremely rapid vibrations. The problem of producing light has been likened to that of maintaining a certain high-pitch note by means of a bell. It should be said a barely audible note; and even these words would not express it, so wonderful is the sensitiveness of the eye. We may deliver powerful blows at long intervals, waste a good deal of energy, and still not get what we want; or we may keep up the note by delivering frequent gentle taps, and get nearer to the object sought by the expenditure of much less energy. In the production of light, as far as the illuminating device is concerned, there can be only one rule — that is, to use as high frequencies as can be obtained; but the means for the production and conveyance of impulses of such character impose, at present at least, great limitations. Once it is decided to use very high frequencies, the return wire becomes unnecessary, and all the appliances are simplified. By the use of obvious means the same result is obtained as though the return wire were used. It is sufficient for this purpose to bring in contact with the bulb, or merely in the vicinity of the same, an insulated body of some surface. The surface need, of course, be the smaller, the higher the frequency and potential used, and necessarily, also, the higher the economy of the lamp or other device.

This plan of working has been resorted to on several occasions this evening. So, for instance, when the incandescence of a button was produced by grasping the bulb with the hand, the body of the experimenter merely served to intensify the action. The bulb used was similar to that illustrated in Fig. 19, and the coil was excited to a small potential, not sufficient to bring the button to incandescence when the bulb was hanging from the wire; and incidentally, in order to perform the experiment in a more suitable manner, the button was taken so large that a perceptible time had to elapse before, upon grasping the bulb, it could be rendered incandescent. The contact with the bulb was, of course, quite unnecessary. It is easy, by using a rather large bulb with an exceedingly small electrode, to adjust the conditions so that the latter is brought to bright incandescence by the mere approach of the experimenter within a few feet of the bulb, and that the incandescence subsides upon his receding.

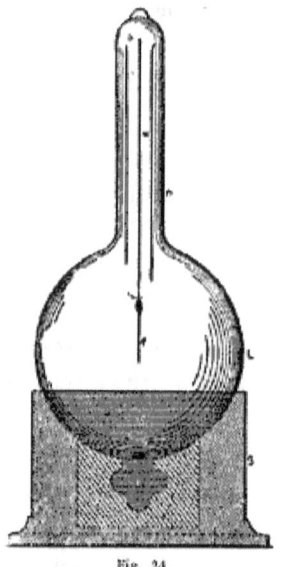
Fig. 24.

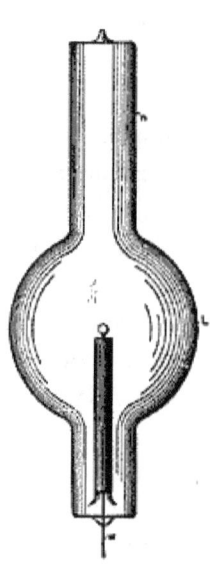
Fig. 25.

In another experiment, when phosphorescence was excited, a similar bulb was used. Here again, originally, the potential was not sufficient to excite phosphorescence until the action was intensified—in this case, however, to present a different feature, by touching the socket with a metallic object held in the hand. The electrode in the bulb was a carbon button so large that it could not be brought to incandescence, and thereby spoil the effect produced by phosphorescence.

Again, in another of the early experiments, a bulb was used as illustrated in

Fig. 12. In this instance, by touching the bulb with one or two fingers, one or two shadows of the stem inside were projected against the glass, the touch of the finger producing the same result as the application of an external negative electrode under ordinary circumstances.

In all these experiments the action was intensified by augmenting the capacity at the end of the lead connected to the terminal. As a rule, it is not necessary to resort to such means, and would be quite unnecessary with still higher frequencies; but when it is desired, the bulb, or tube, can be easily adapted to the purpose.

In Fig. 24, for example, an experimental bulb L is shown, which is provided with a neck n on the top for the application of an external tinfoil coating, which may be connected to a body of larger surface. Sum a lamp as illustrated in Fig. 25 may also be lighted by connecting the tinfoil coating on the neck n to the terminal, and the leading-in wire w to an insulated plate. If the bulb stands in a socket upright, as shown in the cut, a shade of conducting material may be slipped in the neck n, and the action thus magnified.

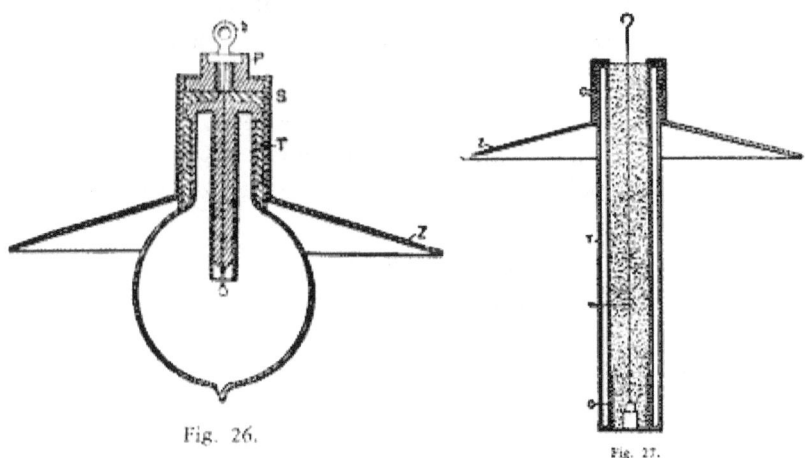

Fig. 26.

Fig. 27.

A more perfected arrangement used in some of these bulbs is illustrated in Fig. 26. In this case the construction of the bulb is as shown and described before, where reference was made to Fig. 19. A zinc sheet Z, with a tubular extension T, is slipped over the metallic socket S. The bulb hangs downward from the terminal t, the zinc sheet Z, performing the double office of intensifier and reflector. The reflector is separated from the terminal t by an extension of the insulating plug P.

A similar disposition with a phosphorescent tube is illustrated in Fig. 27. The tube T is prepared from two short tubes of a different diameter, which are sealed on the ends. On the lower end is placed an outside conducting coating C, which connects to the wire w. The wire has a hook on the upper end for suspension, and passes through the centre of the inside tube, which is filled with some good and tightly packed insulator. On the outside of the upper end of the tube, T, is another conducting coating C_1, upon which is slipped a metallic reflector Z, which should be separated by a thick insulation from the end of wire w.

The economical use of such a reflector or intensifier would require that all energy supplied to an air condenser should be recoverable, or, in other words, that there should not be any losses, neither in the gaseous medium nor through its action elsewhere. This is far from being so, but, fortunately, the losses may be reduced to anything desired. A few remarks are necessary on this subject, in order to make the experiences gathered in the course of these investigations perfectly clear.

Suppose a small helix with many well insulated turns, as in experiment Fig. 17, had one of its ends connected to one of the terminals of the induction coil, and the other to a metal plate, or, for the sake of simplicity, a sphere, insulated in space. When the coil is set to work, the potential of the sphere is alternated, and the small helix now behaves as though its free end were connected to the other terminal of the induction coil. If an iron can be held within the small helix it is quickly brought to a high temperature, indicating the passage of a strong current through the helix. How does the insulated sphere act in this case? It can be a condenser, storing and returning the energy supplied to it, or it can be a mere sink of energy, and the conditions of the experiment determine whether it is more one or the other. The sphere being charged to a high potential, it acts inductively upon the surrounding air, or whatever gaseous medium there might be. The molecules, or atoms, which are near the sphere are of course more attracted, and move through a greater distance than the farther ones. When the nearest molecules strike the sphere they are repelled, and collisions occur at all distances within the inductive action of the sphere. It is now clear that, if the potential be steady, but little loss of energy can be caused in this way, for the molecules which are nearest to the sphere, having had an additional charge imparted to them by contact, are not Attracted until they have parted, if not with all, at least with most of the additional charge, which can be accomplished only after a great many collisions. From the fact that with a steady potential there is but little loss in dry air, one must come to such a conclusion. When the potential of the sphere, instead of being steady, is alternating, the conditions are entirely different. In this case a rhythmical bombardment occurs, no matter whether the molecules after coming in contact with the sphere lose the imparted charge or not; what is more, if the charge is not lost, the impacts are only the more violent. Still if the frequency of the impulses be very small, the loss caused by the impacts and collisions would not be serious unless the potential were excessive. But when extremely high frequencies and more or less high potentials are used, the loss may be very great. The total energy lost per unit of time is proportionate to the product of the number of impacts per second, or the frequency and the energy lost in each impact. But the energy of an impact must be proportionate to the square of the electric density of the sphere, since the charge imparted to the molecule is proportionate to that density. I conclude from this that the total energy lost must be proportionate to the product of the frequency and the square of the electric density; but this law needs experimental confirmation. Assuming the preceding considerations to be true, then, by rapidly alternating the potential of a body immersed in an insulating gaseous medium, any amount of energy may be dissipated into space. Most of that energy then, I believe, is not dissipated in the form of long ether waves, propagated to considerable distance, as is thought most generally, but is consumed — in the case of an insulated sphere, for example — in impact and collisional losses — that is, heat vibrations — on the surface and in the vicinity of the sphere. To reduce the dissipation it is necessary to work with a small electric density the smaller the higher the frequency.

But since, on the assumption before made, the loss is diminished with the square of the density, and since currents of very high frequencies involve considerable waste when transmitted through conductors, it follows that, on the whole, it is better to employ one wire than two. Therefore, if motors, lamps, or devices of any kind are perfected, capable of being advantageously operated by currents of extremely high frequency, economical reasons will make it advisable to use only one wire, especially if the distances are great.

When energy is absorbed in a condenser the same behaves as though its capacity were increased. Absorption always exists more or less, but generally it is small and of no consequence as long as the frequencies are not very great. In using extremely high frequencies, and, necessarily in such case, also high potentials, the absorption — or, what is here meant more particularly by this term, the loss of energy due to the presence of a gaseous medium — is an important factor to be considered, as the energy absorbed it the air condenser may be any fraction of the supplied energy. This would seem to make it very difficult to tell from the measured or computed capacity of an air condenser its actual capacity or vibration period, especially if the condenser is of very small surface and is charged to a very high potential. As many important results are dependent upon the correctness of the estimation of the vibration period, this subject demands the most careful scrutiny of other investigators. To reduce the probable error as much as possible in experiments of the kind alluded to, it is advisable to use spheres or plates of large surface, so as to make the density exceedingly small. Otherwise, when it is practicable, an oil condenser should be used in preference. In oil or other liquid dielectrics there are seemingly no such losses as in gaseous media. It being impossible to exclude entirely the gas in condensers with solid dielectrics, such condensers should be immersed in oil, for economical reasons if nothing else; they can then be strained to the utmost and will remain cool. In Leyden jars the loss due to air is comparatively small, as the tinfoil coatings are large, close together, and the charged surfaces not directly exposed; but when the potentials are very high, the loss may be more or less considerable at, or near, the upper edge of the foil, where the air is principally acted upon. If the jar be immersed in boiled-out oil, it will be capable of performing four times the amount of work which it can for any length of time when used in the ordinary way, and the loss will be inappreciable.

It should not be thought that the loss in heat in an air condenser is necessarily associated with the formation of visible streams or brushes. If a small electrode, inclosed in an unexhausted bulb, is connected to one of the terminals of the coil, streams can be seen to issue from the electrode and the air in the bulb is heated; if, instead of a small electrode, a large sphere is inclosed in the bulb, no streams are observed, still the air is heated.

Nor should it be thought that the temperature of an air condenser would give even an approximate idea of the loss in heat incurred, as in such case heat must be given off much more quickly, since there is, in addition to the ordinary radiation, a very active carrying away of heat by independent carriers going on, and since not only the apparatus, but the air at some distance from it is heated in consequence of the collisions which must occur.

Owing to this, in experiments with such a coil, a rise of temperature can be distinctly observed only when the body connected to the coil is very small. But with apparatus on a larger scale, even a body of considerable bulk would be heated, as, for instance, the body of a person; and I think that skilled physicians might make observations of utility in such experiments, which, if the apparatus were judiciously designed, would not present the slightest danger.

A question of some interest, principally to meteorologists, presents itself here. How does the earth behave? The earth is an air condenser, but is it a perfect or a very imperfect one — a mere sink of energy? There can be little doubt that to such small disturbance as might be caused in an experiment the earth behaves as an almost perfect condenser. But it might be different when its charge is set in vibration by some sudden disturbance occurring in the heavens. In such case, as before stated, probably only little of the energy of the vibrations set up would be lost into space in the form of long ether radiations, but most of the energy, I think, would spend itself in molecular impacts and collisions, and pass off into space in the form of short heat, and possibly light, waves. As both the frequency of the vibrations of the charge and the potential are in all probability excessive, the energy converted into heat may be considerable. Since the density must be unevenly distributed, either in consequence of the irregularity of the earth's surface, or on account of the condition of the atmosphere in various places, the effect produced would accordingly vary from place to place. Considerable variations in the temperature and pressure of the atmosphere may in this manner be caused at any point of the surface of the earth. The variations may be gradual or very sudden, according to the nature of the general disturbance, and may produce rain and storms, or locally modify the weather in any way.

From the remarks before made one may see what an important factor of loss the air in the neighborhood of a charged surface becomes when the electric density is great and the frequency of the impulses excessive. But the action as explained implies that the air is insulating—that is, that it is composed of independent carriers immersed in an insulating medium. This is the case only when the air is at something like ordinary or greater, or at extremely small, pressure. When the air is slightly rarefied and conducting, then true conduction losses occur also. In such case, of course, considerable energy may be dissipated into space even with a steady potential, or with impulses of low frequency, if the density is very great.

When the gas is at very low pressure, an electrode is heated more because higher speeds can be reached. If the gas around the electrode is strongly compressed, the displacements, and consequently the speeds, are very small, and the heating is insignificant. But if in such case the frequency could be sufficiently increased, the electrode would be brought to a high temperature as well as if the gas were at very low pressure; in fact, exhausting the bulb is only necessary because we cannot produce (and possibly not convey) currents of the required frequency.

Returning to the subject of electrode lamps, it is obviously of advantage in such a lamp to confine as much as possible the heat to the electrode by preventing the circulation of the gas in the bulb. If a very small bulb be taken, it would confine the heat better than a large one, but it might not be of sufficient capacity to be operated from the coil, or, if so, the glass might get too hot. A simple way to improve in this direction is to employ a globe of the required size, but to place a small bulb, the diameter of which is properly estimated, over the refractory button contained in the globe. This arrangement is illustrated in Fig. 28.

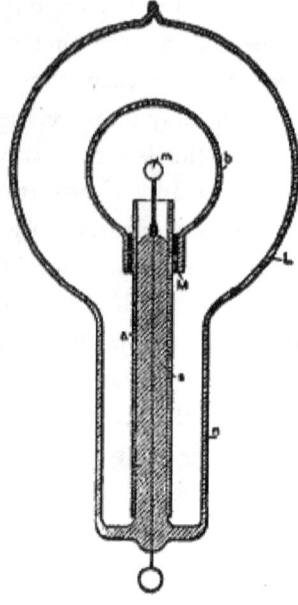

Fig. 28.

The globe L has in this case a large neck n, allowing the small bulb b to slip through. Otherwise the construction is the same as shown in Fig. 18, for example. The small bulb is conveniently supported upon the stem s, carrying the refractory button m. In tube a by several layers of mica M, in order to prevent the cracking of the neck by the rapid heating of the aluminium tube upon a sudden turning on of the current. The inside bulb should be as small as possible when it is desired to obtain light only by incandescence of the electrode. If it is desired to produce phosphorescence, the bulb should be larger, else it would be apt to get too hot, and the phosphorescence would cease. In this arrangement usually only the small bulb shows phosphorescence, as there is practically no bombardment against the outer globe. In some of these bulbs constructed as illustrated in Fig. 28 the small tube was coated with phosphorescent paint, and beautiful effects were obtained. Instead of making the inside bulb large, in order to avoid undue heating, it answers the purpose to make the electrode m larger. In this case the bombardment is weakened by reason of the smaller electric density.

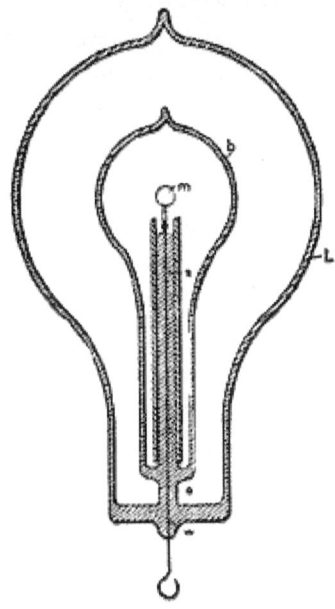

Fig. 29

Many bulbs were constructed on the plan illustrated in Fig. 29. Here a small bulb b, containing the refractory button m, upon being exhausted to a very high degree was sealed in a large globe L, which was then moderately exhausted and sealed off. The principal advantage of this construction was that it allowed of reaching extremely high vacua, and, at the same time use a large bulb. It was found, in the course of experiences with bulbs such as illustrated in Fig. 29, that it was well to make the stem s near the seal at c very thick, and the leading-in wire w thin, as it occurred sometimes that the stem at e was heated and the bulb was cracked. Often the outer globe L was exhausted only just enough to allow the discharge to pass through, and the space between the bulbs appeared crimson, producing a curious effect. In some cases, when the exhaustion in globe L was very low, and the air good conducting, it was found necessary, in order to bring the button m to high incandescence, to place, preferably on the upper part of the neck of the globe, a tinfoil coating which was connected to an insulated body, to the ground, or to the other terminal of the coil, as the highly conducting air weakened the effect somewhat, probably by being acted upon inductively from the wire w, where it entered the bulb at e. Another difficulty—which, however, is always present when the refractory button is mounted in a very small bulb—existed in the construction illustrated in Fig. 29, namely, the vacuum in the bulb b would be impaired in a comparatively short time.

The chief idea in the two last described constructions was to confine the heat to the central portion of the globe by preventing the exchange of air. An advantage is secured, but owing to the heating of the inside bulb and slow evaporation of the glass the vacuum is hard to maintain, even if the construction illustrated in Fig. 28 be chosen, in which both bulbs communicate.

But by far the better way—the ideal way—would be to reach sufficiently high frequencies. The higher the frequency the slower would be the exchange of the air, and I think that a frequency may be reached at which there would be no exchange whatever of the air molecules around the terminal. We would then produce a flame in which there would be no carrying away of material, and a queer flame it would be, for it would be rigid! With such high frequencies the inertia of the particles, would come into play. As the brush, or flame, would gain rigidity in virtue of the inertia of the particles, the exchange of the latter would be prevented. This would necessarily occur, for, the number ~f the impulses being augmented, the potential energy of each would diminish, so that finally only atomic vibrations could be set up, and the motion of translation through measurable space would cease. Thus an ordinary gas burner connected to a source of rapidly alternating potential might have its efficiency augmented to a certain limit, and this for two reasons—because of the additional vibration imparted, and because of a slowing down of the process of carrying off. But the renewal being rendered difficult, and renewal being necessary to maintain the burner, a continued increase of the frequency of the impulses, assuming they could be transmitted to and impressed upon the flame, would result in the "extinction" of the latter, meaning by this term only the cessation of the chemical process.

I think, however, that in the case of an electrode immersed in a fluid insulating medium, and surrounded by independent carriers of electric charges, which can be acted upon inductively, a sufficiently high frequency of the impulses would probably result in a gravitation of the gas all around toward the electrode. For this it would be only necessary to assume that the independent bodies are irregularly shaped; they would then turn toward the electrode their side of the greatest electric density, and this would be a position in which the fluid resistance to approach would be smaller than that offered to the receding.

The general opinion, I do not doubt, is that it is out of the question to reach any such frequencies as might—assuming some of the views before expressed to be true produce any of the results which I have pointed out as mere possibilities. This may be so, but in the course of these investigations, from the observation of many phenomena I have gained the conviction that these frequencies would be much lower than one is apt to estimate at first.

In a flame we set up light vibrations by causing molecules, of atoms, to collide. But what is the ratio of the frequency of the collisions and that of the vibrations set up? Certainly it must be incomparably smaller than that of the knocks of the bell and the sound vibrations, or that of the discharges and the oscillations of the condenser. We may cause the molecules of the gas to collide by the use of alternate electric impulses of high frequency, and so we may imitate the process in a flame; and from experiments with frequencies which we are now able to obtain, I think that the result is producible with impulses which are transmissible through a conductor.

In connection with thoughts of a similar nature, it appeared to me of great interest to demonstrate the rigidity of a vibrating gaseous column. Although with such low frequencies as, say 10,000 per second, which I was able to obtain without difficulty from a specially constructed alternator, the task looked discouraging at first, I made a series of experiments. The trials with air at ordinary pressure led to no result, but with air moderately rarefied I obtain what I think to be an unmistakable experimental evidence of the property sought for. As a result of this kind might lead able investigators to conclusions of importance I will describe one of the experiments performed.

It is well known that when a tube is slightly exhausted the discharge may be passed through it in the form of a thin luminous thread. When produced with currents of low frequency, obtained from a coil operated as usual, this thread is inert. If a magnet be approached to it, the part near the same is attracted or repelled, according to the direction of the lines of force of the magnet. It occurred to me that if such a thread would be produced with currents of very high frequency, it should be more or less rigid, and as it was visible it could be easily studied. Accordingly I prepared a tube about 1 inch in diameter and 1 metre long, with outside coating at each end. The tube was exhausted to a point at which, by a little working the thread discharge could be obtained. It must be remarked here that the general aspect of the tube, and the degree of exhaustion, are quite different than when ordinary low frequency currents are used. As it was found preferable to work with one terminal, the tube prepared was suspended from the end of a wire connected to the terminal, the tinfoil coating being connected to the wire, and to the lower coating sometimes a small insulated plate was attached. When the thread was formed it extended through the upper part of the tube and lost itself in the lower end. If it possessed rigidity it resembled, not exactly an elastic cord stretched tight between two supports, but a cord suspended from a height with a small weight attached at the end. When the finger or a magnet was approached to the upper end of the luminous thread, it could be brought locally out of position by electrostatic or magnetic action; and when the disturbing object was very quickly removed, an analogous result was produced, as though a suspended cord would be displaced and quickly released near the point of suspension. In doing this the luminous thread was set in vibration, and two very sharply marked nodes, and a third indistinct one, were formed. The vibration, once set up, continued for fully eight minutes, dying gradually out. The speed of the vibration often varied perceptibly, and it could be observed that the electrostatic attraction of the glass affected the vibrating thread; but it was clear that the electrostatic action was not the cause of the vibration, for the thread was most generally stationary, and could always be set in vibration by passing the finger quickly near the upper part of the tube. With a magnet the thread could be split in two and both parts vibrated. By approaching the hand to the lower coating of the tube, or insulated plate if attached, the vibration was quickened; also, as far as I could see, by raising the potential of frequency. Thus, either increasing the frequency or passing a stronger discharge of the same frequency corresponded to a tightening of the cord. I did not obtain any experimental evidence with condenser discharges. A luminous band excited in a bulb by repeated discharges of a Leyden jar must possess rigidity, and if deformed and suddenly released should vibrate. But probably the amount of vibrating matter is so small that in spite of the extreme speed the inertia cannot prominently assert itself. Besides, the observation in such a case is rendered extremely difficult on account of the fundamental vibration.

The demonstration of the fact—which still needs better experimental confirmation—that a vibrating gaseous column possesses rigidity, might greatly modify the views of thinkers. When with low frequencies and insignificant potentials indications of that property may be noted, how must a gaseous medium behave under the influence of enormous electrostatic stresses which may be active in the interstellar space, and which may alternate with inconceivable rapidity! The existence of such an electrostatic, rhythmically throbbing force—of a vibrating electrostatic field—would show a possible way how solids might have formed from the ultra-gaseous uterus, and how transverse and all kinds of vibrations may be transmitted through a gaseous medium filling all space. Then, ether might be a true fluid, devoid of rigidity, and at rest, it being merely necessary as a connecting link to enable interaction. What determines the rigidity of a body? It must be the speed and the amount of moving matter. In a gas the speed may be considerable, but the density is exceedingly small; in a liquid the speed would be likely to be small, though the density may be considerable; and in both cases the inertia resistance offered to displacement is practically nil. But place a gaseous (or liquid) column in an intense, rapidly alternating electrostatic field, set the particles vibrating with enormous speeds, then the inertia resistance asserts itself. A body might move with more or less freedom through the vibrating mass, but as a whole it would be rigid.

There is a subject which I must mention in connection with these experiments: it is that of high vacua. This is a subject the study of which is not only interesting, but useful, for it may lead to results of great practical importance. In commercial apparatus such as incandescent lamps, operated from ordinary systems of distribution, a much higher vacuum than obtained at present would not secure a very great advantage. In such a case the work is performed on the filament and the gas is little concerned; the improvement, therefore, would be but trifling. But when we begin to use very high frequencies and potentials, the action of the gas becomes all important, and the degree of exhaustion materially modifies the results. As long as ordinary coils, even very large ones, were used, the study of the subject was limited, because just at a point when it became most interesting it had to be interrupted on account of the "non-striking" vacuum being reached. But presently we are able to obtain from a small disruptive discharge coil potentials much higher than even the largest coil was capable of giving, and, what is more, we can make the potential alternate with great rapidity. Both of these results enable us now to pass a luminous discharge through almost any vacua obtainable, and the field of our investigations is greatly extended. Think we as we may, of all the possible directions to develop a practical illuminant, the line of high vacua seems to be the most promising at present. But to reach extreme vacua the appliances must be much mote improved, and ultimate perfection will not be attained until we shall have discarded the mechanical and perfected an electrical vacuum pump. Molecules and atoms can be thrown out of a bulb under the action of an enormous potential: this will be the principle of the vacuum pump of the future. For the present, we must secure the best results we can with mechanical appliances. In this respect, it might not be out of the way to say a few words about the method of, and apparatus for, producing excessively high degrees of exhaustion of which I have availed myself in the course of these investigations. It is very probable that other experimenters have used similar arrangements; but as it is possible that there may be an item of interest in their description, a few remarks, which will render this investigation more complete, might be permitted.

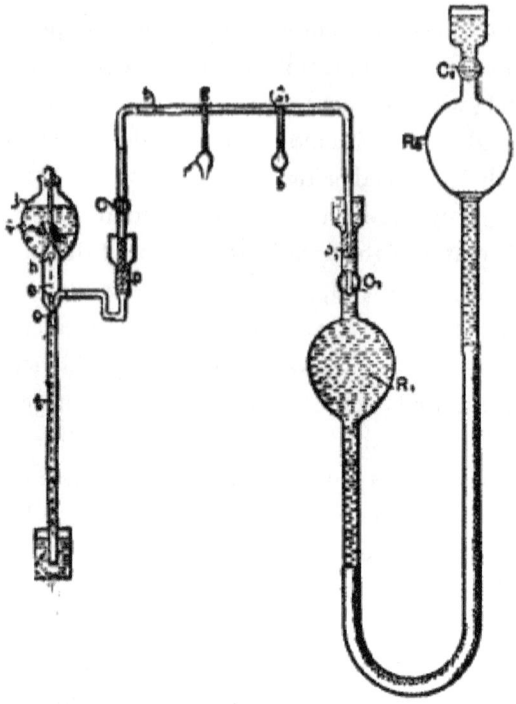

Fig. 30.

The apparatus is illustrated in a drawing shown in Fig. 30. S represents a Sprengel pump, which has been specially constructed to better suit the work required. The stopcock which is usually employed has been omitted, and instead of it a hollow stopper has been fitted in the neck of the reservoir R. This stopper has a small hole h, through which the mercury descends; the size of the outlet o being properly determined with respect to the section of the fall tube t, which is sealed to the reservoir instead of being connected to it in the usual manner. This arrangement overcomes the imperfections and troubles, which often arise from the use of the stopcock on the reservoir and the connection of the latter with the fall tube.

The pump is connected through a U-shaped tube t to a very large reservoir R_1. Especial care was taken in fitting the grinding surfaces of the stoppers p and p_1 and both of these and the mercury caps above them were made exceptionally long. After the U-shaped tube was fitted and put in place, it was heated, so as to soften and take off the strain resulting from imperfect fitting. The U-shaped tube was provided with a stopcock C, and two ground connections g and g_1, — one for a small bulb b, usually containing caustic potash, and the other for the receiver r, to be exhausted.

The reservoir R_1 was connected by means of a rubber tube to a slightly larger reservoir R_2, each of the two reservoirs being provided with a stopcock C_1 and C_2 respectively. The reservoir R_2 could be raised and lowered by a wheel and rack, and the range of its motion was so determined that when it was filled with mercury and the stopcock C, closed, so as to form a Torricellian vacuum in it when raised, it could be lifted so high that the mercury in reservoir R_1 would stand a little above stopcock C_1: and when this stopcock was closed and the reservoir R_2 descended, so as to form a Torricellian vacuum in reservoir R_1, it could be lowered so far as to completely empty the latter, the mercury filling the reservoir R_2 up to a little above stopcock C_2.

The capacity of the pump and of the connections was taken as small as possible relatively to the volume of reservoir, R_1, since, of course, the degree of exhaustion depended upon the ratio of these quantities.

With this apparatus I combined the usual means indicated by former experiments for the production of very high vacua. In most of the experiments it was convenient to use caustic potash. I may venture to say, in regard to its use, that much time is saved and a more perfect action of the pump insured by fusing and boiling the potash w soon as, or even before, the pump settles down. If this course is not followed the sticks, as ordinarily employed, may give moisture off at a certain very slow rate, and the pump may work for many hours without reaching a very high vacuum. The potash was heated either by a spirit lamp or by passing a discharge through it, or by passing a current through a wire contained in it. The advantage in the latter case was that the heating could be more rapidly repeated.

Generally the process of exhaustion was the following:—at the start, the stopcocks C and C_1 being open, and all other connections closed, the reservoir R_2 was raised so far that the mercury filled the reservoir R_1 and a part: of the narrow connecting U-shaped tube. When the pump was set to work, the mercury would, of course, quickly rise in the tube, and reservoir R_2 was lowered, the experimenter keeping the mercury at about the same level. The reservoir R_2 was balanced by a long spring which facilitated the operation, and the friction of the parts was generally sufficient to keep it almost in any position. When the Sprengel pump had done its work, the reservoir R_2 was further lowered and the mercury descended in R_1 and filled R_2, whereupon stopcock C_2 was closed. The air adhering to the walls of R, and that absorbed by the mercury was carried off, and to free the mercury of all air the reservoir R_2 was for a long time worked up and down. During this process some air, which would gather below stopcock C_2, was expelled from R_2 by lowering it far enough and opening the stopcock, closing the latter again before raising the reservoir. When all the air had been expelled from the mercury, and no air would gather in R_2 when it was lowered, the caustic potash was resorted to. The reservoir R_2 was now again raised until the mercury in R_1 stood above stopcock C_1. The caustic potash was fused and boiled, and the moisture partly carried off by the pump and partly re-absorbed; and this process of heating and cooling was repeated many times, and each time, upon the moisture being absorbed or carried off, the reservoir R_2 was for a long time raised and lowered. In this manner all the moisture was carried off from the mercury, and both the reservoirs were in proper condition to be used. The reservoir R_2 was then again raised to the top, and the pump was kept working for a long time. When the highest vacuum obtainable with the pump had been reached the potash bulb was usually wrapped with cotton which was sprinkled with ether so as to keep the potash at a very low temperature, then the reservoir R_2 was lowered, and again reservoir R_1 being emptied the receiver r was quickly sealed up.

When a new bulb was put on, the mercury was always raised above stopcock C_1, which was closed, so as to always keep the mercury and both the reservoirs in fine condition, and the mercury was never withdrawn from R_1 except when the pump had reached the highest degree of exhaustion. It is necessary to observe this rule if it is desired to use the apparatus to advantage.

By means of this arrangement I was able to proceed very quickly, and when the apparatus was in perfect order it was possible to reach the phosphorescent stage in a small bulb in less than fifteen minutes, which is certainly very quick work for a small laboratory arrangement requiring all in all about 100 pounds of mercury. With ordinary small bulbs the ratio of the capacity of the pump, receiver, and connections, and that of reservoir R was about 1 to 20, and the degrees of exhaustion reached were necessarily very high, though I am unable to make a precise and reliable statement how far the exhaustion was carried.

What impresses the investigator most in the course of these experiences is the behavior of gases when subjected to great rapidly alternating electrostatic stresses. But he must remain in doubt as to whether the effects observed are due wholly to the molecules, or atoms, of the gas which chemical analysis discloses to us, or whether there enters into play another medium of a gaseous nature, comprising atoms, or molecules, immersed in a fluid pervading the space. Such a medium, surely must exist, and I am convinced that, for instance, even if air were absent, the surface and neighborhood of a body in space would be heated by rapidly alternating the potential of the body; but no such heating of the surface or neighborhood could occur if all free atoms were removed and only a homogeneous, incompressible, and elastic fluid—such as ether is supposed to be—would remain, for then there would be no impacts, no collisions. In such a case, as far as the body itself is concerned, only frictional losses in the inside could occur.

It is a striking fact that the discharge through a gas is established with ever increasing freedom as the frequency of the impulses is augmented. It behaves in this respect quite contrarily to a metallic conductor. In the latter the impedance enters prominently into play as the frequency is increased, but the gas acts much as a series of condensers would: the facility with which the discharge passes through seems to depend on the rate of change of potential. If it act so, then in a vacuum tube even of great length, and no matter how strong the current, self-induction could not assert itself: to any appreciable degree. We have, then, as far as we can now see, in the gas a conductor which is capable of transmitting electric impulses of any frequency which we may be able to produce. Could the frequency be brought high enough, then a queer system of electric distribution, which would be likely to interest gas companies, might be realized : metal pipes filled with gas — the metal being the insulator, the gas the conductor — supplying phosphorescent bulbs, or perhaps devices as yet uninvented. It is certainly possible to take a hollow core of copper, rarefy the gas in the same, and by passing impulses of sufficiently high frequency through a circuit around it, bring the gas inside to a high degree of incandescence; but as to the nature of the forces there would be considerable uncertainty, for it would be doubtful whether with such impulses the copper core would act as a static screen. Such paradoxes and apparent impossibilities we encounter at every step in this line of work, and therein lies, to a great extent, the charm of the study.

I have here a short and wide tube which is exhausted to a high degree and covered with a substantial coating of bronze, the coating allowing barely the light to shine through. A metallic clasp, with a hook for suspending the tube, is fastened around the middle portion of the latter, the clasp being in contact with the bronze coating. I now want to light the gas inside by suspending the tube on a wire connected to the coil. Any one who would try the experiment for the first time, not having any previous experience, would probably take care to be quite alone when making the trial, for fear that he might become the joke of his assistants. Still, the bulb lights in spite of the metal coating, and the light can be distinctly perceived through the latter. A long tube covered with aluminium bronze lights when held in one hand — the other touching the terminal of the coil — quite powerfully. It might be objected that the coatings are not sufficiently conducting; still, even if they were highly resistant, they ought to screen the gas. They certainly screen it perfectly in a condition of rest, but not by far perfectly when the charge is surging in the coating. But the loss of energy which occurs within the tube, notwithstanding the screen, is occasioned principally by the presence of the gas. Were we to take a large hollow metallic sphere and fill it with a perfect incompressible fluid dielectric, there would be no loss inside of the sphere, and consequently the inside might be considered as perfectly screened, though the potential be very rapidly alternating. Even were the sphere filled with oil, the loss would be incomparably smaller than when the fluid is replaced by a gas, for in the latter case the force produces displacements; that means impact and collisions in the inside.

No matter what the pressure of the gas may be, it becomes an important factor in the bearing of a conductor when the electric density is great and the frequency very high. That in the heating of conductors by lightning discharges air is an element of great importance, is almost as certain as an experimental fact. I may illustrate the action of the air by the following experiment: I take a short tube which is exhausted to a moderate degree and has a platinum wire running through the middle from one end to the other. I pass a steady or low frequency current through the wire, and it is heated uniformly in all parts. The heating here is due to conduction, or frictional losses, and the gas around the wire has—as far as we can see—no function to perform. But now let me pass sudden discharges, or a high frequency current, through the wire. Again the wire is heated, this time principally on the ends and least in the middle portion; and if the frequency of the impulses, or the rate of change, is high enough, the wire might as well be cut in the middle as not, for practically all the heating is due to the rarefied gas: Here the gas might only act as a conductor of no impedance diverting the current from the wire as the impedance of the latter is enormously increased, and merely heating the ends of the wire by reason of their resistance to the passage of the discharge. But it is not at all necessary that the gas in the tube should be conducting; it might be at an extremely low pressure, still the ends of the wire would be heated—as, however, is ascertained by experience—only the two ends would in such case not be electrically connected through the gaseous medium. Now what with these frequencies and potentials occurs in an exhausted tube occurs in the lightning discharges at ordinary pressure. We only need to remember one of the facts arrived at in the course of these investigations, namely, that to impulses of very high frequency the gas at ordinary pressure behaves much in the same manner as though it were at moderately low pressure. I think that in lightning discharges frequently wires or conducting objects are volatilized merely because air is present, and that, were the conductor immersed in an insulating liquid, it would be safe, for then the energy would have to spend itself somewhere else. From the behavior of gases to sudden impulses of high potential I am led to conclude that there can be no surer way of diverting a lightning discharge than by affording it a passage through a volume of gas, if such a thing can be done in a practical manner.

There are two more features upon which I think it necessary to dwell in connection with these experiments—the "radiant state" and the non-striking vacuum."

Any one who has studied Crookes work must have received the impression that the "radiant state" is a property of the gas inseparably connected with an extremely high degree of exhaustion. But it should be remembered that the phenomena observed in an exhausted vessel are limited to the character and capacity of the apparatus which is made use of. I think that in a bulb a molecule, or atom, does not precisely move in a straight line because it meets no obstacle, but because the velocity imparted to it is sufficient to propel it in a sensibly straight line. The mean free path is one thing, but the velocity—the energy associated with the moving body—is another, and under ordinary circumstances I believe that it is mere question of potential or speed. A disruptive discharge coil, when the potential is pushed very far, excites phosphorescence and projects shadows, at comparatively low degrees of exhaustion. In a lightning discharge, matter moves in straight lines at ordinary pressure when the mean free path is exceedingly small, and frequently images of wires or other metallic objects have been produced by the particles thrown off in straight lines.

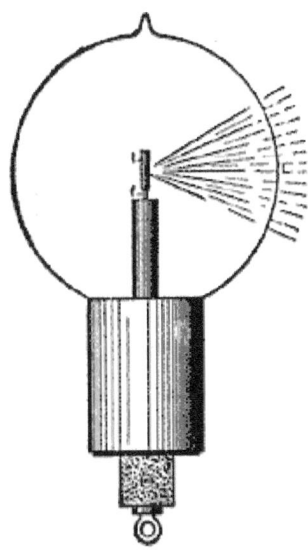

Fig. 31.

I have prepared a bulb to illustrate by an experiment the correctness of these assertions. In a globe L (Fig. 31), I have mounted upon a lamp filaments of a piece of lime l. The lamp filament is connected with a wire which leads into the bulb, and the general construction of the latter is as indicated in Fig. 19, before described. The bulb being suspended from a wire connected to the terminal of the coil, and the latter being set to work, the lime piece l and the projecting parts of the filament f are bombarded. The degree of exhaustion is just such that with the potential the coil is capable of giving phosphorescence of the glass is produced, but disappears as soon as the vacuum is impaired. The lime containing moisture, and moisture being given off as soon as heating occurs, the phosphorescence lasts only for a few moments. When the lime has been sufficiently heated, enough moisture has been given off to impair materially the vacuum of the bulb. As the bombardment goes on, one point of the lime piece is more heated than other points, and the results is that finally practically all the discharge passes through that point which is intensely heated, and a white stream of lime particles (Fig. 31) then breaks forth from that point. This stream is composed of "radiant" matter, yet the degree of exhaustion is low. But the particles move in straight lines because the velocity imparted to them is great, and this is due to three causes—to the great electric density, the high temperature of the small point, and the fact that the particles of the lime are easily torn and thrown off—far more easily than those of carbon. With frequencies such as we are able to obtain, the particles are bodily thrown off and projected to a considerable distance, but with sufficiently high frequencies no such thing would occur: in such case only a stress would spread or a vibration would be propagated through the bulb. It would be out of the question to reach any such frequency on the assumption that the atoms move with the speed of light; but I believe that such a thing is impossible; for this an enormous potential would be required. With potentials which we are able to obtain, even with a disruptive discharge coil, the speed must be quite insignificant.

As to the "non-striking vacuum," the point to be noted is that it can occur only with low frequency impulses, and it is necessitated by the impossibility of carrying off enough energy with such impulses in high vacuum since the few atoms which are around the terminal upon coming in contact with the same are repelled and kept at a distance for a comparatively long period of time, and not enough work can be performed to render the effect perceptible to the eye. If the difference of potential between the terminals is raised, the dielectric breaks down. But with very high frequency impulses there is no necessity for such breaking down, since any amount of work can be performed by continually agitating the atoms in the exhausted vessel, provided the frequency is high enough. It is easy to reach—even with frequencies obtained from an alternator as here used—a stage at which the discharge does not pass between two electrodes in a narrow tube, each of these being connected to one of the terminals of the coil, but it is difficult to reach a point at which a luminous discharge would not occur around each electrode.

A thought which naturally presents itself in connection with high frequency currents, is to make use of their powerful electro-dynamic inductive action to produce light effects in a sealed glass globe. The leading-in wire is one of the defects of the present incandescent lamp, and if no other improvement were made, that imperfection at least should be done away with. Following this thought, I have carried on experiments in various directions, of which some were indicated in my former paper. I may here mention one or two more lines of experiment which have been followed up.

Many bulbs were constructed as shown in Fig. 32 and Fig. 33.

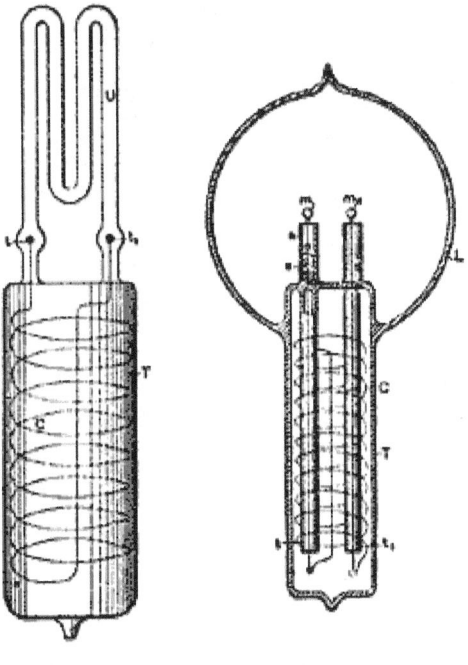

Fig. 32.   Fig. 33.

In Fig. 32 a wide tube T was sealed to a smaller W-shaped tube U, of phosphorescent glass. In the tube T was placed a coil C of aluminium wire, the ends of which were provided with small spheres t and t_1 of aluminium, and reached into the U tube. The tube T was slipped into a socket containing a primary coil through which usually the discharges of Leyden jars were directed, and the rarefied gas in the small U tube was excited to strong luminosity by the high-tension currents induced in the coil C. When Leyden jar discharges were used to induce currents in the coil C, it was found necessary to pack the tube T tightly with insulating powder, as a discharge would occur frequently between the turns of the coil, especially when the primary was thick and the air gap, through which the jars discharged, large, and no little trouble was experienced in this way.

In Fig. 33 is illustrated another form of the bulb constructed. In this case a tube T is sealed to a globe L. The tube contains a coil C, the ends of which pass through two small glass tubes t and t_1, which are sealed to the tube T. Two refractory buttons m and m_1 are mounted on lamp filaments which are fastened to the ends of the wires passing through the glass tubes t and t_1. Generally in bulbs made on this plan the globe L communicated with the tube T. For this purpose the ends of the small tubes t and t_1 were just a trifle heated in the burner, merely to hold the wires, but not to interfere with the communication. The tube T, with the small tubes, wires through the same, and the refractory buttons m and m_1 was first prepared, and then sealed to globe L, whereupon the coil C was slipped in and the connections made to its ends. The tube was then packed with insulating powder, jamming the latter as tight as possible up to very nearly the end, then it was closed and only a small hole left through which the remainder of the powder was introduced, and finally the end of the tube was closed. Usually in bulbs constructed as shown in Fig. 33 an aluminium tube a was fastened to the upper end s of each of the tubes t and t_1, in order to protect that end against the heat. The buttons m and m_1 could be brought to any degree of incandescence by passing the discharges of Leyden jars around the coil C. In such bulbs with two buttons a very curious effect is produced by the formation of the shadows of each of the two buttons.

Another line of experiment, which has been assiduously followed, was to induce by electro-dynamic induction a current or luminous discharge in an exhausted tube or bulb. This matter has received such able treatment at the hands of Prof. J. J. Thomson that I could add but little to what he has made known, even had I made it the special subject of this lecture. Still, since experiences in this line have gradually led me to the present views and results, a few words must be devoted here to this subject.

It has occurred, no doubt, to many that as a vacuum tube is made longer the electromotive force per unit length of the tube, necessary to pass a luminous discharge through the latter, gets continually smaller; therefore, if the exhausted tube be made long enough, even with low frequencies a luminous discharge could be induced in such a tube closed upon itself. Such a tube might be placed around a hall or on a ceiling, and at once a simple appliance capable of giving considerable light would be obtained. But this would be an appliance hard to manufacture and extremely unmanageable. It would not do to make the tube up of small lengths, because there would be with ordinary frequencies considerable loss in the coatings, and besides, if coatings were used, it would be better to supply the current directly to the tube by connecting the coatings to a transformer. But even if all objections of such nature were removed, still, with low frequencies the light conversion itself would be inefficient, as I have before stated. In using extremely high frequencies the length of the secondary—in other words, the site of the vessel—can be reduced as far as desired, and the efficiency of the light conversion is increased; provided that means are invented for efficiently obtaining such high frequencies. Thus one is led, from theoretical and practical considerations, to the use of high frequencies, and this means high electromotive forces and small currents in the primary. When he works with condenser charges—and they are the only means up to the present known for reaching these extreme frequencies—one gets to electromotive forces of several thousands of volts per turn of the primary. He cannot multiply the electro-dynamic inductive effect by taking more turns in the primary, for he arrives at the conclusion that the best way is to work with one single turn—though we must sometimes depart from this rule—and we must get along with whatever inductive effect we can obtain with one turn. But before he has long experimented with the extreme frequencies required to set up in a small bulb an electromotive force of several thousands of volts he realizes the great importance of electrostatic effects, and these effects grow relatively to the electro-dynamic in significance as the frequency is increased.

Now, if anything is desirable in this case, it is to increase the frequency, and this would make it still worse for the electro-dynamic effects. On the other hand, it is easy to exalt the electrostatic action as far as one likes by taking more turns on the secondary, or combining self-induction and capacity to raise the potential. It should also be remembered that, in reducing the current to the smallest value and increasing the potential, the electric impulses of high frequency can be more easily transmitted through a conductor.

These and similar thoughts determined me to devote more attention to the electrostatic phenomena, and to endeavor to produce potentials as high as possible, and alternating as fast as they could be made to alternate. I then found that I could excite vacuum tubes at considerable distance from a conductor connected to a properly constructed coil, and that I could, by converting the oscillatory current of a condenser to a higher potential, establish electrostatic alternating fields which acted through the whole extent of a room, lighting up a tube no matter where it was held in space. I thought I recognized that I had made a step in advance, and I have persevered in this line; but I wish to say that I share with all lovers of science and progress the one and only desire — to reach a result of utility to men in any direction to which thought or experiment may lead me. I think that this departure is the right one, for I cannot see, from the observation of the phenomena which manifest themselves as the frequency is increased, what there would remain to act between two circuits conveying, for instance, impulses of several hundred millions per second, except electrostatic forces. Even with such stifling frequencies the energy would be practically all potential, and my conviction has grown strong that, to whatever kind of motion light may be due, it is produced by tremendous electrostatic stresses vibrating with extreme rapidity.

Of all these phenomena observed with currents, or electric impulses, of high frequency, the most fascinating for an audience are certainly those which are noted in an electrostatic field acting through considerable distance, and the best an unskilled lecturer can do is to begin and finish with the exhibition of these singular effects. I take a tube in the hand and move it about, and it is lighted wherever I may hold it; throughout space the invisible forces act. But I may take another tube and it might not light, the vacuum being very high. I excite it by means of a disruptive discharge coil, and now it will light in the electrostatic field. I may put it away for a few weeks or months, still it retains the faculty of being excited. What change have I produced in the tube in the act of exciting it? If a motion imparted to the atoms, it is difficult to perceive how it can persist so long without being arrested by frictional losses; and if a strain exerted in the dielectric, such as a simple electrification would produce, it is easy to see how it may persist indefinitely but very difficult to understand why such a condition should aid the excitation when we have to deal with potentials which are rapidly alternating.

Since I have exhibited these phenomena for the first time, I have obtained some other interesting effects. For instance, I have produced the incandescence of a button, filament, or wire enclosed in a tube. To get to this result it was necessary to economize the energy which is obtained from the field and direct most of it on the small body to be rendered incandescent. At the beginning the task appeared difficult, but the experiences gathered permitted me to teach the result easily. In Fig. 34 and Fig. 35 two such tubes are illustrated which are prepared for the occasion. In Fig. 34 a short tube T_1, sealed to another long tube T, is provided with a stem s, with a platinum wire sealed in the latter. A very thin lamp filament l is fastened to this wire, and connection to the outside is made through a thin copper wire w. The tube is provided with outside and inside coatings, C and C_1 respectively, and is filled as far as the coatings reach with conducting, and the space above with insulating powder. These coatings are merely used to enable me to perform two experiments with the tube—namely, to produce the effect desired either by direct connection of the body of the experimenter or of another body to the wire w, or by acting inductively through the glass. The stem s is provided with an aluminium tube a for purposes before explained, and only a small part of the filament reaches out of this tube. By holding the tube T_1 anywhere in the electrostatic field the filament is rendered incandescent.

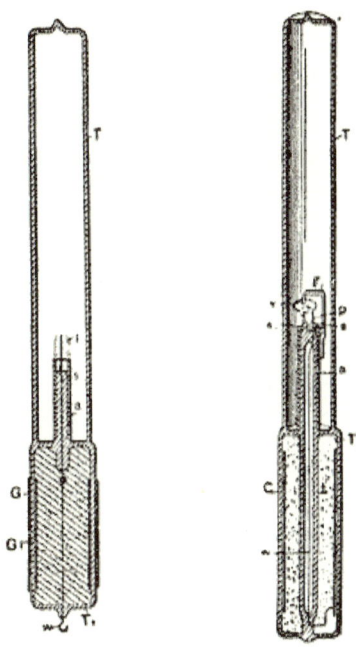

Fig. 34.   Fig. 35.

A more interesting piece of apparatus is illustrated in Fig. 35. The construction is the same as before, only instead of the lamp filament a small platinum wire p, sealed in a stem s, and bent above it in a circle, is connected to the copper wire w, which is joined to an inside coating C. A small stem s_1 is provided with a needle, on the point of which is arranged to rotate very freely a very light fan of mica v. To prevent the fan from falling out, a thin stem of glass g is bent properly and fastened to the aluminium tube. When the glass tube is held anywhere in the electrostatic field the platinum wire becomes incandescent, and the mica vanes are rotated very fast.

Intense phosphorescence may be excited in a bulb by merely connecting it to a plate within the field, and the plate need not be any larger than an ordinary lamp shade. The phosphorescence excited with these currents is incomparably more powerful than with ordinary apparatus. A small phosphorescent bulb, when attached to a wire connected to a coil, emits sufficient light to allow reading ordinary print at a distance of five to six paces. It was of interest to see how some of the phosphorescent bulbs of Professor Crookes would behave with these currents, and he has had the kindness to lend me a few for the occasion. The effects produced are magnificent, especially by the sulphide of calcium and sulphide of zinc. From the disruptive discharge coil they glow intensely merely by holding them in the hand and connecting the body to the terminal of the coil.

To whatever results investigations of this kind may lead, their chief interest lies for the present in the possibilities they offer for the production of an efficient illuminating device. In no branch of electric industry is an advance more desired than in the manufacture of light. Every thinker, when considering the barbarous methods employed, the deplorable losses incurred in our best systems of light production, must have asked himself, What is likely to be the light of the future? Is it to be an incandescent solid, as in the present lamp, or an incandescent gas, or a phosphorescent body, or something like a burner, but incomparably more efficient?

There is little chance to perfect a gas burner; not, perhaps, because human ingenuity has been bent upon that problem for centuries without a radical departure having been made—though this argument is not devoid of force—but because in a burner the higher vibrations can never be reached except by passing through all the low ones. For how is a flame produced unless by a fall of lifted weights? Such process cannot be maintained without renewal, and renewal is repeated passing from low to high vibrations. One way only seems to be open to improve a burner, and that is by trying to reach higher degrees of incandescence. Higher incandescence is equivalent to a quicker vibration; that means more light from the same material, and that, again, means more economy. In this direction some improvements have been made, but the progress is hampered by many limitations. Discarding, then, the burner, there remain the three ways first mentioned, which are essentially electrical.

Suppose the light of the immediate future to be a solid rendered incandescent by electricity. Would it not seem that it is better to employ a small button than a frail filament? From many considerations it certainly must be concluded that a button is capable of a higher economy, assuming, of course, the difficulties connected with the operation of such a lamp to be effectively overcome. But to light such a lamp we require a high potential; and to get this economically we must use high frequencies.

Such considerations apply even more to the production of light by the incandescence of a gas, or by phosphorescence. In all cases we require high frequencies and high potentials. These thoughts occurred to me a long time ago.

Incidentally we gain, by the use of very high frequencies, many advantages, such as a higher economy in the light production, the possibility of working with one lead, the possibility of doing away with the leading-in wire, etc.

The question is, how far can we go with frequencies? Ordinary conductors rapidly lose the facility of transmitting electric impulses when the frequency is greatly increased. Assume the means for the production of impulses of very great frequency brought to the utmost perfection, every one will naturally ask how to transmit them when the necessity arises. In transmitting such impulses through conductors we must remember that we have to deal with pressure and flow, in the ordinary interpretation of these terms. Let the pressure increase to an enormous value, and let the flow correspondingly diminish, then such impulses — variations merely of pressure, as it were — can no doubt be transmitted through a wire even if their frequency be many hundreds of millions per second. It would, of course, be out of question to transmit such impulses through a wire immersed in a gaseous medium, even if the wire were provided with a thick and excellent insulation for most of the energy would be lost in molecular bombardment and consequent heating. The end of the wire connected to the source would be heated, and the remote end would receive but a trifling part of the energy supplied. The prime necessity, then, if such electric impulses are to be used, is to find means to reduce as much as possible the dissipation.

The first thought is, employ the thinnest possible wire surrounded by the thickest practicable insulation. The next thought is to employ electrostatic screens. The insulation of the wire may be covered with a thin conducting coating and the latter connected to the ground. But this would not do, as then all the energy would pass through the conducting coating to the ground and nothing would get to the end of the wire. If a ground connection is made it can only be made through a conductor offering an enormous impedance, or through a condenser of extremely small capacity. This, however, does not do away with other difficulties.

If the wave length of the impulses is much smaller than the length of the wire, then corresponding short waves will be sent up in the conducting coating, and it will be more or less the same as though the coating were directly connected to earth. It is therefore necessary to cut up the coating in sections much shorter than the wave length. Such an arrangement does not still afford a perfect screen, but it is ten thousand times better than none. I think it preferable to cut up the conducting coating in small sections, even if the current waves be much longer than the coating.

If a wire were provided with a perfect electrostatic screen, it would be the same as though all objects were removed from it at infinite distance. The capacity would then be reduced to the capacity of the wire itself, which would be very small. It would then be possible to send over the wire current vibrations of very high frequencies at enormous distance without affecting greatly the character of the vibrations. A perfect screen is of course out of the question, but I believe that with a screen such as I have just described telephony could be rendered practicable across the Atlantic. According to my ideas, the gutta-percha covered wire should be provided with a third conducting coating subdivided in sections. On the top of this should be again placed a layer of gutta-percha and other insulation, and on the top of the whole the armor. But such cables will not be constructed, for ere long intelligence — transmitted without wires will throb through the earth like a pulse through a living organism. The wonder is that, with the present state of knowledge and the experiences gained, no attempt is being made to disturb the electrostatic or magnetic condition of the earth, and transmit, if nothing else, intelligence.

It has been my chief aim in presenting these results to point out phenomena or features of novelty, and to advance ideas which I am hopeful will serve as starting points of new departures. It has been my chief desire this evening to entertain you with some novel experiments. Your applause, so frequently and generously accorded has told me that I have succeeded.

In conclusion, let me thank you most heartily for your kindness and attention, and assure you that the honor I have had in addressing such a distinguished audience, the pleasure I have had in presenting these results to a gathering of so many able men and among them also some of those in whose work for many years past I have found enlightenment and constant pleasure — I shall never forget.

ON LIGHT AND OTHER HIGH-FREQUENCY PHENOMENA (4/4)

—*Delivered before the Franklin Institute, Philadelphia, February 1893, and before the National Electric Light Association, St. Louis, March 1893.*

Introductory — some thoughts on the eye

When we look at the world around us, on Nature, we are impressed with its beauty and grandeur. Each thing we perceive, though it may be vanishingly small, is in itself a world, that is, like the whole of the universe, matter and force governed by law,—a world, the contemplation of which fills us with feelings of wonder and irresistibly urges us to ceaseless thought and inquiry. But in all this vast world, of all objects our senses reveal to us, the most marvelous, the most appealing to our imagination, appears no doubt a highly developed organism, a thinking being. If there is anything fitted to make us admire Nature's handiwork, it is certainly this inconceivable structure, which performs its innumerable motions of obedience to external influence. To understand its workings, to get a deeper insight into this Nature's masterpiece, has ever been for thinkers a fascinating aim, and after many centuries of arduous research men have arrived at a fair understanding of the functions of its organs and senses. Again, in all the perfect harmony of its parts, of the parts which constitute the material or tangible of our being, of all its organs and senses, the eye is the most wonderful. It is the most precious, the most indispensable of our perceptive or directive organs, it is the great gateway through which all knowledge enters the mind. Of all our organs, it is the one, which is in the most intimate relation with that which we call intellect. So intimate is this relation, that it is often said, the very soul shows itself in the eye.

It can be taken as a fact, which the theory of the action of the eye implies, that for each external impression, that is, for each image produced upon the retina, the ends of the visual nerves, concerned in the conveyance of the impression to the mind, must be under a peculiar stress or in a vibratory state, It now does not seem improbable that, when by the power of thought an image is evoked, a distinct reflex action, no matter how weak, is exerted upon certain ends of the visual nerves, and therefore upon the retina. Will it ever be within human power to analyse the condition of the retina when disturbed by thought or reflex action, by the help of some optical or other means of such sensitiveness, that a clear idea of its state might be gained at any time? If this were possible, then the problem of reading one's thoughts with precision, like the characters of an open book, might be much easier to solve than many problems belonging to the domain of positive physical science, in the solution of which many, if not the majority: of scientific men implicitly believe. Helmholtz has shown that the fundi of the eye are themselves, luminous, and he was able to see, in total darkness, the movement of his arm by the light of his own eyes. This is one of the most remarkable experiments recorded in the history of science, and probably only a few men could satisfactorily repeat it, for it is very likely, that the luminosity of the eyes is associated with uncommon activity of the brain and great imaginative power. It is fluorescence of brain action, as it were.

Another fact having a bearing on this subject which has probably been noted by many, since it is stated in popular expressions, but which I cannot recollect to have found chronicled as a positive result of observation is, that at times, when a sudden idea or image presents itself to the intellect, there is a distinct and sometimes painful sensation of luminosity produced in the eye, observable even in broad daylight.

The saying then, that the soul shows itself in the eye, is deeply founded, and we feel that it expresses a great truth. It has a profound meaning even for one who, like a poet or artist, only following; his inborn instinct or love for Nature, finds delight in aimless thoughts and in the mere contemplation of natural phenomena, but a still more profound meaning for one who, in the spirit of positive scientific investigation, seeks to ascertain the causes of the effects. It is principally the natural philosopher, the physicist, for whom the eye is the subject of the most intense admiration.

Two facts about the eye must forcibly impress the mind of the physicist, notwithstanding he may think or say that it is an imperfect optical instrument, forgetting, that the very conception of that which is perfect or seems so to him, has been gained through this same instrument. First, the eye is, as far as our positive knowledge goes, the only organ which is directly affected by that subtle medium, which as science teaches us, must fill all space; secondly, it is the most sensitive of our organs, incomparably more sensitive to external impressions than any other.

The organ of hearing implies the impact of ponderable bodies, the organ of smell the transference of detached material particles, and the organs of taste, and of touch or force, the direct contact, or at least some interference of ponderable matter, and this is true even in those instances of animal organisms, in which some of these organs are developed to a degree of truly marvelous perfection. This being so, it seems wonderful that the organ of sight solely should be capable of being stirred by that, which all our other organs are powerless to detect, yet which plays an essential part in all natural phenomena, which transmits all energy and sustains all motion and, that most intricate of all, life, but which has properties such that even a scientifically trained mind cannot help drawing a distinction between it and all that is called matter. Considering merely this, and the fact that the eye, by its marvelous power, widens our otherwise very narrow range of perception far beyond the limits of the small world which is our own, to embrace myriads of other worlds, suns and stars in the infinite depths of the universe, would make it justifiable to assert, that it is an organ of a higher order. Its performances are beyond comprehension. Nature as far as we know never produced anything more wonderful. We can get barely a faint idea of its prodigious power by analyzing what it does and by comparing. When ether waves impinge upon the human body, they produce the sensations of warmth or cold, pleasure or pain, or perhaps other sensations of which we are not aware, and any degree or intensity of these sensations, which degrees are infinite in number, hence an infinite number of distinct sensations. But our sense of touch, or our sense of force, cannot reveal to us these differences in degree or intensity, unless they are very great. Now we can readily conceive how an organism, such as the human, in the eternal process of evolution, or more philosophically speaking, adaptation to Nature, being constrained to the use of only the sense of touch or force, for instance, might develop this sense to such a degree of sensitiveness or perfection, that it would be capable of distinguishing the minutest differences in the temperature of a body even at some distance, to a hundredth, or thousandth, or millionth part of a degree. Yet, even this apparently impossible performance would not begin to compare with that of the eye, which is capable of distinguishing and conveying to the mind in a single instant innumerable peculiarities of the body, be it in form, or color, or other respects. This power of the eye rests upon two thins, namely, the rectilinear propagation of the disturbance by which it is effected, and upon its sensitiveness. To say that the eye is sensitive is not saying anything. Compared with it, all other organs are monstrously crude. The organ of smell which guides a dog on the trail of a deer, the organ of touch or force which guides an insect in its wanderings, the organ of hearing, which is affected by the slightest disturbances of the air, are sensitive organs, to be sure, but what are they compared with the human eye! No doubt it responds to the faintest echoes or reverberations of the medium; no doubt, it brings us tidings from other worlds, infinitely remote, but in a language we cannot as yet always understand. And why not? Because we live in a medium filled with air and other gases, vapors and a dense mass of solid particles flying about. These play an important part in many phenomena; they fritter away the energy of the vibrations before they can reach the eye; they too, are the carriers of germs of destruction, they get into our lungs and other organs, clog up the channels and imperceptibly, yet inevitably, arrest the stream of life. Could we but do away with all ponderable matter in the line of sight of the telescope, it would reveal to us undreamt of marvels. Even the unaided eye, I think; would he capable of distinguishing in the pure medium, small objects at distances measured probably by hundreds or perhaps thousands of miles.

But there is something else about the eye which impresses us still more than these wonderful features which we observed, viewing it from the standpoint of a physicist, merely as an optical instrument,—something which appeals to us more than its marvelous faculty of being directly affected by the vibrations of the medium, without interference of gross matter, and more than its inconceivable sensitiveness and discerning power. It is its significance in the processes of life. No matter what one's views on nature and life may be, he must stand amazed when, for the first time in his, thoughts, he realizes the importance of the eye in the physical processes and mental performances of the human organism. And how could it be otherwise, when he realizes, that the eye is the means through which the human race has acquired the entire knowledge it possesses, that it controls all our motions, more still, and our actions.

There is no way of acquiring knowledge except through the eye. What is the foundation of all philosophical systems of ancient and modern times, in fact, of all the philosophy of men? I am I think; I think, therefore I am. But how could I think and how would I know that I exist, if I had not the eye? For knowledge involve consciousness; consciousness involves ideas, conceptions; conceptions involve pictures or images, and images the sense of vision, and therefore the organ of sight. But how about blind men, will be asked? Yes, a blind man may depict in magnificent poems, forms and scenes from real life, from a world he physically does not see. A blind man may touch the keys of an instrument with unerring precision, may model the fastest boat, may discover and invent, calculate and construct, may do still greater wonders—but all the blind men who have done such thinks have descended from those who had seeing eyes. Nature may reach the same result in many ways. Like a wave in the physical world, in the infinite ocean of the medium which pervades all, so in the world of organism:, in life, an impulse started proceeds onward, at times, may be, with the speed of light, at times, again, so slowly that for ages and ages it seems to stay; passing through processes of a complexity inconceivable to men, but in;ill its forms, in all its stages, its energy, ever and ever integrally present. A single ray of light from a distant star falling upon the eye of a tyrant in by-gone times, may have altered the course of his life, may have changed the destiny of nations, may have transformed the surface of the globe, so intricate, so inconceivably complex are the processes in Nature. In no way can we get such an overwhelming idea of the grandeur of Nature, as when we consider, that in accordance with the law of the conservation of energy, throughout the infinite, the forces are in a perfect balance, and hence the energy of a single thought may determine the motion of a Universe. It is not necessary that every individual, not even that every generation or many generations, should have the physical instrument of sight, in order to be able to form images and to think, that is, form ideas or conceptions; but sometime or other, during the process of evolution, the eye certainly must have existed, else thought, as we understand it, would be impossible; else conceptions, like spirit, intellect, mind, call it as you may, could not exist. It is conceivable, that in some other world, in some other beings, the eye is replaced by a different organ, equally or more perfect, but these beings cannot be men.

Now what prompts us all to voluntary motions and actions of any kind? Again the eye. If I am conscious of the motion, I must have an idea or conception, that is, an image, therefore the eye. If I am not precisely conscious of the motion, it is, because the images are vague or indistinct, being blurred by the superimposition of many. But when I perform the motion, does the impulse which prompts me to the action come from within or from without? The greatest physicists have not disdained to endeavour to answer this and similar questions and have at tunes abandoned themselves to the delights of pure and unrestrained thought. Such questions are generally considered not to belong to the realm of positive physical science, but will before long be annexed to its domain. Helmholtz has probably thought more on life than any modern scientist. Lord Kelvin expressed his belief that life's process is electrical and that there is a force inherent to the organism and determining its motions. just as much as I am convinced of any physical truth I am convinced that the motive impulse must come from the outside. For, consider the lowest organism we know—and there are probably many lower ones—an aggregation of a few cells only. If it is capable of voluntary motion it can perform an infinite number of motions, all definite and precise. But now a mechanism consisting of a finite number of parts and few at that, cannot perform are infinite number of definite motions, hence the impulses which govern its movements must come from the environment. So, the atom, the ulterior element of the Universe's structure, is tossed about in space eternally, a play to external influences, like a boat in a troubled sea. Were it to stop its motion it would die: hatter at rest, if such a thin; could exist, would be matter dead. Death of matter! Never has a sentence of deeper philosophical meaning been uttered. This is the way in which Prof. Dewar forcibly expresses it in the description of his admirable experiments, in which liquid oxygen is handled as one handles water, and air at ordinary pressure is made to condense and even to solidify by the intense cold: Experiments, which serve to illustrate, in his language, the last feeble manifestations of life, the last quiverings of matter about to die. But human eyes shall not witness such death. There is no death of matter, for throughout the infinite universe, all has to move, to vibrate, that is, to live.

I have made the preceding statements at the peril of treading upon metaphysical ground; in my desire to introduce the subject of this lecture in a manner not altogether uninteresting, I may hope, to an audience such as I have the honor to address. But now, then, returning to the subject, this divine organ of sight, this indispensable instrument for thought and all intellectual enjoyment, which lays open to us the marvels of this universe, through which we have acquired what knowledge we possess, and which prompts us to, and controls, all our physical and mental activity. By what is it affected? By light! What is light?

We have witnessed the great strides which have been made in all departments of science in recent years. So great lave been the advances that we cannot refrain from asking ourselves, Is this all true; or is it but a dream? Centuries ago men have lived, have thought, discovered, invented, and have believed that they were soaring, while they were merely proceeding at a snail's pace. So we too may be mistaken. But taking the truth of the observed events as one of the implied facts of science, we must rejoice in the, immense progress already made and still more in the anticipation of what must come, judging from the possibilities opened up by modern research. There is, however, an advance which we have been witnessing, which must be particularly gratifying to every lover of progress. It is not a discovery, or an invention, or an achievement in any particular direction. It is an advance in all directions of scientific thought and experiment I mean the generalization of the natural forces and phenomena, the looming up of a certain broad idea on the scientific horizon. It is this idea which has, however, long ago taken possession of the most advanced minds, to which I desire to call your attention, and which I intend to illustrate in a general way, in these experiments, as the first step in answering the question "What is light?" and to realize the modern meaning of this word.

It is beyond the scope of my lecture to dwell upon the subject of light in general, my object being merely to bring presently to your notice a certain class of light effects and a number of phenomena observed in pursuing the study of these effects. But to he consistent in my remarks it is necessary to state that, according to that idea, now, accepted by the majority of scientific men as a positive result of theoretical and experimental investigation, the various forms or manifestations of energy which were generally designated as "electric" or more precisely "electromagnetic" are energy manifestations of the same nature as those of radiant heat and light. Therefore the phenomena of light and heat and others besides these, may be called electrical phenomena. Thus electrical science has become the mother science of all and its study has become all important. The day when we shall know exactly what "electricity" is, will chronicle an event probably greater, more important than any other recorded in the history of the human race. The time will come when the comfort, the very existence, perhaps, or man will depend upon that wonderful agent. For our existence and comfort we require heat, light and mechanical power. How do we now get all these? We get them from fuel, we get them by consuming material. What will man do when the forests disappear, when the coal fields are exhausted? Only one thing according to our present knowledge will remain; that is, to transmit power at great distances. Men will go to the waterfalls, to the tides, which are the stores of an infinitesimal part of Nature's immeasurable energy. There will they harness the energy and transmit the same to their settlements, to warm their homes by, to give them light, and to keep their obedient slaves, the machines, toiling. But how will they transmit this energy if not by electricity? Judge then, if the comfort, nay, the very existence, of man will not depend on electricity. I am aware that this view is not that of a practical engineer, but neither is it that of an illusionist, for it is certain, that power transmission, which at present is merely a stimulus to enterprise, will some day be a dire necessity.

It is more important for the student, who takes up the study of light phenomena, to make himself thoroughly acquainted with certain modern views, than to peruse entire books on the subject of light itself, as disconnected from these views. Were I therefore to make these demonstrations before students seeking information — and for the sake of the few of those who may be present, give me leave to so assume — it would be my principal endeavor to impress these views upon their minds in this series of experiments.

It might be sufficient for this purpose to perform a simple and well-known experiment. I might take a familiar appliance, a Leyden jar, charge it from a frictional machine, and then discharge it. In explaining to you its permanent state when charged, and its transitory condition when discharging, calling your attention to the forces which enter into play and to the various phenomena they produce, and pointing out the relation of the forces and phenomena, I might fully succeed in illustrating that modern idea. No doubt, to the thinker, this simple experiment would appeal as much as the most magnificent display. But this is to be an experimental demonstration, and one which should possess, besides instructive, also entertaining features and as such, a simple experiment, such as the one cited, would not go very far towards the attainment of the lecturer's aim. I must therefore choose another way of illustrating, more spectacular certainly, but perhaps also more instructive. Instead of the frictional machine and Leyden jar, I shall avail myself in these experiments of an induction coil of peculiar properties, which was described in detail by me in a lecture before the London Institution of Electrical Engineers, in Feb., 1892. This induction coil is capable of yielding currents of enormous potential differences, alternating with extreme rapidity. With this apparatus I shall endeavor to show you three distinct classes of effects, or phenomena, and it is my desire that each experiment, while serving for the purposes of illustration, should at the same time teach us some novel truth, or show us some novel aspect of this fascinating science. But before doing this, it seems proper and useful to dwell upon the apparatus employed, and method of obtaining the high potentials and high-frequency currents which are made use of in these experiments.

On the apparatus and method of conversion

These high-frequency currents are obtained in a peculiar manner. The method employed was advanced by me about two years ago in an experimental lecture before the *American Institute of Electrical Engineers*. A number of ways, as practiced in the laboratory, of obtaining these currents either from continuous or low frequency alternating currents, is diagrammatically indicated in Fig. 1/165, which will be later described in detail. The general plan is to charge condensers, from a direct or alternate-current source, preferably of high-tension, and to discharge them disruptively while observing well-known conditions necessary to maintain the oscillations of the current. In view of the general interest taken in high-frequency currents and effects producible by them, it seems to me advisable to dwell at some length upon this method of conversion. In order to give you a clear idea of the action, I will suppose that a continuous-current generator is employed, which is often very convenient. It is desirable that the generator should possess such high tension as to be able to break through a small air space. If this is not the case, then auxiliary means have to be resorted to, some of which will be indicated subsequently. When the condensers are charged to a certain potential, the air, or insulating space, gives way and a disruptive discharge: occurs. There is then a sudden rush of current and generally a large portion of accumulated electrical energy spends itself. The condensers are thereupon quickly charged and the same process is repeated in more or less rapid succession. To produce such sudden rushes of current it is necessary to observe certain conditions. If the rate at which the condensers are discharged is the same as that at which they are charged, then, clearly, in the assumed case the condensers do not come into play. If the rate of discharge be smaller than the rate of charging, then, again, the condensers cannot play an important part. But if, on the contrary, the rate of discharging is greater than that of charging, then a succession of rushes of current is obtained. It is evident that, if the rate at which the energy is dissipated by the discharge is very much greater than the rate of supply to the condensers, the sudden rushes will be comparatively few, with long-time intervals between. This always occurs when a condenser of considerable capacity is charged by means of a comparatively small machine. If the rates of supply and dissipation are not widely different, then the rushes of current will be in quicker succession, and this the more, the more nearly equal both the rates are, until limitations incident to each case and depending upon a number of causes are reached. Thus we are able to obtain from a continuous-current generator as rapid a succession of discharges as we like. Of course, the higher the tension of the generator, the smaller need be the capacity of the condensers, and for this reason, principally, it is of advantage to employ a generator of very high tension. Besides, such a generator permits the attaining of greater rates of vibration.

The rushes of current may be of the same direction under the conditions before assumed, but most generally there is an oscillation superimposed upon the fundamental vibration of the current. When the conditions are so determined that there are no oscillations, the current impulses are unidirectional and thus a means is provided of transforming a continuous current of high tension, into a direct current of lower tension, which I think may find employment in the arts.

This method of conversion is exceedingly interesting and I was much impressed by its beauty when I first conceived it. It is ideal in certain respects. It involves the employment of no mechanical devices of any kind, and it allows of obtaining currents of any desired frequency from an ordinary circuit, direct or alternating. The frequency of the fundamental discharges depending on the relative rates of supply and dissipation can be readily varied within wide limits, by simple adjustments of these quantities, and the frequency of the superimposed vibration by the determination of the capacity, self-induction and resistance of the circuit. The potential of the currents, again, may be raised as high as any insulation is capable of withstanding safely by combining capacity and self-induction or by induction in a secondary, which need have but comparatively few turns.

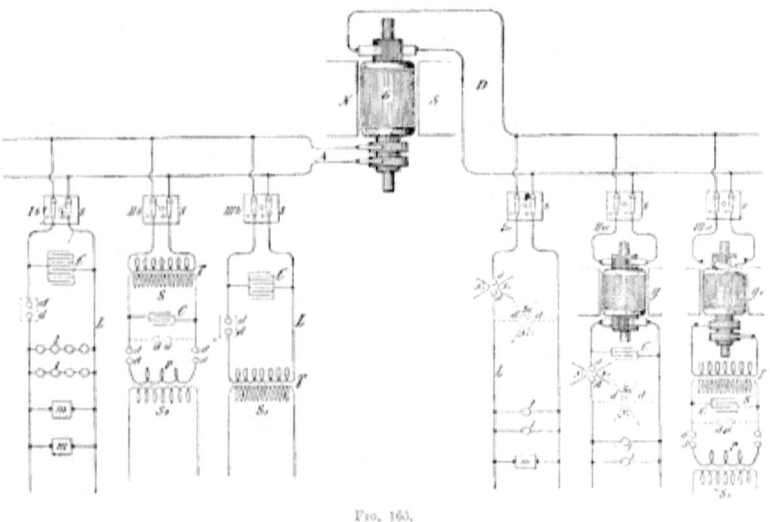

FIG. 165.

As the conditions are often such that the intermittence or oscillation does not readily establish itself, especially when a direct current source is employed, it is of advantage to associate an interrupter with the arc, as I have, some time ago, indicated the use of an air-blast or magnet, or other such device readily at hind. The magnet is employed with special advantage in the conversion of direct currents, as it is then very effective. If the primary source is an alternate current generator. it is desirable, as I have stated on another occasion, that the frequency should be low, and that tile current forming the arc be large, in order to render the magnet more effective.

A form of such discharger with a magnet which has been found convenient, and adopted after some trials, in the conversion of direct currents particularly, is illustrated in Fig. 166. N S are the pole pieces of a very strong magnet which is excited by a coil c. The pole pieces are slotted for adjustment and can be fastened in any position by screws s s_1. The discharge rods d d_1, thinned down on the ends in order to allow a closer approach of the magnetic pole pieces, pass through the columns of brass b b_1 and are fastened in position by screws s_2 s_2. Springs r r_1 and collars c c_1 are slipped on the rods, the latter serving to set the points of the rods at a certain suitable distance by means of screws s_3 s_3 and the former to draw the points apart. When it is desired to start the arc, one of the large rubber handles h h_1 is tapped quickly with the hand, whereby the points of the rods are brought in contact but are instantly separated by the springs r r_1. Such an arrangement has been found to be often necessary, namely in cases when the E. M. F. was not large enough to cause the discharge to break through the gap, and also when it was desirable to avoid short circuiting of the generator by the metallic contact of the rods. The rapidity of the interruptions of the current with a magnet depends on the intensity of the magnetic field and on the potential difference at the end of the arc. The interruptions are generally in such quick succession as to produce a musical sound. Years ago it was observed that when a powerful induction coil is discharged between the poles of a strong magnet, the discharge produces a loud noise not unlike a small pistol shot. It was vaguely stated that the spark was intensified by the presence of the magnetic field. It is now clear that the discharge current, flowing for some time, was interrupted a great number of times by the magnet, thus producing the sound. The phenomenon is especially marked when the field circuit of a large magnet or dynamo is broken in a powerful magnetic field.

When the current through the gap is comparatively large, it is of advantage to slip on the points of the discharge rods pieces of very hard carbon and let the arc play between the carbon pieces. This preserves the rods, and besides has the advantage of keeping the air space hotter, as the heat is not conducted away as quickly through the carbons, and the result is that a smaller E. M. F. in the arc gap is required to maintain a succession of discharges.

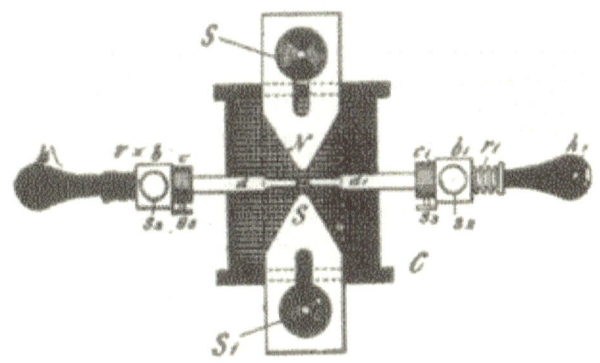

Fig. 166

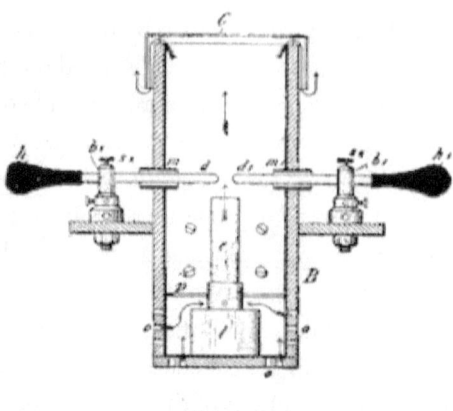

Fig. 167.

Another form of discharger, which may be employed with advantage in some cases, is illustrated in Fig. 167. In this form the discharge rods d d_1 pass through perforations in a wooden box B, which is thickly coated with mica on the inside, as indicated by the heavy lines. The perforations are provided with mica tubes m m_1 of some thickness, which are preferably not in contact with the rods d d_1. The box has a cover c which is a little larger and descends on the outside of the box. The spark gap is warmed by a small lamp l contained in the box. A plate p above the lamp allows the draught to pass only through the chimney a of the lamp, the air entering through holes o o in or near the bottom of the box and following the path indicated by the arrows. When the discharger is in operation, the door of the box is closed so that the light of the arc is not visible outside. It is desirable to exclude the light as perfectly as possible, as it interferes with some experiments. This form of discharger is simple and very effective when properly manipulated. The air being warmed to a certain temperature, has its insulating power impaired; it becomes dielectrically weak, as it were, and the consequence is that the arc can be established at much greater distance. The arc should, of course, be sufficiently insulating to allow the discharge to pass through the gap disruptively. The arc formed under such conditions, when long, may be made extremely sensitive, and the weak draught through the lamp chimney a is quite sufficient to produce rapid interruptions. The adjustment is made by regulating the temperature and velocity of the draught. Instead of using the lamp, it answers the purpose to provide for a draught of warm air in other ways. A very simple way which has been practiced is to enclose the arc in a long vertical tube, with plates on the top and bottom for regulating the temperature and velocity of the air current. Some provision had to be made for deadening the sound.

The air may be rendered dielectrically weak also by rarefaction. Dischargers of this kind have likewise been used by me in connection with a magnet. A large tube is for this purpose provided with heavy electrodes of carbon or metal, between which the discharge is made to pass, the tube being placed in a powerful magnetic field The exhaustion of the tube is carried to a point at which the discharge breaks through easily, but the pressure should be more than 75 millimetres, at which the ordinary thread discharge occurs. In another form of discharger, combining the features before mentioned, the discharge was made to pass between two adjustable magnetic pole pieces, the space between them being kept at an elevated temperature.

It should be remarked here that when such, or interrupting devices of any kind, are used and the currents are passed through the primary of a disruptive discharge coil, it is not, as a rule, of advantage to produce a number of interruptions of the current per second greater than the natural frequency of vibration of the dynamo supply circuit, which is ordinarily small. It should also be pointed out here, that while the devices mentioned in connection with the disruptive discharge are advantageous under certain conditions, they may be sometimes a source of trouble, as they produce intermittences and other irregularities in the vibration which it would be very desirable to overcome.

There is, I regret to say, in this beautiful method of conversion a defect, which fortunately is not vital, .and which I have been gradually overcoming. I will best call attention to this defect and indicate a fruitful line of work, by comparing the electrical process with its mechanical analogue. The process may be illustrated in this manner. Imagine a tank; with a wide opening at the bottom, which is kept closed by spring pressure, but so that it snaps off sudden/y when the liquid in the tank has reached a certain height. Let the fluid be supplied to the tank by means of a pipe feeding at a certain rate. When the critical height of the liquid is reached, the spring gives way and the bottom of the tank drops out. Instantly the liquid falls through the wide opening, and the spring, reasserting itself, closes the bottom again. The tank is now filled, and after a certain time interval the same process is repeated. It is clear, that if the pipe feeds the fluid quicker than the bottom outlet is capable of letting it pass through, the bottom will remain off and the tank; will still overflow. If the rates of supply are exactly equal, then the bottom lid will remain partially open and no vibration of the same and of the liquid column will generally occur, though it might, if started by some means. But if the inlet pipe does not feed the fluid fast enough for the outlet, then there will be always vibration. Again, in such case, each time the bottom flaps up or down, the spring and the liquid column, if the pliability of the spring and the inertia of the moving parts are properly chosen, will perform independent vibrations. In this analogue the fluid may be likened to electricity or electrical energy, the tank to the condenser, the spring to the dielectric, and the pipe to the conductor through which electricity is supplied to the condenser. To make this analogy quite complete it is necessary to make the assumption, that the bottom, each time it gives way, is knocked violently against a non-elastic stop, this impact involving some loss of energy; and that, besides, some dissipation of energy results due to frictional losses. In the preceding analogue the liquid is supposed to be under a steady pressure. If the presence of the fluid be assumed to vary rhythmically, this may be taken as corresponding to the case of an alternating current. The process is then not quite as simple to consider, but the action is the same in principle.

It is desirable, in order to maintain the vibration economically, to reduce the impact and frictional losses as much as possible. As regards the latter, which in the electrical analogue correspond to the losses due to the resistance of the circuits, it is impossible to obviate them entirely, but they can be reduced to a minimum by a proper selection of the dimensions of the circuits and by the employment of thin conductors in the form of strands. But the loss of energy caused by the first breaking through of the dielectric—which in the above example corresponds to the violent knock of the bottom against the inelastic stop—would be more important to overcome. At the moment of the breaking through, the air space has a very high resistance, which is probably reduced to a very small value when the current has reached some strength, and the space is brought to a high temperature. It would materially diminish the loss of energy if the space were always kept at an extremely high temperature, but then there would be no disruptive break. By warming the space moderately by means of a lamp or otherwise, the economy as far as the arc is concerned is sensibly increased. But the magnet or other interrupting device does not diminish the loss in the arc. Likewise, a jet of air only facilitates the carrying off of the energy. Air, or a gas in general, behaves curiously in this respect. When two bodies charged to a very high potential, discharge disruptively through an air space, any amount of energy may be carried off by the air. This energy is evidently dissipated by bodily carriers, in impact and collisional losses of the molecules. The exchange of the molecules in the space occurs with inconceivable rapidity. A powerful discharge taking place between two electrodes, they may remain entirely cool, and yet the loss in the air may represent any amount of energy. It is perfectly practicable, with very great potential differences in the gap, to dissipate several horse-power in the arc of the discharge without even noticing a small increase in the temperature of the electrodes. All the frictional losses occur then practically in the air. If the exchange of the air molecules is prevented, as by enclosing the air hermetically, the bas inside of the vessel is brought quickly to a high temperature, even with a very small discharge. It is difficult to estimate how much of the energy is lost in sound waves, audible or not, in a powerful discharge. When the currents through the gap are large, the electrodes may become rapidly heated, but this is not a reliable measure of the energy wasted in the arc, as the loss through the yap itself may be comparatively small. The air or a gas in general is at ordinary pressure at least, clearly not the best medium through which a disruptive discharge should occur. Air or other gas under great pressure is of curse a much more suitable medium for the discharge gap. I have carried on long-continued experiments in this direction, unfortunately less practicable on account of the difficulties and expense in getting air under great pressure. But even if the medium in the discharge space is solid or liquid, still the same losses take place, though they are generally smaller, for just as soon as the arc is established, the solid or liquid is volatilized. Indeed, there is no body known which would not be disintegrated by the arc, and it is an open question among scientific men, whether an arc discharge could occur at all in the air itself without the particles of the electrodes being torn off. When the current through the gap is very small and the arc very long, I believe that a relatively considerable amount of heat is taken up in the disintegration of the electrodes, which partially on this account may remain quite cold.

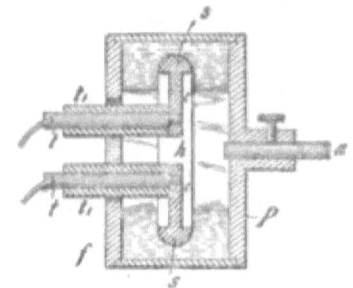 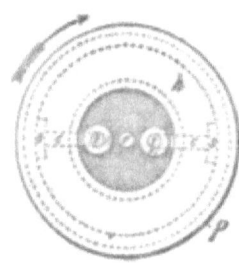

Fig. 168a. Fig. 168b.

The ideal medium for a discharge gap should only crack, and the ideal electrode should be of some material which cannot be disintegrated. With small currents through the gap it is best to employ aluminum, but not when the currents are large. The disruptive break in the air, or more or less in any ordinary medium, is not of the nature of a crack, but it is rather comparable to the piercing of innumerable bullets through a mass offering great frictional resistances to the motion of the bullets, this involving considerable loss of energy. A medium which would merely crack when strained electrostatically—and this possibly might be the case with a perfect vacuum, that is, pure ether—would involve a very small loss in the gap, so small as to be entirely negligible, at least theoretically, because a crack may be produced by an infinitely small displacement. In exhausting an oblong bulb provided with two aluminum terminals, with the greatest care, I have succeeded in producing such a vacuum that the secondary discharge of a disruptive discharge coil would break disruptively through the bulb in the form of fine spark streams. The curious point was that the discharge would completely ignore the terminals and start far behind the two aluminum plates which served as electrodes. This extraordinary high vacuum could only be maintained for a very short while. To return to the ideal medium; think, for the sake of illustration, of a piece of glass or similar body clamped in a vice, and the latter tightened more and more. At a certain point a minute increase of the pressure will cause the ;glass to crack. The loss of energy involved in splitting the glass may be practically nothing, for though the force is great, the displacement need be but extremely small. Now imagine that the glass would possess the property of closing again perfectly the crack upon a minute diminution of the pressure. This is the way the dielectric in the discharge space should behave. But inasmuch as there would be always some loss in the gap, the medium, which should be continuous should exchange through the gap at a rapid rate. In the preceding example, the glass being perfectly closed, it would mean that the dielectric in the discharge space possesses a great insulating power; the glass being cracked, it would signify that the medium in the space is a good conductor. The dielectric should vary enormously in resistance by minute variations of the E. M. F. across the discharge space. This condition is attained, but in an extremely imperfect manner, by warming the air space to a certain critical temperature, dependent on the E. M. F. across the gap, or by otherwise impairing the insulating power of the air. But as a matter of fact the air does never break down disruptively, if this term be rigorously interpreted, for before the sudden rush of the current occurs, there is always a weak current preceding it, which rises first gradually and then with comparative suddenness. That is the reason why the rate of change is very much greater when glass, for instance, is broken through, than when the break takes place through an air space of equivalent dielectric strength. As a medium for the discharge space, a solid, or even a liquid, would be preferable therefore. It is somewhat difficult to conceive of a solid body which would possess the property of closing instantly after it has been cracked. But a liquid, especially under great pressure, behaves practically like a solid, while it possesses the property of closing the crack. Hence it was thought that a liquid insulator might be more suitable as a dielectric than air. Following out this idea, a number of different forms of dischargers in which a variety of such insulators, sometimes under great pressure, were employed, have been experimented upon. It is thought sufficient to dwell in a few words upon one of the forms experimented upon. One of these dischargers is illustrated in Figs. 168a and 168b.

A hollow metal pulley P (Fig. 168a), was fastened upon an arbor a, which by suitable means was rotated at a considerable speed. On the inside of the pulley, but disconnected from the same, was supported a thin disc h (which is shown thick for the sake of clearness), of hard rubber in which there were embedded two metal segments s s with metallic extensions e e into which were screwed conducting terminals t t covered with thick tubes of hard rubber t t. The rubber disc b with its metallic segments s s, was finished in a lathe, and its entire surface highly polished so as to offer the smallest possible frictional resistance to the motion through a fluid. In the hollow of the pulley an insulating liquid such as a thin oil was poured so as to reach very nearly to the opening left in the flange f, which was screwed tightly on the front side of the pulley. The terminals t t, were connected to the opposite coatings of a battery of condensers so that the discharge occurred through the liquid. When the pulley was rotated, the liquid was forced against the rim of the pulley and considerable fluid pressure resulted. In this simple way the discharge gap was filled with a medium which behaved practically like a solid, which possessed the duality of closing instantly upon the occurrence of the break, and which moreover was circulating through the gap at a rapid rate. Very powerful effects were produced by discharges of this kind with liquid interrupters, of which a number of different forms were made. It was found that, as expected, a longer spark for a given length of wire was obtainable in this way than by using air as an interrupting device. Generally the speed, and therefore also the fluid pressure, was limited by reason of the fluid friction, in the form of discharger described, but the practically obtainable speed was more than sufficient to produce a number of breaks suitable for the circuits ordinarily used. In such instances the metal pulley P was provided with a few projections inwardly, and a definite number of breaks was then produced which could be computed from the speed of rotation of the pulley. Experiments were also carried on with liquids of different insulating power with the view of reducing the loss in the arc. When an insulating liquid is moderately warmed, the loss in the arc is diminished.

A point of some importance was noted in experiments with various discharges of this kind. It was found, for instance, that whereas the conditions maintained in these forms were favorable for the production of a great spark length, the current so obtained was not best suited to the production of light effects. Experience undoubtedly has shown, that for such purposes a harmonic rise and fall of the potential is preferable. Be it that a solid is rendered incandescent, or phosphorescent, or be it that energy is transmitted by condenser coating through the glass, it is quite certain that a harmonically rising and falling potential produces less destructive action, and that the vacuum is more permanently maintained. This would be easily explained if it were ascertained that the process going on in an exhausted vessel is of an electrolytic nature.

In the diagrammatical sketch, Fig. 165, which has been already referred to, the cases which are most likely to be met with in practice are illustrated. One has at his disposal either direct or alternating currents from a supply station. It is convenient for an experimenter in an isolated laboratory to employ a machine G, such as illustrated, capable of giving both kinds of currents. In such case it is also preferable to use a machine with multiple circuits, as in many experiments it is useful and convenient to have at one's disposal currents of different phases. In the sketch, D represents the direct and A the alternating circuit. In each of these, three branch circuits are shown, all of which are provided with double line switches s s s s s s. Consider first the direct current conversion; la represents the simplest case. If the E. M. F. of the generator is sufficient to break through a small air space, at least when the latter is warmed or otherwise rendered poorly insulating, there is no difficulty in maintaining a vibration with fair economy by judicious adjustment of the capacity, self-induction and resistance of the circuit L. containing the devices l l m. The magnet N, S, can be in this case advantageously combined with the air space, The discharger d d with the magnet may be placed either way, as indicated by the full or by the dotted lines. The circuit la with the connections and devices is supposed to possess dimensions such as are suitable for the maintenance of a vibration. But usually the E. M. F. on the circuit or branch la will be something like a 100 volts or so, and in this case it is not sufficient to break through the gap. Many different means may be used to remedy this by raising the E. M. F. across the gap. The simplest is probably to insert a large self-induction coil in series with the circuit L. When the arc is established, as by the discharger illustrated in Fig. 166, the magnet blows the arc out the instant it is formed. Now the extra current of the break, being of high E. M. F., breaks through the gap, and a path of low resistance for the dynamo current being again provided, there is a sudden rush of current from the dynamo upon the weakening or subsidence of the extra current. This process is repeated in rapid succession, and in this manner I have maintained oscillation with as low as 50 volts, or even less, across the gap. But conversion on this plan is not to be recommended on account of the too heavy currents through the gap and consequent heating of the electrodes; besides, the frequencies obtained in this way are low, owing to the high self-induction necessarily associated with the circuit. It is very desirable to have the E. M. F. as high as possible, first, in order to increase the economy of the conversion, and secondly, to obtain high frequencies. The difference of potential in this electric oscillation is, of course, the equivalent of the stretching force in the mechanical vibration of the spring. To obtain very rapid vibration in a circuit of some inertia, a great stretching force or difference of potential is necessary. Incidentally, when the E. M. F. is very great, the condenser which is usually employed in connection with the circuit need but have a small capacity, and many other advantages are gained. With a view of raising the E. M. F. to a many times greater value than obtainable from ordinary distribution circuits, a rotating transformer g is used, as indicated at IIa, Fig. 1, or else a separate high potential machine is driven by means of a motor operated from the generator G. The latter plan is in fact preferable, as changes are easier made. The connections from the high tension winding are quite similar to those in branch Ia with the exception that a condenser C, which should be adjustable, is connected to the high tension circuit. Usually, also, an adjustable self-induction coil in series with the circuit has been employed in these experiments. When the tension of the currents is very high, the magnet ordinarily used in connection with the discharger is of comparatively small value, as it is quite easy to adjust the dimensions of the circuit so that oscillation is maintained. The employment of a steady E. M. F. in the high frequency conversion affords some advantages over the employment of alternating E. M. F., as the adjustments are much simpler and the action can be easier controlled. But unfortunately one is limited by the obtainable potential difference. The winding also breaks down easily in consequence of the sparks which form between the sections of the armature or commutator when a vigorous oscillation takes place. Besides, these transformers are expensive to build. It has been found by experience that it is best to follow the plan illustrated at Met, In this arrangement a rotating transformer g, is employed to convert the low tension direct currents into low frequency alternating currents, preferably also of small tension. The tension of the currents is then raised in a stationary transformer T. The secondary s of this transformer is connected to an adjustable condenser C which discharges through the gap or discharger d d, placed in either of the ways indicated, through the primary P of a disruptive discharge coil, the high frequency current being obtained from the secondary s of this coil, as described on previous occasions. This will undoubtedly be found the cheapest and most convenient way of converting direct currents.

The three branches of the circuit A represent the usual cases met in practice when alternating currents are converted. In Fig. 165, Ib a condenser C, generally of large capacity, is connected to the circuit L containing the devices l l, m m. The devices m m are supposed to be of high self-induction so as to bring the frequency of the circuit more or less to that of the dynamo. In this instance the discharger d d should best have a number of makes and breaks per second equal to twice the frequency of the dynamo. If not so, then it should have at least a number equal to a multiple or even fraction of the dynamo frequency. It should be observed, referring to Ib, that the conversion to a high potential is also effected when the discharger d d, which is shown in the sketch, is omitted. But the effects which are produced by currents which rise instantly to high values, as in a disruptive discharge, are entirely different from those produced by dynamo currents which rise and fall harmonically. So, for instance, there might be in a given case a number of makes and breaks at d d equal to just twice the frequency of the dynamo, or in other words, there may be the same number of fundamental oscillations as would be produced without the discharge gap, and there might even not be any quicker superimposed vibration; yet the differences of potential at the various points of the circuit, the impedance and other phenomena, dependent upon the rate of change, will bear no similarity in the two cases. Thus, when working with currents discharging disruptively, the element chiefly to be considered is not the frequency, as a student might be apt to believe, but the rate of change per unit of time. With low frequencies in a certain measure the same effects may be obtained as with high frequencies, provided the rate of change is sufficiently great. So if a low frequency current is raised to a potential of, say, 75,000 volts, and the high tension current passed through a series of high resistance lamp filaments, the importance of the rarefied gas surrounding the filament is clearly noted, as will be seen later; or, if a low frequency current of several thousand amperes is passed through a metal bar, striking phenomena of impedance are observed, just as with currents of high frequencies. But it is, of course, evident that with low frequency currents it is impossible to obtain such rates of change per unit of time as with high frequencies, hence the effects produced by the latter are much more prominent. It is deemed advisable to make the preceding remarks, inasmuch as many more recently described effects have been unwittingly identified with high frequencies. Frequency alone in reality does not mean anything, except when an undisturbed harmonic oscillation is considered.

In the branch IIIb a similar disposition to that in Ib is illustrated, with the difference that the currents discharging through the gap d d are used to induce currents in the secondary s of a transformer T. In such case the secondary should be provided with an adjustable condenser for the purpose of tuning it to the primary.

IIb illustrates a plan of alternate current high frequency conversion which is most frequently used and which is found to be most convenient. This plan has been dwelt upon in detail on previous occasions and need not be described here.

Some of these results were obtained by the use of a high frequency alternator. A description of such machines will be found in my original paper before the *American Institute of Electrical Engineers*, and in periodicals of that period, notably in The Electrical Engineer of March 18, 1891.

I will now proceed with the experiments.

On phenomena produced by electrostatic force

The first class of effects I intend to show you are effects produced by electrostatic force. It is the force which governs the motion of the atoms, which causes them to collide and develop the life-sustaining energy of heat and light, and which causes them to aggregate in an infinite variety of ways, according to Nature's fanciful designs, and to form all these wondrous structures we perceive around us; it is, in fact, if our present views be true, the most important force for us to consider in. Nature. As the term electrostatic might imply a steady electric condition, it should be remarked, that in these experiments the force is not constant, but varies at a rate which may be considered moderate, about one million times a second, or thereabouts. This enables me to produce many effects which are not producible with an unvarying force.

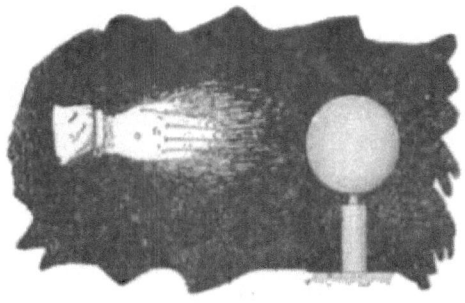

Fig. 169.

When two conducting bodies are insulated and electrified, we say that an electrostatic force is acting between them. This force manifests itself in attractions, repulsions and stresses in the bodies and space or medium without. So great may be the strain exerted in the air, or whatever separates the two conducting bodies, that it may break down, and we observe sparks or bundles of light or streamers, as they are called. These streamers form abundantly when the force through the air is rapidly varying. I will illustrate this action of electrostatic force in a novel experiment in which I will employ the induction coil before referred to. The coil is contained in a trough filled with oil, and placed under the table. The two ends of .the secondary wire pass through the two thick columns of hard rubber which protrude to solve height above the table. It is necessary to insulate the ends or terminals of the secondary heavily with hard rubber, because even dry wood is by far ton poor an insulator for these currents of enormous potential differences. On one of the terminals of the coil, I ,have placed a large sphere of sheet brass, which is connected to a larger insulated brass plate, in order to enable me to perform the experiments under conditions, which, as you will see, are more suitable for this experiment. I now set the coil to work and approach the free terminal with a metallic object held in my hand, this simply to avoid burns As I approach the metallic object to a distance of eight or tell inches, a torrent of furious sparks breaks forth from the end of the secondary wire, which passes through the rubber column. The sparks cease when the metal in my hand touches the wire. My arm is now traversed by a powerful electric current, vibrating at about the rate of one million times a second. All around me the electrostatic force makes itself felt, and the air molecules and particles of dust flying about are acted upon and are hammering violently against my body. So great is this agitation of the particles, that when the lights are turned out you may see streams of feeble light appear on some parts of my body. When such a streamer breaks out on any part of the body, it produces a sensation like the pricking of a needle. Were the potentials sufficiently high and the frequency of the vibration rather low, the skin would probably be ruptured under the tremendous strain, and the blood would rush out with great force in the form of fine spray or jet so thin as to be invisible, just as oil will when placed on the positive terminal of a Holtz machine. The breaking through of the skin though it may seem impossible at first, would perhaps occur, by reason of the tissues finder the skin being incomparably better conducting. This, at least, appears plausible, judging from some observations.

I can make these streams of light visible to all, by touching with the metallic object one of the terminals as before, and approaching my free hand to the brass sphere, which is connected to the second terminal of the coil. As the hand is approached, the air between it and the sphere, or in the immediate neighborhood, is more violently agitated, and you see streams of light now break forth from my finger tips and from the whole hand (Fig. 169). Were I to approach the hand closer, powerful sparks would jump from the brass sphere to my hand, which might be injurious. The streamers offer no particular inconvenience, except that in the ends of the finger tips a burning sensation is felt. They should not be confounded with those produced by an influence machine, because in many respects they behave differently. I have attached the brass sphere and plate to one of the terminals in order to prevent the formation of visible streamers on that terminal, also in order to prevent sparks from jumping at a considerable distance. Besides, the attachment is favorable for the working of the coil.

The streams of light which you have observed issuing from my hand are due to a potential of about 200,000 volts, alternating in rather irregular intervals, sometimes like a million times a second. A vibration of the same amplitude, but four times as fast, to maintain which over 3,000,000 volts would be required, would be mare than sufficient to envelop my body in a complete sheet of flame. But this flame would not burn me up; quite contrarily, the probability is ,that I would not be injured in the least. Yet a hundredth part of that energy, otherwise directed; would be amply sufficient to kill a person.

The amount of energy which may thus be passed into the body of a person depends on the frequency and potential of the currents, and by making both of these very great, a vast amount of energy may be passed into the body without causing any discomfort, except perhaps, in the arm, which is traversed by a true conduction current. The reason why no pain in the body is felt, and no injurious effect noted, is that everywhere, if a current be imagined to flow through the body, the direction of its flow would be at right angles to the surface; hence the body of the experimenter offers an enormous section to the current, and the density is very small, with the exception of the arm, perhaps, where the density may be considerable. Lout if only a small fraction of that energy would be applied in such a way that a current would traverse the body in the same manner as a low frequency current, a shock would be received which might be fatal. A direct or low frequency alternating current is fatal, I think, principally because its distribution through the body is not uniform, as it must divide itself in minute streamlets of great density, whereby some organs are vitally injured. That such a process occurs I have not the least doubt, though no evidence might apparently exist, or be found upon examination. The surest to injure and destroy life, is a continuous current, but the most painful is an alternating current of very low frequency. The expression of these views, which are the result of long continued experiment and observation, both with steady and varying currents, is elicited by the interest which is at present taken in this subject, and by the manifestly erroneous ideas which are daily propounded in journals on this subject.

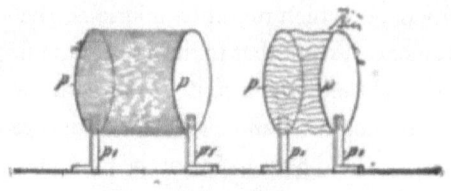

FIG. 170a.   FIG. 170b.

I may illustrate an effect of the electrostatic force by another striking experiment, but before, I must call your attention to one or two facts. I have said before, that whet; the medium between two oppositely electrified bodies is strained beyond a certain limit it gives way and, stated in popular language, the opposite electric charges unite and neutralize each other. This breaking down of the medium occurs principally when the force acting between the bodies is steady, or varies at a moderate rate. Were the variation sufficiently rapid, such a destructive break would not occur, no matter how great the force, for all the energy would be spent in radiation, convection and mechanical and chemical action. Thus the spark length, or greatest distance which a spark will jump between the electrified bodies is the smaller, the greater the variation or time rate of change. But this rule may be taken to be true only in a general way, when comparing rates which are widely different.

I will show you by an experiment the difference in the effect produced by a rapidly varying and a steady or moderately varying force. I have here two large circular brass plates p p (Fig. 170a and Fig. 170b), supported on movable insulating stands on the table, connected to the ends of the secondary of a coil similar to the one used before. I place the plates ten or twelve inches apart and set the coil to work. You see the whole space between the plates, nearly two cubic feet, filled with uniform light, Fig. 6a. This light is due to the streamers you have seen in the first experiment, which are now much more intense. I have already pointed out the importance of these streamers in commercial apparatus and their still greater importance in some purely scientific investigations, Often they are to weak to be visible, but they always exist, consuming energy and modifying the action of .the apparatus. When intense, as they are at present, they produce ozone in great quantity, and also, as Professor Crookes has pointed out, nitrous acid. So quick is the chemical action that if a coil, such as this one, is worked for a very long time it will make the atmosphere of a small room unbearable, for the eyes and throat are attacked. But when moderately produced, the streamers refresh the atmosphere wonderfully, like a thunder-storm, and exercises unquestionably a beneficial effect.

In this experiment the force acting between the plates changes in intensity and direction at a very rapid rate. I will now make the rate of change per unit time much smaller. This I effect by rendering the discharges through the primary of the induction coil less frequent, and also by diminishing the rapidity of the vibration in the secondary. The former result is conveniently secured by lowering the E. M. F. over the air gap in the primary circuit, the latter by approaching the two brass plates to a distance of about three or four inches. When the coil is set to work, you see no streamers or light between the plates, yet the medium between them is under a tremendous strain. I still further augment the strain by raising the E. M. F. in the primary circuit, and soon you see the air give away and the hall is illuminated by a shower of brilliant and noisy sparks, Fig. 170b. These sparks could be produced also with unvarying force; they have been for many years a familiar phenomenon, though they were usually obtained from an entirely different apparatus. In describing these two phenomena so radically different in appearance, I have advisedly spoken of a "force" acting between the plates. It would be in accordance with the accepted views to say, that there was an "alternating E. M. F.", acting between the plates. This term is quite proper and applicable in all cases where there is evidence of at least a possibility of an essential inter-dependence of the electric state of the plates, or electric action in their neighborhood. But if the plates were removed to an infinite distance, or if at a finite distance, there is no probability or necessity whatever for such dependence. I prefer to use the term "electrostatic force," and to say that such a force is acting around each plate or electrified insulated body in general. There is an inconvenience in usin this expression as the term incidentally, means a steady electric condition; but a proper nomenclature will eventually settle this difficulty.

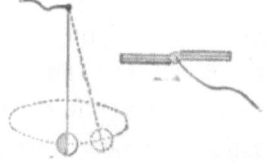

Fig. 171.   Fig. 172a.   Fig. 172b.

I now return to the experiment to which I have already alluded, and with which I desire to illustrate a striking effect produced by a rapidly varying electrostatic force. I attach to the end of the wire, l (Fig. 171), which is in connection with one of the terminals of the secondary of the induction coil, an exhausted bulb b. This bulb contains a thin carbon filament f, which is fastened to a platinum wire w, sealed in the glass and leading outside of the bulb, where it connects to the wire l. The bulb may be exhausted to any degree attainable with ordinary apparatus. Just a moment before, you have witnessed the breaking down of the air between the charged brass plates. You know that a plate of glass, or any other insulating material, would break down in like manner. Had I therefore a metallic coating attached to the outside of the bulb, or placed near the same, and were this coating connected to the other terminal of the coil, you would be prepared to see the glass give way if the strain were sufficiently increased. Even were the coating not connected to the other terminal, but to an insulated plate, still, if you have followed recent developments, you would naturally expect a rupture of the glass.

But it will certainly surprise you to note that under the action of the varying electrostatic force, the glass gives way when all other bodies are removed from the bulb. In fact, all the surrounding bodies we perceive might be removed to an infinite distance without affecting the result in the slightest. When the coil is set to work, the glass is invariably broken through at the seal, or other narrow channel, and the vacuum is quickly impaired. Such a damaging break would not occur with a steady force, even if the same were many times greater. The break is due to the agitation of the molecules of the gas within the bulb, and outside of the same. This agitation, which is generally most violent in the narrow pointed channel near the seal, causes a heating and rupture of the glass. This rupture would, however, not occur, not even with a varying force, if the medium filling the inside of the bulb, and that surrounding it, were perfectly homogeneous. The break occurs much quicker if the top of the bulb is drawn out into a fine fibre. In bulbs used with these coils such narrow, pointed channels must therefore be avoided.

When a conducting body is immersed in air, or similar insulating medium, consisting of, or containing, small freely movable particles capable of being electrified, and when the electrification of the body is made to undergo a very rapid change—which is equivalent to saying that the electrostatic force acting around the body is varying in intensity,—the small particles are attracted and repelled, and their violent impacts against the body may cause a mechanical motion of the latter. Phenomena of this kind are noteworthy, inasmuch as they have not been observed before with apparatus such as has been commonly in use. If a very light conducting sphere be suspended on an exceedingly fine wire, and charged to a steady potential, however high, the sphere will remain at rest. Even if the potential would be rapidly varying, provided that the small particles of matter, molecules or atoms, are evenly distributed, no motion of the sphere should result. But if one side of the conducting sphere is covered with a thick insulating layer, the impacts of the particles will cause the sphere to move about, generally in irregular curves, Fig. 172a. In like manner, as I have shown on a previous occasion, a fan of sheet metal, Fig. 172b, covered partially with insulating material as indicated, and placed upon the terminal of the coil so as to turn freely on it, is spun around.

All these phenomena you have witnessed and others which will be shown later, are due to the presence of a medium like air, and would not occur in a continuous medium. The action of the air may be illustrated still better by the following experiment. I take a glass tube t, Fig. 173, of about an inch in diameter, which has a platinum wire w sealed in the lower end, and to which is attached a thin lamp filament f. I connect the wire with the terminal of the coil and set the coil to work. The platinum wire is now electrified positively and negatively in rapid succession and the wire and air inside of the tube is rapidly heated by the impacts of the particles, which may be so violent as to render the filament incandescent. But if I pour oil in the tube, just as soon as the wire is covered with the oil, all action apparently ceases and there is no marked evidence of heating. The reason of this is that the oil is a practically continuous medium. The displacements in such a continuous medium are, with these frequencies, to all appearance incomparably smaller than in air, hence the work performed in such a medium is insignificant. But oil would behave very differently with frequencies many times as great, for even though the displacements be small, if the frequency were much greater, considerable work might be performed in the oil.

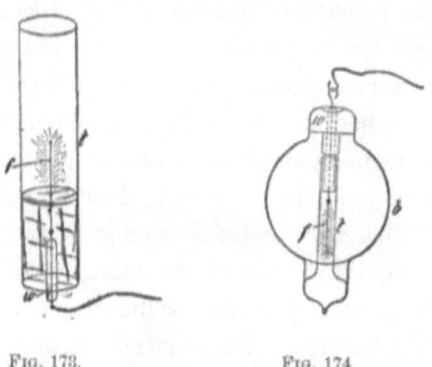

Fig. 173.   Fig. 174.

The electrostatic attractions and repulsions between bodies of measurable dimensions are, of all the manifestations of this force, the first so-called electrical phenomena noted. But though they have been known to us for many centuries, the precise nature of the mechanism concerned in these actions is still unknown to us, and has not been even quite satisfactorily explained. What kind of mechanism must that be? We cannot help wondering when we observe two magnets attracting and repelling each other with a force of hundreds of pounds with apparently nothing between them. We have in our commercial dynamos magnets capable of sustaining in mid-air tons of weight. But what are even these forces acting between magnets when compared with the tremendous attractions and repulsions produced by electrostatic force, to which there is apparently no limit as to intensity. In lightning discharges bodies are often charged to so high a potential that they are thrown away with inconceivable force and torn asunder or shattered into fragments. Still even such effects cannot compare with the attractions and repulsions which exist between charged molecules or atoms, and which are sufficient to project them with speeds of many kilometres a second, so that under their violent impact bodies are rendered highly incandescent and are volatilized. It is of special interest for the thinker who inquires into the nature of these forces to note that whereas the actions between individual molecules or atoms occur seemingly under any conditions, the attractions and repulsions of bodies of measurable dimensions imply a medium possessing insulating properties. So, if air; either by being rarefied or heated, is rendered more or less conducting, these actions between two electrified bodies practically cease, while the actions between the individual atoms continue to manifest themselves.

An experiment may serve as an illustration and as a means of bringing out other features of interest: Some time ago I showed that a lamp filament or wire mounted in a bulb and connected to one of the terminals of a high tension secondary coil is set spinning, the top of the filament generally describing a circle. This vibration was very energetic when the air in the bulb was at ordinary pressure and became less energetic when the air in the bulb was strongly compressed. It ceased altogether when the air was exhausted so as to become comparatively good conducting. I found at that time that no vibration tool: place when the bulb was very highly exhausted. But I conjectured that the vibration which I ascribed to the electrostatic action between the walls of the bulb and the filament should take place also in a highly exhausted bulb. To test this under conditions which were more favorable, a bulb like the one in Fig. 174, was constructed. It comprised a globe b, in the neck of which was sealed a platinum wire w, carrying a thin lamp filament f. In the lower part of the globe a tube t was sealed so as to surround the filament. The exhaustion was carried as far as it was practicable with the apparatus employed.

This bulb verified my expectation, for the filament was set spinning when the current was turned on, and became incandescent. It also showed another interesting feature, bearing upon the preceding remarks, namely, when the filament had been kept incandescent some time, the narrow tube and the space inside were brought to an elevated temperature, and as the gas in the tube then became conducting, the electrostatic attraction between the glass and the filament became very weak or ceased, and the filament came to rest, When it came to rest it would glow far more intensely. This was probably due to its assuming the position in the centre of the tube where the molecular bombardment was most intense, and also partly to the fact that the individual impacts were more violent and that no part of the supplied energy was converted into mechanical movement. Since, in accordance with accepted views, in this experiment the incandescence must be attributed to the impacts of the particles, molecules or atoms in tire heated space, these particles must therefore, in order to explain such action, be assumed to behave as independent carriers of electric charges immersed in an insulating medium; yet there is no attractive force between the glass tube and the filament because the space in the tube is, as a whole, conducting.

It is of some interest to observe in this connection that whereas the attraction between two electrified bodies may cease owing to the impairing of the insulating power of the medium in which they are immersed, the repulsion between the bodies may still be observed. This may be explained in a plausible way. When the bodies are placed at some distance in a poorly conducting medium, such as slightly warmed or rarefied air, and are suddenly electrified, opposite electric charges being imparted to them, these charges equalize more or less by leakage through the air. But if the bodies are similarly electrified, there is less opportunity afforded for such dissipation, hence the repulsion observed in such case is greater than the attraction. Repulsive actions in a gaseous medium are however, as Prof. Crookes has shown, enhanced by molecular bombardment.

On current or dynamic electricity phenomena

So far, I have considered principally effects produced by a varying electrostatic force in an insulating medium, such as air. When such a force is acting upon a conducting body of measurable dimensions, it causes within the same, or on its surface, displacements of the electricity, and gives rise to electric currents, and these produce another kind of phenomena, some of which I shall presently endeavor to illustrate. In presenting this second class of electrical effects, I will avail myself principally of such as are producible without any return circuit, hoping to interest you the more by presenting these phenomena in a more or less novel aspect.

It has been a long time customary, owing to .the limited experience with vibratory currents, to consider an electric current as something circulating in a closed conducting path. It was astonishing at first to realize that a current may flow through tile conducting path even if the latter be interrupted; and it was still more surprising to learn, that sometimes it may be even easier to make a current flow under such conditions than through a closed path. But that old idea is gradually disappearing, even among practical men, and will soon be entirely forgotten.

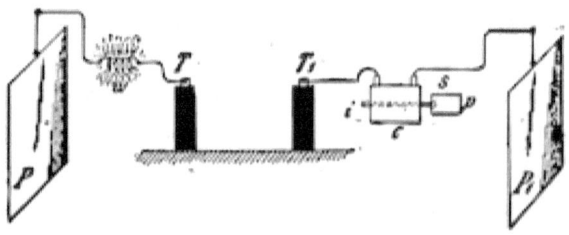

FIG. 175.

If I connect an insulated metal plate P, Fig. 175, to one of the terminals T of the induction coil by means of a wire, though this plate be very well insulated, a current passes through the wire when the coil is set to work. First I wish to give you evidence that there is a current passing through the connecting wire. An obvious way of demonstrating this is to insert between the terminal of the coil and the insulated plate a very thin platinum or german silver wire w and bring the latter to incandescence or fusion by the current. This requires a rather large plate or else current impulses of very high potential and frequency. Another way is to take a coil C, Fig. 175, containing many turns of thin insulated wire and to insert the same in the path of the current to the plate. When I connect one of the ends of the coil to the wire leading to another insulated plate P_1, and its other end to the terminal T_1 of the induction coil, and set the latter to work, a current passes through the inserted coil C and the existence of the current may be made manifest in various ways. For instance, I insert an iron core i within the coil. The current being one of very high frequency, will, if it be of some strength, soon bring the iron core to a noticeably higher temperature, as the hysteresis and current losses are great with such high frequencies. One might take a core of small size, laminated or not, it would matter little; but ordinary iron wire 1/16th or 1/8th of an inch thick is suitable for the purpose. While the induction coil is working, a current traverses the inserted coil and only a few moments are sufficient to bring the iron wire i to an elevated temperature sufficient to soften the sealing wax s and cause a paper washer p fastened by it to the iron wire to fall off. But with the apparatus such as I have here, other, much more interesting, demonstrations of this kind can be made. I have a secondary s, Fig. 176, of coarse wire, wound upon a coil similar to the first. In the preceding experiment the current through the coil C, Fig. 175, was very small, but there being many turns a strong heating effect was, nevertheless, produced in the iron wire. Had I passed that current through a conductor in order to show the heating of the latter, the current might have been too small to produce the effect desired. But with this coil provided with a secondary winding, I can now transform the feeble current of high tension which passes through the primary P into a strong secondary current of low tension, and this current will quite certainly do what I expect. In a small glass tube (t, Fig. 176), I have enclosed a coiled platinum wire, w, this merely in order to protect the wire. On each end of the glass tube is sealed a terminal of stout wire to which one of the ends of the platinum wire w, is connected. I join the terminals of the secondary coil to these terminals and insert the primary p, between the insulated plate P_1, and the terminal T_1, of the induction coil as before. The latter being set to work, instantly the platinum wire w is rendered incandescent and can be fused, even if it be very thick.

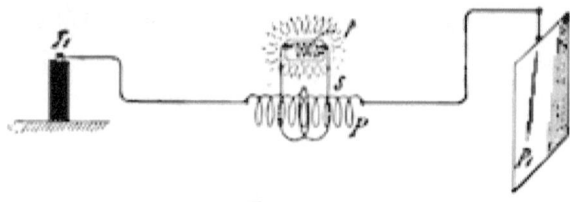

FIG. 176.

Instead of the platinum wire I now take an ordinary 50-volt 16 c. p. lamp. When I set the induction coil in operation the lamp filament is brought to high incandescence. It is, however, not necessary to use the insulated plate, for the lamp (l Fig. 177) is rendered incandescent even if the plate P_1 be disconnected. The secondary may also he connected to the primary as indicated by the dotted line in Fig. 177, to do away more or less with the electrostatic induction or to modify the action otherwise.

I may here call attention to a number of interesting observations with the lamp. First, I disconnect one of the terminals of the lamp from the secondary s. When the induction coil plays, a glow is noted which fills the whole bulb. This glow is due to electrostatic induction. It increases when the bulb is grasped with the hand, and the capacity of the experimenter's body thus added to the secondary circuit. The secondary, in effect, is equivalent to a metallic coating, which would be placed near the primary. If the secondary, or its equivalent, the coating, were placed symmetrically to the primary, the electrostatic induction would be nil under ordinary conditions, that is, when a primary return circuit is used, as both halves would neutralize each other. The secondary is in fact placed symmetrically to the primary, but the action of both halves of the latter, when only one of its ends is connected to the induction coil, is not exactly equal; hence electrostatic induction takes place, and hence the glow in the bulb. I can nearly equalize the action of both halves of the primary by connecting the other, free end of the same to the insulated plate, as in the preceding experiment. When the plate is connected, the glow disappears. With a smaller plate it would not entirely disappear and then it would contribute to the brightness of the filament when the secondary is closed, by warming the air in the bulb.

To demonstrate another interesting feature, I have adjusted the coils used in a certain way. I first connect both the terminals of the lamp to the secondary, one end of the primary being connected to the terminal T_1 of the induction coil and the other to tae insulated plate P, as before. When the current is turned on, the lamp glows brightly, as shown in Fig. 178b, in which C is a fine wire coil and s a coarse wire secondary wound upon it. If the insulated plate P₁ is disconnected, leaving one of the ends a of the primary insulated, the filament becomes dark or generally it diminishes in brightness (Fig. 178a). Connecting again the plate P1 and raising the frequency of the current, I make the filament quite dark or barely red (Fig. 178b).

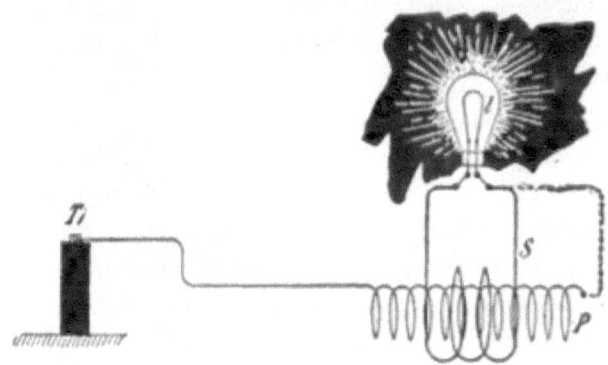

FIG. 177.

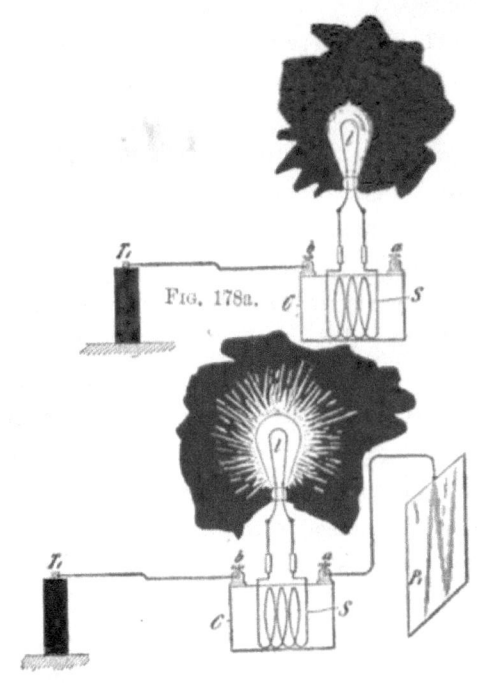

FIG. 178a.

FIG. 178b.

Once more I will disconnect the plate. One will of course infer that when the plate is disconnected, the current through the primary will be weakened, that therefore the E. M. F. will fall in the secondary .s and that the brightness of the lamp will diminish. This might be the case and the result can be secured by an easy adjustment of the coils; also by varying the frequency and potential of the currents. But it is perhaps of greater interest to note, that the lamp increases in brightness when the plate is disconnected (Fig. 179b). In this case all the energy the primary receives is now sunk info it, like the charge of a battery in an ocean cable, but most of that energy is recovered through the secondary and used to light the lamp. The current traversing the primary is strongest at the end b which is connected to the terminal T, of the induction coil, and diminishes in strength towards the remote end a. But the dynamic inductive effect exerted upon the secondary s is now greater than before, when the suspended plate was connected to the primary. These results might have been produced by a number of causes. For instance, the plate P_1 being connected, the reaction from the coil C may be such as to diminish the potential at the terminal T_1 of the induction coil, and therefore weaken the current through the primary of the coil C. Or the disconnecting of the plate may diminish the capacity effect with relation to the primary of the latter coil to such an extent that the current through it is diminished, though the potential at the terminal T_1 of the induction coil may be the same or even higher. Or the result might have been produced by the change of please of the primary and secondary currents and consequent reaction. But the chief determining factor is the relation of the self-induction and capacity of coil C and plate P_1 and the frequency of the currents. The greater brightness of the filament in Fig. 179a is, however, in part due to the heating of the rarefied gas in the lamp by electrostatic induction; which, as before remarked, is greater when the suspended plate is disconnected.

Still another feature of some interest I tray here bring to your attention. When the insulated date is disconnected and the secondary of the coil opened, by approaching a small object to the secondary, but very small sparks can be drawn from it, showing that the electrostatic induction is small in this case. But upon the secondary being closed upon itself or through the lamp, the filament, glowing brightly, strong sparks are obtained from the secondary. The electrostatic induction is now much greater, because the closed secondary determines a greater flow of current through the primary and principally through that half of it which is connected to the induction coil. If now the bulb be grasped with the hand, the capacity of the secondary with reference to the primary is augmented by the experimenter's body and the luminosity of the filament is increased, the Incandescence now being due partly to the flow of current through the filament and partly to the molecular bombardment of the rarefied gas in the bulb.

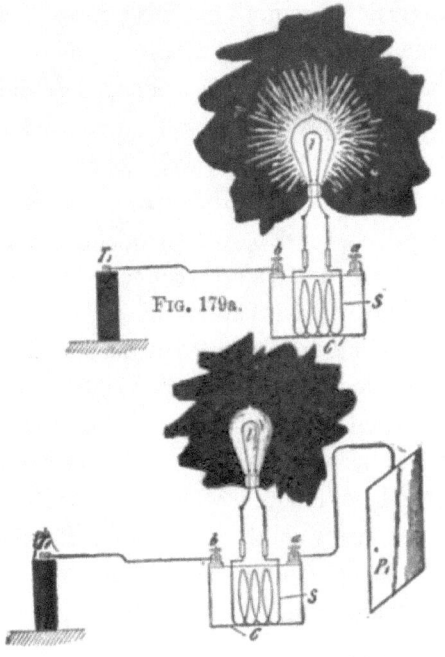

Fig. 179a.

Fig. 179b.

The preceding experiments will have prepared one for the next following results of interest, obtained in the course of these investigations. Since I can pass a current through an insulated wire merely by connecting one of its ends to the source of electrical energy, since I can induce by it another current, magnetize all iron core, and, in short, perform all operations as though a return circuit were used, clearly I can also drive a motor by the aid of only one wire. On a former occasion I have described a simple form of motor comprising a single exciting coil, an iron core and disc. Fig. 180 illustrates a modified way of operating such an alternate current motor by currents induced in a transformer connected to one lead, and several other arrangements of circuits for operating a certain class of alternating motors founded on the action of currents of differing phase. In view of the present state of the art it is thought sufficient to describe these arrangements in a few words only. The diagram, Fig. 180 II., shows a primary coil P, connected with one of its ends to the line L leading from a high tension transformer terminal T_1. In inductive relation to this primary P is a secondary s of coarse wire in the circuit of which is a coil C. The currents induced in the secondary energize the iron core i, which is preferably, but not necessarily, subdivided, and set the metal disc d in rotation. Such a motor M_2 as diagrammatically shown in Fig. 180 II., has been called a "magnetic lag motor", but this expression may be objected to by those who attribute the rotation of the disc to eddy currents circulating in minute paths when the core i is finally subdivided. In order to operate such a motor effectively on the plan indicated, the frequencies should not he too high, not more than four or five thousand, though the rotation is produced even with ten thousand per second, or more.

In Fig. 180 I., a motor M_1 having two energizing circuits, A and B, is diagrammatically indicated. The circuit A is connected to the line L and in series with it is a primary P, which may have its free end connected to an insulated plate P_1, such connection being indicated by the dotted lines. The other motor circuit B is connected to the secondary s which is in inductive relation to the primary P. When the transformer terminal T_1 is alternately electrified, currents traverse the open line L and also circuit A and primary P. The currents through the latter induce secondary currents in the circuit S, which pass through the energizing coil B of the motor. The currents through the secondary S and those through the primary P differ in phase 90 degrees, or nearly so, and are capable of rotating an armature placed in inductive relation to the circuits A and B.

In Fig. 180 III., a similar motor M_3 with two energizing circuits A_1 and B_1 is illustrated. A primary P, connected with one of its ends to the line L has a secondary S, which is preferably wound for a tolerably high E. M. F., and to which the two energizing circuits of the motor are connected, one directly to the ends of the secondary and the other through a condenser C, by the action of which the currents traversing the circuit A_1 and B_1 are made to differ in phase.

In Fig. 180 IV., still another arrangement is shown. In this case two primaries P_1 and P_2 are connected to the line L, one through a condenser C of small capacity, and the other directly. The primaries are provided with secondaries S_1 and S_2 which are in series with the energizing circuits, A_2 and B_2 and a motor M_3 the condenser C again serving to produce the requisite difference in the phase of the currents traversing the motor circuits. As such phase motors with two or more circuits are now well known in the art, they have been here illustrated diagrammatically. No difficulty whatever is found in operating a motor in the manner indicated, or in similar ways; and although such experiments up to this day present only scientific interest, they may at a period not far distant, be carried out with practical objects in view.

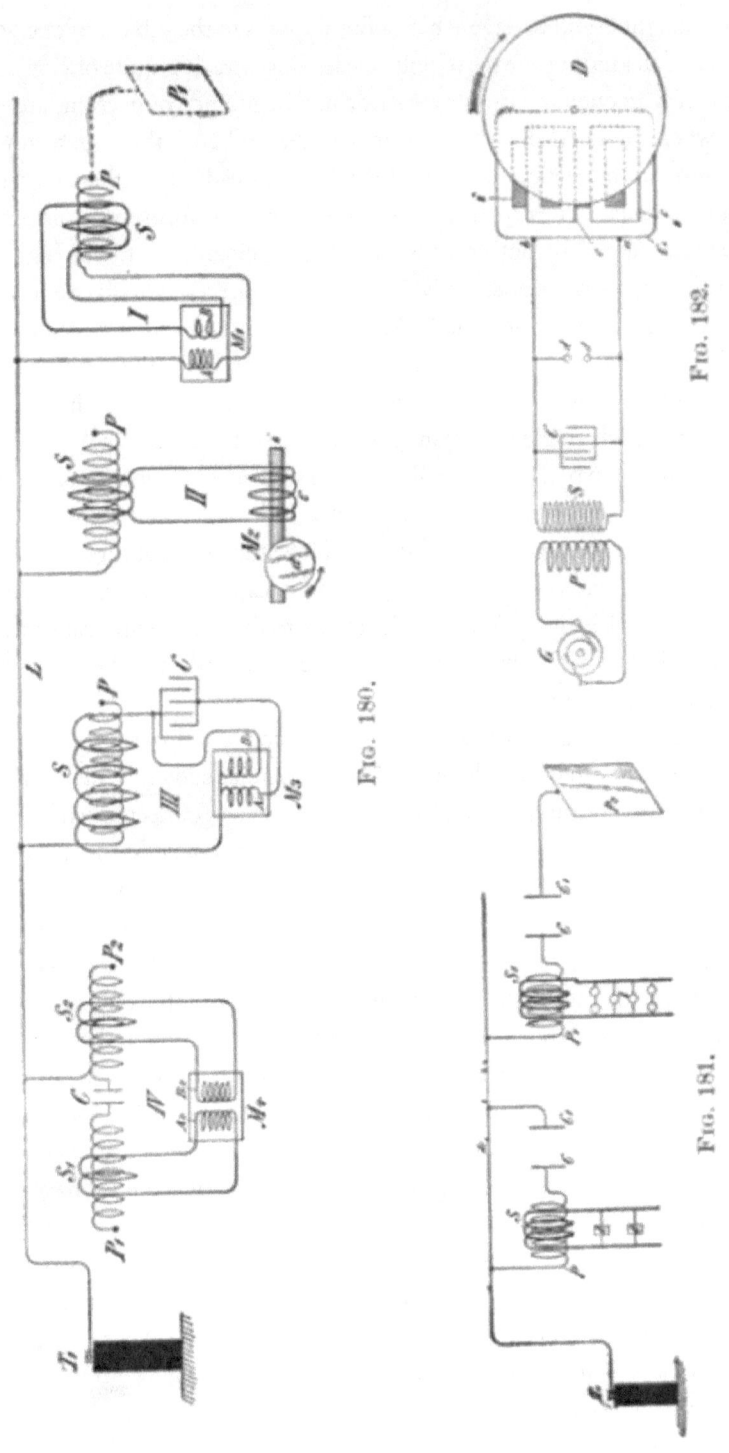

It is thought useful to devote here a few remarks to the subject of operating devices of all kinds by means of only one leading wire. It is quite obvious, that when high-frequency currents are made use of, ground connections are—at least when the E. M. F. of the currents is great—better than a return wire. Such ground connections are objectionable with steady or low frequency currents on account of destructive chemical actions of the former and disturbing influences exerted by both on the neighboring circuits; but with high frequencies these actions practically do not exist. Still, even ground connections become, superfluous when the E. M. F. is very high, for soon a condition is reached, when the current may be passed more economically through open, than through closed, conductors. Remote as might seem an industrial application of such single wire transmission of energy to one not experienced in such lines of experiment, it will not seem so to anyone who for some time has carried on investigations of such nature. Indeed I cannot see why such a plan, should not be practicable. Nor should it be thought that for carrying out such a plan currents of very high frequency are expressly required, for just as soon as potentials of say 30,000 volts are used, the single wire transmission may be effected with low frequencies, and experiments have been made by me from which these inferences are trade.

When the frequencies are very high it has been found in laboratory practice quite easy to regulate the effects in the manner shown in diagram Fig. 181. Here two primaries P and P_1 are shown, each connected with one of its ends to the line L and with the other end to the condenser plates C and C, respectively. Near these are placed other condenser plates C_1 and C_1, the former being connected to the line L and the latter to an insulated larger plate P_2. On the primaries are wound secondaries S and S_1, of coarse wire, connected to the devices d and l respectively. By varying the distances of the condenser plates C and C_1, and C and C_1 the currents through the secondaries S and S_1 are varied in intensity. The curious feature is the great sensitiveness, the slightest change in the distance of the plates producing considerable variations in the intensity or strength of the currents. The sensitiveness may be rendered extreme by making the frequency such, that the primary itself, without any plate attached to its free end, satisfies, in conjunction with the closed secondary, the condition of resonance. In such condition an extremely small change in the capacity of the free terminal produces great variations. For instance, I have been able to adjust the conditions so that the mere approach of a person to the coil produces a considerable change in the brightness of the lamps attached to the secondary. Such observations and experiments possess, of course, at present, chiefly scientific interest, but they may soon become of practical importance.

Very high frequencies are of course not practicable with motors on account of the necessity of employing iron cores. But one may use sudden discharges of low frequency and thus obtain certain advantages of high-frequency currents without rendering the iron core entirely incapable of following the changes and without entailing a very great expenditure of energy in the core. I have found it quite practicable to operate with such low frequency disruptive discharges of condensers, alternating-current motors. A certain class of such motors which I advanced a few years ago, which contain closed secondary circuits, will rotate quite vigorously when the discharges are directed through the exciting coils. One reason that such a motor operates so well with these discharges is that the difference of phase between the primary and secondary currents is 90 degrees, which is generally not the case with harmonically rising and falling currents of low frequency. It might not be without interest to show an experiment with a simple motor of this kind, inasmuch as it is commonly thought that disruptive discharges are unsuitable for such purposes. The motor is illustrated in Fig. 182. It comprises a rather large iron core i with slots on the top into which are embedded thick copper washers c c. In proximity to the core is a freely-movable metal disc D. The core is provided with a primary exciting coil C_1 the ends a and b of which are connected to the terminals of the secondary S of an ordinary transformer, the primary P of the latter being connected to an alternating distribution circuit or generator G of low or moderate frequency. The terminals of the secondary S are attached to a condenser C which discharges through an air gap d d which may be placed in series or shunt to the coil C_1. When the conditions are properly chosen the disc D rotates with considerable effort and the iron core i does not get very perceptibly hot. With currents from a high-frequency alternator, on the contrary, the core gets rapidly hot and the disc rotates with a much smaller effort. To perform the experiment properly it should be first ascertained that the disc D is not set in rotation when the discharge is not occurring at d d. It is preferable to use a large iron core and a condenser of large capacity so as to bring the superimposed quicker oscillation to a very low pitch or to do away with it entirely. By observing certain elementary rules I have also found it practicable to operate ordinary series or shunt direct-current motors with such disruptive discharges, and this can be done with or without a return wire.

Impedance phenomena

Among the various current phenomena observed, perhaps the most interesting are those of impedance presented by conductors to currents varying at a rapid rate. In my first paper before the *American Institute of Electrical Engineers*, I have described a few striking observations of this kind. Thus I showed that when such currents or sudden discharges are passed through a thick metal bar there may be points on the bar only a few inches apart, which have a sufficient potential difference between them to maintain at bright incandescence an ordinary filament lamp. I have also described the curious behavior of rarefied gas surrounding a conductor, due to such sudden rushes of current. These phenomena have since been more carefully studied and one or two novel experiments of this kind are deemed of sufficient interest to be described here.

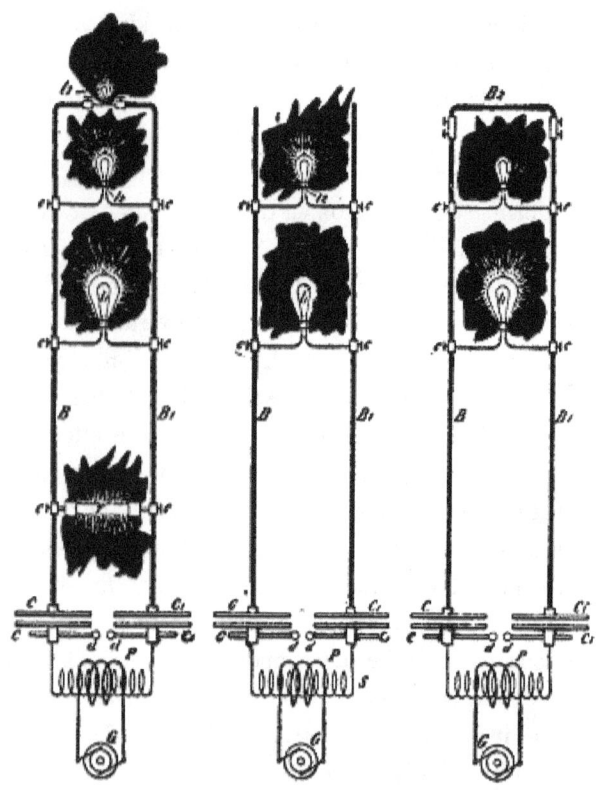

FIGS. 183a, 183b and 183c.

Referring to Fig. 183a, B and B_1 are very stout copper bars connected at their lower ends to plates C and C_1, respectively, of a condenser, the opposite plates of the latter being connected to the terminals of the secondary S of a high-tension transformer, the primary P of which is supplied with alternating currents from an ordinary low-frequency dynamo G or distribution circuit. The condenser discharges through an adjustable gap d d as usual. By establishing a rapid vibration it was found quite easy to perform the following curious experiment. The bars B and B_1 were joined at the top by a low-voltage lamp l_3 a little lower was placed by means of clamps C C, a 50-volt lamp l_2; and still lower another 100-volt lamp l_1; and finally, at a certain distance below the latter lamp, an exhausted tube T. By carefully determining the positions of these devices it was found practicable to maintain them all at their proper illuminating power. Yet they were all connected in multiple arc to the two stout copper bars and required widely different pressures. This experiment requires of course some time for adjustment but is quite easily performed.

In Figs. 183b and 183c, two other experiments are illustrated which, unlike the previous experiment, do not require very careful adjustments. In Fig. 183b, two lamps, l_1 and l_2, the former a 100-volt and the latter a 50-volt are placed in certain positions as indicated, the 100-volt lamp being below the 50-volt lamp. When the arc is playing at d d and the sudden discharges are passed through the bars B B_1, the 50-volt lamp will, as a rule, burn brightly, or at least this result is easily secured, while the 100-volt lamp will burn very low or remain quite dark, Fig. 183b. Now the bars B B1 may be joined at the top by a thick cross bar B2 and it is quite easy to maintain the 100-volt lamp at full candle-power while the 50-volt lamp remains dark, Fig. 183c. These results, as I have pointed out previously, should not be considered to be due exactly to frequency but rather to the time rate of change which may be great, even with low frequencies. A great many other results of the same kind, equally interesting, especially to those who are only used to manipulate steady currents, may be obtained and they afford precious clues in investigating the nature of electric currents.

In the preceding experiments I have already had occasion to show some light phenomena and it would now be proper to study these in particular; but to make this investigation more complete I think it necessary to make first a few remarks on the subject of electrical resonance which has to be always observed in carrying out these experiments.

On electrical resonance

The effects of resonance are being more and more noted by engineers and are becoming of great importance in the practical operation of apparatus of all kinds with alternating currents. A few general remarks may therefore be made concerning these effects. It is clear, that if we succeed in employing the effects of resonance practically in the operation of electric devices the return wire will, as a matter of course, become unnecessary, for the electric vibration may be conveyed with one wire just as well as, and sometimes even better than, with two. The question first to answer is, then, whether pure resonance effects are producible. Theory and experiment both show that such is impossible in Nature, for as the oscillation becomes more and more vigorous, the losses in the vibrating bodies and environing media rapidly increase and necessarily check the vibration which otherwise would go on increasing forever. It is a fortunate circumstance that pure resonance is not producible, for if it were there is no telling what dangers might not lie in wait for the innocent experimenter. But to a certain degree resonance is producible, the magnitude of the effects being limited by the imperfect conductivity and imperfect elasticity of the media or, generally stated, by frictional losses. The smaller these losses, the more striking are the effects. The same is the case in mechanical vibration. A stout steel bar may be set in vibration by drops of water falling, upon it at proper intervals; and with glass, which is more perfectly elastic, the resonance effect is still more remarkable, for a goblet may be burst by singing into it a tone of the proper pitch. The electrical resonance is the more perfectly attained, the smaller the resistance or the impedance of the conducting path and the more perfect the dielectric. In a Leyden jar discharging through a short stranded cable of thin wires these requirements are probably best fulfilled, and the resonance effects are , therefore very prominent. Such is not the case with dynamo machines, transformers and their circuits, or with commercial apparatus in general in which the presence of iron cores complicates the action or renders it impossible. In regard to Leyden jars with which resonance effects are frequently demonstrated, I would say that the effects observed are often attributed but are seldom due to true resonance, for an error is quite easily made in this respect. This may be undoubtedly demonstrated by the following experiment. Take, for instance, two large insulated metallic plates or spheres which I shall designate A and B; place them at a certain small distance apart and charge them from a frictional or influence machine to a potential so high that just a slight increase of the difference of potential between them will cause the small air or insulating space to break down. This is easily reached by making a few preliminary trials. If now another plate—fastened on an insulating handle and connected by a wire to one of the terminals of a high tension secondary of an induction coil, which is maintained in action by an alternator (preferably high frequency)—is approached to one of the charged bodies A or B, so a s to be nearer to either one of them, the discharge will invariably occur between them; at least it will, if the potential of the coil in connection with the plate is sufficiently high. But the explanation of this will soon be found in the fact that the approached plate acts inductively upon the bodies A and B and causes a spark to pass between them. When this spark occurs, the charges which were previously imparted to these bodies from the influence machine, must needs be lost, since the bodies are brought in electrical connection through the arc formed. Now this arc is formed whether there be resonance or not. But even if the spark would not be produced, still there is an alternating E. M. F. set up between the bodies when the plate is brought near one of them; therefore the approach of the plate, if it does not always actually, will, at any rate, tend to break down the air space by inductive action. Instead of the spheres or plates A and B we may take the coatings of a Leyden jar with the same result, and in place of the machine,—which is a high frequency alternator preferably, because it is more suitable for the experiment and also for the argument,—we may take another Leyden jar or battery of jars. When such jars are discharging through a circuit of low resistance the same is traversed by currents of very high frequency. The plate may now be connected to one of the coatings of the second jar, and when it is brought near to the first jar just previously charged to a high potential from an influence machine, the result is the same as before, and the first jar will discharge through a small air space upon the second being caused to discharge. But both jars and their circuits need not be tuned any closer than a basso profundo is to the note produced by a mosquito, as small sparks will be produced through the air space, or at least the latter will be considerably more strained owing to the setting up of an alternating E. M. F. by induction, which takes place when one of the jars begins to discharge. Again another error of a similar nature is quite easily made. If the circuits of the two jars are run parallel and close together, and the experiment has been performed of discharging one by the other, and now a coil of wire be added to one of the circuits whereupon the experiment does not succeed, the conclusion that this is due to the fact that the circuits are now not tuned, would be far from being safe. For the two circuits act as condenser coatings and the addition of the coil to one of them is equivalent to bridging them, at the point where the coil is placed, by a small condenser, and the effect of the latter might be to prevent the spark from jumping through the discharge space by diminishing the alternating E. M. F. acting across the same. All these remarks, and many more which might be added but for fear of wandering too far from the subject, are made with the pardonable intention of cautioning the unsuspecting student, who might gain an entirely unwarranted opinion of his skill at seeing every experiment succeed; but they are in no way thrust upon the experienced as novel observations.

In order to make reliable observations of electric resonance effects it is very desirable, if not necessary, to employ an alternator giving currents which rise and fall harmonically, as in working with make and break currents the observations are not always trustworthy, since many phenomena, which depend on the rate of change, may be produced with widely different frequencies. Even when making such observations with an alternator one is apt to be mistaken. When a circuit is connected to an alternator there are an indefinite number of values for capacity and self-induction which, in conjunction, will satisfy the condition of resonance. So there are in mechanics an infinite number of tuning forks which will respond to a note of a certain pitch, or loaded springs which have a definite period of vibration. But the resonance will be most perfectly attained in that case in which the motion is effected with the greatest freedom. Now in mechanics, considering the vibration in the common medium — that is, air — it is of comparatively little importance whether one tuning fork be somewhat larger than another, because the losses in the air are not very considerable. One may, of course, enclose a tuning fork in an exhausted vessel and by thus reducing the air resistance to a minimum obtain better resonant action. Still the difference would not be very great. But it would make a great difference if the tuning fork were immersed in mercury. In the electrical vibration it is of enormous importance to arrange the conditions so that the vibration is effected with the greatest freedom. The magnitude of the resonance effect depends, under otherwise equal conditions, on the quantity of electricity set in motion or on the strength of the current driven through the circuit. But the circuit opposes the passage of .the currents by reason of its impedance and therefore, to secure the best action it is necessary to reduce the impedance to a minimum. It is impossible to overcome it entirely, but merely in part, for the ohmic resistance cannot be overcome. But when the frequency of the impulses is very great, the flow of the current is practically determined by self-induction. Now self-induction can be overcome by combining it with capacity. If the relation between these is such, that at the frequency used they annul each other, that is, have such values as to satisfy the condition of resonance, and the greatest quantity of electricity is made to flow through the external circuit, then the best result is obtained. It is simpler and safer to join the condenser in series with the self-induction. It is clear that in such combinations there will be, for a given frequency, and considering only the fundamental vibration, values which will give the best result, with the condenser in shunt to the self-induction coil; of course more such values than with the condenser in series. But practical conditions determine the selection. In the latter case in performing the experiments one may take a small self-induction and a large capacity or a small capacity and a large self-induction, but the latter is preferable, because it is inconvenient to adjust a large capacity by small steps. By taking a coil with a very large self-induction the critical capacity is reduced to a very small value, and the capacity of the coil itself may be sufficient. It is easy, especially by observing certain artifices, to wind a coil through which the impedance will be reduced to the value of the ohmic resistance only; and for any coil there is, of course, a frequency at which the maximum current will be made to pass through the coil. The observation of the relation between self-induction, capacity and frequency is becoming important in the operation of alternate current apparatus, such as transformers or motors, because by a judicious determination of the elements the employment of an expensive condenser becomes unnecessary. Thus it is possible to pass through the coils of an alternating current motor under the normal working conditions the required current with a low E. M. F. and do away entirely with the false current, and the larger the motor, the easier such a plan becomes practicable; but it is necessary for this to employ currents of very high potential or high frequency.

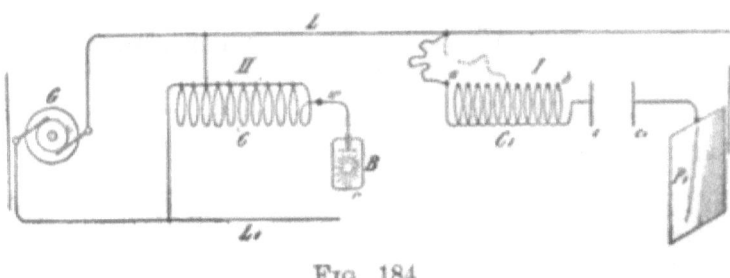

FIG. 184.

In Fig. 184 I. is shown a plan which has been followed in the study of the resonance effects by means of a high frequency alternator. C_1 is a coil of many turns, which is divided into small separate sections for the purpose of adjustment. The final adjustment was made sometimes with a few thin iron wires (though this is not always advisable) or with a closed secondary. The coil C_1 is connected with one of its ends to the line L from the alternator G and with the other end to one of the plates C of a condenser C C_1, the plate (C_1) of the latter being connected to a much larger plate P_1. In this manner both capacity and self-induction were adjusted to suit the dynamo frequency.

As regards the rise of potential through resonant action, of course, theoretically, it may amount to anything since it depends on self-induction and resistance and since these may have any value. But in practice one is limited in the selection of these values and besides these, there are other limiting causes. One may start with, say, 1,000 volts and raise the E. M. F. to 50 times that value, but one cannot start with 100,000 and raise it to ten times that value because of the losses in the media which are great, especially if the frequency is high. It should be possible to start with, for instance, two volts from a high or low frequency circuit of a dynamo and raise the E. M. F. to many hundred times that value. Thus coils of the proper dimensions might be connected each with only one of its ends to the mains from a machine of low E. M. F., and though the circuit of the machine would not be closed in the ordinary acceptance of the term, yet the machine might be burned out if a proper resonance effect would be obtained. I have not been able to produce, nor have I observed with currents from a dynamo machine, such great rises of potential. It is possible, if not probable, that with currents obtained from apparatus containing iron the disturbing influence of the latter is the cause that these theoretical possibilities cannot be realized. But if such is the case I attribute it solely to the hysteresis and Foucault current losses in the core. Generally it was necessary to transform upward, when the E. M. F. was very low, and usually an ordinary form of induction coil was employed, but sometimes the arrangement illustrated in Fig. 184 II., has been found to be convenient. In this case a coil C is made in a great many sections, a few of these being used as a primary. In this manner both primary and secondary are adjustable. One end of the coil is connected to the line L_1 from the alternator, and the other line L is connected to the intermediate point of the coil. Such a coil with adjustable primary and secondary will be found also convenient in experiments with the disruptive discharge. When true resonance is obtained the top of the wave must of course be on the free end of the coil as, for instance, at the terminal of the phosphorescence bulb B. This is easily recognized by observing the potential of a point on the wire w near to the coil.

In connection with resonance effects and the problem of transmission of energy over a single conductor which was previously considered, I would say a few words on a subject which constantly fills my thoughts and which concerns the welfare of all. I mean the transmission of intelligible signals or perhaps even power to any distance without the use of wires. I am becoming daily more convinced of the practicability of the scheme; and though I know full well that the great majority of scientific men will not believe that such results can be practically and immediately realized, yet I think that all consider the developments in recent years by a number of workers to have been such as to encourage thought and experiment in this direction. My conviction has grown so strong, that I no longer look upon this plan of energy or intelligence transmission as a mere theoretical possibility, but as a serious problem in electrical engineering, which must be carried out some day. The idea of transmitting intelligence without wires is the natural outcome of the most recent results of electrical investigations. Some enthusiasts have expressed their belief that telephony to any distance by induction through the air is possible. I cannot stretch my imagination se far, but I do firmly believe that it is practicable to disturb by means of powerful machines the electrostatic condition of the earth and thus transmit intelligible signals and perhaps power. In fact, what is there against the carrying out of such a scheme? We now know that electric vibration may be transmitted through a single conductor. Why then not try to avail ourselves of the earth for this purpose? We need not be frightened by the idea of distance. To the weary wanderer counting the mile-posts the earth may appear very large but to that happiest of all men, the astronomer, who gazes at the heavens and by their standard judges the magnitude of our globe, it appears very small. And so I think it must seem to the electrician, for when he considers the speed with which an electric disturbance is propagated through the earth all his ideas of distance must completely vanish.

A point of great importance would be first to know what is the capacity of the earth? and what charge does it contain if electrified? Though we have no positive evidence of a charged body existing in space without other oppositely electrified bodies being near, there is a fair probability that the earth is such a body, for by whatever process it was separated from other bodies—and this is the accepted view of its origin—it must have retained a charge, as occurs in all processes of mechanical separation. If it be a charged body insulated in space its capacity should be extremely small, less than one-thousandth of a farad. But the upper strata of the air are conducting, and so, perhaps, is the medium in free space beyond the atmosphere, and these may contain an opposite charge. Then the capacity might be incomparably greater. In any case it is of the greatest importance to get an idea of what quantity of electricity the earth contains. It is difficult to say whether we shall ever acquire this necessary knowledge, but there is hope that we may, and that is, by means of electrical resonance. If ever we can ascertain at what period the earth's charge, when disturbed, oscillates with respect to an oppositely electrified system or known circuit, we shall know a fact possibly of the greatest importance to the welfare of the human race. I propose to seek for the period by means of an electrical oscillator, or a source of alternating electric currents. One of the terminals of the source would be connected to earth as, for instance, to the city water mains, the other to an insulated body of large surface. It is possible that the outer conducting air strata, or free space, contain an opposite charge and that, together with the earth, they form a condenser of very large capacity. In such case the period of vibration may be very low and an alternating dynamo machine might serve for the purpose of the experiment. I would then transform the current to a potential as high as it would be found possible and connect the ends of the high tension secondary to the ground and to the insulated body. By varying the frequency of the currents and carefully observing the potential of the insulated body and watching for the disturbance at various neighboring points of the earth's surface resonance might be detected. Should, as the majority of scientific men in all probability believe, the period be extremely small, then a dynamo machine would not do and a proper electrical oscillator would have to be produced and perhaps it might not be possible to obtain such rapid vibrations. But whether this be possible or not, and whether the earth contains a charge or not, and whatever may be its period of vibration, it certainly is possible—for of this we have daily evidence—to produce some electrical disturbance sufficiently powerful to be perceptible by suitable instruments at any point of the earth's surface.

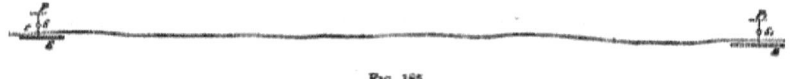

FIG. 185.

Assume that a source of alternating currents be connected, as in Fig. 185, with one of its terminals to earth (conveniently to the water mains) and with the other to a body of large surface P. When the electric oscillation is set up there will be a movement of electricity in and out of P, and alternating currents will pass through the earth, converging to, or diverging from, the point C where the ground connection is made. In this manner neighboring points on the earth's surface within a certain radius will be disturbed. But the disturbance will diminish with the distance, and the distance at which the effect will still be perceptible will depend on the quantity of electricity set in motion. Since the body P is insulated, in order to displace a considerable quantity, the potential of the source must be excessive, since there would be limitations as to the surface of P. The conditions might be adjusted so that the generator or source S will set up the same electrical movement as though its circuit were closed. Thus it is certainly practicable to impress an electric vibration at least of a certain low period upon the earth by means of proper machinery. At what distance such a vibration might be made perceptible can only be conjectured. I have on another occasion considered the question how the earth might behave to electric disturbances. There is no doubt that, since in such an experiment the electrical density at the surface could be but extremely small considering the size of the earth, the air would not act as a very disturbing factor, and there would be not much energy lost through the action of the air, which would be the case if the density were great. Theoretically, then; it could not require a great amount of energy to produce a disturbance perceptible at great distance, or even all over the surface of the globe.

Now, it is quite certain that at any point within a certain radius of the source S a properly adjusted self-induction and capacity device can be set in action by resonance. But not only can this be done, but another source S_1 Fig. 185, similar to S, or any number of such sources, can be set to work in synchronism with the latter, and the vibration thus intensified and spread over a large area, or a flow of electricity produced to or from the source S_1 if the same be of opposite phase to the source S. I think that beyond doubt it is possible to operate electrical devices in a city through the ground or pipe system by resonance from an electrical oscillator located at a central point. But the practical solution of this problem would be of incomparably smaller benefit to man than the realization of the scheme of transmitting intelligence, or perhaps power, to any distance through the earth or environing medium. If this is at all possible, distance does not mean anything. Proper apparatus must first be produced by means of which the problem can be attacked and I have devoted much thought to this subject. I am firmly convinced that it can be done and hope that we shall live to see it done.

On the light phenomena produced by high-frequency currents of high potential and general remarks relating to the subject

Returning now to the light effects which it has been the chief object to investigate, it is thought proper to divide these effects into four classes: 1. Incandescence of a solid. 2. Phosphorescence. 3. Incandescence or phosphorescence of a rarefied gas; and 4. Luminosity produced in a gas at ordinary pressure. The first question is: How are these luminous effects produced? In order to answer this question as satisfactorily as I am able to do in the light of accepted views and with the experience acquired, and to add some interest to this demonstration, I shall dwell here upon a feature which I consider of great importance, inasmuch as it promises, besides, to throw a better light upon the nature of most of the phenomena produced by high-frequency electric currents. I have on other occasions pointed out the great importance of the presence of the rarefied gas, or atomic medium in general, around the conductor through which alternate currents of high frequency are passed, as regards the heating of the conductor by the currents. My experiments, described some time ago, have shown that, the higher the frequency and potential difference of the currents, the more important becomes the rarefied gas in which the conductor is immersed, as a factor of the heating. The potential difference, however, is, as I then pointed out, a more important element than the frequency. When both of these are sufficiently high, the heating may be almost entirely due to the presence of the rarefied gas. The experiments to follow will show the importance of the rarefied gas, or, generally, of gas at ordinary or other pressure as regards the incandescence or other luminous effects produced by currents of this kind.

I take two ordinary 50-volt 16 c. p. lamps which are in every respect alike, with the exception, that one has been opened at the top and the air has filled the bulb, while the other is at the ordinary degree of exhaustion of commercial lamps. When I attach the lamp which is exhausted to the terminal of the secondary of the coil, which I have already used, as in experiments illustrated in Fig. 179a for instance, and turn on the current, the filament, as you have before seen, comes to high incandescence. When I attach the second lamp, which is filled with air, instead of the former, the filament still glows, but much less brightly. This experiment illustrates only in part the truth of the statements before made. The importance of the filament's being immersed in rarefied gas is plainly noticeable but not to such a degree as might be desirable. The reason is that the secondary of this coil is wound for low tension, having only 150 turns, and the potential difference at the terminals of the lamp is therefore small. Were I to take another coil with many more turns in the secondary, the effect would be increased, since it depends partially on the potential difference, as before remarked. But since the effect likewise depends on the frequency, it may be properly stated that it depends on the time rate of the variation of the potential difference. The greater this variation, the more important becomes the gas as an element of heating. I can produce a much greater rate of variation in another way, which, besides, has the advantage of doing away with the objections, which might be made in the experiment just shown, even if both the lamps were connected in series or multiple arc to the coil, namely, that in consequence of the reactions existing between the primary and secondary coil the conclusions are rendered uncertain. This result I secure by charging, from an ordinary transformer which is fed from the alternating current supply station, a battery of condensers, and discharging the latter directly through a circuit of small self-induction, as before illustrated in Figs. 183a, 183b and 183c.

 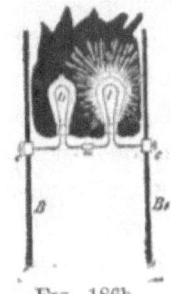 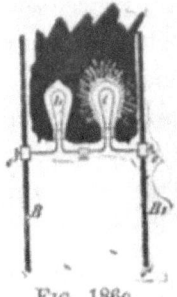

FIG. 186a.　　　　　　FIG. 186b.　　　　　　FIG. 186c.

In Figs. 186a, 186b and 186c, the heavy copper bars BB_1 are connected to the opposite coatings of a battery of condensers, or generally in such way, that the high frequency or sudden discharges are made to traverse them. I connect first an ordinary 50-volt incandescent lamp to the bars by means of the clamps C C. The discharges being; passed through the lamp, the filament is rendered incandescent, though the current through it is very small, and would not be nearly sufficient to produce a visible effect under the conditions of ordinary use of the lamp. Instead of this I now attach to the bars another lamp exactly like the first, but with the seal broken off, the bulb being therefore filled with air at ordinary pressure. When the discharges are directed through the filament, as before, it does not become incandescent. But the result might still be attributed to one of the many possible reactions. I therefore connect both the lamps in multiple arc as illustrated in Fig. 186a. Passing tile discharges through both the lamps, again the filament in the exhausted lamp l glows very brightly while that in the non-exhausted lamp l_1 remains dark, as previously. But it should not be thought that the latter lamp is' taking only a small fraction of the energy supplied to both the lamps; on the contrary, it may consume a considerable portion of the energy and it may become even hotter than the one which burns brightly. In this experiment the potential difference at the terminals of the lamps varies in sign theoretically three to four million times a second. The ends of the filaments are correspondingly electrified, and the gas in the bulbs is violently agitated and a large portion of the supplied energy is thus converted into heat. In the non-exhausted bulb, there being a few million times more gas molecules than in the exhausted one, the bombardment, which is most violent at the ends of the filament, in the neck of the bulb, consumes a large portion of the energy without producing any visible effect. The reason is that, there being many molecules, the bombardment is quantitatively considerable, but the individual impacts are not very violent, as the speeds of the molecules are comparatively small owing to the small free path. In the exhausted bulb, on the contrary, the speeds are very great, and the individual impacts are violent and therefore better adapted to produce a visible effect. .Besides, the convection of heat is greater in the former bulb. In both the bulbs the current traversing the filaments is very small, incomparably smaller than that which they require on an ordinary low-frequency circuit. The potential difference, however, at the ends of the filaments is very great and might be possibly 20,000 volts or more, if the filaments were straight and their ends far apart. In the ordinary lamp a spark generally occurs between the ends of the filament or between the platinum wires outside, before such a difference of potential can be reached.

It might be objected that in the experiment before shown the lamps, being in multiple arc, the exhausted lamp might take a much larger current and that the effect observed might not be exactly attributable to the action of the gas in the bulbs. Such objections will lose much weight if I connect the lamps in series, with the wine result. When this is done and the discharges are directed through the filaments, it is again noted that the filament in the non-exhausted bulb l, remains dark, while that in the exhausted one (l) glows even snore intensely than under its normal conditions of working, Fig. 186b. According to general ideas the current through the filaments should now be the same, were it not modified by the presence of the gas around the filaments.

At this juncture I may point out another interesting feature, which illustrates the effect of the rate of change of potential of the currents. I will leave the two lamps connected in series to the bars BB_1, as in the previous experiment, Fig. 186b, but will presently reduce considerably the frequency of the currents, which was excessive in the experiment just before shown. This I may do by inserting a self-induction coil in the path of the discharges, or by augmenting the capacity of the condensers. When I now pass these low-frequency discharges through the lamps, the exhausted lamp l again is as bright as before, but it is noted also that the non-exhausted lamp l_1 glows, though not quite as intensely as the other. Reducing the current through the lamps, I may bring the filament in the latter lamp to redness, and, though the filament in the exhausted lamp l is bright, Fig. 186c, the degree of its incandescence is much smaller than in Fig. 186b, when the currents were of a much higher frequency.

In these experiments the gas acts in two opposite ways in determining the degree of the incandescence of the filaments, that is, by convection and bombardment. The higher the frequency and potential of the currents, the more important becomes the bombardment. The convection on the contrary should be the smaller, the higher the frequency. When the currents are steady there is practically no bombardment, and convection may therefore with such currents also considerably modify the degree of incandescence and produce results similar to those just before shown. Thus, it two lamps exactly alike, one exhausted and one not exhausted, are connected in multiple arc or series to a direct-current machine, the filament in the non-exhausted lamp will require a considerably greater current to be rendered incandescent. This result is entirely due to convection, and the effect is the more prominent the thinner the filament. Professor Ayrton and Mr. Kilgour some time ago published quantitative results concerning the thermal emissivity by radiation and convection in which the effect with thin wires was clearly shown. This effect may be strikingly illustrated by preparing a number of small, short, glass tubes, each containing through its axis the thinnest obtainable platinum wire. If these tubes be highly exhausted, a number of them may be connected in multiple arc to a direct-current machine and all of the wires may be kept at incandescence with a smaller current than that required to render incandescent a single one of the wires if the tube be not exhausted. Could the tubes be so highly exhausted that convection would be nil, then the relative amounts of heat given off by convection and radiation could be determined without the difficulties attending thermal quantitative measurements. If a source of electric impulses of high frequency and very high potential is employed, a still greater number of the tubes may be taken and the wires rendered incandescent by a current not capable of warming perceptibly a wire of the same size immersed in air at ordinary pressure, and conveying the energy to all of them.

I may here describe a result which is still more interesting, and to which I have been led by the observation of these phenomena. I noted that small differences in the density of the air produced a considerable difference in the degree of incandescence of the wires, and I thought that, since in a tube, through which a luminous discharge is passed, tile gas is generally not of uniform density, a very thin wire contained in the tube might be rendered incandescent at certain places of smaller density of the gas, while it would remain dark at the places of greater density, where the convection would be greater and the bombardment less intense. Accordingly a tube t was prepared, as illustrated in Fig. 187, which contained through the middle a very fine platinum wire w. The tube was exhausted to a moderate degree and it was found that when it was attached to the terminal of a high-frequency coil the platinum wire w, would indeed, become incandescent in patches, as illustrated in Fig. 187. Later a number of these tubes with one or more wires were prepared, each showing this result. The effect was best noted when the striated discharge occurred in the tube, but was also produced when the striae were not visible, showing that, even then, the gas in the tube was, not or uniform density. The position of the striae was generally such, that the rarefactions corresponded to the places of incandescence or greater brightness on the wire w. But in a few instances it was noted, that the bright spots on the wire were covered by the dense parts of the striated discharge as indicated by l in Fig. 187, though the effect was barely perceptible. This was explained in a plausible way by assuming that the convection was not widely different in the dense and rarefied places, and that the bombardment was greater on the dense places of the striated discharge. It is, in fact, often observed in bulbs, that under certain conditions a thin wire is brought to higher incandescence when the air is not too highly rarefied. This is the case when the potential of the coil is not high enough for the vacuum, but the result may be attributed to many different causes. In all cases this curious phenomenon of incandescence disappears when the tube, or rather the wire, acquires throughout a uniform temperature.

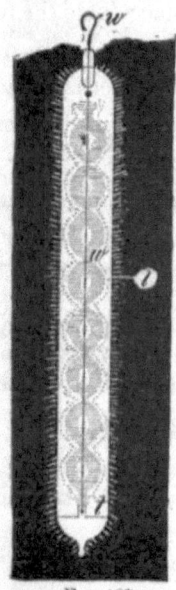

Fig. 187.

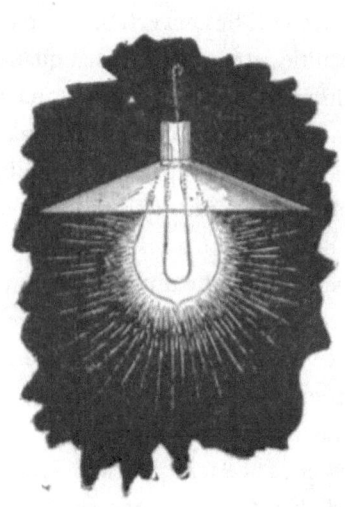

Fig. 188.

Disregarding now the modifying effect of convection there are then two distinct causes which determine the incandescence of a wire or filament with varying currents, that is, conduction current and bombardment. With steady currents we have to deal only with the former of these two causes, and the heating effect is a minimum, since the resistance is least to steady flow. When the current is a varying one the resistance is greater, and hence the heating effect is increased. Thus if the rate of change of the current is very great, the resistance may increase to such an extent that the filament is brought to incandescence with inappreciable currents, and we are able to take a short and thick block of carbon or other material and brim, it to bright incandescence with a current incomparably smaller than that required to bring to the same deuce of incandescence an ordinary thin lamp filament with a steady or low frequency current. This result is important, and illustrates how rapidly our views on these subjects are changing, and how quickly our field of knowledge is extending. In the art of incandescent lighting, to view this result in one aspect only, it has been commonly considered as an essential requirement for practical success, that the lamp filament should be thin and of high resistance. But now we know that the resistance of the filament to the steady flow does not mean anything; the filament might as well be short and thick; for if it be immersed in rarefied gas it will become incandescent by the passage of a small current. It all depends on the frequency and potential of the currents. We may conclude from this, that it would be of advantage, so far as the lamp is considered, to employ high frequencies for lighting, as they allow the use of short and thick filaments and smaller currents.

If a wire or filament be immersed in a homogeneous medium, all the heating is due to true conduction current, but if it be enclosed in an exhausted vessel the conditions are entirely different. Here the gas begins to act and the heating effect of the conduction current, as is shown in many experiments, may be very small compared with that of the bombardment. This is especially the case if the circuit is not closed and the potentials are of course very high. Suppose that a fine filament enclosed in an exhausted vessel be connected with one of its ends to the terminal of a high tension coil and with its other end to a large insulated plate. Though the circuit is not closed, the filament, as I have before shown, is brought to incandescence. If the frequency and potential be comparatively low, the filament is heated by the current passing through it. If the frequency and potential, and principally the latter, be increased, the insulated plate need be but very small, or may be done away with entirely; still: the filament will become incandescent, practically all the heating being then due to the bombardment. A practical way of combining both the effects of conduction currents and bombardment is illustrated in Fig. 188, in which an ordinary lamp is shown provided with a very thin filament which has one of the ends of the latter connected to a shade serving the purpose of the insulated plate, and the other end to the terminal of a high tension source. It should not be thought that only rarefied gas is an important factor in the heating of a conductor by varying currents, but gas at ordinary pressure may become important, if the potential difference and frequency of the currents is excessive. On this subject I have already stated, that when a conductor is fused by a stroke of lightning, the current through it may be exceedingly small, not even sufficient to heat the conductor perceptibly, were the latter immersed in a homogeneous medium.

From the preceding it is clear that when a conductor of high resistance is connected to the terminals of a source of high frequency currents of high potential, there may occur considerable dissipation of energy, principally at the ends of the conductor, in consequence of the action of the gas surrounding the conductor. Owing to this, the current through a section of the conductor at a point midway between its ends may be much smaller than through a section near the ends. Furthermore, the current passes principally through the outer portions of the conductor, but this effect is to be distinguished from the skin effect as ordinarily interpreted, for the latter would, or should, occur also in a continuous incompressible medium. If a great many incandescent lamps are connected in series to a source of such currents, the lamps at the ends may burn brightly, whereas those in the middle may remain entirely dark. This is due principally to bombardment, as before stated. But even if the currents be steady, provided the difference of potential is very great, the lamps at the end will burn more brightly than those in the middle. In such case there is no rhythmical bombardment, and the result is produced entirely by leakage. This leakage or dissipation into space when the tension is high, is considerable when incandescent lamps are used, and still more considerable with arcs, for the latter act like flames. Generally, of course, the dissipation is much smaller with steady, than with varying, currents.

I have contrived an experiment which illustrates in an interesting manner the effect of lateral diffusion. If a very long tube is attached to the terminal of a high frequency coil, the luminosity is greatest near the terminal and falls off gradually towards the remote end. This is more marked if the tube is narrow.

A small tube about one-half inch in diameter and twelve inches long (Fig. 189), has one of its ends drawn out into a fine fibre f nearly three feet long. The tube is placed in a brass socket T which can be screwed on the terminal T_1 of the induction coil. The discharge passing through the tube first illuminates the bottom of the same, which is of comparatively large section; but through the long glass fibre the discharge cannot pass. But gradually the rarefied gas inside becomes warmed and more conducting and the discharge spreads into the glass fibre. This spreading is so slow, that it may take half a minute or more until the discharge has worked through up to the top of the glass fibre, then presenting the appearance of a strongly luminous thin thread. By adjusting the potential at the terminal the light may be made to travel upwards at any speed. Once, however, the glass fibre is heated, the discharge breaks through its entire length instantly. The interesting point to be noted is that, the higher the frequency of the currents, or in other words, the greater relatively the lateral dissipation, at a slower rate may the light be made to propagate through the fibre. This experiment is best performed with a highly exhausted and freshly made tube. When the tube has been used for some time the experiment often fails. It is possible that the gradual and slow impairment of the vacuum is the cause. This slow propagation of the discharge through a very narrow glass tube corresponds exactly to the propagation of heat through a bar warmed at one end. The quicker the heat is carried away laterally the longer time it will take for the heat to warm the remote end. When the current of a low frequency coil is passed through the fibre from end to end, then the lateral dissipation is small and the discharge instantly breaks through almost without exception.

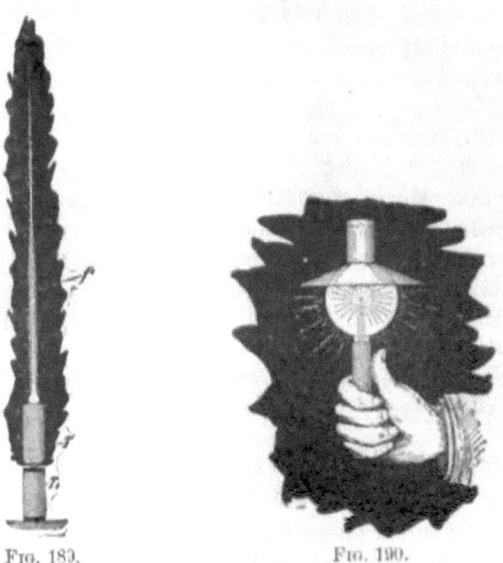

Fig. 189.   Fig. 190.

After these experiments and observations which have shown the importance of the discontinuity or atomic structure of the medium and which will serve to explain, in a measure at least, the nature of the four kinds of light effects producible with these currents, I may now give you an illustration of these effects. For the sake of interest I may do this in a manner which to many of you might be novel. You have seen: before that we may now convey the electric vibration to a body by means of a single wire or conductor of any kind. Since the human frame is conducting I may convey the vibration through my body.

Fig. 191. Fig. 192.

First, as in some previous experiments, I connect my body with one of the terminals of a high-tension transformer and take in my hand an exhausted bulb which contains a small carbon button mounted upon a platinum wire leading to the outside of the bulb, and the button is rendered incandescent as soon as the transformer is set to work (Fig. 190). I may place a conducting shade on the bulb which serves to intensify the action, but it is not necessary. Nor is it required that the button should be in conducting connection with the hand through a wire leading through the glass, for sufficient energy may be transmitted through the glass itself by inductive action to render the button incandescent.

Next I take a highly exhausted bulb containing a strongly phosphorescent body, above which is mounted a small plate of aluminum on a platinum wire leading to the outside, and the currents flowing through my body excite intense phosphorescence in the bulb (Fig. 191). Next again I take in my hand a simple exhausted tube, and in the same manner the gas inside the tube is rendered highly incandescent or phosphorescent (Fig. 192). Finally, I may take in my hand a wire, bare or covered with thick insulation, it is quite immaterial; the electrical vibration is so intense as to cover the wire with a luminous film (Fig. 193). A few words must now be devoted to each of these phenomena. In the first place, I will consider the incandescence of a button or of a solid in general, and dwell upon some facts which apply equally to all these phenomena. It was pointed out before that when a thin conductor, such as a lamp filament, for instance, is connected with one of its ends to the terminal of a transformer of high tension the filament is brought to incandescence partly by a conduction current and partly by bombardment.

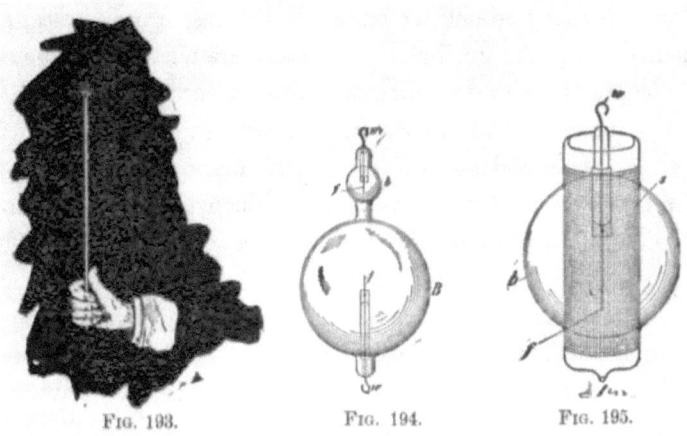

Fig. 193.   Fig. 194.   Fig. 195.

The shorter and thicker the filament the more important becomes the latter, and finally, reducing the filament to a mere button, all the heating must practically be attributed to the bombardment. So in the experiment before shown, the button is rendered incandescent by the rhythmical impact of freely movable small bodies in the bulb. These bodies may be the molecules of the residual gas, particles of dust or lumps torn from the electrode; whatever they are, it is certain that the heating of the button is essentially connected with the pressure of such freely movable particles, or of atomic matter in general in the bulb. The heating is the more intense the greater the number of impacts per second and the greater the energy of each impact. Yet the button would be heated also if it were connected to a source of a steady potential. In such a case electricity mould be carried away from the button by the freely movable carriers or particles flying about, and the quantity of electricity thus carried away might be sufficient to bring the button to incandescence by its passage through the latter. But the bombardment could not be of great importance in such case. For this reason it would require a comparatively very great supply of energy to the button to maintain it at incandescence with a steady potential. The higher the frequency of the electric impulses the more economically can the button be maintained at incandescence. One of the chief reasons why this is so, is, I believe, that with impulses of very high frequency there is less exchange of the freely movable carriers around the electrode and this means, that in the bulb the heated matter is better confined to the neighborhood of the button. If a double bulb, as illustrated in Fig. 194 be made, comprising a large globe B and a small one b, each containing as usual a filament f mounted on a platinum wire w and w_1 it is found, that if the filaments f f be exactly alike, it requires less energy to keep the filament in the globe b at a certain degree of incandescence, than that in the globe B. This is due to the confinement of the movable particles around the button. In this case it is also ascertained, that the filament in the small globe b is less deteriorated when maintained a certain length of time at incandescence. This is a necessary consequence of the fact that the gas in the small bulb becomes strongly heated and therefore a very good conductor, and less work is then performed on the button, since the bombardment becomes less intense as the conductivity of the gas increases. In this construction, of course, the small bulb becomes very hot and when it reaches an elevated temperature the convection and radiation on the outside increase. On another occasion I have shown bulbs in which this drawback was largely avoided. In these instances a very small bulb, containing a refractory button, was mounted in a large globe and the space between the walls of both was highly exhausted. The outer large globe remained comparatively cool in such constructions. When the large globe was on the pump and the vacuum between the walls maintained permanent by the continuous action of the pump, the outer globe would remain quite cold, while the button in tile small bulb was kept at incandescence. But when the seal was made, and the button in the small bulb maintained incandescent some length of time, the large globe too would become warmed. Prom this I conjecture that if vacuous space (as Prof. Dewar finds) cannot convey heat, it is so merely in virtue of our rapid motion through space or, generally speaking, by the motion of the medium relatively to us, for a permanent condition could not be maintained without the medium being constantly renewed. A vacuum cannot, according to all evidence, be permanently maintained around a hot body.

In these constructions, before mentioned, the small bulb inside would, at least in the first stages, prevent all bombardment against the outer large globe. It occurred to me then to ascertain how a metal sieve would behave in this respect, and several bulbs, as illustrated in Fig. 195, were prepared for this purpose. In a globe b, was mounted a thin filament f (or button) upon a platinum wire w passing through a glass stem and leading to the outside of the globe. The filament f was surrounded by a metal sieve s. It was found in experiments with such bulbs that a sieve with wide meshes apparently did not in the slightest affect the bombardment against the globe b. When the vacuum was high, the shadow of the sieve was clearly projected against the globe and the latter would get hot in a short while. In some bulbs the sieve ,v was connected to a platinum wire sealed in the glass. When this wire was connected to the other terminal of the induction coil (the E. M. F. being kept low in this case), or to an insulated plate, the bombardment against the outer globe b was diminished. By taking a sieve with fine meshes the bombardment against the globe b was always diminished, but even then if the exhaustion was carried very far, and when the potential of the transformer was very high, the globe b would be bombarded and heated quickly, though no shadow of the sieve was visible, owing to the smallness of the meshes. But a glass tube or other continuous body mounted so as to surround the filament, did entirely cut off the bombardment and for a while the outer globe b would remain perfectly cold. Of course when the glass tube was sufficiently heated the bombardment against the outer globe could be noted at once. The experiments with these bulbs seemed to show that the speeds of the projected molecules or particles must be considerable (though quite insignificant when compared with that of light), otherwise it would be difficult to understand how they could traverse a fine metal sieve without being affected, unless it were found that such small particles or atoms cannot be acted upon directly at measurable distances. In regard to the speed of the projected atoms, Lord Kelvin has recently estimated it at about one kilometre a second or thereabouts in an ordinary Crookes bulb. As the potentials obtainable with a disruptive discharge coil are much higher than with ordinary coils, the speeds must, of course, be much greater when the bulbs are lighted from such a coil. Assuming the speed to be as high as five kilometres and uniform through the whole trajectory, as it should be in a very highly exhausted vessel, then if the alternate electrifications of the electrode would be of a frequency of five million, the greatest distance a particle could get away from the electrode would be one millimetre, and if it could be acted upon directly at that distance, the exchange of electrode matter or of the atoms would be very slow and there would be practically no bombardment against the bulb. This at least should be so, if the action of an electrode upon the atoms of the residual gas would be such as upon electrified bodies which we can perceive. A hot body enclosed in a n exhausted bulb produces always atomic bombardment, but a hot body has no definite rhythm, for its molecules perform vibrations of all kinds.

If a bulb containing a button or filament be exhausted as high as is possible with tile greatest care and by the use of the best artifices, it is often observed that the discharge cannot, at first, break through, but after some time, probably in consequence of some changes within the bulb, the discharge finally passes through and the button is rendered incandescent. In fact, it appears that the higher the degree of exhaustion the easier is the incandescence produced. There seem to be no other causes to which the incandescence might be attributed in such case except to the bombardment or similar action of the residual gas, or of particles of matter in general. But if the bulb be exhausted with the greatest care can these play an important part? Assume the vacuum in the bulb to be tolerably perfect, the great interest then centres in the question: Is the medium which pervades all space continuous or atomic? If atomic, then the heating of a conducting button or filament in an exhausted vessel might be due largely to ether bombardment, and then the heating of a conductor in general through which currents of high frequency or high potential are passed must be modified by the behavior of such medium; then also the skin effect, the apparent increase of the ohmic resistance, etc., admit, partially at least, of a different explanation.

It is certainly more in accordance with many phenomena observed with high frequency currents to hold that all space is pervaded with free atoms, rather than to assume that it is devoid of these, and dark and cold, for so it must be, if filled with a continuous medium, since in such there can be neither heat nor light. Is then energy transmitted by independent carriers or by the vibration of a continuous medium? This important question is by no means as yet positively answered. But most of the effects which are here considered, especially the light effects, incandescence, or phosphorescence, involve the presence of free atoms and would be impossible without these.

In regard to the incandescence of a refractory button (or filament) in an exhausted receiver, which has been one of the subjects of this investigation, the chief experiences, which may serve as a guide in constructing such bulbs, may be summed up as follows: 1. The button should be as small as possible, spherical, of a smooth or polished surface, and of refractory material which withstands evaporation best. 2. The support of the button should be very thin and screened by an aluminum and mica sheet, as I have described on another occasion. 3. The exhaustion of the bulb should be as high as possible. 4. The frequency of the currents should be as high as practicable. 5. The currents should be of a harmonic rise and fall, without sudden interruptions. 6. The heat should be confined to the button by inclosing the same in a small bulb or otherwise. 7. The space between the walls of the small bulb and the outer globe should be highly exhausted.

Most of the considerations which apply to the incandescence of a solid just considered may likewise be applied to phosphorescence. Indeed, in an exhausted vessel the phosphorescence is, as a rule, primarily excited by the powerful beating of the electrode stream of atoms against the phosphorescent body. Even in many cases, where there is no evidence of such a bombardment, I think that phosphorescence is excited by violent impacts of atoms, which are not necessarily thrown off from the electrode but are acted upon from the same inductively through the medium or through chains of other atoms. That mechanical shocks play an important part in exciting phosphorescence in a bulb may be seen from the following experiment. If a bulb, constructed as that illustrated in Fig. 174, be taken and exhausted with the greatest care so that the discharge cannot pass, the filament facts by electrostatic induction upon the tube t and the latter is set in vibration. If the tube o be rather wide, about an inch or so, the filament may be so powerfully vibrated that whenever it hits the glass tube it excites phosphorescence. But the phosphorescence ceases when the filament comes to rest. The vibration can be arrested and again started by varying the frequency of the currents. Now the filament has its own period of vibration, and if the frequency of the currents is such that there is resonance, it is easily set vibrating, though the potential of the currents be small. I have often observed that the filament in the bulb is destroyed by such mechanical resonance. The filament vibrates as a rule so rapidly that it cannot be seen and the experimenter may at first be mystified. When such an experiment as the one described is carefully performed, the potential of the currents need be extremely small, and for this reason I infer that the phosphorescence is then due to the mechanical shock of the filament against the glass, just as it is produced by striking a loaf of sugar with a knife. The mechanical shock produced by the projected atoms is easily noted when a bulb containing a button is grasped in the hand and the current turned on suddenly. I believe that a bulb could be shattered by observing the conditions of resonance.

In the experiment before cited it is, of course, open to say, that the glass tube, upon coming in contact with the filament, retains a charge of a certain sign upon the point of contact. If now the filament main touches the glass at the same point while it is oppositely charged, the charges equalize under evolution of light. But nothing of importance would be gained by such an explanation. It is unquestionable that the initial charges given to the atoms or to the glass play some part in exciting phosphorescence. So for instance, if a phosphorescent bulb be first excited by a high frequency coil by connecting it to one of the terminals of the latter and the degree of luminosity be noted and then the bulb be highly charged from a Holtz machine by attaching it preferably to the positive terminal of the machine, it is found that when the bulb is again connected to the terminal of the high frequency coil, the phosphorescence is far more intense. On another occasion I have considered the possibility of some phosphorescent phenomena in bulbs being produced by the incandescence of an infinitesimal layer on the surface of the phosphorescent body. Certainly the impact of the atoms is powerful enough to produce intense incandescence by the collisions, since they bring quickly to a high temperature a body of considerable bulk. If any such effect exists, then the best appliance for producing phosphorescence in a bulb, which we know so far, is a disruptive discharge coil giving an enormous potential with but few fundamental discharges, say 25-30 per second, just enough to produce a continuous impression upon the eye. It is a fact that such a coil excites phosphorescence under almost any condition and at all degrees of exhaustion, and I have observed effect; which appear to be due to phosphorescence even at ordinary pressures of the atmosphere, when the potentials are extremely high. But if phosphorescent light is produced by the equalization of charges of electrified atoms (whatever this may mean ultimately), then the higher the frequency of the impulses or alternate electrifications, the more economical will be the light production. It is a long known and noteworthy fact that all the phosphorescent bodies are poor conductors of electricity and heat, and that all bodies cease to emit phosphorescent light when they are brought to a certain temperature. Conductors on the contrary do not possess this quality. There are but few exceptions to the rule. Carbon is one of them. Becquerel noted that carbon phosphoresces at a certain elevated temperature preceding the dark red. This phenomenon may be easily observed in bulbs provided with a rather large carbon electrode (say, a sphere of six millimetres diameter). If the current is turned on after a few seconds, a snow white film covers the electrode, just before it bets dark red. Similar effects are noted with other conducting bodies, but many scientific men will probably not attribute them to true phosphorescence. Whether true incandescence has anything to do with phosphorescence excited by atomic impact or mechanical shocks still remains to be decided, but it is a fact that all conditions, which tend to localize and increase the heating effect at the point of impact, are almost invariably the most favorable for the production of phosphorescence. So, if the electrode be very small, which is equivalent to saying in general, that the electric density is great; if the potential be high, and if the gas be highly rarefied, all of which things imply high speed of the projected atoms, or matter, and consequently violent impacts — the phosphorescence is very intense. If a bulb provided with a large and small electrode be attached to the terminal of an induction coil, the small electrode excites phosphorescence while the large one may not do so, because of the smaller electric density and hence smaller speed of the atoms. A bulb provided with a large electrode may be grasped with the hand while the electrode is connected to the terminal of the coil and it may not phosphoresce; but if instead of grasping the bulb with the hand, the same be touched with a pointed wire, the phosphorescence at once spreads through the bulb, because of the great density at the point of contact. With low frequencies it seems that gases of great atomic weight excite more intense phosphorescence than those of smaller weight, as for instance, hydrogen. With high frequencies the observations are not sufficiently reliable to draw a conclusion. Oxygen, as is well-known, produces exceptionally strong effects, which may be in part due to chemical action. A bulb with hydrogen residue seems to be most easily excited. Electrodes which are most easily deteriorated produce more intense phosphorescence in bulbs, but the condition is not permanent because of the impairment of the vacuum and the deposition of the electrode matter upon the phosphorescent surfaces. Some liquids, as oils, for instance, produce magnificent effects of phosphorescence (or fluorescence?), but they last only a few seconds. So if a bulb has a trace of oil on the walls and the current is turned on, the phosphorescence only persists for a few moments until the oil is carried away. Of all bodies so far tried, sulphide of zinc seems to be the most susceptible to phosphorescence. Some samples, obtained through the kindness of Prof. Henry in Paris, were employed in many of these bulbs. One of the defects of this sulphide is, that it loses its quality of emitting light when brought to a temperature which is by no means high. It can therefore, be used only for feeble intensities. An observation which might deserve notice is, that when violently bombarded from an aluminum electrode it assumes a black color, but singularly enough, it returns to the original condition when it cools down.

The most important fact arrived at in pursuing investigations in this direction is, that in all cases it is necessary, in order to excite phosphorescence with a minimum amount of energy, to observe certain conditions. Namely, there is always, no matter what the frequency of the currents, degree of exhaustion and character of the bodies in the bulb, a certain potential (assuming the bulb excited from one terminal) or potential difference (assuming the bulb to be excited with both terminals) which produces the most economical result. If the potential be increased, considerable energy may be wasted without producing any more light, and if it be diminished, then again the light production is not as economical. The exact condition under which the best result is obtained seems to depend on many things of a different nature, and it is to be yet investigated by other experimenters, but it will certainly have to be observed when such phosphorescent bulbs are operated, if the best results are to be obtained.

Coming now to the most interesting of these phenomena, the incandescence or phosphorescence of gases, at low pressures or at the ordinary pressure of the atmosphere, we must seek the explanation of these phenomena in the same primary causes, that is, in shocks or impacts of the atoms. Just as molecules or atoms beating upon a solid body excite phosphorescence in the same or render it incandescent, so when colliding among themselves they produce similar phenomena. But this is a very insufficient explanation and concerns only the crude mechanism. Light is produced by vibrations which go on at a rate almost inconceivable. If we compute, from the energy contained in the form of known radiations in a definite space the force which is necessary to set up such rapid vibrations, we find, that though the density of the ether be incomparably smaller than that of any body we know, even hydrogen, the force is something surpassing comprehension. What is this force, which in mechanical measure may amount to thousands of tons per square inch? It is electrostatic force in the light of modern views. It is impossible to conceive how a body of measurable dimensions could be charged to so high a potential that the force would be sufficient to produce these vibrations. Long before any such charge could be imparted to the body it would be shattered into atoms. The sun emits light and heat, and so does an ordinary flame or incandescent filament, but in neither of these can the force be accounted for if it be assumed that it is associated with the body as a whole. Only in one way may we account for it, namely, by identifying it with the atom. An atom is so small, that if it be charged by coming in contact with an electrified body and the charge be assumed to follow the same law as in the case of bodies of measurable dimensions, it must retain a quantity of electricity which is fully capable of accounting for these forces and tremendous rates of vibration. But the atom behaves singularly in this respect—it always takes the same "charge".

It is very likely that resonant vibration plays a most important part in all manifestations of energy in nature. Throughout space all matter is vibrating, and all rates of vibration are represented, from the lowest musical note to the highest pitch of the chemical rays, hence an atom, or complex of atoms, no matter what its period, must find a vibration with which it is in resonance. When we consider the enormous rapidity of the light vibrations, we realize the impossibility of producing such vibrations directly with any apparatus of measurable dimensions, and we are driven to the only possible means of attaining the object of setting up waves of light by electrical means and economically, that is, to affect the molecules or atoms of a gas, to cause them to collide and vibrate. We then must ask ourselves—How can free molecules or atoms be affected?

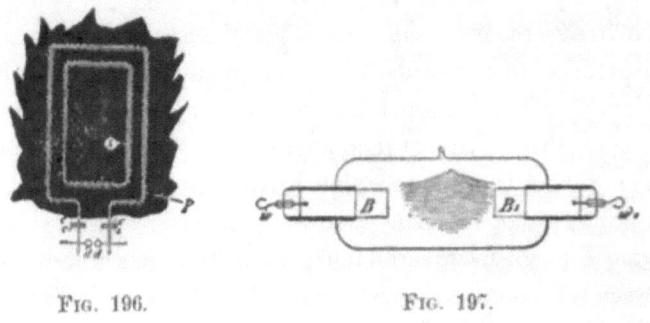

Fig. 196. Fig. 197.

It is a fact that they can be affected by electrostatic force, as is apparent in many of these experiments. By varying the electrostatic force we can agitate the atoms, and cause them to collide accompanied by evolution of heat and light. It is not demonstrated beyond doubt that vie can affect them otherwise. If a luminous discharge is produced in a closed exhausted tube, do the atoms arrange themselves in obedience to any other but to electrostatic force acting in straight lines from atom to atom? Only recently I investigated the mutual action between two circuits with extreme rates of vibration. When a battery of a few jars (c c c c, Fig. 196), is discharged through a primary P of low resistance (the connections being as illustrated in Figs. 183a, 183b and 183c), and the frequency of vibration is many millions there are great differences of potential between points on the primary not more than a few inches apart. These differences may be 10,000 volts per inch, if not more, taking the maximum value of the E. M. F. The secondary S is therefore acted upon by electrostatic induction, which is in such extreme cases of much greater importance than the electro-dynamic. To such sudden impulses the primary as well as the secondary are poor conductors, and therefore great differences of potential may be produced by electrostatic induction between adjacent points on the secondary. Then sparks may jump between the wires and streamers become visible in the dark if the light of the discharge through the spark gap d d be carefully excluded. If now we substitute a closed vacuum tube for the metallic secondary S, the differences of potential produced in the tube by electrostatic induction from the primary are fully sufficient to excite portions of it; but as the points of certain differences of potential on the primary are not fixed, but are generally constantly changing in position, a luminous band is produced in the tube, apparently not touching the glass, as it should, if the points of maximum and minimum differences of potential were fixed on the primary. I do not exclude the possibility of such a tube being excited only by electro-dynamic induction, for very able physicists hold this view; but in my opinions, there is as yet no positive proof given that atoms of a gas in a closed tube may arrange themselves in chains under the action of an : electromotive impulse produced by electro-dynamic induction in the tube. I have been unable so far to produce striae in a tube, however long, and at whatever degree of exhaustion, that is, striae at right angles to the supposed direction of the discharge or the axis of the tube; but I have distinctly observed in a large bulb, in which a wide luminous band was produced by passing a discharge of a battery through a wire surrounding the bulb, a circle of feeble luminosity between two luminous bands, one of which was more intense than the other. Furthermore, with my present experience I do not think that such a gas discharge in a closed tube can vibrate, that is, vibrate as a whole. I am convinced that no discharge through a gas can vibrate. The atoms of a gas behave very curiously in respect to sudden electric impulses. The gas does not seem to possess any appreciable inertia to such impulses, for it is a fact, that the higher the frequency of the impulses, with the greater freedom does the discharge pass through the gas. If the gas possesses no inertia then it cannot vibrate, for some inertia is necessary for the free vibration. I conclude from this that if a lightning discharge occurs between two clouds, there can be no oscillation, such as would be expected, considering the capacity of the clouds. But if the lightning discharge strike the earth, there is always vibration—in the earth, but not in the cloud. In a gas discharge each atom vibrates at its oven rate, but there is no vibration of the conducting gaseous mass as a whole. This is an important consideration in the great problem of producing light economically, for it teaches us that to reach this result we must use impulses of very high frequency and necessarily also of high potential. It is a fact that oxygen produces a more intense light in a tube. Is it because oxygen atoms possess some inertia and the vibration does not die out instantly? But then nitrogen should be as good, and chlorine and vapors of many other bodies much better than oxygen, unless the magnetic properties of the latter enter prominently into play. Or, is the process in the tube of an electrolytic nature? Many observations certainly speak for it, the most important being that matter is always carried away from the electrodes and the vacuum in a bulb cannot be permanently maintained. If such process takes place in reality, then again must we take refuge in high frequencies, for, with such, electrolytic action should be reduced to a minimum, if not rendered entirely impossible. It is an undeniable fact that with very high frequencies, provided the impulses be of harmonic nature, like those obtained from an alternator, there is less deterioration and the vacua are more permanent. With disruptive discharge coils there are sudden rises of potential and the vacua are more quickly impaired, for the electrodes are deteriorated in a very short time. It was observed in some large tubes, which were provided with heavy carbon blocks B B_1, connected to platinum wires w w_1 (as illustrated in Fig. 197), and which were employed in experiments with the disruptive discharge instead of the ordinary air gap, that the carbon particles under the action of the powerful magnetic field in which the tube was placed, were deposited in regular fine lines in the middle of the tube, as illustrated. These lines were attributed to the deflection or distortion of the discharge by the magnetic field, but why the deposit occurred principally where the field was most intense did not appear quite clear. A fact of interest, likewise noted, was that the presence of a strong magnetic field increases the deterioration of the electrodes, probably by reason of the rapid interruptions it produces, whereby there is actually a higher E. M. F. maintained between the electrodes.

Much would remain to be said about the luminous effects produced in gases at low or ordinary pressures. With the present experiences before us we cannot say that the essential nature of these charming phenomena is sufficiently known. But investigations in this direction are being pushed with exceptional ardor. Every line of scientific pursuit has its fascinations, but electrical investigation appears to possess a peculiar attraction, for there is no experiment or observation of any kind in the domain of this wonderful science which would not forcibly appeal to us. Yet to me it seems, that of all the many marvelous things we observe, a vacuum tube, excited by an electric impulse from a distant source, bursting forth out of the darkness and illuminating the room with its beautiful light, is as lovely a phenomenon as can greet our eyes. More interesting still it appears when, reducing the fundamental discharges across the gap to a very small number and waving the tube about we produce all kinds of designs in luminous lines. So by way of amusement I take a straight long tube, or a square one, or a square attached to a straight tube, and by whirling them about in the hand, I imitate the spokes of a wheel, a Gramme winding, a drum winding, an alternate current motor winding, etc. (Fig. 198). Viewed from a distance the effect is weak and much of its beauty is lost, but being near or holding the tube in the hand, one cannot resist its charm.

Fig. 198.

In presenting these insignificant results I have not attempted to arrange and coordinate them, as would be proper in a strictly scientific investigation, in which every succeeding result should be a logical sequence of the preceding, so that it might be guessed in advance by the careful reader or attentive listener. I have preferred to concentrate my energies chiefly upon advancing novel facts or ideas which might serve as suggestions to others, and this may serve as an excuse for the lack of harmony. The explanations of the phenomena have been given in good faith and in the spirit of a student prepared to find that they admit of a better interpretation. There can be no great harm in a student taking an erroneous view, but when great minds err, the world must dearly pay for their mistakes.

March 1893

ON THE DISSIPATION OF THE ELECTRICAL ENERGY OF THE HERTZ RESONATOR

—*The Electrical Engineer*—*December 21, 1892*

Anyone who, like myself, has had the pleasure of witnessing the beautiful demonstrations with vibrating diaphragms which Prof. Bjerknes, exhibited in person at the Paris Exposition in 1880, must have admired his ability and painstaking care to such a degree, as to have an almost implicit faith in the correctness of observations made by him. His experiments "On the Dissipation of the Electrical Energy of the Hertz Resonator," which are described in the issue of Dec. 14, of *The Electrical Engineer*, are prepared in the same ingenious and skillful manner, and the conclusions drawn from them are all the more interesting as they agree with the theories put forth by the most advanced thinkers. There can not be the slightest doubt as to the truth of these conclusions, yet the statements which follow may serve to explain in part the results arrived at in a different manner; and with this object in view I venture to call attention to a condition with which, in investigations such as those of Prof. Bjerknes, the experimenter is confronted.

The apparatus, oscillator and resonator, being immersed in air, or other discontinuous medium, there occurs—as I have pointed out in the description of my recent experiments before the English and French scientific societies—dissipation of energy by what I think might be appropriately called electric sound waves or sound-waves of electrified air. In Prof. Bjerknes's experiments principally this dissipation in the resonator need be considered, though the sound-waves—if this term be permitted—which emanate from the surfaces at the oscillator may considerably affect the observations made at some distance from the latter. Owing to this dissipation the period of vibration of an air-condenser can not be accurately determined, and I have already drawn attention to this important fact. These waves are propagated at right angles from the charged surfaces when their charges are alternated, and dissipation occurs, even if the surfaces are covered with thick and excellent insulation. Assuming that the "charge" imparted to a molecule or atom either by direct contact or inductively is proportionate to the electric density of the surface, the dissipation should be proportionate to the square of the density and to the number of waves per second. The above assumption, it should be stated, does not agree with some observations from which it appears that an atom can not take but a certain maximum charge; hence, the charge imparted may be practically independent of the density of the surface, but this is immaterial for the present consideration. This and other points will be decided when accurate quantitative determinations, which are as yet wanting, shall be trade. At present it appears certain from experiments with, high-frequency currents, that this dissipation of energy from a wire, for instance, is not very far from being proportionate to the frequency of the alternations, and increases very rapidly when the diameter of the wire is made exceedingly small. On the latter point the recently published results of Prof. Ayrton and H. Kilgour on "The Thermal Emissivity of Thin Wires in Air" throw a curious light. Exceedingly thin wires are capable of dissipating a comparatively very great amount of energy by the agitation of the surrounding air, when they are connected to a source of rapidly alternating potential. So in the experiment cited, a thin hot wire is found to be capable of emitting an extraordinarily great amount of heat, especially at elevated temperatures. In the case of a hot wire it must of course be assumed that the increased emissivity is due to the more rapid convection and not, to any, appreciable degree, to an increased radiation. Were the latter demonstrated, it would show that a wire, made hot by the application of heat in ordinary ways, behaves in some respects like one, the charge of which is rapidly alternated, the dissipation of energy per unit of surface kept at a certain temperature depending on the curvature of the surface. I do not recall any record of experiments intended to demonstrate this, yet this effect, though probably very small, should certainly be, looked for.

A number of observations showing the peculiarity, of very thin wires were made in the course of my experiments. I noted, for instance, that in the well-known Crookes instrument the mica vanes are repelled with comparatively greater force when the incandescent platinum wire is exceedingly thin. This observation enabled me to produce the spin of such vanes mounted in a vacuum tube when the latter was placed in an alternating electrostatic field. This however does not prove anything in regard to radiation, as in a highly exhausted vessel tile phenomena are principally due to molecular bombardment or convection.

When I first undertook to produce the incandescence of a wire enclosed in a bulb, by connecting it to only one of the terminals of a high tension transformer, I could not succeed for a long time. On one occasion I had mounted in a bulb a thin platinum wire, but my apparatus was not adequate to produce the incandescence. I made other bulbs, reducing the length of the wire to a small fraction; still I did not succeed. It then occurred to me that it would be desirable to have the surface of the wire as large as possible, yet the bulk small, and I provided a bulb with an exceedingly thin wire of a bulk about equal to that of the short but much thicker wire. On turning the current on the bulb the wire was instantly fused. A series of subsequent experiments showed that when the diameter of the wire was exceedingly small, considerably more energy would be dissipated per unit surface at all degrees of exhaustion than was to be expected, even on the assumption that the energy given off was in proportion to the square of the electric density. There is likewise evidence which, though not possessing the certainty of an accurate quantitative determination, is nevertheless reliable because it is the result of a great many observations, namely, that with the increase of the density the dissipation is more rapid for thin than for thick wires.

The effects noted in exhausted vessels with high-frequency currents are merely diminished in degree when the air is at ordinary pressure, but heating and dissipation occurs, as I have demonstrated, under the ordinary atmospheric conditions. Two very thin wires attached to the terminals of a high-frequency coil are capable of giving off an appreciable amount of energy. When the density is very great, the temperature of the wires may be perceptibly raised, and in such case probably the greater portion of the energy which is dissipated owing to the presence of a discontinuous medium is transformed into heat at the surface or in close proximity to the wires. Such heating could not occur in a medium possessing either of the two qualities, namely, perfect incompressibility or perfect elasticity. In fluid insulators, such as oils, though they are far from being perfectly incompressible or elastic to electric displacement, the heating is much smaller because of the continuity of the fluid.

When the electric density of the wire surfaces is small, there is no appreciable local heating, nevertheless energy is dissipated in air, by waves, which differ from ordinary sound-waves only because the air is electrified. These waves are especially conspicuous when the discharges of a powerful battery are directed through a short and thick metal bar, the number of discharges per second being very small. The experimenter may feel the impact of the air at distances of six feet or more from the bar, especially if be takes the precaution to sprinkle the face or hands with ether. These waves cannot be entirely stopped by the interposition of an insulated metal plate.

Most of the striking phenomena of mechanical displacement, sound, heat and light which have been observed, imply the presence of a medium of a gaseous structure that is one consisting of independent carriers capable of free motion.

When a glass plate is placed near a condenser the charge of which is alternated, the plate emits a sound. This sound is due to the rhythmical impact of the air against the plate. I have also found that the ringing of a condenser, first noted by Sir William Thomson, is due to the presence of the air between or near the charged surfaces.

When a disruptive discharge coil is immersed in oil contained in a tank, it is observed that the surface of the oil is agitated. This may be thought to be due to the displacements produced in the oil by the changing stresses, but such is not the case. It is the air above the oil which is agitated and causes the motion of the latter; the oil itself would remain at rest. The displacements produced in it by changing electrostatic stresses are insignificant; to such stresses it may be said to be compressible to but a very small degree. The action of the air is shown in a curious manner for if a pointed metal bar is taken in the hand and held with the point close to the oil, a hole two inches deep is formed in the oil by the molecules of the air, which are violently projected from the point.

The preceding statements may have a general bearing upon investigations in which currents of high frequency and potential are made use of, but they also have a more direct bearing upon the experiments of Prof. Bjerknes which are here considered, namely, the "skin effect," is increased by the action of the air. Imagine a wire immersed in a medium, the conductivity of which would be some function of the frequency and potential difference but such, that the conductivity increases when either or bout of these elements are increased. In such a medium, the higher the frequency and potential difference, the greater wilt be the current which will find its way through the surrounding medium, and the smaller the part which will pass through the central portion of the wire: In the case of a wire immersed in air and traversed by a high-frequency current, the facility with which the energy is dissipated may be considered as the equivalent of the conductivity; and the analogy would be quite complete, were it not that besides the air another medium is present, the total dissipation being merely modified by the presence of the air to an extent as yet not ascertained. Nevertheless, I have sufficient evidence to draw the conclusion, that the results obtained by Prof. Bjerknes are affected by the presence of air in the following manner: 1. The dissipation of energy is more rapid when the resonator is immersed in air than it would be in a practically continuous medium, for instance, oil. 2. The dissipation owing to the presence of air renders the difference between magnetic and non-magnetic metals more striking. The first conclusion follows directly from the preceding remarks; the second follows front the two facts that the resonator receives always the same amount of energy, independent of the nature of the metal, and that the magnetism of the metal increases the impedance of the circuit. A resonator of magnetic metal behaves virtually as though its circuit were longer. There is a greater potential difference set up per unit of length; although this rosy not show itself in the deflection of the electrometer owing to the lateral dissipation. The effect of the increased impedance is strikingly illustrated in the two experiments of Prof. Bjerknes when copper is deposited upon an iron wire, and next iron upon a copper wire. Considerable thickness of copper deposit was required in the former experiment, but very little thickness of iron in the latter, as should be expected.

Taking the above views, I believe, that in the experiments of Prof. Bjerknes which lead him to undoubtedly correct conclusions, the air is a factor fully as important, if not more so, than the resistance of the metals.

December 21, 1892

TESLA'S OSCILLATOR AND OTHER INVENTIONS

—By Thomas Commerford Martin. Century Magazine. April 1895, pp. 916-933.

An Authoritative Account of some of his Recent Electrical Work[1]

Kobeleff, the great Russian general, once said of the political conditions in Central Asia, that they changed every moment; hence the necessity for vigilance, no less the price of empire than of liberty. Thus changeful, also, is the aspect of that vast new electrical domain which the thought and invention of our age have subdued. They who would inform themselves expertly about it, in whatever respect, must ever keep up an attitude of strained attention. Its theoretical problems assume novel phases daily. Its old appliances ceaselessly give way to successors. Its methods of production, distribution, and utilization vary from year to year. Its influence on the times is ever deeper, yet one can never be quite sure into what part of the social or industrial system it is next to thrust a revolutionary force. Its fanciful dreams of yesterday are the magnificent triumphs of to-morrow, and its advance toward domination in the twentieth century is as irresistible as that of steam in the nineteenth.

Throughout this change there has prevailed a consistency of purpose: a steady aim has been leveled at definite goals; while useful arts in multitude attest the solidity of the work done. If, therefore, we find a tremendous outburst of activity at the very moment when, after twenty-five years of superlative productiveness, electricians were ready, with the reforming English statesman, to rest and be thankful, we may safely assume that electricity has reached another of those crucial points at which it becomes worth the while of the casual outside observer to glance at what is going on. To the timid and the conservative, even to many initiated, these new departures have indeed become exasperating. They demand the unlearning of established facts, and insist on right-about-faces that disregard philosophical dignity. The sensations of a dog attempting to drink sea-water after a lifetime spent on inland lakes are feeble compared with those of men who discover that electricity is quite other than the fluid which they have believed it to be from their youth up, and that actually there is no such thing as electricity or an electric current.

Electricity has, indeed, taken distinctively new ground of late years; and its present state of unrest—unsurpassed, perhaps, in other regions of research—is due to recent theory and practice, blended in a striking manner in the discoveries of Mr. Nikola Tesla,[2] who, though not altogether alone, has come to be a foremost and typical figure of the era now begun. He invites attention today, whether for profound investigations into the nature of electricity, or for beautiful inventions in which is offered a concrete embodiment of the latest means for attaining the ends most sought after in the distribution of light, heat, and power, and in the distant communication of intelligence. Any one desirous of understanding the trend and scope of modem electrical advance will find many clues in the work of this inventor. The present article discloses a few of the more important results which he has attained, some of the methods and apparatus which he employs, and one or two of the theories to which he resorts for an explanation of what is accomplished.

By a brief preliminary survey, we may determine our historical longitude and latitude, and thus ascertain a little more precisely where we are. It is necessary to recapitulate facts known and accepted. Let it, then, be remarked that aside from the theories and interpretations that have beset the science, electricity as an art has for three hundred years been directed chiefly to securing an abundant, cheap, efficient, and economical supply of the protean agency, be it what it may. Frictional machine, Leyden jar, coil, battery, magnet, dynamo, oscillator,—these are but the steps in a process as regular and well-defined as those which take us from the aboriginal cradling of gold out of river sands up to the refining of ore with all the appliances of modem mechanism and chemistry. Each stage in electrical evolution has seen the conquest of some hitherto unknown art—electrotherapy, telegraphy, telephony, electric lighting, electric heating, power transmission; yet each has had limitations set on it by the conditions prevailing. With a mere battery much can be done; with a magnet, still more; with a dynamo, we touch possibilities of all kinds, for we compel the streams, the coalfields, and the winds to do us service: but with Mr. Tesla's new oscillator we may enlist even the ether-waves, and turn our wayward recruits into resistless trained forces, sweeping across continents of unimagined opportunity.

The dynamo, slowly perfected these fifty years, has rendered enormous benefits, and is destined to much further usefulness. But all that we learn now about it of any intrinsic value is to build it bigger, or to specialize it; and the moment a device reaches that condition of development, the human intellect casts about for something else in which the elements are to be subtler and less gross. Based upon currents furnished by modern dynamo-electric machines, the arc-light and the trolley-car seek to monopolize street illumination and transportation, while the incandescent lamp has preempted for exclusive occupancy the interiors of our halls and homes. Yet the abandonment of gas, horses, and sails is slow, because the dynamo and its auxiliaries have narrow boundaries, trespassing which, they cease to offer any advantage. We can all remember the high hopes with which, for example, incandescent lighting was introduced some fifteen years ago. Even the most cynical detractor of it will admit that its adoption has been quick and widespread; but as a simple matter of fact, today, all the lamps and all the lighting dynamos in the country would barely meet the needs of New York and Chicago if the two cities were to use no other illuminant than electricity. In all England there are only 1,750,000 incandescent lamps contesting for supremacy with probably 75,000,000 gas-burners, and the rate of increase is small, if indeed it exceeds that of gas. Evidently, some factor is wanting, and a new point of departure, even in mere commercial work, is to be sought, so that with longer circuits, better current-generating apparatus, and lamps that will not burn out, the popular demand for a pure and perfect light can be met. In power transmission, also, unsatisfied problems of equal magnitude crop up. "Is there any load that water cannot lift?" asked Emerson. "If there be, try steam; or if not that, try electricity. Is there any exhausting of these means?" None, provided that our mechanics be right.

It must not be supposed that the new electricity is iconoclastic. In the minds of a great many people of culture the idea prevails that invention is as largely a process of pulling down as of building up; and electricity, in spreading from one branch of industry to another, encounters the prejudice that always rebuffs the innovator. The assumption is false. It may be true that in the gladiatorial arena where the principles of science contend, one party or the other always succumbs and drags out its dead; but in the arts long survival is the law for all the appliances that have been found of any notable utility. It simply becomes a question of the contracting sphere within which the old apparatus is hedged by the advent of the new; and that relation once established by processes complex and long continued, capable even of mathematical determination, the two go on together, complementary in their adjustment to specific human needs. In its latest outgrowths, electrical application exemplifies this. After many years' use of dynamo-electric machinery giving what is known as a "continuous current," the art has reached the conclusion that only with the "alternating current" can it fulfil the later duties laid upon it, and accomplish the earlier tasks that remain untouched. With the continuous current we have learned the rudiments of lighting and power distribution. With the alternating current, manipulated and coaxed to yield its highest efficiency, we may solve the problems of aerial and marine navigation by electricity, operate large railway systems, transmit the energy of Niagara hundreds of miles, and, in Mr. Tesla's own phrase, "hook our machinery directly to that of Nature."

The generation of current

Let us see wherein lies the difference between these two kinds of currents. In all dynamos the generation of what we call electric current is effected by the whirling of coils of wire in front of magnets, or conversely. The wires that lead away from the machine and back to it to complete the necessary circuit, may be compared to a circle of troughs or to a pipe-line; the coils and magnets are comparable to pump mechanism; and the lamps or motors driven by the current, to fountains or faucets spaced out on the trough circle. This comparison is crudity itself, but it gives a fairly exact idea. The current travels along the surface of the wire rather than inside, its magnetic or ether whorls resembling rubber bands sliding along a lead-pencil. A machine that produces continuous current, dipping its wire coils or buckets into the magnetic field of force, has all its jets, as they come around to discharge themselves, headed one way, and complicated devices called "commutators" have been unavoidable for the purpose of "rectifying" them. A machine that produces alternating currents, on the contrary, has its jets thrown first into one end of the trough system, and then into the other, and therefore dispenses with the rectifying or commutating valves. On the other hand, it requires peculiar adjustment of its fountains and faucets to the streams rushing in either way. It is an inherent disadvantage of the continuous-current system that it cannot deliver energy successfully at any great distance at high pressure, and that therefore the pipe-line must be relatively as bulky as were the hollow wooden logs which were once employed for water-conduits in New York. The advantage of the alternating current is that it can be delivered at exceedingly high pressures over very slender wires, and used either at that pressure or at lower or higher ones, obtained by means of a "transformer," which, according to its use, answers both to the idea of a magnetic reducing valve, and to that of a spring-board accelerating the rapidity of motion of any object alighting on it. Obviously a transformer cannot return more than is put into it, so that it gives out the current received with less pressure but in greater volume, or raised in pressure but diminished in the volume of the stream. In some like manner a regiment of soldiers may be brought by express to any wharf, and transferred, Indian file, to a sailing barge or an ocean liner indifferently; but throughout the trip the soldiers will constitute the same regiment, and when picked up by another train across the ferry, the body, though there be loss by desertion and sickness, will retain its identity, even if the ranks are broken in filling the cars, and are reformed four abreast at the end of the journey.

Alternating currents

Let us, still recapitulating familiar facts, make the next step in our review of what is involved in the resort to alternating currents. It was stated above that the current-consuming devices such as motors, likened to fountains, needed peculiar adjustments to the inflow first from one side and then from the other. Not to put it at all too strongly, they would not work, and have largely remained inoperative to the present time. Lamps would burn, but motors would not run, and this fact limited seriously the adoption and range of the otherwise flexible and useful alternating current until Mr. Tesla discovered a beautiful and unsuspected solution of the problem, and thus embarked on one part of the work now revealing grander possibilities every day. The transmission of the power of Niagara has become possible since the discovery of the method. In his so-called "rotating magnetic field," a pulley mounted upon a shaft is perpetually running after a magnetic "pole" without ever being able to catch it. The fundamental idea is to produce magnetism shifting circularly, in contrast with the old and known phenomenon of magnetism in a fixed position. Those who have seen the patient animal inside the treadmill wheel of the well at Carisbrooke Castle can form an idea of the ingenuity of Mr. Tesla's plan.

Ordinarily, alternating-current generators, such as are now in common use, have a great number of projecting poles to cause the alternations of current, and hence their "frequency" is high—that is, the current makes a great many to-and-fro motions per second, and each ebb-and-flow in the circuit is termed the "period" or "frequency," one alternation being the rise from zero to maximum value and down to nothing again, and the other the same thing backward. If we ruled a horizontal straight line, and then drew a round-bellied Hogarth curve of beauty across it, the half of the curve above the line would be illustrative of the positive flow, the lower half of the negative flow; the top of one oval and the bottom of the other oval would be the maxima respectively; positive and negative n and the point where the curve crossed the straight line would mark the instant when the current changes its direction. A swinging pendulum is an analogy favored by scientists in their endeavors to illustrate popularly the process of the generation of the alternating current. Each time the copper wire in the coils on the dynamo armature is rotated past the pole of the dynamo field, the currents in each coil follow this rise and fall; so that the number of the magnets and coils determines the period or frequency, as stated. The more numerous the magnets, and the faster the rotation of the coils, the quicker will be the ebbs and flows of current. But the character of the work to be done, and existing conditions, govern the rate at which the current is thus to be set vibrating; and no small amount of skill and knowledge enters here. The men who can predicate the right thing to do are still few and far between. The field has as yet been little explored. Moreover, in one of the deepest problems now engaging the thought of electrical engineers,—namely, the production of cheap light and cheap power by these new means,—opposite conditions pull different ways. Mr. Tesla made up his mind some time ago that for motor work it was better to have few frequencies; and the whole drift of power transmission is on that path, the frequency adopted for the work at Niagara being only twenty-five. But, as was natural, he ran through the whole scale of low and high frequencies, and soon discovered that for obtaining light, one great secret lay in the utilization of currents of high frequency and high potential. Some years ago, after dealing with the power problem as above described, Mr. Tesla attacked the light problem by building a number of novel alternating-current generators for the purpose, and attained with them alternations up to 30,000 per second. These machines transcended anything theretofore known in the art, and their currents were further raised in pressure by "step up" transformers and condensers. But these dynamos had their shortcomings. The number of the poles and coils could not be indefinitely increased, and there was a limit to the speed. To go to the higher frequencies, therefore, Mr. Tesla next invented his "disruptive discharge coil," which permitted him to reach remarkably high frequency and high pressure, and, what is more, to obtain these qualities from any ordinary current, whether alternating or continuous. With this apparatus he surprised the scientists both of this country and of Europe in a series of most interesting demonstrations. It is not too much to say that these experiments marked an epoch in electricity, yielding results which lie at the root of his later work with the oscillator in an inconceivably wider range of phenomena.

The Tesla Oscillator

Up to this point we have been considering both continuous-current and alternating-current dynamos as driven by the ordinary steam engine. Perhaps nine tenths of all the hundreds of thousands of dynamos in the world today are so operated, the remainder being driven by water-wheels, gas-engines, and compressed air. Now, each step from consuming the coal under the boilers that deliver steam to the engines, up to the glow of the filament in an incandescent lamp, is attended with loss. As in every other cycle that has to do with heat transformation, the energy is more or less frittered away, just as in July the load in an iceman's cart crumbles and melts in transit along the street. Actual tests prove that the energy manifesting itself as light in an incandescent lamp is barely 5% of that received as current. In the luminosity of a gas flame the efficiency is even smaller. Professor Tyndall puts the useful light-waves of a gas flame at less than 1% of all the waves caused by the combustion going on in it. If we were dealing with a corrupt city government, such wretched waste and inefficiency would not be tolerated; and in sad reality the extravagance is but on a par with the wanton destruction of whole forests for the sake of a few sticks of lumber. Armies of inventors have flung themselves on the difficulties involved in these barbaric losses occurring at every stage of the calorific, mechanical, and electric processes; and it is indeed likely that many lines of improvement have already been compelled to yield their utmost, reaching terminal forms. A moment's thought will show that one main object must be the elimination of certain steps in the transfer of the energy; and obviously, if engine and dynamo both have large losses, it will be a gain to merge the two pieces of apparatus. The old-fashioned electric-light station or street-railway power-house is a giddy maze of belts and shafting; in the later plants engine and dynamo are coupled directly together on one base. This is a notable stride, but it still leaves us with a dynamo in which some part of the wire wound on it is not utilized at every instant, and with an engine of complicated mechanism. The steam-cylinder, with its piston, is the only thing actually doing work, and all the rest of the imposing collection of fly-wheel, governor-balls, eccentrics, valves, and what not, is for the purpose of control and regulation.

In his oscillator Mr. Tesla, to begin with, has stripped the engine of all this governing mechanism. By giving also to the coils in which the current is created as they cut the "lines of force" of the magnets, a to-and-fro or reciprocating motion, so that the influence on them is equal in every direction, he has overcome the loss of the idle part of the wire experienced in rotating armatures; and, moreover, greatest achievement of all, he has made the currents regulate the mechanical motions. No matter how close the governing of the engine that drives the ordinary dynamo, with revolving armature, there is some irregularity in the generation of current. In the Tesla oscillator, if its inventor and the evidence of one's eyes may be believed, the vibrations of the current are absolutely steady and uniform, so that one could keep the time of day with the machine about as well as with a clock. It was this superlative steadiness of the vibration or frequency that Mr. Tesla aimed at, for one thing. The variations caused by the older apparatus might be slight, but minute errors multiplied by high rates of occurrence soon become perceptible, and militate against desirable uniformity and precision of action. Back of the tendencies to irregularity in the old-fashioned electrical apparatus were the equal or greater tendencies in the steam-engine; and over and above all were the frightful losses due to the inefficient conversion in both of the power released from the fuel under the boiler generating the steam.

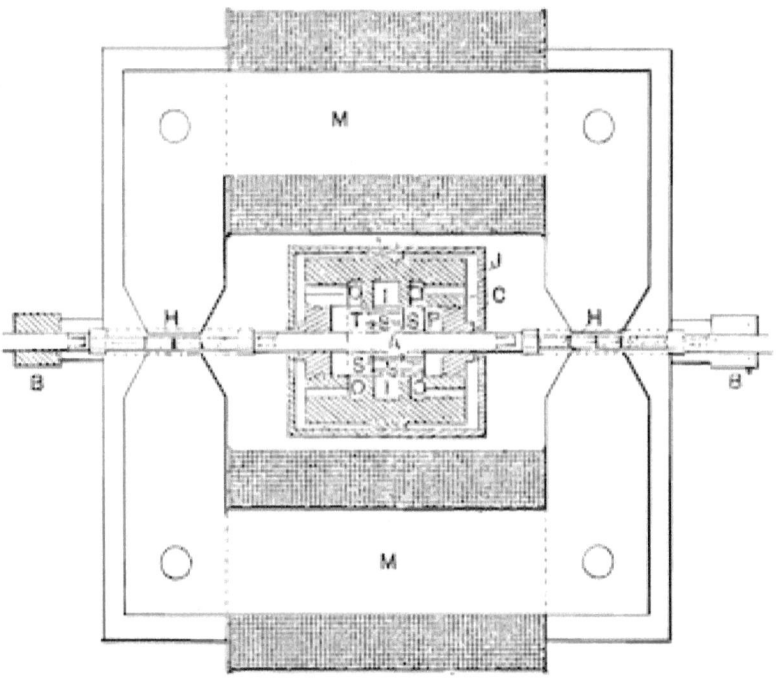

Fig. 1. Diagram of working parts of early form of tesla oscillator, as if seen from above, in section. (From "The Electrical Engineer," by permission.)

Gain in one direction with a radical innovation usually means gain in many others, through a growing series. I confess I do not know which of the advantages of the oscillator to place first; and I doubt whether its inventor has yet been able to sit down and sum up all the realities and possibilities to which it is a key. One thing he does: he presses forward. Our illustration, Fig. 2, shows one of his latest forms of oscillator in perspective, while the diagram, Fig. 1, exhibits the internal mechanism of one of the early forms. Fig. 2 will serve as a text for the subsequent heads of discourse. The steam-chest is situated on the bed-plate between the two electromagnetic systems, each of which consists of field coils between which is to move the armature or coil of wire. There are two pistons to receive the impetus of the incoming steam in the chest, and in the present instance steam is supplied at a pressure of 350 pounds, although as low as 80 is also used in like oscillators, where steam of the higher pressure is not obtainable. We note immediately the absence of all the governing appliances of the ordinary engine. They are non-existent. The steam chest is the engine, bared to the skin like a prizefighter, with every ounce counting. Besides easily utilizing steam at a remarkably high pressure, the oscillator holds it under no less remarkable control, and, strangest of all, needs no packing to prevent leak. It is a fair inference, too, that, denuded in this way of superfluous weight and driven at high pressure, the engine must have an economy far beyond the common. With an absence of friction due to the automatic cushioning of the light working parts, it is also practically indestructible. Moreover, for the same pressure and the same piston speed the engine has about one thirtieth or one fortieth of the usual weight, and occupies a proportionately smaller space. This diminution of bulk and area is equally true of the electrical part. The engine-pistons carry at their ends the armature coils, and these they thrust reciprocatively in and out of the magnetic field of the field coils, thus generating current by their action.

If one watches any dynamo, it will be seen that the coils constituting the "armature" are swung around in front of magnets, very much as a turnstile revolves inside the barricading posts; and the current that goes out to do work on the line circuit is generated inductively in the coils, because they cut lines of influence emanating from the ends of the magnets, and forming what has been known since Faraday's time as the "field of force." In the Tesla oscillator, the rotary motion of the coils is entirely abandoned, and they are simply darted to and fro at a high speed in front of the magnets, thus cutting the lines of the "field of force" by shooting in and out of them very rapidly, shuttle-fashion. The great object of cutting as many lines of an intense field of force as swiftly, smoothly, regularly, and economically as possible is thus accomplished in a new and, Mr. Tesla believes, altogether better way. The following description of remarkable new phenomena in electricity will justify him in regarding the oscillator as an extremely valuable instrument of research, while time will demonstrate its various commercial and industrial benefits.

Incidentally it may be remarked that the crude idea of obtaining currents by means of a coil or a magnetic core attached to the piston of a reciprocating steam-engine, is not in itself an entire novelty. It may also be noted that steam-turbines of extremely high rotative velocity are sometimes used instead of slow-moving engines to drive dynamos. But in the first class of long-abandoned experiments no practical result of any kind was ever reached before by any sort of device; and in the second class there is the objection that the turbine is driven by means of isolated shocks that cannot be overcome by any design of the blades, and which frustrate any attempts to perform work of the kind now under survey. What we are dealing with here is a dual, interacting machine, half mechanical, half electrical, of smallest bulk, extremely simple, utilizing steam under conditions unquestionably of the highest efficiency, its vibrations independent of load and pressure, delivering currents of the greatest regularity ever known for practical work or research. That such a combination should produce electricity for half the consumption of steam previously necessary with familiar apparatus in equivalent results, need not surprise us; yet think how much a saving of that kind would mean in well-nigh every industry consuming power!

Fig. 2. Latest form of tesla oscillator, combining in one mechanism dynamo and steam engine.

The oscillator and the production of light

Having obtained with the oscillator currents of high potential, high frequency, and high regularity, what shall be done with them? Mr. Tesla having already grappled successfully with the great difficulties of long distance power transmission, as narrated above, has first answered that question by boldly assailing the problem of the production of light in a manner nearer, perhaps, to that which gives us sunshine than was ever attempted before. Between us and the sun stretches the tenuous, sensitive ether, and every sensation of light that the eye experiences is caused by the effect of five hundred trillions of waves every second impressed on the ether by the molecular energy of the sun traveling along it rhythmically. If the waves have a lower frequency than this 500,000,000,000,000, they will chiefly engender heat. In our artificial methods of getting light we imitatively agitate the ether so poorly that the waves our bonfires set up rarely get above the rate at which they become sensible to us in heat, and only a few waves attain the right pitch or rapidity to cause the sensation of light. At the upper end of the keyboard of vibration of the ether is a high, shrill, and yet inaudible note,—light,—which we want to strike and to keep on striking; but we fumble at the lower, bass end of the instrument all the time, and never touch that topmost note without wasting the largest part of our energy on the intermediate ones, which we do not at all wish to touch. Light (the high note) without heat (the lower notes) is the desideratum. The inefficiency of the gas flame has been mentioned. In the ordinary incandescent lamp the waste is not so great; but even there the net efficiency of any one hundred units of energy put into it as electric current is at the most five or six of light, the waste occurring in the process of setting the molecules of the filament and the little air left in the bulb into the state of vibration under which they must work before they can throw out energy-waves on the ether, which will be conveyed to us through the glass of the bulb the ether as light rather than as heat. The glass is as unconfining to the ether as a coarse sieve is to water.

Now Mr. Tesla takes his currents of high frequency and high potential, subjects the incandescent lamp to them, and, skipping some of those intermediate wasteful heat stages of lower wave vibration experienced in the old methods, gets the ether-charged molecules more quickly into the intensely agitated condition necessary to yield light. Using his currents, produced electromagnetically, as we have seen, to load each fugitive molecule with its charge, which it receives and exercises electrostatically, he gets the ether medium into a state of excitement in which it seems to become capable of almost anything. In one of his first lectures, Mr. Tesla said:

Electrostatic effects are in many ways available for the production of light. For instance, we may place a body of some refractory material in a closed, and preferably in a more or less air exhausted, globe, connect it to a source of high, rapidly alternating potential, causing the molecules of the gas to strike it many times a second at enormous speeds, and in this way, with trillions of invisible hammers, pound it until it gets incandescent. Or we may place a body in a very highly exhausted globe, and by employing very high frequencies and potentials maintain it at any degree of incandescence. Or we may disturb the ether carried by the molecules of a gas, or their static charges, causing them to vibrate or emit light.

These anticipatory statements are confirmed today by what Mr. Tesla has actually done in one old way revolutionized, and in three new ways: (1) the incandescence of a solid; (2) phosphorescence; (3) incandescence or phosphorescence of a rarefied gas; and (4) luminosity produced in a gas at ordinary pressure.

Lamps with buttons or bars in place of filaments

Taking lamps in the first category, it may be stated that it had been commonly supposed that the light-giving conductor in the lamp, to be efficient and practical, should be fine; hence the name "filament" given to the carbon loop some in such lamps. But with the Tesla currents, the resistance or friction of the filament to the of flow of current does not count for anything: the filament may just as well be short and thick, for it will rapidly reach and steadily maintain proper incandescence by the passage of a small current of the right high frequency and potential. An action is set up as the result of which the filament is hit millions of times a second by the bombardment of the molecules around it in a merciless ring of tormentors. The vibrations of the current in similar manner will cause the infinite jostling of the molecules of solid and gas against a small polished carbon or metallic button or bar in a lamp, and brilliant light is also obtainable in this way.

Fig. 3. First photograph ever taken by phosphorescent light. The face is that of Mr. Tesla, and the source of light is one of his phosphorescent bulbs. Time of exposure: eight minutes. Date of photograph January, 1894.

Fig. 4. Phosphograph of Mr. Clemens (Mark Twain), taken in the Tesla Laboratory January, 1894. Time of exposure: ten minutes.

Fig. 5. Three phosphorescent bulbs under test for actinic value, photographed by their own light.

Light and photographs with tesla phosphorescent bulbs

In the field of lighting by phosphorescence we reach hitherto un-trodden ground. Phosphorescent light has been associated with the idea of "cold light," or the property of becoming luminous with the omission of the intermediate step of combustion, as commonly understood. As a physical action, we know it in the light of the firefly, which Professor S. P. Langley rates at an efficiency of 100%, all its radiations lying within the limits of the visible spectrum. By means of the Tesla currents phosphorescent light strong enough even to photograph by has been obtained; and Fig. 3, representing the inventor himself, is the first portrait or photograph of any kind ever taken by phosphorescent light. A bulb whose light-giving member is coated with sulphide of zinc treated in a special way was rendered phosphorescent by means of current obtained from a high-frequency transformer coil. The current used was alternated or oscillated about 10,000 times per second. The exposure was about eight minutes.

Fig. 4, of Mr. Clemens (Mark Twain), was taken a few weeks later—early in 1894—with the aid of the same bulb, and with an exposure of about ten minutes. In order to test more closely the actinic value of phosphorescent light, some bulbs subject to high-frequency currents were photographed, or, if we may coin a new word, "phosphographed," with a somewhat longer exposure. They are shown in Fig. 5. The right-hand, bright pair utilize sulphide of zinc in form for luminosity. The third bulb, seen faintly to the left of them, has a coating of sulphide of calcium. Although, judged by eye, it glowed with a brightness fully equal to that of the other two, the actinic value was evidently much less. It is, perhaps, needless to say that these demonstrations invite to an endless variety of experiments, in which investigators will find a host of novel phenomena awaiting them as to phosphorescence and fluorescence produced with electrical currents.

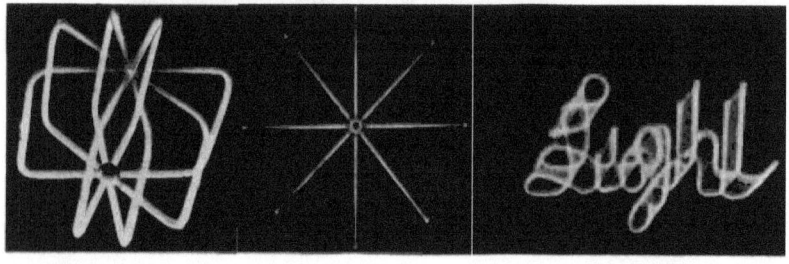

Fig. 6.   Fig. 7.   Fig. 8.

Figs. 6, 7, And 8 are tesla tubes in different forms in which light is obtained without filament or combustion. (Photographed by their own light.)

Light from empty bulbs in free space

The third and fourth classes of lighting enumerated above as obtained by Tesla currents are those caused by the incandescence or phosphorescence of a rarefied gas and the luminosity of a gas at ordinary pressure. We get pure, beautiful light without any filament or any combustion. In Figs. 6, 7, and 8 we have tubes or bulbs by means of which some of these interesting phenomena are obtained and illustrated. The bulbs shown are more or less exhausted of air. In the case of Figs. 6 and 7 the glass of the tubes is the ordinary German glass. In Fig. 8, uranium glass "green" was employed. This last was held in the hand while a photograph was taken of it by its own light; whence the unsteadiness of the negative. To obtain the beautiful illumination seen in all three, the bulbs were simply approached within a few inches of the terminal of a high-frequency coil or transformer. Just here it may be pointed out that the lamps are spoken of as unattached, in free space. Ordinary incandescent lighting is done, as everybody knows, with the lamps' bases firmly attached to the two current-bearing wires. Even where the lamps have been used on the ordinary alternating circuits in which the transformer is employed to — step down, — or reduce, for safe use, the higher-tension current brought to it by the wire from the dynamo, the lamps have to be attached to the "secondary" wires of the coil so as to make a closed circuit for them. But as we rise in the frequency of the current, as we leave behind the electrodynamic conditions for the electrostatic ones, so we free ourselves from the restrictions and limitations of solid wires for the conveyance of the effects sought, until at last we reach a point where all the old ideas of the necessity of a tangible circuit vanish. It is all circuit if we can properly direct the right kind of impulses through it. As Mr. Tesla long ago pointed out, most of the experiments usually performed with a static machine of glass plates can also be performed with an induction-coil of wire if the currents are alternated rapidly enough; and it is in reality here that Mr. Tesla parts company with other distinguished workers who have fixed their attention merely on the results attainable with electrodynamic apparatus. Before passing on, let us quote the inventor himself:

"Powerful electrostatic effects are a sine qua non of light production on the lines indicated by theory. Electromagnetic effects are primarily unavailable, for the reason that to produce the required effects we would have to pass the current impulses through a conductor which, long before the required frequency of the impulses could be reached, would cease to transmit them. On the other hand, electromagnetic waves many times longer than those of light, and producible by sudden discharge of a condenser, could not be utilized, it would seem, unless we availed ourselves of their effect upon conductors as in the present methods, which are wasteful. We could not affect by means of such waves the static molecular or atomic charges of a gas, and cause them to vibrate and to emit light. Long transverse waves cannot, apparently, produce such effects, since excessively small electromagnetic disturbances may pass readily through miles of air. Such dark waves, unless they are of the length of true light-waves, cannot, it would seem, excite luminous radiation in a Geissler tube, and the luminous effects which are producible by induction in a tube devoid of electrodes, I am inclined to consider as being of an electrostatic nature. To produce such luminous effects straight electrostatic thrusts are required; these, whatever be their frequency, may disturb the molecular charges and produce light."

Effects with attuned but widely separated circuits

A few experiments performed in Mr. Tesla's laboratory work shop afford an idea of the flexibility of the methods by which powerful electrostatic effects are produced across many feet of intervening space. The workshop is a room about forty by eighty feet, and ten or twelve feet high. A circuit of small cable is carried around it from the terminals of the oscillator. In the center of the clear, open space is placed a coil, wound drum fashion, three or four feet high, and unconnected with the current source save through the medium of the atmosphere. The coil is provided, as shown in the picture, with two condenser plates for adjustment, standing up like cymbals. The plates act after the manner of a spring, and the coil is comparable to an electromagnetic weight. The system of apparatus in the middle of the room has therefore a certain period of vibration, just as though it were a tuning-fork, or a sheet of thin resonant glass. Around the room, over the cable, there are sent from the oscillator electrical current vibrations. By carefully adjusting the condenser plates so that the periodicity or swing of the induced current is brought into step with that of the cable currents, powerful sparks are made to pour across between the plates in the dense streams shown in Fig. 9. In this manner it is easy to reach tensions as high as 200,000 and 300,000 volts.

Fig. 9 Experiment showing play of electric sparks between condenser plates, produced by electric charge. The coil, standing in the center of a large room, is unconnected with the energizing circuit. (From flashlight photograph.)

No one who has witnessed these significant experiments can fail to be impressed with the evidence of the actuality of a medium, call it ether or what you will, which in spite of its wonderful tenuity is as capable of transmitting energy as though it were air or water. Still more impressive to a layman, perhaps, is the confidence and easy precision with which these fine adjustments are brought about.

Fig. 10. Experiment showing the lighting up of an ordinary incandescent lamp, at a distance, through the influence of electrified ether-waves. (From flash-light photograph.)

In Fig. 10 there is a similar coil, in the middle of the same room, which has been so adjusted to the vibrations sent around the shop that an ordinary sixteen-candle-power incandescent lamp is well lighted up.

Fig. 11 pursues this a little further. Above the coil a circle of wire is held by an observer, an incandescent lamp is attached to the circle. As before, the vibration of the ether in the coil is brought into harmony with the vibrations emitted from the cable. The inductive effect upon the circle held loosely in free space by the observer is so pronounced that lamp is immediately lighted up, though it may connected with but one terminal wire, or with two. A 100-volt lamp is used, requiring when employed ordinarily more than one tenth of a horse-power right off the connecting circuit wires direct from the dynamo to bring it up to proper illuminating value. Hence, as will be seen, there is actual proof here of the transmission of at least that amount of energy across a space of some twenty feet and into the bulb by actually no wire at all. This need not surprise us when we remember that on a bright day the ether delivers steadily from the sun a horse-power of energy to every seven square feet of the earth's surface toward it: so great is its capacity for transmitting energy. Mr. Tesla with his "electrostatic thrusts" has simply learned the knack of loading electrically on the good-natured ether a little of the protean energy of which no amount has yet sufficed to break it down or put it out of temper. We may assume either an enormous speed in what may be called the transmitting wheelwork of the ether, since the weight is inconceivably small; or else that the ether is a mere transmitter of energy by its well-nigh absolute incompressibility.

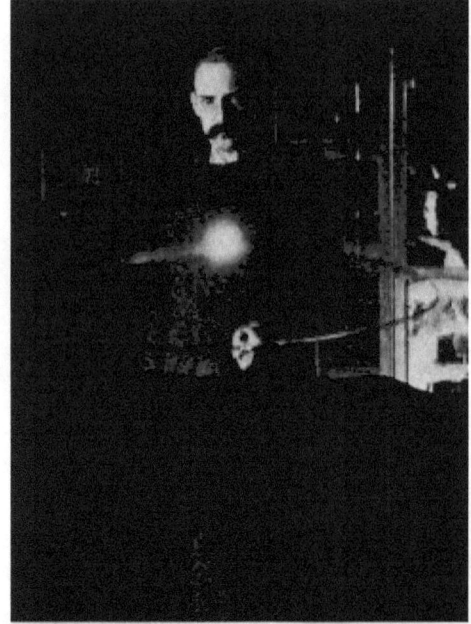

Fig. 11. Experiment illustrating the lighting of an incandescent lamp in free space by induction from coil below, energized by distant circuit around the room. The loop of wire carrying the lamp is held by Mr. Marion Crawford. (From flash-light photograph.)

Fig. 12. Similar experiment, illustrating the phenomenon of impedance. The loop of wire, carrying two lamps, is held by Mr. Joseph Jefferson. (From flash-light photograph.)

Curious impedance phenomenon

In Fig. 12 we have another remarkable experiment illustrated. Standing over the coil in the center of the room, the observer holds a hoop of stout wire in his hand. One or more lamps are connected with two points on the wire, so that the lamps are "short-circuited" by the short bar of wire. The vibrations are, however, so extremely rapid that in spite of the opposite terminals being united in this way, the current does not flow past them neglectfully, in the apparently easier path, as it should, but brings them to a bright incandescence. We have here an example of what is known as "impedance" phenomena, in which the current is oddly choked back at certain points and not at others. Under the conditions of "impedance," the best electrical conductor loses its property of conducting, and behaves like a highly resisting substance. Elaborating further these experimental results, Mr. Tesla shows that a gas — a perfect non-conductor under ordinary circumstances — may be more conductive than the best copper wire, provided the currents vibrate rapidly enough. The fantastic side of this phenomenon he touched on playfully once by suggesting that perchance in such wise we might some day utilize gas to convey electricity, and the old gas-pipe to insulate it.

Fig. 13. Similar experiment, the high-tension current bring passed through the body before it beings the lamps to incandescence. The loop is held over the resonating coil by Mr. Clemens (Mark Twain). (From a flash-light photograph.)

Lamps lighted by currents passed through the human body

In Fig. 13 a most curious and weird phenomenon is illustrated. A few years ago electricians would have considered it quite remarkable, if indeed they do not now. The observer holds a loop of bare wire in his hands. The currents induced in the loop by means of the "resonating" coil over which it is held, traverse the body of the observer, and at the same time, as they pass between his bare hands, they bring two or three lamps held there to bright incandescence. Strange as it may seem, these currents, of a voltage one or two hundred times as high as that employed in electrocution, do not inconvenience the experimenter in the slightest. The extremely high tension of the currents which Mr. Clemens is seen receiving prevents them from doing any harm to him.

Transmission of intelligence by attuned or — resonating — currents

Reference has been made to the "resonating" quality of the circuits and coils. It would be wearisome, and indeed is not necessary, here to dwell on the difficulty often experienced in establishing the relation of "resonance," and the instantaneity with which it can be disturbed. It may be stated, in order to give some idea of the conditions to be observed in these experiments, that when an electric circuit is traversed by a rapidly oscillating current which sets tip waves in the ether around the wire, the effect of these waves upon another circuit situated at some distance from the first can be largely varied by proper adjustments. The effect is most pronounced when the second circuit is so adjusted that its period of vibration is the same as that of the first. This harmonizing is deftly accomplished by varying either of the two elements which chiefly govern the rapidity of the vibration, viz., the so-called "capacity" and the "self-induction." Whatever the exact process may be, it is clear that these two quantities in their effect answer almost directly to what are known in mechanics as pliability and as weight or inertia. Attach to a spring a weight, and it will vibrate at a certain rate. By changing the weight, or modifying the pliability of the spring, any period of vibration is obtainable. In very exact adjustments, minute changes will completely upset the balance, and the very last straw of fine wire, for example, in the induction-coil which gives the self-induction will break the spell. As Mr. Tesla has said, it is really a lucky thing that pure resonance is not obtainable; for if it were, all kinds of dangers might lie in store for us by the increasing oscillations of every kind that would be set up. It will, however, have been gathered that if one electrical circuit can be tuned to another effectively, we shall need no return wire, as heretofore, for motors or for lights, the one wire being, if anything, better than two, provided we have vibration of the right value; and if we have that, we might get along without any wires or any "currents." Here again we must quote Mr. Tesla:

In connection with resonance effects and the problem of transmission of energy over a single conductor, I would say a few words on a subject which constantly fills my thoughts, and which concerns the welfare of all. I mean the transmission of intelligible signals, or perhaps even power, to any distance without the use of wires. I am becoming daily more convinced of the practicability of the scheme; and though I know full well that the majority of scientific men will not believe that such results can be practically and immediately realized, yet I think that all consider the developments of recent years by a number of workers to have been such as to encourage thought and experiment in this direction. My conviction has grown so strong that I no longer look upon this plan of energy or intelligence transmission as a mere theoretical possibility, but as a serious problem in electrical engineering which must be carried out some day. The idea of transmitting intelligence without wire is the natural outcome of the most recent results of electrical investigations, Some enthusiasts have expressed their belief that telephony to any distance by induction through the air is possible. I cannot stretch my imagination so far; but I do firmly believe that it is practicable to disturb by means of powerful machines the electrostatic condition of the earth, and thus transmit intelligible signals and perhaps power. In fact, what is there against the carrying out of such a scheme? We now know that electric vibration may be transmitted through a single conductor. Why, then, not try to avail ourselves of the earth for this purpose? We need not be frightened by the idea of distance. To the weary wanderer counting the of mile-posts the earth may appear very large; but to that happiest of all men, the astronomer who gazes at the heavens, and by their standard judges the magnitude of our globe, it appears very small. And so I think it must seem to the electrician; for when he considers the speed with which an electric disturbance is propagated through the earth, all his ideas of distance must completely vanish. A point of great importance would be first to know what is the capacity of the earth, and what charge does it contain of electricity.

Disturbance and demonstration of the earth's electrical charge

Part of Mr. Tesla's more recent work has been in the direction here indicated; for in his oscillator he has not simply a new practical device, but a new implement of scientific research. With the oscillator, if he has not as yet actually determined the earth's electrical charge or "capacity," he has obtained striking effects which conclusively demonstrate that he has succeeded in disturbing it. He connects to the earth, by one of its ends, a coil (see Fig. 15) in which rapidly vibrating currents are produced, the other end being free in space. With this coil he does actually what one would he doing with a pump forcing air into an elastic football. At each alternate stroke the ball would expand and contract. But it is evident that such a ball, if filled with air, would, when suddenly expanded or contracted, vibrate at its own rate. Now if the strokes of the pump be so timed that they are in harmony with the individual vibrations of the ball, an intense vibration or surging will be obtained. The purple streamers electricity thus elicited from the earth and pouring out to the ambient air are marvelous. Such a display is seen in Fig. 14, where the crown of the coil, tapering upward in a Peak of Teneriffe, flames with the outburst of a solar photosphere.

The currents which are made to pass in and out of the earth by means of this coil can also be directed upon the human body. An observer mounted on a chair, and touching the coil with a metal rod, can, by careful adjustments, divert enough of it upon himself to cause its manifestation from and around him in splinters of light. This halo effect, obtained by sending the electricity of the earth through a human being,— the highest charge positively ever given in safety,— is, to say the least, curious, and deeply suggestive. Mr. Tesla's temerity in trying the effect first upon his own person can be justified only by his close and accurate calculation of what the amount of the discharge from the earth would be.

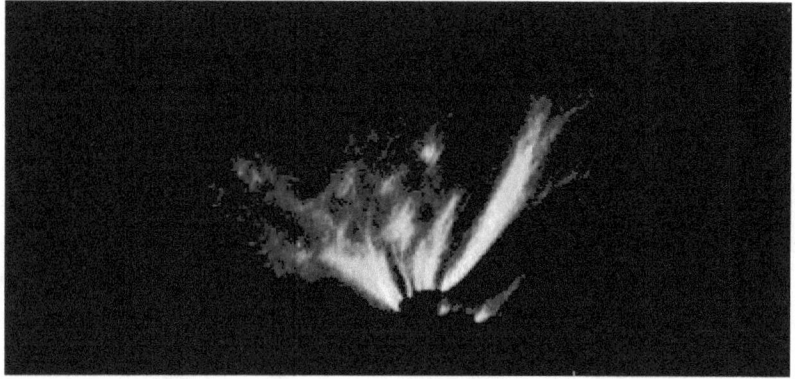

Fig. 14. Effect of electrical discharge from the earth by tesla coil.
(Photographed by its own light.)

Considering that in the adjustments necessary here, a small length of wire or a small body of any kind added to the coil or brought into its vicinity may destroy entirely all effect, one can imagine the pleasure which the investigator feels when thus rewarded by unique phenomena. After searching with patient toil for two or three years after a result calculated in advance, he is compensated by being able to witness a most magnificent display of fiery streams and lightning discharges breaking out from the tip of the wire with the roar of a gas-well. Aside from their deep scientific import and their wondrous fascination as a spectacle, such effects point to many new realizations making for the higher welfare of the human race. The transmission of power and intelligence is but one thing; the modification of climatic conditions maybe another. Perchance we shall "call up" Mars in this way some day, the electrical charge of both planets being utilized in signals.

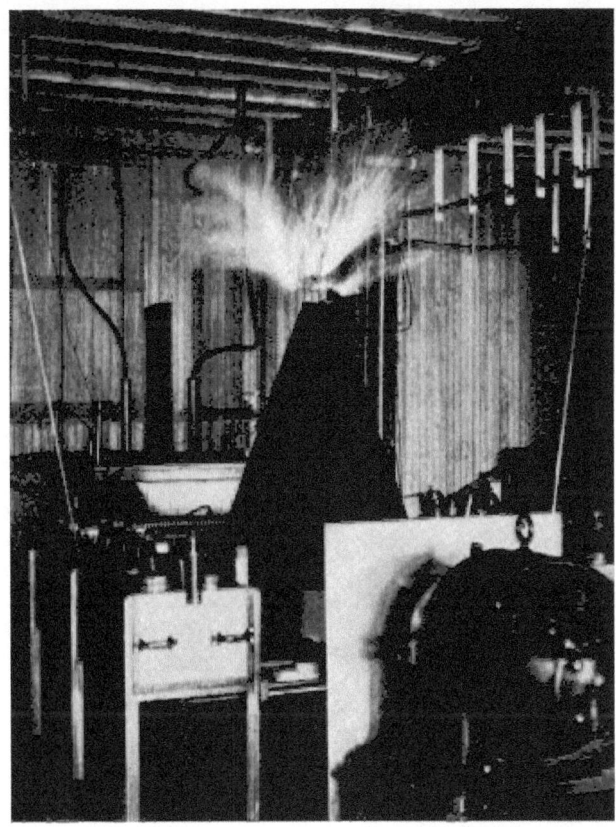

Fig. 15. Tesla coil for ascertaining and discharging the electricity of the earth. The streamers at top of coil are of purple hue, and in form resemble filaments of seaweed, the effect of mass being caused by prolonged exposure of flash-light negative.

Here are great results, lofty aims, and noble ideas; and yet they are but a beggarly few of all those with which Mr. Tesla, by his simple, modest work, has associated his name during recent years. He is not an impracticable visionary, but a worker who, with solid achievements behind him, seeks larger and better ones that lie before, as well as fuller knowledge. I have ventured to supplement data as to his late inventions by some of his views as to the ether, which throughout this presentation of his work has been treated familiarly as the maid-of-all-work of the universe. All our explanations of things are but half-way houses to the ultimate facts. It may be said, then, in conclusion, that while Mr. Tesla does not hold Professor Oliver Lodge's ingenious but intricate notion of two electricities and two ethers, and of the ether as itself electricity, he does belong to what Lord Kelvin has spoken of as the nineteenth-century school of plenum, accepting one ether for light, heat, electricity, and magnetism, outward manifestations of an inward unity whose secret we shall some day learn.

Thomas Commerford Martin.
April 1895

Notes:

- The photographs reproduced in this article were taken, under the special direction of the inventor, by Tonnele & Co.
- A biographical sketch of Mr. Tesla, by the present writer, with portrait, appeared in THE CENTURY for February, 1894.

EARTH ELECTRICITY TO KILL MONOPOLY

—The World Sunday Magazine—March 8, 1896. A Way to Harness Free Electric Currents Discovered by Nikola Tesla

The World is on the eve of an astounding revelation. The conditions under which we exist will be changed. The end has come to telegraph and telephone monopolies with a crash. Incidentally, all the other monopolies that depend on power of any kind will come to a sudden stop. The earth currents of electricity are to be harnessed. Nature supplies them free of charge. The cost of power and light and heat will be practically nothing.

Tesla's Marvellous Achievement.

COMMUNICATING FOUR MILES THROUGH PIKE'S PEAK WITHOUT WIRES.

The scientist-electricians who have for years been trying to master the mystery of electrical earth currents with which the ground beneath your feet is filled, are on the threshold of success. The success of the experiments they have under way means much to them, but vastly more to the people. It means that if Nikola Tesla succeeds in harnessing the electrical earth currents and putting then to work for man there will be an end to oppressive, extortionate monopolies in steam, telephone, telegraphs and the other commercial uses of electricity, and that the grasping millionaires who have for two decades milked the people's purse with electrical fingers will have to relinquish their monopoly.

Nikola Tesla has discovered the secret of the electric earth currents of nature, and they will be adapted to the use of man. He has succeeded in transmitting sound by the currents that make an electric riot of the earth. The transmission of power will follow. His experiments reduced to commercially practicable uses will be able to tap the electric currents of the earth and make them serve the purposes of industry and of trade just as a well digger over on Long Island taps water or a Pennsylvania miner opens a vein of coal. The mighty electrical energy that has been stored up in the earth for ages will be harnessed and made to move the machinery of men.

Electricity will be as free as the air. For the privilege of its use legislatures will not have to be bribed or men corrupted at the polls, and public boards will not have to be "seen" to bestow exclusive franchises upon corporations organized to use public property for purposes of private gain, and make the people pay the original cost of their investment and excessive charges for service in order to squeeze dividends out of copiously watered shares.

Monopolies for purveying steams power too will be forced to capitulate to free electricity, for with the latter manufactures will only have to connect their dynamos with the earth currents to set their machinery in motion. The successful adaptation of Tesla's discovery will administer a death-blow to the most galling slavery that has ever yoked the activities of men to the treadmill of monopoly. Tesla is the wizard who is going to emancipate modern industries from the shackles of corrupting, dividend-grabbing, monopolistic corporations.

Sound travels with amazing speed, but electrical vibrations travel so swiftly that it is difficult to conjure up a figure which will graphically illustrate their speed. Here is one that will perhaps convey a vivid and lucid impression. In fancy place yourself at a table with a revolver in one hand and a finger of the other hand on the key of a telegraph instrument connected with a wire that girdles the globe seven times and laps over on the eighth turn a distance equal to 11,000 miles. Pull the trigger of the pistol and simultaneously press the telegraph key. While the sound of the report of the revolver is traveling 1,100 feet the electrical impulse imparted by the pressure on the key will pass seven and a half times around the world through the wire with which the key is connected.

Sound travels 1,250 feet a second and electrical impulse 186,000 miles a second. If the electrical currents with which the earth is filled can be harnessed and put to work a new era in electricity will have dawned. It is to the mastering of the mystery of these earth currents and their adaptation that scientist like Tesla have been striving.

In the course of Tesla's experiments it is reported he found that in the vicinity of large cities there were so many conflicting earth currents that satisfactory results could not be obtained. So he went out to Denver and near there found a better field for experimenting. There he met a friend interested in electrical research. They went to Pike's Peak. Conspicuous among their baggage were two autoharps.

Tesla and his friend scaled the rugged sides of the peak. At an elevation agreed upon they separated. Tesla skirted the peak and on reaching a point precisely opposite the place at which he left his friend he stopped. The two experimenters, on a line drawn straight through the peak, were thus separated by four miles of stratified rock. The two autoharps had been very delicately attuned before the scientists parted, and a time fixed for Mr. Tesla's comrade to play an air (also agreed upon) on the autoharp.

Tesla waited patiently the arrival of the appointed time. Then he connected his harp with the ground in such a way as to secure harmonic resonance with the earth current. The manner and medium of this connection are secrets. The receiving autoharp was equipped with a microphone. As the time approached for his friend on the other side of the peak to strum the appointed tune Tesla listened with rapt attention.

At last, as a tuning-fork responds to its harmonic note sounded on the strings of a piano, the autoharp in Mr. Tesla's hands gave out the harmonic tunes of "Ben Bolt" which his companion at his station four miles away straight through the peak was plucking from the tense wires of his instrument. The experiment was a success. After many tunes had been played Tesla and his companion descended the peak. A statement of the facts and results of the experiment was written and attested before a notary public as a matter of scientific record.

The electric currents are in the earth. Their strength is great enough to furnish all the power and light man needs. Mr. Tesla has overcome the initial difficulty, and has located and tapped the earth's currents. The rest will follow, as followed the telephone, Prof. Bell's discovery of how to transmit speech over a wire.

March 8, 1896

ON ELECTRICITY

—The Address On the Occasion of the Commemoration of the Introduction of Niagara Falls Power In Buffalo At the Ellicot Club, January 12, 1897. Electrical Review, January 27, 1897

I have scarcely had courage enough to address an audience on a few unavoidable occasions, and the experience of this evening, even as disconnected from the cause of our meeting, is quite novel to me. Although in those few instances, of which I have retained agreeable memory, my words have met with a generous reception, I never deceived myself, and knew quite well that my success was not due to any excellency in the rhetorical or demonstrative art. Nevertheless, my sense of duty to respond to the request with which I was honored a few days ago was strong enough to overcome my very grave apprehensions in regard to my ability of doing justice to the topic assigned to me. It is true, at times—even now, as I speak—my mind feels full of the subject, but I know that, as soon as I shall attempt expression, the fugitive conceptions will vanish, and I shall experience certain well known sensations of abandonment, chill and silence. I can see already your disappointed countenances and can read in them the painful regret of the mistake in your choice.

These remarks, gentlemen, are not made with selfish desire of winning your kindness and indulgence on my shortcomings, but with the honest intention of offering you an apology for your disappointment. Nor are they made—as you might be disposed to think—in that playful spirit which, to the enjoyment of the listeners is often displayed by belated speakers. On the contrary, I am deeply earnest in my wish that I were capable of having the fire of eloquence kindled in me, that I might dwell in adequate terms on this fascinating science of electricity, on the marvelous development which electrical annals have recorded and which, as one of the speakers justly remarked, stamp this age as the Electrical Age, and particularly on the great event we are commemorating this day. Unfortunately, this my desire must remain unfulfilled, but I am hopeful that in my formless and incomplete statements, among the few ideas and facts I shall mention there may be something of interest and usefulness, something befitting this unique occasion.

Gentlemen, there are a number of features clearly discernible in, and characteristic of, human intellectual progress in more recent times—features which afford great comfort to the minds of all those who have really at heart the advancement and welfare of mankind.

First of all the inquiry, by the aid of the microscope and electrical instruments of precision, into the nature of our organs and senses, and particularly of those through which we commune directly with the outside world and through which knowledge is conveyed to our minds, has revealed their exact construction and mode of action, which is in conformity with simple and well established physical principles and laws. Hence the observations we make and the facts we ascertain by their help are real facts and observations, and our knowledge is true knowledge. To illustrate: our knowledge of form, for instance, is dependent upon the positive fact that light propagates in straight lines, and, owing to this, the image formed by a lens is exactly similar to the object seen. Indeed, my thoughts in such fields and directions have led me to the conclusion that most all human knowledge is based on this simple truth, since practically every idea or conception—and therefore all knowledge—presupposes visual impressions. But if light would not propagate in accordance with the law mentioned, but in conformity with any other law which we might presently conceive, whereby not only the image might not bear any likeness to the object seen, but even the images of the same object at different times or distances might not resemble each other, then our knowledge of form would be very defective, for then we might see, for example, a three-cornered figure as a six or twelve-cornered one. With the clear understanding of the mechanism and mode of action of our organs, we remove all doubts as to the reality and truth of the impressions received from the outside, and thus we bar out—forever, we may hope—that unhealthy speculation and skepticism into which formerly even strong minds were apt to fall.

Let me tell you of another comforting feature. The progress in a measured time is nowadays more rapid and greater than it ever was before. This is quite in accordance with the fundamental law of motion, which commands acceleration and increase of momentum or accumulation of energy under the action of a continuously acting force and tendency, and is the more true as every advance weakens the elements tending to produce friction and retardation. For, after all, what is progress, or—more correctly—development, or evolution, if not a movement, infinitely complex and often unscrutinizable, it is true, but nevertheless exactly determined in quantity as well as in quality of motion by the physical conditions and laws governing? This feature of more recent development is best shown in the rapid merging together of the various arts and sciences by the obliteration of the hard and fast lines of separation, of borders, some of which only a few years ago seemed unsurpassable, and which, like veritable Chinese walls, surrounded every department of inquiry and barred progress. A sense of connectedness of the various apparently widely different forces and phenomena we observe is taking possession of our minds, a sense of deeper understanding of nature as a whole, which, though not yet quite clear and defined, is keen enough to inspire us with the confidence of vast realizations in the near future.

But these features chiefly interest the scientific man, the thinker and reasoner. There is another feature which affords us still more satisfaction and enjoyment, and which is of still more universal interest, chiefly because of its bearing upon the welfare of mankind. Gentlemen, there is an influence which is getting strong and stronger day by day, which shows itself more and more in all departments of human activity, and influence most fruitful and beneficial — the influence of the artist. It was a happy day for the mass of humanity when the artist felt the desire of becoming a physician, an electrician, an engineer or mechanician or — whatnot — a mathematician or a financier; for it was he who wrought all these wonders and grandeur we are witnessing. It was he who abolished that small, pedantic, narrow-grooved school teaching which made of an aspiring student a galley-slave, and he who allowed freedom in the choice of subject of study according to one's pleasure and inclination, and so facilitated development.

Some, who delight in the exercise of the powers of criticism, call this an asymmetrical development, a degeneration or departure from the normal, or even a degradation of the race. But they are mistaken. This is a welcome state of things, a blessing, a wise subdivision of labors, the establishment of conditions most favorable to progress. Let one concentrate all his energies in one single great effort, let him perceive a single truth, even though he be consumed by the sacred fire, then millions of less gifted men can easily follow. Therefore it is not as much quantity as quality of work which determines the magnitude of the progress.

It was the artist, too, who awakened that broad philanthropic spirit which, even in old ages, shone in the teachings of noble reformers and philosophers, that spirit which makes men in all departments and positions work not as much for any material benefit or compensation — though reason may command this also — but chiefly for the sake of success, for the pleasure there is in achieving it and for the good they might be able to do thereby to their fellow-men. Through his influence types of men are now pressing forward, impelled by a deep love for their study, men who are doing wonders in their respective branches, whose chief aim and enjoyment is the acquisition and spread of knowledge, men who look far above earthly things, whose banner is Excelsior! Gentlemen, let us honor the artist; let us thank him, let us drink his health!

Now, in all these enjoyable and elevating features which characterize modern intellectual development, electricity, the expansion of the science of electricity, has been a most potent factor. Electrical science has revealed to us the true nature of light, has provided us with innumerable appliances and instruments of precision, and has thereby vastly added to the exactness of our knowledge. Electrical science has disclosed to us the more intimate relation existing between widely different forces and phenomena and has thus led us to a more complete comprehension of Nature and its many manifestations to our senses. Electrical science, too, by its fascination, by its promises of immense realizations, of wonderful possibilities chiefly in humanitarian respects, has attracted the attention and enlisted the energies of the artist; for where is there a field in which his God-given powers would be of a greater benefit to his fellow-men than this unexplored, almost virgin, region, where, like in a silent forest, a thousand voices respond to every call?

With these comforting features, with these cheering prospects, we need not look with any feeling of incertitude or apprehension into the future. There are pessimistic men, who, with anxious faces, continuously whisper in your ear that the nations are secretly arming—arming to the teeth; that they are going to pounce upon each other at a given signal and destroy themselves; that they are all trying to outdo that victorious, great, wonderful German army, against which there is no resistance, for every German has the discipline in his very blood—every German is a soldier, But these men are in error. Look only at our recent experience with the British in that Venezuela difficulty. Two other nations might have crashed together, but not the Anglo-Saxons; they are too far ahead. The men who tell you this are ignoring forces which are continually at work, silently but resistlessly—forces which say Peace!

There is the genuine artist, who inspires us with higher and nobler sentiments, and makes us abhor strife and carnage. There is the engineer, who bridges gulfs and chasms, and facilitates contact and equalization of the heterogeneous masses of humanity. There is the mechanic, who comes with his beautiful time and energy-saving appliances, who perfects his flying machine, not to drop a bag of dynamite on a city or vessel, but to facilitate transport and travel. There, again, is the chemist, who opens new resources and makes existence more pleasant and secure; and there is the electrician, who sends his messages of peace to all parts of the globe. The time will not be long in coming when those men who are turning their ingenuity to inventing quick-firing guns, torpedoes and other implements of destruction—all the while assuring you that it is for the love and good of humanity—will find no takers for their odious tools, and will realize that, had they used their inventive talent in other directions; they might have reaped a far better reward than the sestertia received. And then, and none too soon the cry will be echoed everywhere. Brethren, stop these high-handed methods of the strong, these remnants of barbarism so inimical to progress! Give that valiant warrior opportunities for displaying a more commendable courage than that he shows when, intoxicated with victory, he rushes to the destruction of his fellowmen. Let him toil day and night with a small chance of achieving and yet be unflinching; let him challenge the dangers of exploring the heights of the air and the depths of the sea; let him brave the dread of the plague, the heat of the tropic desert and the ice of the polar region. Turn your energies to warding off the common enemies and danger, the perils that are all around you, that threaten you in the air you breathe, in the water you drink, in the food you consume. It is not strange, is it not shame, that we, beings in the highest state of development in this our world, beings with such immense powers of thought and action, we, the masters of the globe, should be absolutely at the mercy of our unseen foes, that we should not know whether a swallow of food or drink brings joy and life or pain and destruction to us! In this most modern and sensible warfare, in which the bacteriologist leads, the services electricity will render will prove invaluable. The economical production of high-frequency currents, which is now an accomplished fact, enables us to generate easily and in large quantities ozone for the disinfection of the water and the air, while certain novel radiations recently discovered give hope of finding effective remedies against ills of microbic origin, which have heretofore withstood all efforts of the physician. But let me turn to a more pleasant theme.

I have referred to the merging together of the various sciences or departments of research, and to a certain perception of intimate connection between the manifold and apparently different forces and phenomena. Already we know, chiefly through the efforts of a bold pioneer, that light, radiant heat, electrical and magnetic actions are closely related, not to say identical. The chemist professes that the effects of combination and separation of bodies he observes are due to electrical forces, and the physician and physiologist will tell you that even life's progress is electrical. Thus electrical science has gained a universal meaning, and with right this age can claim the name "Age of Electricity."

I wish much to tell you on this occasion—I may say I actually burn for desire of telling you—what electricity really is, but I have very strong reasons, which my coworkers will best appreciate, to follow a precedent established by a great and venerable philosopher, and I shall not dwell on this purely scientific aspect of electricity.

There is another reason for the claim which I have before stated which is even more potent than the former, and that is the immense development in all electrical branches in more recent years and its influence upon other departments of science and industry. To illustrate this influence I only need to refer to the steam or gas engine. For more than half a century the steam engine has served the innumerable wants of man. The work it was called to perform was of such variety and the conditions in each case were so different that, of necessity, a great many types of engines have resulted. In the vast majority of cases the problem put before the engineer was not as it should have been, the broad one of converting the greatest possible amount of heat energy into mechanical power, but it was rather the specific problem of obtaining the mechanical power in such form as to be best suitable for general use. As the reciprocating motion of the piston was not convenient for practical purposes, except in very few instances, the piston was connected to a crank, and thus rotating motions was obtained, which was more suitable and preferable, though it involved numerous disadvantages incident to the crude and wasteful means employed. But until quite recently there were at the disposal of the engineer, for the transformation and transmission of the motion of the piston, no better means than rigid mechanical connections. The past few years have brought forcibly to the attention of the builder the electric motor, with its ideal features. Here was a mode of transmitting mechanical motion simpler by far, and also much more economical. Had this mode been perfected earlier, there can be no doubt that, of the many different types of engine, the majority would not exist, for just as soon as an engine was coupled with an electric generator a type was produced capable of almost universal use. From this moment on there was no necessity to endeavor to perfect engines of special designs capable of doing special kinds of work. The engineer's task became now to concentrate all his efforts upon one type, to perfect one kind of engine — the best; the universal, the engine of the immediate future; namely, the one which is best suitable for the generation of electricity. The first efforts in this direction gave a strong impetus to the development of the reciprocating high speed engine, and also to the turbine, which latter was a type of engine of very limited practical usefulness, but became, to a certain extent, valuable in connection with the electric generator and motor. Still, even the former engine, though improved in many particulars, is not radically changed, and even now has the same objectionable features and limitations. To do away with these as much as possible, a new type of engine is being perfected in which more favorable conditions for economy are maintained, which expands the working fluid with utmost rapidity and loses little heat on the walls, an engine stripped of all usual regulating mechanism—packings, oilers and other appendages—and forming part of an electric generator; and in this type, I may say, I have implicit faith.

The gas or explosive engine has been likewise profoundly affected by the commercial introduction of electric light and power, particularly in quite recent years. The engineer is turning his energies more and more in this direction, being attracted by the prospect of obtaining a higher thermodynamic efficiency. Much larger engines are now being built, the construction is constantly improved, and a novel type of engine, best suitable for the generation of electricity, is being rapidly evolved.

There are many other lines of manufacture and industry in which the influence of electrical development has been even more powerfully felt. So, for instance, the manufacture of a great variety of articles of metal, and especially of chemical products. The welding of metals by electricity, though involving a wasteful process, has, nevertheless, been accepted as a legitimate art, while the manufacture of metal sheet, seamless tubes and the like affords promise of much improvement. We are coming gradually, but surely, to the fusion of bodies and reduction of all kinds of ores—even of iron ores—by the use of electricity, and in each of these departments great realizations are probable. Again, the economical conversion of ordinary currents of supply into high-frequency currents opens up new possibilities, such as the combination of the atmospheric nitrogen and the production of its compounds; for instance, ammonia and nitric acid, and their salts, by novel processes.

The high-frequency currents also bring us to the realization of a more economical system of lighting; namely by means of phosphorescent bulbs or tubes, and enable us to produce with these appliances light of practically any candle-power. Following other developments in purely electrical lines, we have all rejoiced in observing the rapid strides made, which, in quite recent years, have been beyond our most sanguine expectations. To enumerate the many advances recorded is a subject for the reviewer, but I can not pass without mentioning the beautiful discoveries of Lenart and Roentgen, particularly the latter, which have found such a powerful response throughout the scientific world that they have made us forget, for a time, the great achievement of Linde in Germany, who has effected the liquefaction of air on an industrial scale by a process of continuous cooling: the discovery of argon by Lord Rayleigh and Professor Ramsay, and the splendid pioneer work of Professor Dewar in the field of low temperature research. The fact that the

United States have contributed a very liberal share to this prodigious progress must afford to all of us great satisfaction. While honoring the workers in other countries and all those who, by profession or inclination, are devoting themselves to strictly scientific pursuits, we have particular reasons to mention with gratitude the names of those who have so much contributed to this marvelous development of electrical industry in this country. Bell, who, by his admirable invention enabling us to transmit speech to great distances, has profoundly affected our commercial and social relations, and even our very mode of life; Edison, who, had he not done anything else beyond his early work in incandescent lighting, would have proved himself one of the greatest benefactors of the age; Westinghouse, the founder of the commercial alternating system; Brush, the great pioneer of arc lighting; Thomson, who gave us the first practical welding machine, and who, with keen sense, contributed very materially to the development of a number of scientific and industrial branches; Weston, who once led the world in dynamo design, and now leads in the construction of electric instruments; Sprague, who, with rare energy, mastered the problem and insured the success of practical electrical railroading; Acheson, Hall, Willson and others, who are creating new and revolutionizing industries here under our very eyes at Niagara. Nor is the work of these gifted men nearly finished at this hour. Much more is still to come, form fortunately, most of them are still full of enthusiasm and vigor. All of these men and many more are untiringly at work investigating new regions and opening up unsuspected and promising fields. Weekly, if not daily, we learn through the journals of a new advance into some unexplored region, where at every step success beckons friendly, and leads the toiler on to hard and harder tasks.

But among all these many departments of research, these many branches of industry, new and old, which are being rapidly expanded, there is one dominating all others in importance—one which is of the greatest significance for the comfort and welfare, not to say for the existence, of mankind, and that is the electrical transmission of power. And in this most important of all fields, gentlemen, long afterwards, when time will have placed the events in their proper perspective, and assigned men to their deserved places, the great event we are commemorating today will stand out as designating a new and glorious epoch in the history of humanity—an epoch grander than that marked by the advent of the steam engine. We have many a monument of past ages: we have the palaces and pyramids, the temples of the Greek and the cathedrals of Christendom. In them is exemplified the power of men, the greatness of nations, the love of art and religious devotion. But that monument at Niagara has something of its own, more in accord with our present thoughts and tendencies. It is a monument worthy of our scientific age, a true monument of enlightenment and of peace. It signifies the subjugation of natural forces to the service of man, the discontinuance of barbarous methods, the relieving of millions from want and suffering. No matter what we attempt to do, no matter to what fields we turn our efforts, we are dependent on power. Our economists may propose more economical systems of administration and utilization of resources, our legislators may make wiser laws and treaties, it matters little; that kind of help can be only temporary. If we want to reduce poverty and misery, if we want to give to every deserving individual what is needed for a safe existence of an intelligent being, we want to provide more machinery, more power. Power is our mainstay, the primary source of our many-sided energies. With sufficient power at our disposal we can satisfy most of our wants and offer a guaranty for safe and comfortable existence to all, except perhaps to those who are the greatest criminals of all — the voluntarily idle.

The development and wealth of a city, the success of a nation, the progress of the whole human race, is regulated by the power available. Think of the victorious march of the British, the like of which history has never recorded. Apart from the qualities of the race, which have been of great moment, they own the conquest of the world to coal. For with coal they produce their iron; coal furnishes them light and heat; coal drives the wheels of their immense manufacturing establishments, and coal propels their conquering fleets. But the stores are being more and more exhausted; the labor is getting dearer and dearer, and the demand is continuously increasing. It must be clear to every one that soon some new source of power supply must be opened up, or that at least the present methods must be materially improved. A great deal is expected from a more economical utilization of the stored energy of the carbon in a battery; but while the attainment of such a result would be hailed as a great achievement; it would not be as much of an advance towards the ultimate and permanent method of obtaining power as some engineers seem to believe. By reasons both of economy and convenience we are driven to the general adoption of a system of energy supply from central stations, and for such purposes the beauties of the mechanical generation of electricity can not be exaggerated. The advantages of this universally accepted method are certainly so great that the probability of replacing the engine dynamos by batteries is, in my opinion, a remote one, the more so as the high-pressure steam engine and gas engine give promise of a considerably more economical thermodynamic conversion. Even if we had this day such an economical coal battery, its introduction in central stations would by no means be assured, as its use would entail many inconveniences and drawbacks. Very likely the carbon could not be burned in its natural form as in a boiler, but would have to be specially prepared to secure uniformity in the current generation. There would be a great many cells needed to make up the electro-motive force usually required. The process of cleaning and renewal, the handling of nasty fluids and gases and the great space necessary for so many batteries would make it difficult, if not commercially unprofitable, to operate such a plant in a city or densely populated district. Again if the station be erected in the outskirts, the conversion by rotating transformers or otherwise would be a serious and unavoidable drawback. Furthermore, the regulating appliances and other accessories which would have to be provided would probably make the plant fully as much, if not more, complicated than the present. We might, of course, place the batteries at or near the coal mine, and from there transmit the energy to distant points in the form of high-tension alternating currents obtained from rotating transformers, but even in this most favorable case the process would be a barbarous one, certainly more so than the present, as it would still involve the consumption of material, while at the same time it would restrict the engineer and mechanic in the exercise of their beautiful art. As to the energy supply in small isolated places as dwellings, I have placed my confidence in the development of a light storage battery, involving the use of chemicals manufactured by cheap water power, such as some carbide or oxygen-hydrogen cell.

But we shall not satisfy ourselves simply with improving steam and explosive engines or inventing new batteries; we have something much better to work for, a greater task to fulfill. We have to evolve means for obtaining energy from stores which are forever inexhaustible, to perfect methods which do not imply consumption and waste of any material whatever. Upon this great possibility, which I have long ago recognized, upon this great problem, the practical solution of which means so much for humanity, I have myself concentrated my efforts since a number of years, and a few happy ideas which came to me have inspired me to attempt the most difficult, and given me strength and courage in adversity. Nearly six years ago my confidence had become strong enough to prompt me to an expression of hope in the ultimate solution of this all dominating problem. I have made progress since, and have passed the stage of mere conviction such as is derived from a diligent study of known facts, conclusions and calculations. I now feel sure that the realization of that idea is not far off. But precisely for this reason I feel impelled to point out here an important fact, which I hope will be remembered. Having examined for a long time the possibilities of the development I refer to, namely, that of the operation of engines on any point of the earth by the energy of the medium, I find that even under the theoretically best conditions such a method of obtaining power can not equal in economy, simplicity and many other features the present method, involving a conversion of the mechanical energy of running water into electrical energy and the transmission of the latter in the form of currents of very high tension to great distances. Provided, therefore, that we can avail ourselves of currents of sufficiently high tension, a waterfall affords us the most advantageous means of getting power from the sun sufficient for all our wants, and this recognition has impressed me strongly with the future importance of the water power, not so much because of its commercial value, though it may be very great, but chiefly because of its bearing upon our safety and welfare. I am glad to say that also in this latter direction my efforts have not been unsuccessful, for I have devised means which will allow us the use in power transmission of electromotive forces much higher than those practicable with ordinary apparatus. In fact, progress in this field has given me fresh hope that I shall see the fulfillment of one of my fondest dreams; namely, the transmission of power from station to station without the employment of any connecting wire. Still, whatever method of transmission be ultimately adopted, nearness to the source of power will remain an important advantage.

Gentlemen, some of the ideas I have expressed may appear to many of you hardly realizable; nevertheless, they are the result of long-continued thought and work. You would judge them more justly if you would have devoted your life to them, as I have done. With ideas it is like with dizzy heights you climb: At first they cause you discomfort and you are anxious to get down, distrustful of your own powers; but soon the remoteness of the turmoil of life and the inspiring influence of the altitude calm your blood; your step gets firm and sure and you begin to look—for dizzier heights. I have attempted to speak to you on "Electricity," its development and influence, but I fear that I have done it much like a boy who tries to draw a likeness with a few straight lines. But I have endeavored to bring out one feature, to speak to you in one strain which I felt sure would find response in the hearts of all of you, the only one worthy of this occasion—the humanitarian. In the great enterprise at Niagara we see not only a bold engineering and commercial feat, but far more, a giant stride in the right direction as indicated both by exact science and philanthropy. Its success is a signal for the utilization of water powers all over the world, and its influence upon industrial development is incalculable. We must all rejoice in the great achievement and congratulate the intrepid pioneers who have joined their efforts and means to bring it about. It is a pleasure to learn of the friendly attitude of the citizens of Buffalo and of the encouragement given to the enterprise by the Canadian authorities. We shall hope that other cities, like Rochester on this side and Hamilton and Toronto in Canada, will soon follow Buffalo's lead. This fortunate city herself is to be congratulated. With resources now unequaled, with commercial facilities and advantages such as few cities in the world possess, and with the enthusiasm and progressive spirit of its citizens, it is sure to become one of the greatest industrial centers of the globe.

January 27, 1897

HIGH FREQUENCY OSCILLATORS FOR ELECTRO-THERAPEUTIC AND OTHER PURPOSES

—*The Electrical Engineer. Vol. XXVI. November 17, 1898. No. 550.*

Some theoretical possibilities offered by currents of very high frequency and observations which I casually made while pursuing experiments with alternating currents, as well as the stimulating influence of the work of Hertz and of views boldly put forth by Oliver Lodge, determined me some time during 1889 to enter a systematic investigation of high frequency phenomena, and the results soon reached were such as to justify further efforts towards providing the laboratory with efficient means for carrying on the research in this particular field, which has proved itself so fruitful since. As a consequence alternators of special design were constructed and various arrangements for converting ordinary into high frequency currents perfected, both of which were duly described and are now—I assume—familiar.

One of the early observed and remarkable features of the high frequency currents, and one which was chiefly of interest to the physician, was their apparent harmlessness which made it possible to pass relatively great amounts of electrical energy through the body of a person without causing pain or serious discomfort. This peculiarity which, together with other mostly unlooked-for properties of these currents I had the honor to bring to the attention of scientific men first in an article in a technical journal in February, 1891, and in subsequent contributions to scientific societies, made it at once evident that these currents would lend themselves particularly to electro-therapeutic uses.

With regard to the electrical actions in general, and by analogy, it was reasonable to infer that the physiological effects, however complex, might be resolved in three classes. First the statical, that is, such as are chiefly dependent on the magnitude of electrical potential; second, the dynamical, that is, those principally dependent on the quality of electrical movement or current's strength through the body, and third, effects of a distinct nature due to electrical waves or oscillations, that is, impulses in which the electrical energy is alternately passing in more or less rapid succession through the static and dynamic forms.

Most generally in practice these different actions are coexistent, but by a suitable selection of apparatus and observance of conditions the experimenter may make one or other of these effects predominate. Thus he may pass through the body, or any part of the same, currents of comparatively large volume under a small electrical pressure, or he may subject the body to a high electrical pressure while the current is negligibly small, or he may put the patient under the influence of electrical waves transmitted, if desired, at considerable distance through space.

While it remained for the physician to investigate the specific actions on the organism and indicate proper methods of treatment, the various ways of applying these currents to the body of a patient suggested themselves readily to the electrician.

As one cannot be too clear in describing a subject, a diagrammatic illustration of the several modes of connecting the circuits which I will enumerate, though obvious for the majority, is deemed of advantage.

The first and simplest method of applying the currents was to connect the body of the patient to two points of the generator, be it a dynamo or induction coil. Fig. 1 is intended to illustrate this case. The alternator G may be one giving from five to ten thousand complete vibrations per second, this number being still within the limit of practicability. The electromotive force — as measured by a hot wire instrument — may be from fifty to one hundred volts. To enable strong currents to be passed through the tissues, the terminals T T, which serve to establish contact with the patient's body should, of course, be of large area, and covered with cloth saturated with a solution of electrolyte harmless to the skin, or else the contacts are made by immersion. The regulation of the currents is best effected by means of an insulating trough A provided with two metal terminals T' T' of considerable surface, one of which, at least, should be movable. The trough is filled with water, and an, electrolytic solution is added to the same, until a degree of conductivity is obtained suitable for the experiments.

When it is desired to use small currents of high tension, a secondary coil is resorted to, as illustrated in Fig. 2. I have found it from the outset convenient to make a departure from the ordinary ways of winding the coils with a considerable number of small turns. For many reasons the physician will find it better to provide a large hoop H of not less than, say three feet in diameter and preferably more, and to wind upon it a few turns of stout cable P. The secondary coil S is easily prepared by taking two wooden hoops h h and joining them with stiff cardboard. One single layer of ordinary magnet wire, and not too thin at that, will be generally sufficient, the number of turns necessary for the particular use for which the coil is intended being easily ascertained by a few trials. Two plates of large surface, forming an adjustable condenser, may be used for the purpose of synchronizing the secondary with the primary circuit, but this is generally not necessary. In this manner a cheap coil is obtained, and one which cannot be easily injured. Additional advantages, however, will be found in the perfect regulation which is effected merely by altering the distance between the primary and secondary, for which adjustment provision should be made, and, furthermore, in the occurrence of harmonics which are more pronounced in such large coils of thick wire, situated at some distance from the primary.

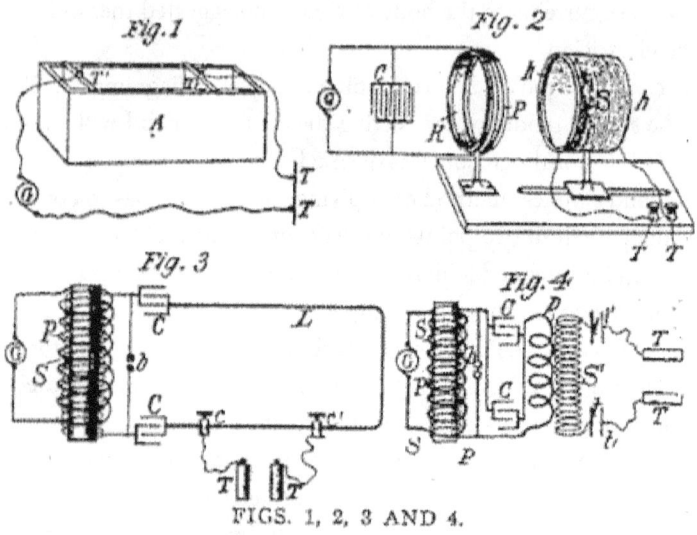

FIGS. 1, 2, 3 AND 4.

The preceding arrangements may also be used with alternating or interrupted currents of low frequency, but certain peculiar properties of high frequency currents make it possible to apply the latter in ways entirely impracticable with the former.

One of the prominent characteristics of high frequency or, to be more general, of rapidly varying currents, is that they pass with difficulty through stout conductors of high self-induction. So great is the obstruction which self-induction offers to their passage that it was found practicable, as shown in the early experiments to which reference has been made, to maintain differences of potential of many thousands of volts between two points-not more than a few inches apart-of a thick copper bar of inappreciable resistance. This observation naturally suggested the disposition illustrated in Fig. 3. The source of high frequency impulses is in this instance a familiar type of transformer which may be supplied from a generator G of ordinary direct or alternating currents. The transformer comprises a primary P, a secondary S, two condensers C C which are joined in series, a loop or coil of very thick wire L and a circuit interrupting device at break b. The currents are derived from the loop

L by two contacts c c', one or both of which are capable of displacement along the wire L. By varying the distance between these contacts, any difference of potential, from a few volts to many thousands, is readily obtained on the terminals or handles T T. This mode of using the currents is entirely safe and particularly convenient, but it requires a very uniform working of the break b employed for charging and discharging the condenser.

Another equally remarkable feature of high frequency impulses was found in the facility with which they are transmitted through condensers, moderate electromotive forces and very small capacities being required to enable currents of considerable volume to pass. This observation made it practicable to resort to a plan such as indicated in Fig. 4. Here the connections are similar to those shown in the preceding case, except that the condensers C C are joined in parallel. This lowers the frequency of the currents, but has the advantage of allowing the working with a much smaller difference of potential on the terminals of the secondary S. Since the later is the chief item of expense of such apparatus and since its price rapidly increases with the number of turns required, the experimenter will find it generally cheaper to make a sacrifice in the frequency, which, however, will be high enough for most purposes. However, he only needs to reduce proportionately the number of turns or the length of primary p to obtain the same frequency as before, but the economy of transformation will be somewhat reduced in so doing and the break h will require more attention. The secondary S, of the high frequency coil has two metal plates t t of considerable surface connected to its terminals, and the current for use is derived from two similar plates t' t' in proximity to the former. Both the tension and volume of the currents taken from terminals T T may be easily regulated and in a continuous manner by simply varying the distance between the two pair of plates t t and t' t' respectively.

A facility is also afforded in this disposition for raising or lowering the potential of one of the terminals T, irrespective of the changes produced on the other terminal, this making it possible to cause a stronger action on one or other part of the patient's body.

The physician may find it for some or other reason convenient to modify the arrangements in Figs. 2, 3 and 4 by connecting one terminal of the high frequency source to the ground. The effects will be in most respects the same, but certain peculiarities will be noted in each case. When a ground connection is made it may be of some consequence which of the terminals of the secondary is connected to the ground, as in high frequency discharges the impulses of one direction are generally preponderating.

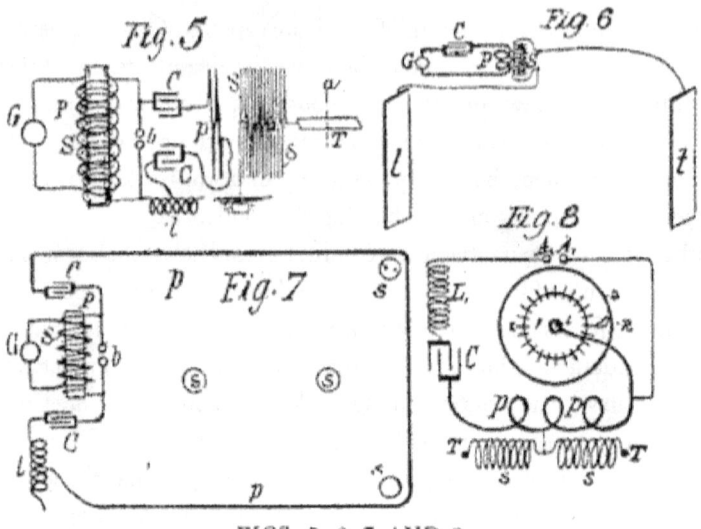

FIGS. 5, 6, 7 AND 8.

Among the various noteworthy features of these currents there is one which lends itself especially to many valuable uses. It is the facility which they afford for conveying large amounts of electrical energy to a body entirely insulated in space. The practicability of this method of energy transmission, which is already receiving useful applications and promises to become of great importance in the near future, has helped to dispel the old notion assuming the necessity of a return circuit for the conveyance of electrical energy in any considerable amount. With novel appliances we are enabled to pass through a wire, entirely insulated on one end, currents strong enough to fuse it, or to convey through the wire any amount of energy to an insulated body. This mode of applying high frequency currents in medical treatment appears to me to offer the greatest possibilities at the hands of the physician. The effects produced in this manner possess features entirely distinct from those observed when the currents are applied in any of the before mentioned or similar ways.

The circuit connections as usually made are illustrated schematically in Fig. 5, which, with reference to the diagrams before shown, is self-explanatory. The condensers C C, connected in series, are preferably charged by a step-up transformer, but a. high frequency alternator, static machine, or a direct current generator, if it be of sufficiently high tension to enable the use of small condensers, may be used with more or less success. The primary p, through which the high frequency discharges of the condensersare passed, consists of very few turns of cable of as low resistance as possible, and the secondary s, preferably at some distance from the primary to facilitate free oscillation, has one of its ends — that is the one which is nearer to the primary — connected to the ground, while the other end leads to an insulated terminal T, with which the body of the patient is connected. It is of importance in this case to establish synchronism between the oscillations in the primary and secondary circuits p and s respectively. This will be as a rule best effected by varying the self-induction of the circuit including the primary loop or coil p, for which purpose an adjustable self-induction e is provided; but in cases when the electromotive force of the generator is exceptionally high, as when a static machine is used and a condenser consisting of merely two plates offers sufficient capacity, it will be simpler to attain the same object by varying the distance of the plates.

The primary and secondary oscillations being in close synchronism, the points of highest potential will be on a part of terminal T, and the consumption of energy will occur chiefly there. The attachment of the patient's body to the terminal will in most cases very materially affect the period of oscillation in the secondary, making it longer, and a readjustment of the primary circuit will have to be made in each case to suit the capacity of the body connected with terminal T. Synchronism should always be preserved, and the intensity of the action varied by moving the secondary coil to or from the primary, as may be desired. I know of no method which would make it possible to subject the human body to such excessive electrical pressures as are practicable with this, or of one which would enable the conveying to and giving off from the body without serious injury amounts of electrical energy approximating even in a remote degree those which are entirely practicable when this manner of applying the energy is resorted to. This is evidently due to the fact that the action is chiefly superficial, the largest possible section being offered to the transfer of the current, or, to say more correctly, of the energy. With a very rapidly and smoothly working break I would not think it impossible to convey to the body of a person and to give off into the space energy at the rate of several horse power with impunity, while a small part of this amount applied in other ways could not fail to produce injury.

When a person is subjected to the action of such a coil, the proper adjustments being carefully observed, luminous streams are seen in the dark issuing from all parts of the body. These streams are short and of delicate texture when the number of breaks is very great and the action of the device b (Fig. 5) free of any irregularities, but when the number of breaks is small or the action of the device imperfect, long and noisy streams appear which cause some discomfort. The physiological effects produced with apparatus of this kind may be graduated from a hardly perceptible action when the secondary is at a great distance from the primary, to a most violent one when both coils are placed at a small distance. In the latter case only a few seconds are sufficient to cause a feeling of warmth all over the body, and soon after the person perspires freely. I have repeatedly, in demonstrations to friends, exposed myself longer to the action of the oscillations, and each time, after the lapse of an hour or so, an immense fatigue, of which it is difficult to give an idea, would take hold of me. It was greater than I experienced on some occasions after the most straining and prolonged bodily exertion. I could scarcely make a step and could keep the eyes open only with the greatest difficulty. I slept soundly afterward, and the after-effect was certainly beneficial, but the medicine was manifestly too strong to be used frequently.

One should be cautious in performing such experiments for more than one reason. At or near the surface of the skin, where the most intense action takes place, various chemical products are formed, the chief being ozone and nitrogen compounds. The former is itself very destructive, this feature being illustrated by the fact that the rubber insulation of a wire is destroyed so quickly as to make the use of such insulation entirely impracticable. The compounds of nitrogen, when moisture is present, consists largely of nitric acid which might, by excessive application, prove hurtful to the skin. So far, I have not noted injuries which could be traced directly to this cause, though on several occasions burns were produced in all respects similar to those which were later observed and attributed to the Rontgen rays. This view is seemingly being abandoned, having not been substantiated by experimental facts, and so also is the notion that these rays are transverse vibrations. But while investigation is being turned in what appears to be the right direction, scientific men are still at sea. This state of things impedes the progress of the physicist in these new regions and makes the already hard task of the physician still more difficult and uncertain.

One or two observations made while pursuing experiments with the apparatus described might be found as deserving mention here. As before stated, when the oscillations in the primary and secondary circuits are in synchronism, the points of highest potential are on some portion of the terminal T. The synchronism being perfect and the length of the secondary coil just equal to one-quarter of the wave length, these points will be exactly on the free end of terminal T, that is, the one situated farthest from the end of the wire attached to the terminal. If this be so and if now the period of the oscillations in the primary be shortened, the points of highest potential will recede towards the secondary coil, since the wave-length is reduced and since the attachment of one end of the secondary coil to the ground determines the position of the nodal points, that is, the points of least potential. Thus, by varying the period of vibration of the primary circuit in any manner, the points of highest potential may be shifted accordingly along the terminal T, which has been shown, designedly, long to illustrate this feature. The same phenomenon is, of course, produced if the body of a patient constitutes the terminal, and an assistant may by the motion of a handle cause the points of highest potential to shift along the body with any speed he may desire. When the action of the coil is vigorous, the region of highest potential is easily and unpleasantly located by the discomfort or pain experienced, and it is most curious to feel how the pain wanders up and down, or eventually across the body, from hand to hand, if the connection to the coil is accordingly made — in obedience to the movement of the handle controlling the oscillations. Though I have not observed any specific action in experiments of this kind, I have always felt that this effect might be capable of valuable use in electro-therapy.

Another observation which promises to lead to much more useful results is the following: As before remarked, by adopting the method described, the body of a person may be subjected without danger to electrical pressures vastly in excess of any producible by ordinary apparatus, for they may amount to several million volts, as has been shown in actual practice. Now, when a conducting body is electrified to so high a degree, small particles, which may be adhering firmly to its surface, are torn off with violence and thrown to distances which can be only conjectured. I find that not only firmly adhering matter, as paint, for instance, is thrown off, but even the particles of the toughest metals are torn off. Such actions have been thought to be restricted to a vacuous enclosure, but with a powerful coil they occur also in the ordinary atmosphere. The facts mentioned would make it reasonable to expect that this extraordinary effect which, in other ways, I have already usefully applied, will likewise prove to be of value in electro-therapy. The continuous improvement of the instruments and the study of the phenomenon may shortly lead to the establishment of a novel mode of hygienic treatment which would permit an instantaneous cleaning of the skin of a person, simply by connecting the same to, or possibly, by merely placing the person in the vicinity of a source of intense electrical oscillations, this having the effect of throwing off, in a twinkle of the eye, dust or particles of any extraneous matter adhering to the body.

Such a result brought about in a practicable manner would, without doubt, be of incalculable value in hygiene and would be an efficient and time-saving substitute for a water bath, and particularly appreciated by those whose contentment consists in undertaking more than they can accomplish.

High frequency impulses produce powerful inductive actions and in virtue of this feature they lend themselves in other ways to the uses of the electro-therapist. These inductive effects are either electrostatic or electrodynamic. The former diminish much more rapidly with the distance — with the square of the same — the latter are reduced simply in proportion to the distance. On the other hand, the former grow with the square of intensity of the source, while the latter increase in a simple proportion with the intensity. Both of these effects may be utilized for establishing a field of strong action extending through considerable space, as through a large hall, and such an arrangement might be suitable for use in hospitals or institutions of this kind, where it is desirable to treat a number of patients at the same time.

Fig. 6 illustrates the manner, as I have shown it originally, in which such a field of electrostatic action is established. In this diagram G is a generator of currents of very high frequency, C a condenser for counteracting the self-induction of the circuit which includes the primary P of an induction coil, the secondary S of which has two plates t t of large surface connected to its terminals. Well known adjustments being observed, a very strong action occurs chiefly in the space between the plates, and the body of a person is subjected to rapid variations of potential and surgings of current, which produce, even at a great distance, marked physiological effects. In my first experiments I used two metal plates as shown, but later I found it preferable to replace them by two large hollow spheres of brass covered with wax of a thickness of about two inches. The cables leading to the terminals of the secondary coil were similarly covered, so that any of them could be approached without danger of the insulation breaking down. In this manner the unpleasant shocks, to which the experimenter was exposed when using the plates, were prevented.

In Fig. 7 a plan for similarly utilizing the dynamic inductive effects of high frequency currents is illustrated. As the frequencies obtainable from an alternator are not as high as is desired, conversion by means of condensers is resorted to. The diagram will be understood at a glance from the foregoing description. It only need be stated that the primary p, through which the condensers are made to discharge, is formed by a thick stranded cable of low self-induction and resistance, and passes all around the hall. Any number of secondary coils s s s, each consisting generally of a single layer of rather thick wire, may be provided. I have found it practicable to use as many as one hundred, each being adjusted for a definite period and responding to a particular vibration passed through the primary. Such a plant I have had in use in my laboratory since 1892, and many times it has contributed to the pleasure of my visitors and also proved itself of practical utility. On a latter occasion I had the pleasure of entertaining some of the members with experiments of this kind, and this opportunity I cannot let pass without expressing my thanks for the interest which was awakened in me by their visit, as well as for the generous acknowledgment of the courtesy by the Association. Since that time my apparatus has been very materially improved, and now I am able to create a field of such intense induction in the laboratory that a coil three feet in diameter, by careful adjustment, will deliver energy at the rate of one-quarter of a horse power, no matter where it is placed within the area enclosed by the primary loops. Long sparks, streamers and all other phenomena obtainable with induction coils are easily producible anywhere within the space, and such coils, though not connected to anything, may be utilized exactly as ordinary coils, and what is still more remarkable, they are more effective. For the past few years I have often been urged to show experiments in public, but, though I was desirous to comply with such requests, pressing work has so far made it 'impossible. These advances have been the result of slow but steady improvement in the details of the apparatus which I hope to be able to describe connectedly in the near future.

However remarkable the electrodynamic inductive effects, which I have mentioned, may appear, they may be still considerably intensified by concentrating the action upon a very small space. It is evident that since, as before stated, electromotive forces of many thousand volts are maintained between two points of a conducting bar or loop only a few inches long, electromotive forces of approximately the same magnitude will be set up in conductors situated near by. Indeed, I found that it was practicable in this manner to pass a discharge through a highly exhausted bulb, although the electromotive force required amounted to as much as ten or twenty thousand volts, and for a long time I followed up experiments in this direction with the object of producing light in a novel and more economical way. But the tests left no doubt that there was great energy consumption attendant to this mode of illumination, at least with the apparatus I had then at command, and, finding another method which promised a higher economy of transformation, my efforts turned in this new direction. Shortly afterward (some time in June, 1891) Prof. J. J. Thomson described experiments which were evidently the outcome of long investigation, and in which he supplied much novel and interesting information, and this made me return with renewed zeal to my own experiments. Soon my efforts were centered upon producing in a small space the most intense inductive action, and by gradual improvement in the apparatus I obtained results of a surprising character. For instance, when the end of a heavy bar of iron was thrust within a loop powerfully energized, a few moments were sufficient to raise the bar to a high temperature. Even heavy lumps of other metals were heated as rapidly as though they were placed in a furnace. When a continuous band formed of a sheet of tin was thrust into the loop, the metal was fused instantly, the action being comparable to an explosion, and no wonder, for the frictional losses accumulated in it at the rate of possibly ten horse power. Masses of poorly conducting material behaved similarly, and when a highly exhausted bulb was pushed into the loop, the glass was heated in a few seconds nearly to the point of melting.

When I first observed these astonishing actions, I was interested to study their effects upon living tissues. As may be assumed, I proceeded with all the necessary caution, and well I might, for I had the evidence that in a turn of only a few inches in diameter an electromotive force of more than ten thousand volts was produced, and such high pressure would be more than sufficient to generate destructive currents in the tissue. This appeared all the more certain as bodies of comparatively poor conductivity were rapidly heated and even partially destroyed. One may imagine my astonishment when I found that I could thrust my hand or any other part of the body within the loop and hold it there with impunity. More than on one occasion, impelled by a desire to make some novel and useful observation, I have willingly or unconsciously performed an experiment connected with some risk, this being scarcely avoidable in laboratory experience, but I have always believed, and do so now, that I have never undertaken anything in which, according to my own estimation, the chances of being injured were so great as when I placed my head within the space in which such terribly destructive forces were at work. Yet I have done so, and repeatedly, and have felt nothing. But I am firmly convinced that there is great danger attending such an experiment, and some one going just a step farther than I have gone may be instantly destroyed. For, conditions may exist similar to those observable with a vacuum bulb. It may be placed in the field of the loop, however intensely energized, and so long as no path for the current is formed, it will remain cool and consume practically no energy. But the moment the first feeble current passes, most of the energy of the oscillations rushes to the place of consumption. If by any action whatever, a conducting path were formed within the living tissue or bones of the head, it would result in the instant destruction of these and death of the foolhardy experimenter. Such a method of killing, if it were rendered practicable, would be absolutely painless. Now, why is it that in a space in which such violent turmoil is going on living tissue remains uninjured? One might say the currents cannot pass because of the great self-induction offered by the large conducting mass. But this it cannot be, because a mass of metal offers a still higher self-induction and is heated just the same. One might argue the tissues offer too great a resistance. But this again cannot be the reason, for all evidence shows that the tissues conduct well enough, and besides, bodies of approximately the same resistance are raised to a high temperature. One might attribute the apparent harmlessness of the oscillations to the high specific heat of the tissue, but even a rough quantitative estimate from experiments with other bodies shows that this view is untenable. The only plausible explanation I have so far found is that the tissues are condensers. This only can account for the absence of injurious action. But it is remarkable that, as soon as a heterogeneous circuit is constituted, as by taking in the hands a bar of metal and forming a closed loop in this manner, the passage of the currents through the arms is felt, and other physiological effects are distinctly noted. The strongest action is, of course, secured when the exciting loop makes only one turn, unless the connections take up a considerable portion of the total length of the circuit, in which case the experimenter should settle upon the least number of turns by carefully estimating what he loses by increasing the number of turns, and what he gains by utilizing thus a greater proportion of the total length of the circuit. It should be borne in mind that, when the exciting coil has a considerable number of turns and is of some length, the effects of electrostatic induction may preponderate, as there may exist a very great difference of potential—a hundred thousand volts or more—between the first and last turn. However, these latter effects are always present even when a single turn is employed.

When a person is placed within such a loop, any pieces of metal, though of small bulk, are perceptibly warmed. Without doubt they would be also heated—particularly if they were of iron—when embedded in living tissue, and this suggests the possibility of surgical treatment by this method. It might be possible to sterilize wounds, or to locate, or even to extract metallic objects, or to perform other operations of this kind within the sphere of the surgeon's duties in this novel manner.

Most of the results enumerated, and many others still more remarkable, are made possible only by utilizing the discharges of a condenser. It is probable that but a very few—even among those who are working in these identical fields—fully appreciate what a wonderful instrument such a condenser is in reality. Let me convey an idea to this effect. One may take a condenser, small enough to go in one's vest pocket, and by skillfully using it he may create an electrical pressure vastly in excess—a hundred times greater if necessary—than any producible by the largest static machine ever constructed. Or, he may take the same condenser and, using it in a different way, he may obtain from it currents against which those of the most powerful welding machine are utterly insignificant. Those who are imbued with popular notions as to the pressures of static machines and currents obtainable with a commercial transformer, will be astonished at this statement—yet the truth of it is easy to see. Such results are obtainable, and easily, because the condenser can discharge the stored energy in an inconceivably short time. Nothing like this property is known in physical science. A compressed spring, or a storage battery, or any other form of device capable of storing energy, cannot do this; if they could, things undreamt of at present might be accomplished by their means. The nearest approach to a charged condenser is a high explosive, as dynamite. But even the most violent explosion of such a compound bears no comparison with the discharge or explosion of a condenser. For, while the pressures which are produced in the detonation of a chemical compound are measured in tens of tons per square inch, those which may be caused by condenser discharges may amount to thousands of tons per square inch, and if a chemical could be made which would explode as quickly as a condenser can be discharged under conditions which are realizable—an ounce of it would quite certainly be sufficient to render useless the largest battleship.

That important realizations would follow from the use of an instrument possessing such ideal properties I have been convinced since long ago, but I also recognized early that great difficulties would have to be overcome before it could replace less perfect implements now used in the arts for the manifold transformations of electrical energy. These difficulties were many. The condensers themselves, as usually manufactured, were inefficient, the conductors wasteful, the best insulation inadequate, and the conditions for the most efficient conversion were hard to adjust and to maintain. One difficulty, however, which was more serious than the others, and to which I called attention when I first described this system of energy transformation, was found in the devices necessarily used for controlling the charges and discharges of the condenser. They were wanting in efficiency and reliability and threatened to prove a decided drawback, greatly restricting the use of the system and depriving it of many valuable features. For a number of years I have tried to master this difficulty. During this time a great number of such devices were experimented upon. Many of them promised well at first, only to prove inadequate in the end. Reluctantly, I came back upon an idea on which I had worked long before. It was to replace the ordinary brushes and commutator segments by fluid contacts. I had encountered difficulties then, but the intervening years in the laboratory were not spent in vain, and I made headway. First it was necessary to provide for a circulation of the fluid, but forcing it through by a pump proved itself impractical. Then the happy idea presented itself to make the pumping device an integral part of the circuit interrupter, inclosing both in a receptacle to prevent oxidation. Next some simple ways of maintaining the circulation, as by rotating a body of mercury, presented themselves. Then I learned how to reduce the wear and losses which still existed. I fear that these statements, indicating how much effort was spent in these seemingly insignificant details will not convey a high idea of my ability, but I confess that my patience was taxed to the utmost. Finally, though, I had the satisfaction of producing devices which are simple and reliable in their operation, which require practically no attention and which are capable of effecting a transformation of considerable amounts of energy with fair economy. It is not the best that can be done, by any means, but it is satisfactory, and I feel that the hardest task is done.

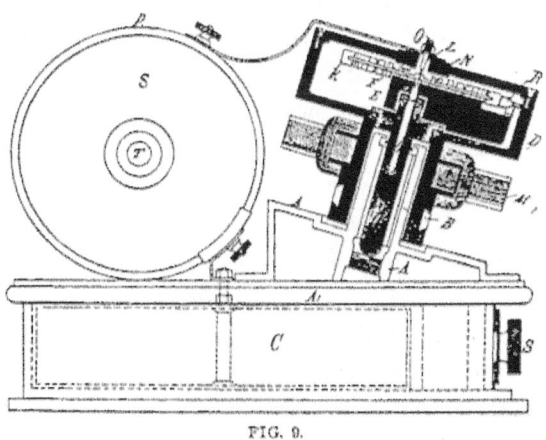

FIG. 9.

The physician will now be able to obtain an instrument suitable to fulfill many requirements. He will be able to use it in electro-therapeutic treatment in most of the ways enumerated. He will have the facility of providing himself with coils such as he may desire to have for any particular purpose, which will give him any current or any pressure he may wish to obtain. Such coils will consist of but a few turns of wire, and the expense of preparing them will be quite insignificant. The instrument will also enable him to generate Rontgen rays of much greater power than obtainable with ordinary apparatus. A tube must still be furnished by the manufacturers which will not deteriorate and which will allow to concentrate larger amounts of energy upon the electrodes. When this is done, nothing will stand in the way of an extensive and efficient application of this beautiful discovery which must ultimately prove itself of the highest value, not only at the hands of the surgeon, but also of the electro-therapist and, what is most important, of the bacteriologist.

To give a general idea of an instrument in which many of the latter improvements are embodied, I would refer to Fig. 9, which illustrates the chief parts of the same in side elevation and partially in vertical cross-section. The arrangement of the parts is the same as in the form of instrument exhibited on former occasions, only the exciting coil with the vibrating interrupter is replaced by one of the improved circuit breakers to which reference has been made.

This device comprises a casting A with a protruding sleeve B, which in a bushing supports a freely rotatable shaft a. The latter carries an armature within a stationary field magnet M and on the top, a hollow iron pulley D, which contains the break proper. Within the shaft a, and concentrically with the same, is placed a smaller shaft b, likewise freely movable on ball-bearings and supporting a weight E. This weight being on one side and the shafts a and b inclined to the vertical, the weight remains stationary as the pulley is rotated. Fastened to the weight E is a device R in the form of a scoop with very thin walls, narrow on the end nearer to the pulley and wider on the other end. A small quantity of mercury being placed in the pulley and the latter rotated against the narrow end of the scoop, a portion of the fluid is taken up and thrown in a thin and wide stream towards the centre of the pulley. The top of the latter is hermetically closed by an iron washer, as shown. this washer supporting on a steel rod L a disk F of the same metal provided with a number of thin contact blades K. The rod L is insulated by washers from the pulley and for the convenience of filling in the mercury a small screw o is provided. The bolt L forming one terminal of the circuit breaker is connected by a copper strip to the primary p. The other end of the primary coil leads to one of the terminals of the condenser C, contained in a compartment of a box A, another compartment of the same being reserved for switch S and terminals of the instrument. The other terminal of the condenser is connected to the casting A and through it to pulley D. When the pulley is rotated, the contact blades K are brought rapidly in and out of contact with the stream of mercury, thus closing and opening the circuit in quick succession. With such a device it is easy to obtain ten thousand makes and breaks per second and even more. The secondary s is made of two separate coils and so arranged that it can be slipped out, and a metal strip in its middle connects it to the primary coil. This is done to prevent the secondary from breaking down when one of the terminals is overloaded, as it often happens in working Rontgen bulbs. This form of coil will withstand a very much greater difference of potential than coils as ordinarily constructed.

The motor has both field and armature built of plates, so that it can be used on alternating as well as direct current supply circuits, and the shafts are as nearly as possible vertical, so as to require the least care in oiling. Thus, the only thing which really requires some attention is the commutator of the motor, but where alternating currents are always available, this source of possible trouble is easily done away with.

The circuit connections of the instrument have been already shown and the mode of operation explained in periodicals. The usual manner of connecting is illustrated in Fig. 8, in which A. A, are the terminals of the supply circuit, L, a self-induction coil for raising the pressure, which is connected in series with condenser C and primary P P. The remaining letters designate the parts correspondingly marked in Fig. 9 and will be understood with reference to the latter.

November 17, 1898

PLANS TO DISPENSE WITH ARTILLERY OF THE PRESENT TYPE

—The Sun, New York, November 21, 1898

... Referring to my latest invention [the telautomaton], I wish to bring out a point which has been overlooked. I arrived, as has been stated, at the idea through entirely abstract speculations on the human organism, which I conceived to be a self-propelling machine, the motions of which are governed by impressions received through the eye.

Endeavoring to construct a mechanical model resembling in its essential, material features the human body, I was led to combine a controlling device, or organ sensitive to certain waves, with a body provided with propelling and directing mechanism, and the rest naturally followed.

Originally the idea interested me only from the scientific point of view, but soon I saw that I had made a departure which sooner or later must produce a profound change in things and conditions presently existing. I hope this change will be for the good only, for, if it were otherwise, I wish that I had never made the invention.

The future may or may not bear out my present convictions, but I can not refrain from saying that it is difficult for me to see at present how, with such a principle brought to great perfection, as it undoubtedly will be in the course of time, guns can maintain themselves as weapons.

We shall be able, by availing ourselves of this advance, to send a projectile at much greater distance, it will not be limited in any way by weight or amount of explosive charge, we shall be able to submerge it at command, to arrest it in its flight, and call it back, and send it out again and explode it at will, and, more than this, it will never make a miss, since all chance in this regard, if hitting the object of attack were at all required, is eliminated. But the chief feature of such a weapon is still to be told; namely, it may be made to respond only to a certain note or tune, it may be endowed with selective power.

Directly such an arm is produced, it becomes almost impossible to meet it with a corresponding development. It is this feature, perhaps more than in its power of destruction, that its tendency to arrest the development of arms and to stop warfare will reside.

With renewed thanks, I remain,
Very truly yours, N. Tesla.

New York, November 19, 1898.

TESLA DESCRIBES HIS EFFORTS IN VARIOUS FIELDS OF WORK

—*Electrical Review. New York. Nov, 30, 1898*
[From The Sun, New York, November 21, 1898]

To the editor of *The Sun*:

"Sir: Had it not been for other urgent duties, I would before this have acknowledged your highly appreciative editorial of November 13. Such earnest comments and the frequent evidences of the highest appreciation of my labors by men who are the recognized leaders of this day in scientific speculation, discovery and invention are a powerful stimulus, and I am thankful for them. There is nothing that gives me so much strength and courage as the feeling that those who are competent to judge have faith in me.

"Permit me on this occasion to make a few statements which will define my position in the various fields of investigation you have touched upon.

"I can not but gratefully acknowledge my indebtedness to earlier workers, as Dr. Hertz and Dr. Lodge, in my efforts to produce a practical and economical lighting system on the lines which I first disclosed in a lecture at Columbia College in 1891. There exists a popular error in regard to this light, inasmuch as it is believed that it can be obtained without generation of heat. The enthusiasm of Dr. Lodge is probably responsible for this error, which I have pointed out early by showing the impossibility of reaching a high vibration without going through the lower or fundamental tones. On purely theoretical grounds such a result is thinkable, but it would imply a device for starting the vibrations of unattainable qualities, inasmuch as it would have to be entirely devoid of inertia and other properties of matter. Though I have conceptions in this regard, I dismiss for the present this proposition as being impossible. We can not produce light without heat, but we can surely produce a more efficient light than that obtained in the incandescent lamp, which, though a beautiful invention, is sadly lacking in the feature of efficiency. As the first step toward this realization, I have found it necessary to invent some method for transforming economically the ordinary currents as furnished from the lighting circuits into electrical vibrations of great rapidity. This was a difficult problem, and it was only recently that I was able to announce its practical and thoroughly satisfactory solution. But this was not the only requirement in a system of this kind. It was necessary also to increase the intensity of the light, which at first was very feeble. In this direction, too, I met with complete success, so that at present I am producing a thoroughly serviceable and economical light of any desired intensity. I do not mean to say that this system will revolutionize those in use at present, which have resulted from the cooperation of many able men. I am only sure that it will have its fields of usefulness.

"As to the idea of rendering the energy of the sun available for industrial purposes, it fascinated me early but I must admit it was only long after I discovered the rotating magnetic field that it took a firm hold upon my mind. In assailing the problem I found two possible ways of solving it. Either power was to be developed on the spot by converting the energy of the sun's radiations or the energy of vast reservoirs was to be transmitted economically to any distance. Though there were other possible sources of economical power, only the two solutions mentioned offer the ideal feature of power being obtained without any consumption of material. After long thought I finally arrived at two solutions, but on the first of these, namely, that referring to the development of power in any locality from the sun's radiations, I can not dwell at present. The system of power transmission without wires, in the form in which I have described it recently, originated in this manner. Starting from two facts that the earth was a conductor insulated in space, and that a body can not be charged without causing an equivalent displacement of electricity in the earth, I undertook to construct a machine suited for creating as large a displacement as possible of the earth's electricity.

"This machine was simply to charge and discharge in rapid succession a body insulated in space, thus altering periodically the amount of electricity in the earth, and consequently the pressure all over its surface. It was nothing but what in mechanics is a pump, forcing water from a large reservoir into a small one and back again. Primarily I contemplated only the sending of messages to great distances in this manner, and I described the scheme in detail, pointing out on that occasion the importance of ascertaining certain electrical conditions of the earth. The attractive feature of this plan was that the intensity of the signals should diminish very little with the distance, and, in fact, should not diminish at all, if it were not for certain losses occurring, chiefly in the atmosphere. As all my previous ideas, this one, too, received the treatment of Marsyas, but it forms, nevertheless, the basis of what is now known as "wireless telegraphy." This statement will bear rigorous examination, but it is not made with the intent of detracting from the merit of others. On the contrary, it is with great pleasure that I acknowledge the early work of Dr. Lodge, the brilliant experiments of Marconi, and of a later experimenter in this line, Dr. Slaby, of Berlin. Now, this idea I extended to a system of power transmission, and I submitted it to Helmholtz on the occasion of his visit to this country. He unhesitatingly said that power could certainly be transmitted in this manner, but he doubted that I could ever produce an apparatus capable of creating the high pressures of a number of million volts, which were required to attack the problem with any chance of success, and that I could overcome the difficulties of insulation. Impossible as this problem seemed at first, I was fortunate to master it in a comparatively short time, and it was in perfecting this apparatus that I came to a turning point in the development of this idea. I, namely, at once observed that the air, which is a perfect insulator for currents produced by ordinary apparatus, was easily traversed by currents furnished by my improved machine, giving a tension of something like 2,500,000 volts. A further investigation in this direction led to another valuable fact; namely, that the conductivity of the air for these currents increased very rapidly with its degree of rarefaction, and at once the transmission of energy through the upper strata of air, which, without such results as I have obtained, would be nothing more than a dream, became easily realizable. This appears all the more certain, as I found it quite practicable to transmit, under conditions such as exist in heights well explored, electrical energy in large amounts. I have thus overcome all the chief obstacles which originally stood in the way, and the success of my system now rests merely on engineering skill.

"Referring to my latest invention, I wish to bring out a point which has been overlooked. I arrived, as has been stated, at the idea through entirely abstract speculations on the human organism, which I conceived to be a self-propelling machine, the motions of which are governed by impressions received through the eye. Endeavoring to construct a mechanical model resembling in its essential, material features the human body, I was led to combine a controlling device, or organ sensitive to certain waves, with a body provided with propelling and directing mechanism, and the rest naturally followed. Originally the idea interested me only from the scientific point of view, but soon I saw that I had made a departure which sooner or later must produce a profound change in things and conditions presently existing. I hope this change will be for the good only, for, if it were otherwise, I wish that I had never made the invention. The future may or may not bear out my present convictions, but I can not refrain from saying that it is difficult for me to see at present how, with such a principle brought to great perfection, as it undoubtedly will be in the course of time, guns can maintain themselves as weapons. We shall be able, by availing ourselves of this advance, to send a projectile at much greater distance, it will not be limited in any way by weight or amount of explosive charge, we shall be able to submerge it at command, to arrest it in its flight, and call it back, and to send it out again and explode it at will, and, more than this, it will never make a miss, since all chance in this regard, if hitting the object of attack were at all required, is eliminated. But the chief feature of such a weapon is still to be told; namely, it may be made to respond only to a certain note or tune, it may be endowed with selective power. Directly such an arm is produced, it becomes almost impossible to meet it with a corresponding development. It is this feature, perhaps, more than in its power of destruction, that its tendency to arrest the development of arms and to stop warfare will reside. With renewed thanks, I remain,

Very truly, yours,
N. TESLA."

New York, November 21, 1898.

ON CURRENT INTERRUPTERS

—*Electrical Review*—*March 15, 1899*

To stimulate the ardor of the zealous experimenters, who believe in the revolutionary character of this discovery, it might be well to suggest one or two such simple devices for interrupting the current. For instance, a very primitive contrivance of this kind comprises a poker—yes, an ordinary poker, connected by means of a flexible cable to one of the mains of the generator, and a bathtub filled with conducting fluid which is connected in any suitable manner, through the primary of an induction coil, to the other pole of the generator. When the experimenter desires to take a Roentgen picture, he brings the end of the poker to white heat, and, thrusting the same into the bathtub, he will at once witness an astonishing phenomenon; the seething and boiling liquid making and breaking the current in rapid succession, and the powerful rays generated will at once convince him of the great practical value of this discovery. I might further suggest that the poker may be conveniently heated by means of a welding machine.

Another device, entirely automatic, and probably suitable for use in suburban districts, comprises two insulated metal plates, supported in any convenient manner, in close proximity to each other. These plates are connected through the primary of an induction coil with the terminals of a generator, and are bridged by two movable contacts joined by a flexible cable. The two contacts are both attached to the legs of a good-sized chicken standing astride on the plates. Heat being applied to the latter, muscular contractions are produced in the legs of the chicken, which thus makes and breaks the current through the induction coil. Any number, of such chickens may be provided and the contacts connected in series or multiple arc, as may be desired, thus, increasing the frequency of the impulses. In this manner fierce sparks, suitable for most purposes, may be obtained, and vacuum tubes may be operated, and these contrivances will be found a notable improvement on certain circuit-breakers of old, with which two enterprising editors undertook some years ago to revolutionize the systems of electric lighting. The enterprising editors, are wiser now. They are to be congratulated, and their readers, scientific societies and the profession, all ought to be congratulated, and—"all is well that ends well." The observant experimenter will not fail to note that the fierce sparks frighten the chickens, which are thus put into more violent spasms and muscular contractions, this again increasing the fierceness of the sparks, which, in return, causes a greater fright of the chickens and increased speed of interruptions; it is, in fact, as Kipling says:

> "*Interdependence absolute, foreseen, ordained, decreed,*
> *To work, ye'll note, as any tilt an' every rate o'speed."*

But to return, in all earnestness, to the "electrolytic interrupter" described, this is a device with which I am perfectly familiar, having carried on extensive experiments with the same two or three years ago. It was one of many devices which I invented in my efforts to produce an economical contrivance of this kind. The name is really not appropriate, inasmuch as any fluid, either conducting or made so in any manner, as by being rendered acid or alkaline, or by being heated, may be used. I have even found it possible, under certain conditions, to operate with mercury. The device is extremely simple, but the great waste of energy attendant upon its operation and certain other defects make it entirely unsuitable for any valuable, practical purpose, and as far as those instances are concerned, in which a small amount of energy is needed, much better results are obtained by a properly designed mechanical circuit-breaker. The experimenters are very likely deceived by finding that an induction coil gives longer sparks when this device is inserted in place of the ordinary break, but this is due merely to the fact that the break is not properly designed. Of the total energy supplied from the mains, scarcely one-fourth is obtainable of that amount, which a well constructed mechanical break furnishes in the secondary, and although I have designed many improved forms, I have found it impossible to increase materially the economy. Two improvements, however, which I found at that time necessary to introduce, I may mention for the benefit of those who are using the device. As will be readily noted, the small terminal is surrounded by a gaseous bubble, in which the makes and breaks are formed, generally in an irregular manner, by the liquid being driven towards the terminal at some point. The force which drives the liquid is evidently the pressure of the fluid column, and by increasing the fluid pressure in any manner the liquid is forced with greater speed towards the terminal and thus the frequency is increased. Another necessary improvement was to make a provision for preventing the acid or alkali from being carried off into the atmosphere, which always happens more or less, even if the liquid column be of some height. During my early experiments with the device, I became so interested in it that I neglected this precaution, and I noted that the acid had attacked all the apparatus in my laboratory.

The experimenter will conveniently carry out both of these improvements by taking a long glass tube of, say, six to eight feet in length, and arranging the interrupting device close to the bottom of the tube, with an outlet for eventually replacing the` liquid. The high column will prevent the fumes from vitiating the atmosphere of the room, and the increased pressure will add materially to the effectiveness of the performance. If the liquid column be,. say, nine times as high, the force driving the fluid towards the contact is nine times as great, and this force is capable, under the same conditions, of driving the fluid three times as fast, hence the frequency is increased in that ratio and, .in fact, in a somewhat greater ratio, as the gaseous bubble, being compressed, is rendered smaller, and therefore the liquid is made to travel through a smaller distance. The electrode, of course, should be very small :to insure the regularity of operation, and it is not necessary to use platinum. The pressure may, however, be increased in other ways, and I have obtained some results of interest, in experiments of this kind.

As before stated, the device is very wasteful, and, while it, may be used in some instances, I consider it of little or no practical value. It will please me to be convinced of the contrary; but I do not think that I am erring. My chief reasons for this statement are that there are many other ways in which by far better results are obtainable with are equally simple, if not more so. I may mention one here, based on a different principle which is incomparably more effective, more efficient and also simpler on the whole. It comprises a fine stream of conducting fluid which is made to issue, with any desired speed, from an orifice connected with one pole of a generator, through the primary of the induction coil, against the other terminal of the generator placed at a small distance. This device gives discharges of a remarkable suddenness, and the frequency may be brought within reasonable limits, almost to anything desired. I have used this device for a long time in connection with ordinary coils and in a form of my own coil with results greatly superior in every respect to those obtainable with the form of device discussed.

March 15, 1899

THE PROBLEM OF INCREASING HUMAN ENERGY

—With special references to the harnessing of the sun's energy Century Illustrated Magazine, June 1900

The onward movement of man — The energy of the movement — The three ways of increasing human energy.

Of all the endless variety of phenomena which nature presents to our senses, there is none that fills our minds with greater wonder than that inconceivably complex movement which, in its entirety, we designate as human life; Its mysterious origin is veiled in the forever impenetrable mist of the past, its character is rendered incomprehensible by its infinite intricacy, and its destination is hidden in the unfathomable depths of the future. Whence does it come? What is it? Whither does it tend? Are the great questions which the sages of all times have endeavored to answer.

Modern science says: The sun is the past, the earth is the present, the moon is the future. From an incandescent mass we have originated, and into a frozen mass we shall turn. Merciless is the law of nature, and rapidly and irresistibly we are drawn to our doom. Lord Kelvin, in his profound meditations, allows us only a short span of life, something like six million years, after which time the suns bright light will have ceased to shine, and its life giving heat will have ebbed away, and our own earth will be a lump of ice, hurrying on through the eternal night. But do not let us despair. There will still be left upon it a glimmering spark of life, and there will be a chance to kindle a new fire on some distant star. This wonderful possibility seems, indeed, to exist, judging from Professor Dewar's beautiful experiments with liquid air, which show that germs of organic life are not destroyed by cold, no matter how intense; consequently they may be transmitted through the interstellar space. Meanwhile the cheering lights of science and art, ever increasing in intensity, illuminate our path, and marvels they disclose, and the enjoyments they offer, make us measurably forgetful of the gloomy future.

Though we may never be able to comprehend human life, we know certainly that it is a movement, of whatever nature it be. The existence of movement unavoidably implies a body which is being moved and a force which is moving it. Hence, wherever there is life, there is a mass moved by a force. All mass possesses inertia, all force tends to persist. Owing to this universal property and condition, a body, be it at rest or in motion, tends to remain in the same state, and a force, manifesting itself anywhere and through whatever cause, produces an equivalent opposing force, and as an absolute necessity of this it follows that every movement in nature must be rhythmical. Long ago this simple truth was clearly pointed out by Herbert Spencer, who arrived at it through a somewhat different process of reasoning. It is borne out in everything we perceive—in the movement of a planet, in the surging and ebbing of the tide, in the reverberations of the air, the swinging of a pendulum, the oscillations of an electric current, and in the infinitely varied phenomena of organic life. Does not the whole of human life attest to it? Birth, growth, old age, and death of an individual, family, race, or nation, what is it all but a rhythm? All life manifestation, then, even in its most intricate form, as exemplified in man, however involved and inscrutable, is only a movement, to which the same general laws of movement which govern throughout the physical universe must be applicable.

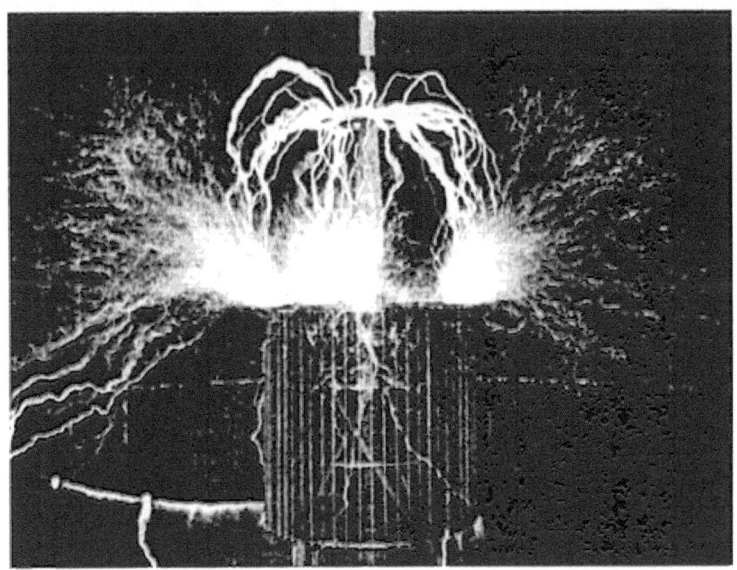

Fig. 1. Burning the nitrogen of the atmosphere.

Note to Fig.1: This result is produced by the discharge of an electrical oscillator giving twelve million volts. The electrical pressure, alternating one hundred thousand times per second, excites the normally inert nitrogen, causing it to combine with the oxygen. The flame-like discharge shown in the photograph measures sixty-five feet across.

When we speak of man, we have a conception of humanity as a whole, and before applying scientific methods to, the investigation of his movement we must accept this as a physical fact. But can anyone doubt today that all the millions of individuals and all the innumerable types and characters constitute an entity, a unit? Though free to think and act, we are held together, like the stars in the firmament, with ties inseparable. These ties cannot be seen, but we can feel them. I cut myself in the finger, and it pains me: this finger is a part of me. I see a friend hurt, and it hurts me, too: my friend and I are one. And now I see stricken down an enemy, a lump of matter which, of all the lumps of matter in the universe, I care least for, and it still grieves me. Does this not prove that each of us is only part of a whole?

For ages this idea has been proclaimed in the consummately wise teachings of religion, probably not alone as a means of insuring peace and harmony among men, but as a deeply founded truth. The Buddhist expresses it in one way, the Christian in another, but both say the same: We are all one. Metaphysical proofs are, however, not the only ones which we are able to bring forth in support of this idea. Science, too, recognizes this connectedness of separate individuals, though not quite in the same sense as it admits that the suns, planets, and moons of a constellation are one body, and there can be no doubt that it will be experimentally confirmed in times to come, when our means and methods for investigating psychical and other states and phenomena shall have been brought to great perfection. Still more: this one human being lives on and on. The individual is ephemeral, races and nations come and pass away, but man remains. Therein lies the profound difference between the individual and the whole. Therein, too, is to be found the partial explanation of many of those marvelous phenomena of heredity which are the result of countless centuries of feeble but persistent influence.

Conceive, then, man as a mass urged on by a force. Though this movement is not of a translatory character, implying change of place, yet the general laws of mechanical movement are applicable to it, and the energy associated with this mass can be measured, in accordance with well-known principles, by half the product of the mass with the square of a certain velocity. So, for instance, a cannon-ball which is at rest possesses a certain amount of energy in the form of heat, which we measure in a similar way. We imagine the ball to consist of innumerable minute particles, called atoms or molecules, which vibrate or whirl around one another. We determine their masses and velocities, and from them the energy of each of these minute systems, and adding them all together, we get an idea of the total heat-energy contained in the ball, which is only seemingly at rest. In this purely theoretical estimate this energy may then be calculated by multiplying half of the total mass — that is half of the sum of all the small masses — with the square of a velocity which is determined from the velocities of the separate particles. In like manner we may conceive of human energy being measured by half the human mass multiplied with the square of the velocity which we are not yet able to compute. But our deficiency in this knowledge will not vitiate the truth of the deductions I shall draw, which rest on the firm basis that the same laws of mass and force govern throughout nature.

Man, however, is not an ordinary mass, consisting of spinning atoms and molecules, and containing merely heat-energy. He is a mass possessed of certain higher qualities by reason of the creative principle of life with which he is endowed. His mass, as the water in an ocean wave, is being continuously exchanged, new taking the place of the old. Not only this, but he grows propagates, and dies, thus altering his mass independently, both in bulk and density. What is most wonderful of all, he is capable of increasing or diminishing his velocity of movement by the mysterious power he possesses by appropriating more or less energy from other substance, and turning it into motive energy. But in any given moment we may ignore these slow changes and assume that human energy is measured by half the product of man's mass with the square of a certain hypothetical velocity. However we may compute this velocity, and whatever we may take as the standard of its measure, we must, in harmony with this conception, come to the conclusion that the great problem of science is, and always will be, to increase the energy thus defined. Many years ago, stimulated by the perusal of that deeply interesting work, Draper's "History of the Intellectual Development of Europe," depicting so vividly human movement, I recognized that to solve this eternal problem must ever be the chief task of the man of science. Some results of my own efforts to this end I shall endeavor briefly to describe here.

Let, then, in diagram a, M represent the mass of man. This mass is impelled in one direction by a force f, which is resisted by another partly frictional and partly negative force R, acting in a direction exactly opposite, and retarding the movement of the mass. Such an antagonistic force is present in every movement and must be taken into consideration. The difference between these two forces is the effective force which imparts a velocity V to the mass M in the direction of the arrow on the line representing the force f. In accordance with the preceding, the human energy will then be given by the product $MV^2 = MV \times V$, in which M is the total mass of man in the ordinary interpretation of the term "mass," and V is a certain hypothetical velocity, which, in the present state of science, we are unable exactly to define and determine. To increase the human energy is, therefore, equivalent to increasing this product, and there are, as will readily be seen, only three ways possible to attain this result, which are illustrated in the above diagram. The first way shown in the top figure, is to increase the mass (as indicated by the dotted circle), leaving the two opposing forces the same. The second way is to reduce the retarding force R to a smaller value r, leaving the mass and the impelling force the same, as diagrammatically shown in the middle figure. The third way, which is illustrated in the last figure, is to increase the impelling force f to a higher value F, while the mass and the retarding force R remain unaltered. Evidently fixed limits exist as regards increase of mass and reduction of retarding force, but the impelling force can be increased indefinitely. Each of these three possible solutions presents a different aspect of the main problem of increasing human energy, which is thus divided into three distinct problems, to be successively considered.

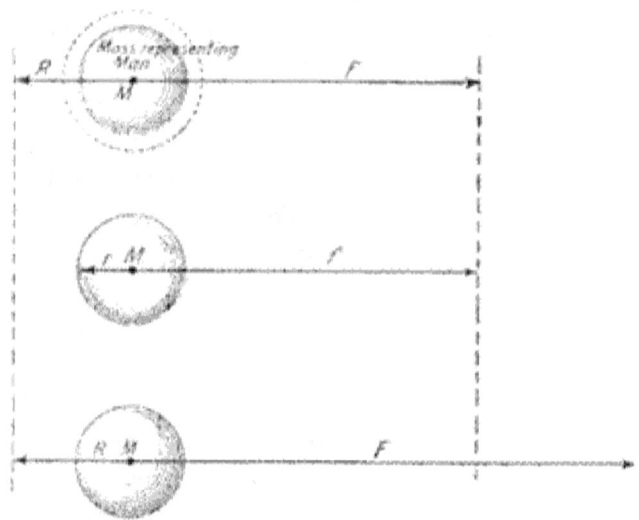

Diagram a. The three ways of increasing human energy.

The first problem: how to increase the human mass — The burning of atmospheric nitrogen

Viewed generally, there are obviously two ways of increasing the mass of mankind: first, by aiding and maintaining those forces and conditions which tend to increase it; and, second, by opposing and reducing those which tend to diminish it. The mass will be increased by careful attention to health, by substantial food, by moderation, by regularity of habits, by promotion of marriage, by conscientious attention to children, and, generally stated, by the observance of all the many precepts and laws of religion and hygiene.

But in adding new mass to the old, three cases again present themselves. Either the mass added is of the same velocity as the old, or it is of a smaller or of a higher velocity. To gain an idea of the relative importance of these cases, imagine a train composed of, say, one hundred locomotives running on a track, and suppose that, to increase the energy of the moving mass, four more locomotives are added to the train. If these four move at the same velocity at which the train is going, the total energy will be increased 4%; if they are moving at only one half of that velocity, the increase will amount to only 1%; if they are moving at twice that velocity, the increase of energy will be 16%. This simple illustration shows that it is of greatest importance to add mass of a higher velocity. Stated more to the point, if, for example, the children be of the same degree of enlightenment as the parents,—that is, mass of the "same velocity,"—the energy will simply increase proportionately to the number added. If they are less intelligent or advanced, or mass of "smaller velocity," there will be a very slight gain in the energy; but if they are further advanced, or mass of "higher velocity," then the new generation will add very considerably to the sum total of human energy. any addition of mass of "smaller velocity," beyond that indispensable amount required by the law expressed in the proverb, *Mens sana in corpore sano*, should be strenuously opposed. For instance, the mere development of muscle, as aimed at in some of our colleges, I consider equivalent to adding mass of "smaller velocity," and I would not commend it, although my views were different when I was a student myself. Moderate exercise, insuring the right balance between mind and body, and the highest efficiency of performance, is, of course, a prime requirement. The above example shows that the most important result to be attained is the education, or the increase of the "velocity," of the mass newly added. Conversely, it scarcely need be stated that everything that is against the teachings of religion and the laws of hygiene is tending to decrease the mass. Whisky, wine, tea coffee, tobacco, and other such stimulants are responsible for the shortening of the lives of many, and ought to be used with moderation. But I do not think that rigorous measures of suppression of habits followed through many generations are commendable. It is wiser to preach moderation than abstinence. We have become accustomed to these stimulants, and if such reforms are to be effected, they must be slow and gradual. Those who are devoting their energies to such ends could make themselves far more useful by turning their efforts in other directions, as, for instance, toward providing pure water. For every person who perishes from the effects of a stimulant, at least a thousand die from the consequences of drinking impure water. This precious fluid, which daily infuses new life into us, is likewise the chief vehicle through which disease and death enter our bodies. The germs of destruction it conveys are enemies all the more terrible as they perform their fatal work unperceived. They seal our doom while we live and enjoy. The majority of people are so ignorant or careless in drinking water, and the consequences of this are so disastrous, that a philanthropist can scarcely use his efforts better than by endeavoring to enlighten those who are thus injuring themselves. By systematic purification and sterilization of the drinking water the human mass would be very considerably increased. It should be made a rigid rule—which might be enforced by law—to boil or to sterilize otherwise the drinking water in every household and public place. The mere filtering does not afford sufficient security against infection. All ice for internal uses should be artificially prepared from water thoroughly sterilized. The importance of eliminating germs of disease from the city water is generally recognized, but little is being done to improve the existing conditions, as no satisfactory method of sterilizing great quantities of water has yet been brought forward. By improved electrical appliances we are now enabled to produce ozone cheaply and in large amounts, and this ideal disinfectant seems to offer a happy solution of the important question.

Gambling, business rush, and excitement, particularly on the exchanges, are causes of much mass reduction, all the more so because the individuals concerned represent units of higher value. Incapacity of observing the first symptoms of an illness, and careless neglect of the same, are important factors of mortality. In noting carefully every new sign of approaching danger, and making conscientiously every possible effort to avert it, we are not only following wise laws of hygiene in the interest of our well-being and the success of our labors, but we are also complying with a higher moral duty. Everyone should consider his body as a priceless gift from one whom he loves above all, as a marvelous work of art, of indescribable beauty and mastery beyond human conception, and so delicate and frail that a word, a breath, a look, nay, a thought, may injure it. Uncleanliness, which breeds disease and death, is not only a self destructive but highly immoral habit. In keeping our bodies free from infection, healthful, and pure, we are expressing our reverence for the high principle with which they are endowed. He who follows the precepts of hygiene in this spirit is proving himself, so far, truly religious. Laxity of morals is a terrible evil, which poisons both mind and body, and which is responsible for a great reduction of the human mass in some countries. Many of the present customs and tendencies are productive of similar hurtful results. For example, the society life, modern education and pursuits of women, tending to draw them away from their household duties and make men out of them, must needs detract from the elevating ideal they represent, diminish the artistic creative power, and cause sterility and a general weakening of the race. A thousand other evils might be mentioned, but all put together, in their bearing upon the problem under discussion, they could not equal a single one, the want of food, brought on by poverty, destitution, and famine. Millions of individuals die yearly for want of food, thus keeping down the mass. Even in our enlightened communities, and not withstanding the many charitable efforts, this is still, in all probability, the chief evil. I do not mean here absolute want of food, but want of healthful nutriment.

How to provide good and plentiful food is, therefore, a most important question of the day. On the general principles the raising of cattle as a means of providing food is objectionable, because, in the sense interpreted above, it must undoubtedly tend to the addition of mass of a "smaller velocity." It is certainly preferable to raise vegetables, and I think, therefore, that vegetarianism is a commendable departure from the established barbarous habit. That we can subsist on plant food and perform our work even to advantage is not a theory, but a well-demonstrated fact. Many races living almost exclusively on vegetables are of superior physique and strength. There is no doubt that some plant food, such as oatmeal, is more economical than meat, and superior to it in regard to both mechanical and mental performance. Such food, moreover, taxes our digestive organs decidedly less, and, in making us more contented and sociable, produces an amount of good difficult to estimate. In view of these facts every effort should be made to stop the wanton and cruel slaughter of animals, which must be destructive to our morals. To free ourselves from animal instincts and appetites, which keep us down, we should begin at the very root from which we spring: we should effect a radical reform in the character of the food.

There seems to be no philosophical necessity for food. We can conceive of organized beings living without nourishment, and deriving all the energy they need for the performance of their life functions from the ambient medium. In a crystal we have the clear evidence of the existence of a formative life-principle, and though we cannot understand the life of a crystal, it is none the less a living being. There may be, besides crystals, other such individualized, material systems of beings, perhaps of gaseous constitution, or composed of substance still more tenuous. In view of this possibility,—nay, probability, we cannot apodictically deny the existence of organized beings on a planet merely because the conditions on the same are unsuitable for the existence of life as we conceive it. We cannot even, with positive assurance, assert that some of them might not be present here, in this our world, in the very midst of us, for their constitution and life-manifestation may be such that we are unable to perceive them.

The production of artificial food as a means for causing an increase of the human mass naturally suggests itself, but a direct attempt of this kind to provide nourishment does not appear to me rational, at least not for the present. Whether we could thrive on such food is very doubtful. We are the result of ages of continuous adaptation, and we cannot radically change without unforeseen and, in all probability, disastrous consequences. So uncertain an experiment should not be tried. By far the best way, it seems to me, to meet the ravages of the evil, would be to find ways of increasing the productivity of the soil. With this object the preservation of forests is of an importance which cannot be overestimated, and in this connection, also, the utilization of water-power for purposes of electrical transmission, dispensing in many ways with the necessity of burning wood, and tending thereby to forest preservation, is to be strongly advocated. But there are limits in the improvement to be effected in this and similar ways.

To increase materially the productivity of the soil, it must be more effectively fertilized by artificial means. The question of food-production resolves itself, then, into the question how best to fertilize the soil. What it is that made the soil is still a mystery. To explain its origin is probably equivalent to explaining the origin of life itself. The rocks, disintegrated by moisture and heat and wind and weather, were in themselves not capable of maintaining life. Some unexplained condition arose, and some new principle came into effect, and the first layer capable of sustaining low organisms, like mosses was formed. These, by their life and death, added more of the life sustaining quality to the soil, and higher organisms could then subsist, and so on and on, until at last highly developed plant and animal life could flourish. But though the theories are, even now, not in agreement as to how fertilization is effected, it is a fact, only too well ascertained, that the soil cannot indefinitely sustain life, and some way must be found to supply it with the substances which have been abstracted from it by the plants. The chief and most valuable among these substances are compounds of nitrogen, and the cheap production of these is, therefore, the key for the solution of the all-important food problem. Our atmosphere contains an inexhaustible amount of nitrogen, and could we but oxidize it and produce these compounds, an incalculable benefit for mankind would follow.

Long ago this idea took a powerful hold on the imagination of scientific men, but an efficient means for accomplishing this result could not be devised. The problem was rendered extremely difficult by the extraordinary inertness of the nitrogen, which refuses to combine even with oxygen. But here electricity comes to our aid: the dormant affinities of the element are awakened by an electric current of the proper quality. As a lump of coal which has been in contact with oxygen for centuries without burning will combine with it when once ignited, so nitrogen, excited by electricity, will burn. I did not succeed, however, in producing electrical discharges exciting very effectively the atmospheric nitrogen until a comparatively recent date, although I showed, in May, 1891, in a scientific lecture, a novel form of discharge or electrical flame named "St. Elmo's hotfire," which, besides being capable of generating ozone in abundance, also possessed, as I pointed out on that occasion, distinctly the quality of exciting chemical affinities. This discharge or flame was then only three or four inches long, its chemical action was likewise very feeble, and consequently the process of oxidation of nitrogen was wasteful. How to intensify this action was the question. Evidently electric currents of a peculiar kind had to be produced in order to render the process of nitrogen combustion more efficient.

The first advance was made in ascertaining that the chemical activity of the discharge was very considerably increased by using currents of extremely high frequency or rate of vibration. This was an important improvement, but practical considerations soon set a definite limit to the progress in this direction. Next, the effects of the electrical pressure of the current impulses, of their wave-form and other characteristic features, were investigated. Then the influence of the atmospheric pressure and temperature and of the presence of water and other bodies was studied, and thus the best conditions for causing the most intense chemical action of the discharge and securing the highest efficiency of the process were gradually ascertained. Naturally, the improvements were not quick in coming; still, little by little, I advanced. The flame grew larger and larger, and its oxidizing action grew more intense. From an insignificant brush-discharge a few inches long it developed into a marvelous electrical phenomenon, a roaring blaze, devouring the nitrogen of the atmosphere and measuring sixty or seventy feet across. Thus slowly, almost imperceptibly, possibility became accomplishment. All is not yet done, by any means, but to what a degree my efforts have been rewarded an idea may be gained from an inspection of Fig. 1, which, with its title, is self explanatory. The flame-like discharge visible is produced by the intense electrical oscillations which pass through the coil shown, and violently agitate the electrified molecules of the air. By this means a strong affinity is created between the two normally indifferent constituents of the atmosphere, and they combine readily, even if no further provision is made for intensifying the chemical action of the discharge. In the manufacture of nitrogen compounds by this method, of course, every possible means bearing upon the intensity of this action and the efficiency of the process will be taken advantage of, and, besides, special arrangements will be provided for the fixation of the compounds formed, as they are generally unstable, the nitrogen becoming again inert after a little lapse of time. Steam is a simple and effective means for fixing permanently the compounds. The result illustrated makes it practicable to oxidize the atmospheric nitrogen in unlimited quantities, merely by the use of cheap mechanical power and simple electrical apparatus. In this manner many compounds of nitrogen may be manufactured all over the world, at a small cost, and in any desired amount, and by means of these compounds the soil can be fertilized and its productiveness indefinitely increased. An abundance of cheap and healthful food, not artificial, but such as we are accustomed to, may thus be obtained. This new and inexhaustible source of food-supply will be of incalculable benefit to mankind, for it will enormously contribute to the increase of the human mass, and thus add immensely to human energy. Soon, I hope, the world will see the beginning of an industry which, in time to come, will, I believe, be in importance next to that if iron.

> The second problem: how to reduce the force retarding the human mass — The art of telautomatics

As before stated, the force which retards the onward movement of man is partly frictional and partly negative. To illustrate this distinction I may name, for example, ignorance, stupidity, and imbecility as some of the purely frictional forces, or resistances devoid of any directive tendency. On the other hand, visionariness, insanity, self-destructive tendency, religious fanaticism, and the like, are all forces of a negative character, acting in definite directions. To reduce or entirely overcome these dissimilar retarding forces, radically different methods must be employed. One knows, for instance, what a fanatic may do, and one can take preventive measures, can enlighten, convince, and, possibly direct him, turn his vice into virtue; but one does not know, and never can know, what a brute or an imbecile may do, and one must deal with him as with a mass, inert, without mind, let loose by the mad elements. A negative force always implies some quality, not infrequently a high one, though badly directed, which it is possible to turn to good advantage; but a directionless, frictional force involves unavoidable loss. Evidently, then, the first and general answer to the above question is: turn all negative force in the right direction and reduce all frictional force.

There can be no doubt that, of all the frictional resistances, the one that most retards human movement is ignorance. Not without reason said that man of wisdom, Buddha: "Ignorance is the greatest evil in the world." The friction which results from ignorance, and which is greatly increased owing to the numerous languages and nationalities, can be reduced only by the spread of knowledge and the unification of the heterogeneous elements of humanity. No effort could be better spent. But however ignorance may have retarded the onward movement of man in times past, it is certain that, nowadays, negative forces have become of greater importance. Among these there is one of far greater moment than any other. It is called organized warfare. When we consider the millions of individuals, often the ablest in mind and body, the flower of humanity, who are compelled to a life of inactivity and unproductiveness, the immense sums of money daily required for the maintenance of armies and war apparatus, representing ever so much of human energy, all the effort uselessly spent in the production of arms and implements of destruction, the loss of life and the fostering of a barbarous spirit, we are appalled at the inestimable loss to mankind which the existence of these deplorable conditions must involve. What can we do to combat best this great evil?

Law and order absolutely require the maintenance of organized force. No community can exist and prosper without rigid discipline. Every country must be able to defend itself, should the necessity arise. The conditions of today are not the result of yesterday, and a radical change cannot be effected to-morrow. If the nations would at once disarm, it is more than likely that a state of things worse than war itself would follow. Universal peace is a beautiful dream, but not at once realizable. We have seen recently that even the nobel effort of the man invested with the greatest worldly power has been virtually without effect. And no wonder, for the establishment of universal peace is, for the time being, a physical impossibility. War is a negative force, and cannot be turned in a positive direction without passing through, the intermediate phases. It is a problem of making a wheel, rotating one way, turn in the opposite direction without slowing it down, stopping it, and speeding it up again the other way.

It has been argued that the perfection of guns of great destructive power will stop warfare. So I myself thought for a long time, but now I believe this to be a profound mistake. Such developments will greatly modify, but not arrest it. On the contrary, I think that every new arm that is invented, every new departure that is made in this direction, merely invites new talent and skill, engages new effort, offers new incentive, and so only gives a fresh impetus to further development. Think of the discovery of gun-powder. Can we conceive of any more radical departure than was effected by this innovation? Let us imagine ourselves living in that period: would we not have thought then that warfare was at an end, when the armor of the knight became an object of ridicule, when bodily strength and skill, meaning so much before, became of comparatively little value? Yet gunpowder did not stop warfare: quite the opposite—it acted as a most powerful incentive. Nor do I believe that warfare can ever be arrested by any scientific or ideal development, so long as similar conditions to those prevailing now exist, because war has itself become a science, and because war involves some of the most sacred sentiments of which man is capable. In fact, it is doubtful whether men who would not be ready to fight for a high principle would be good for anything at all. It is not the mind which makes man, nor is it the body; it is mind and body. Our virtues and our failings are inseparable, like force and matter. When they separate, man is no more.

Another argument, which carries considerable force, is frequently made, namely, that war must soon become impossible be cause the means of defense are outstripping the means of attack. This is only in accordance with a fundamental law which may be expressed by the statement that it is easier to destroy than to build. This law defines human capacities and human conditions. Were these such that it would be easier build than to destroy, man would go on unresisted, creating and accumulating without limit. Such conditions are not of this earth. A being which could do this would not be a man: it might be a god. Defense will always have the advantage over attack, but this alone, it seems to me, can never stop war. By the use of new principles of defense we can render harbors impregnable against attack, but we cannot by such means prevent two warships meeting in battle on the high sea. And then, if we follow this idea to its ultimate development, we are led to the conclusion that it would be better for mankind if attack and defense were just oppositely related; for if every country, even the smallest, could surround itself with a wall absolutely impenetrable, and could defy the rest of the world, a state of things would surely be brought on which would be extremely unfavorable to human progress. It is by abolishing all the barriers which separate nations and countries that civilization is best furthered.

Again, it is contended by some that the advent of the flying-machine must bring on universal peace. This, too, I believe to be an entirely erroneous view. The flying-machine is certainly coming, and very soon, but the conditions will remain the same as before. In fact, I see no reason why a ruling power, like Great Britain, might not govern the air as well as the sea. Without wishing to put myself on record as a prophet, I do not hesitate to say that the next years will see the establishment of an "air-power," and its center may be not far from New York. But, for all that, men will fight on merrily.

The ideal development of the war principle would ultimately lead to the transformation of the whole energy of war into purely potential, explosive energy, like that of an electrical condenser. In this form the war-energy could be maintained without effort; it would need to be much smaller in amount, while incomparably more effective.

As regards the security of a country against foreign invasion, it is interesting to note that it depends only on the relative, and not the absolute, number of the individuals or magnitude of the forces, and that, if every country should reduce the war-force in the same ratio, the security would remain unaltered. An international agreement with the object of reducing to a minimum the war-force which, in view of the present still imperfect education of the masses, is absolutely indispensable, would, therefore, seem to be the first rational step to take toward diminishing the force retarding human movement.

Fortunately, the existing conditions cannot continue indefinitely, for a new element is beginning to assert itself. A change for the better is eminent, and I shall now endeavor to show what, according to my ideas, will be the first advance toward the establishment of peaceful relations between nations, and by what means it will eventually be accomplished.

Let us go back to the early beginning, when the law of the stronger was the only law. The light of reason was not yet kindled, and the weak was entirely at the mercy of the strong. The weak individual then began to learn how to defend himself. He made use of a club, stone, spear, sling, or bow and arrow, and in the course of time, instead of physical strength, intelligence became the chief deciding factor in the battle. The wild character was gradually softened by the awakening of noble sentiments, and so, imperceptibly, after ages of continued progress, we have come from the brutal fight of the unreasoning animal to what we call the "civilized warfare" of today, in which the combatants shake hands, talk in a friendly way, and smoke cigars in the entr'actes, ready to engage again in deadly conflict at a signal. Let pessimists say what they like, here is an absolute evidence of great and gratifying advance.

But now, what is the next phase in this evolution? Not peace as yet, by any means. The next change which should naturally follow from modern developments should be the continuous diminution of the number of individuals engaged in battle. The apparatus will be one of specifically great power, but only a few individuals will be required to operate it. This evolution will bring more and more into prominence a machine or mechanism with the fewest individuals as an element of warfare, and the absolutely unavoidable consequence of this will be the abandonment of large, clumsy, slowly moving, and unmanageable units. Greatest possible speed and maximum rate of energy-delivery by the war apparatus will be the main object. The loss of life will become smaller and smaller, and finally, the number of the individuals continuously diminishing, merely machines will meet in a contest without blood-shed, the nations being simply interested, ambitious spectators. When this happy condition is realized, peace will be assured. But, no matter to what degree of perfection rapid-fire guns, high-power cannon, explosive projectiles, torpedo-boats, or other implements of war may be brought, no matter how destructive they may be made, that condition can never be reached through any such development. All such implements require men for their operation; men are indispensable parts of the machinery. Their object is to kill and to destroy. Their power resides in their capacity for doing evil. So long as men meet in battle, there will be bloodshed. Bloodshed will ever keep up barbarous passion. To break this fierce spirit, a radical departure must be made, an entirely new principle must be introduced, something that never existed before in warfare—a principle which will forcibly, unavoidably, turn the battle into a mere spectacle, a play, a contest without loss of blood. To bring on this result men must be dispensed with: machine must fight machine. But how accomplish that which seems impossible? The answer is simple enough: produce a machine capable of acting as though it were part of a human being—no mere mechanical contrivance, comprising levers, screws, wheels, clutches, and nothing more, but a machine embodying a higher principle, which will enable it to per form its duties as though it had intelligence, experience, judgment, a mind! This conclusion is the result of my thoughts and observations which have extended through virtually my whole life, and I shall now briefly describe how I came to accomplish that which at first seemed an unrealizable dream.

   A long time ago, when I was a boy, I was afflicted with a singular trouble, which seems to have been due to an extraordinary excitability of the retina. It was the appearance of images which, by their persistence, marred the vision of real objects and interfered with thought. When a word was said to me, the image of the object which it designated would appear vividly before my eyes, and many times it was impossible for me to tell whether the object I saw was real or not. This caused me great discomfort and anxiety, and I tried hard to free myself of the spell. But for a long time I tried in vain, and it was not, as I clearly recollect, until I was about twelve years old that I succeeded for the first time, by an effort of the will, in banishing an image which presented itself. My happiness will never be as complete as it was then, but, unfortunately (as I thought at that time), the old trouble returned, and with it my anxiety. Here it was that the observations to which I refer began. I noted, namely, that whenever the image of an object appeared before my eyes I had seen something that reminded me of it. In the first instances I thought this to be purely accidental, but soon I convinced myself that it was not so. A visual impression, consciously or unconsciously received, invariably preceded the appearance of the image. Gradually the desire arose in me to find out, every time, what caused the images to appear, and the satisfaction of this desire soon became a necessity. The next observation I made was that, just as these images followed as a result of something I had seen, so also the thoughts which I conceived were suggested in like manner. Again, I experienced the same desire to locate the image which caused the thought, and this search for the original visual impression soon grew to be a second nature. My mind became automatic, as it were, and in the course of years of continued, almost unconscious performance, I acquired the ability of locating every time and, as a rule, instantly the visual impression which started the thought. Nor is this all. It was not long before I was aware that also all my movements were prompted in the same way, and so, searching, observing, and verifying continuously, year by year, I have, by every thought and every act of mine, demonstrated, and do so daily, to my absolute satisfaction, that I am an automaton endowed with power of movement, which merely responds to external stimuli beating upon my sense organs, and thinks and acts and moves accordingly. I remember only one or two cases in all my life in which I was unable to locate the first impression which prompted a movement or a thought, or even a dream.

Fig. 2. The first practical telautomaton

Note: A machine having all the bodily or translatory movements and the operations of the interior mechanism controlled from a distance without wires. The crewless boat shown in the photograph contains its own motive power, propelling and steering machinery, and numerous other accessories, all of which are controlled by transmitting from a distance, without wires, electrical oscillations to a circuit carried by the boat and adjusted to respond only to these oscillations.

With these experiences it was only natural that, long ago, I conceived the idea of constructing an automaton which would mechanically represent me, and which would respond, as I do myself, but, of course, in a much more primitive manner, to external influences. Such an automaton evidently had to have motive power, organs for locomotion, directive organs, and one or more sensitive organs so adapted as to be excited by external stimuli. This machine would, I reasoned, perform its movements in the manner of a living being, for it would have all the chief mechanical characteristics or elements of the same. There was still the capacity for growth, propagation, and, above all, the mind which would be wanting to make the model complete. But growth was not necessary in this case, since a machine could be manufactured full grown, so to speak. As to the capacity for propagation, it could likewise be left out of consideration, for in the mechanical model it merely signified a process of manufacture. Whether the automation be of flesh and bone, or of wood and steel, it mattered little, provided it could perform all the duties required of it like an intelligent being. To do so, it had to have an element corresponding to the mind, which would effect the control of all its movements and operations, and cause it to act, in any unforeseen case that might present itself, with knowledge, reason, judgment, and experience. But this element I could easily embody in it by conveying to it my own intelligence, my own understanding. So this invention was evolved, and so a new art came into existence, for which the name "telautomatics" has been suggested, which means the art of controlling the movements and operations of distant automatons. This principle evidently was applicable to any kind of machine that moves on land or in the water or in the air. In applying it practically for the first time, I selected a boat (see Fig. 2). A storage battery placed within it furnished the motive power. The propeller, driven by a motor, represented the locomotive organs. The rudder, controlled by another motor likewise driven by the battery, took the place of the directive organs. As to the sensitive organ, obviously the first thought was to utilize a device responsive to rays of light, like a selenium cell, to represent the human eye. But upon closer inquiry I found that, owing to experimental and other difficulties, no thoroughly satisfactory control of the automaton could be effected by light, radiant heat, hertzian radiations, or by rays in general, that is, disturbances which pass in straight lines through space. One of the reasons was that any obstacle coming between the operator and the distant automaton would place it beyond his control. Another reason was that the sensitive device representing the eye would have to be in a definite position with respect to the distant controlling apparatus, and this necessity would impose great limitations in the control. Still another and very important reason was that, in using rays, it would be difficult, if not impossible, to give to the automaton individual features or characteristics distinguishing it from other machines of this kind. Evidently the automaton should respond only to an individual call, as a person responds to a name. Such considerations led me to conclude that the sensitive device of the machine should correspond to the ear rather than the eye of a human being, for in this case its actions could be controlled irrespective of intervening obstacles, regardless of its position relative to the distant controlling apparatus, and, last, but not least, it would remain deaf and unresponsive, like a faithful servant, to all calls but that of its master. These requirements made it imperative to use, in the control of the automaton, instead of light or other rays, waves or disturbances which propagate in all directions through space, like sound, or which follow a path of least resistance, however curved. I attained the result aimed at by means of an electric circuit placed within the boat, and adjusted, or "tuned," exactly to electrical vibrations of the proper kind transmitted to it from a distant "electrical oscillator." This circuit, in responding, however feebly, to the transmitted vibrations, affected magnets and other contrivances, through the medium of which were controlled the movements of the propeller and rudder, and also the operations of numerous other appliances.

By the simple means described the knowledge, experience, judgment — the mind, so to speak — of the distant operator were embodied in that machine, which was thus enabled to move and to perform all its operations with reason and intelligence. It behaved just like a blindfolded person obeying directions received through the ear.

The automatons so far constructed had "borrowed minds," so to speak, as each merely formed part of the distant operator who conveyed to it his intelligent orders; but this art is only in the beginning. I purpose to show that, however impossible it may now seem, an automaton may be contrived which will have its "own mind," and by this I mean that it will be able, independent of any operator, left entirely to itself, to perform, in response to external influences affecting its sensitive organs, a great variety of acts and operations as if it had intelligence. It will be able to follow a course laid out or to obey orders given far in advance; it will be capable of distinguishing between what it ought and what it ought not to do, and of making experiences or, otherwise stated, of recording impressions which will definitely affect its subsequent actions. In fact, I have already conceived such a plan.

Although I evolved this invention many years ago and explained it to my visitors very frequently in my laboratory demonstrations, it was not until much later, long after I had perfected it, that it became known, when, naturally enough, it gave rise to much discussion and to sensational reports. But the true significance of this new art was not grasped by the majority, nor was the great force of the underlying principle recognized. As nearly as I could judge from the numerous comments which appeared, the results I had obtained were considered as entirely impossible. Even the few who were disposed to admit the practicability of the invention saw in it merely an automobile torpedo, which was to be used for the purpose of blowing up battleships, with doubtful success. The general impression was that I contemplated simply the steering of such a vessel by means of Hertzian or other rays. There are torpedoes steered electrically by wires, and there are means of communicating without wires, and the above was, of course an obvious inference. Had I accomplished nothing more than this, I should have made a small advance indeed. But the art I have evolved does not contemplate merely the change of direction of a moving vessel; it affords means of absolutely controlling, in every respect, all the innumerable translatory movements, as well as the operations of all the internal organs, no matter how many, of an individualized automaton. Criticisms to the effect that the control of the automaton could be interfered with were made by people who do not even dream of the wonderful results which can be accomplished by use of electrical vibrations. The world moves slowly, and new truths are difficult to see. Certainly, by the use of this principle, an arm for attack as well as defense may be provided, of a destructiveness all the greater as the principle is applicable to submarine and aerial vessels. There is virtually no restriction as to the amount of explosive it can carry, or as to the distance at which it can strike, and failure is almost impossible. But the force of this new principle does not wholly reside in its destructiveness. Its advent introduces into warfare an element which never existed before—a fighting-machine without men as a means of attack and defense. The continuous development in this direction must ultimately make war a mere contest of machines without men and without loss of life—a condition which would have been impossible without this new departure, and which, in my opinion, must be reached as preliminary to permanent peace. The future will either bear out or disprove these views. My ideas on this subject have been put forth with deep conviction, but in a humble spirit.

The establishment of permanent peaceful relations between nations would most effectively reduce the force retarding the human mass, and would be the best solution of this great human problem. But will the dream of universal peace ever be realized? Let us hope that it will. When all darkness shall be dissipated by the light of science, when all nations shall be merged into one, and patriotism shall be identical with religion, when there shall be one language, one country, one end, then the dream will have become reality.

> The third problem: how to increase the force accelerating the human mass — The harnessing of the sun's energy

Of the three possible solutions of the main problem of increasing human energy, this is by far the most important to consider, not only because of its intrinsic significance, but also because of its intimate bearing on all the many elements and conditions which determine the movement of humanity. In order to proceed systematically, it would be necessary for me to dwell on all those considerations which have guided me from the outset in my efforts to arrive at a solution, and which have led me, step by step, to the results I shall now describe. As a preliminary study of the problem an analytical investigation, such as I have made, of the chief forces which determine the onward movement, would be of advantage, particularly in conveying an idea of that hypothetical "velocity" which, as explained in the beginning, is a measure of human energy; but to deal with this specifically here, as I would desire, would lead me far beyond the scope of the present subject. Suffice it to state that the resultant of all these forces is always in the direction of reason, which therefore, determines, at any time, the direction of human movement. This is to say that every effort which is scientifically applied, rational, useful, or practical, must be in the direction in which the mass is moving. The practical, rational man, the observer, the man of business, he who reasons, calculates, or determines in advance, carefully applies his effort so that when coming into effect it will be in the direction of the movement, making it thus most efficient, and in this knowledge and ability lies the secret of his success. Every new fact discovered, every new experience or new element added to our knowledge and entering into the domain of reason, affects the same and, therefore, changes the direction of movement, which, however, must always take place along the resultant of all those efforts which, at that time, we designate as reasonable, that is, self-preserving, useful, profitable, or practical. These efforts concern our daily life, our necessities and comforts, our work and business, and it is these which drive man onward.

But looking at all this busy world about us, on all this complex mass as it daily throbs and moves, what is it but an immense clock-work driven by a spring? In the morning, when we rise, we cannot fail to note that all the objects about us are manufactured by machinery: the water we use is lifted by steam-power; the trains bring our breakfast from distant localities; the elevators in our dwelling and our office building, the cars that carry us there, are all driven by power; in all our daily errands, and in our very life-pursuit, we depend upon it; all the objects we see tell us of it; and when we return to our machine-made dwelling at night, lest we should forget it, all the material comforts of our home, our cheering stove and lamp, remind us of how much we depend on power. And when there is an accidental stoppage of the machinery, when the city is snowbound, or the life sustaining movement otherwise temporarily arrested, we are affrighted to realize how impossible it would be for us to live the life we live without motive power. Motive power means work. To increase the force accelerating human movement means, therefore, to perform more work.

So we find that the three possible solutions of the great problem of increasing human energy are answered by the three words: food, peace, work. Many a year I have thought and pondered, lost myself in speculations and theories, considering man as a mass moved by a force, viewing his inexplicable movement in the light of a mechanical one, and applying the simple principles of mechanics to the analysis of the same until I arrived at these solutions, only to realize that they were taught to me in my early childhood. These three words sound the key-notes of the Christian religion. Their scientific meaning and purpose now clear to me: food to increase the mass, peace to diminish the retarding force, and work to increase the force accelerating human movement. These are the only three solutions which are possible of that great problem, and all of them have one object, one end, namely, to increase human energy. When we recognize this, we cannot help wondering how profoundly wise and scientific and how immensely practical the Christian religion is, and in what a marked contrast it stands in this respect to other religions. It is unmistakably the result of practical experiment and scientific observation which have extended through the ages, while other religions seem to be the outcome of merely abstract reasoning. Work, untiring effort, useful and accumulative, with periods of rest and recuperation aiming at higher efficiency, is its chief and ever-recurring command. Thus we are inspired both by Christianity and Science to do our utmost toward increasing the performance of mankind. This most important of human problems I shall now specifically consider.

The source of human energy — The three ways of drawing energy from the sun

First let us ask: Whence comes all the motive power? What is the spring that drives all? We see the ocean rise and fall, the rivers flow, the wind, rain, hail, and snow beat on our windows, the trains and steamers come and go; we here the rattling noise of carriages, the voices from the street; we feel, smell, and taste; and we think of all this. And all this movement, from the surging of the mighty ocean to that subtle movement concerned in our thought, has but one common cause. All this energy emanates from one single center, one single source—the sun. The sun is the spring that drives all. The sun maintains all human life and supplies all human energy. Another answer we have now found to the above great question: To increase the force accelerating human movement means to turn to the uses of man more of the sun's energy. We honor and revere those great men of bygone times whose names are linked with immortal achievements, who have proved themselves benefactors of humanity—the religious reformer with his wise maxims of life, the philosopher with his deep truths, the mathematician with his formula—, the physicist with his laws, the discover with his principles and secrets wrested from nature, the artist with his forms of the beautiful; but who honors him, the greatest of all,—who can tell the name of him,—who first turned to use the sun's energy to save the effort of a weak fellow-creature? That was man's first act of scientific philanthropy, and its consequences have been incalculable.

From the very beginning three ways of drawing energy from the sun were open to man. The savage, when he warmed his frozen limbs at a fire kindled in some way, availed himself of the energy of the sun stored in the burning material. When he carried a bundle of branches to his cave and burned them there, he made use of the sun's stored energy transported from one to another locality. When he set sail to his canoe, he utilized the energy of the sun applied to the atmosphere or the ambient medium. There can be no doubt that the first is the oldest way. A fire, found accidentally, taught the savage to appreciate its beneficial heat. He then very likely conceived of the idea of carrying the glowing members to his abode. Finally he learned to use the force of a swift current of water or air. It is characteristic of modern development that progress has been effected in the same order. The utilization of the energy stored in wood or coal, or, generally speaking, fuel, led to the steam-engine. Next a great stride in advance was made in energy-transportation by the use of electricity, which permitted the transfer of energy from one locality to another without transporting the material. But as to the utilization of the energy of the ambient medium, no radical step forward has as yet been made known.

The ultimate results of development in these three directions are: first, the burning of coal by a cold process in a battery; second, the efficient utilization of the energy of the ambient medium; and, third the transmission without wires of electrical energy to any distance. In whatever way these results may be arrived at, their practical application will necessarily involve an extensive use of iron, and this invaluable metal will undoubtedly be an essential element in the further development along these three lines. If we succeed in burning coal by a cold process and thus obtain electrical energy in an efficient and inexpensive manner, we shall require in many practical uses of this energy electric motors — that is, iron. If we are successful in deriving energy from the ambient medium, we shall need, both in the obtainment and utilization of the energy, machinery — again, iron. If we realize the transmission of electrical energy without wires on an industrial scale, we shall be compelled to use extensively electric generators — once more, iron. Whatever we may do, iron will probably be the chief means of accomplishment in the near future, possibly more so than in the past. How long its reign will last is difficult to tell, for even now aluminium is looming up as a threatening competitor. But for the time being, next to providing new resources of energy, it is of the greatest importance to making improvements in the manufacture and utilization of iron. Great advances are possible in these latter directions, which, if brought about, would enormously increase the useful performance of mankind.

> Great possibilities offered by iron for increasing human performance — Enormous waste in iron manufacture

Iron is by far the most important factor in modern progress. It contributes more than any other industrial product to the force accelerating human movement. So general is the use of this metal, and so intimately is it connected with all that concerns our life, that it has become as indispensable to us as the very air we breathe. Its name is synonymous with usefulness. But, however great the influence of iron may be on the present human development, it does not add to the force urging man onward nearly as much as it might. First of all, its manufacture as now carried on is connected with an appalling waste of fuel — that is, waste of energy. Then, again, only a part of all the iron produced is applied for useful purposes. A good part of it goes to create frictional resistances, while still another large part is the means of developing negative forces greatly retarding human movement. Thus the negative force of war is almost wholly represented in iron. It is impossible to estimate with any degree of accuracy the magnitude of this greatest of all retarding forces, but it is certainly very considerable. If the present positive impelling force due to all useful applications of iron be represented by ten, for instance, I should not think it exaggeration to estimate the negative force of war, with due consideration of all its retarding influences and results, at, say, six. On the basis of this estimate the effective impelling force of iron in the positive direction would be measured by the difference of these two numbers, which is four. But if, through the establishment of universal peace, the manufacture of war machinery should cease, and all struggle for supremacy between nations should be turned into healthful, ever active and productive commercial competition, then the positive impelling force due to iron would be measured by the sum of those two, numbers, which is sixteen—that is, this force would have four times its present value. This example is, of course, merely intended to give an idea of the immense increase in the useful performance of mankind which would result from a radical reform of the iron industries supplying the implements of warfare.

A similar inestimable advantage in the saving of energy available to man would be secured by obviating the great waste of coal which is inseparably connected with the present methods of manufacturing iron. In some countries, such as Great Britain, the hurtful effects of this squandering of fuel are beginning to be felt. The price of coal is constantly rising, and the poor are made to suffer more and more. Though we are still far from the dreaded "exhaustion of the coal-fields," philanthropy commands us to invent novel methods of manufacturing iron, which will not involve such barbarous waste of this valuable material from which we derive at present most of our energy. It is our duty to coming generations to leave this store of energy intact for them, or at least not to touch it until we shall have perfected processes for burning coal more efficiently. Those who are coming after us will need fuel more than we do. We should be able to manufacture the iron we require by using the sun's energy, without wasting any coal at all. As an effort to this end the idea of smelting iron ores by electric currents obtained from the energy of falling water has naturally suggested itself to many. I have myself spent much time in endeavoring to evolve such a practical process, which would enable iron to be manufactured at small cost. After a prolonged investigation of the subject, finding that it was unprofitable to use the currents generated directly for smelting the ore, I devised a method which is far more economical.

Economical production of iron by a new process

The industrial project, as I worked it out six years ago, contemplated the employment of the electric currents derived from the energy of a waterfall, not directly for smelting the ore, but for decomposing water for a preliminary step. To lessen the cost of the plant, I proposed to generate the currents in exceptionally cheap and simple dynamos, which I designed for this sole purpose. The hydrogen liberated in the electrolytic decomposition was to be burned or recombined with oxygen, not with that from which it was separated, but with that of the atmosphere. Thus very nearly the total electrical energy used up in the decomposition of the water would be recovered in the form of heat resulting from the recombination of the hydrogen. This heat was to be applied to the smelting of ore. The oxygen gained as a by-product of the decomposition of the water I intended to use for certain other industrial purposes, which would probably yield good financial returns, inasmuch as this is the cheapest way of obtaining this gas in large quantities. In any event, it could be employed to burn all kinds of refuse, cheap hydrocarbon, or coal of the most inferior quality which could not be burned in air or be otherwise utilized to advantage, and thus again a considerable amount of heat would be made available for the smelting of the ore. To increase the economy of the process I contemplated, furthermore, using an arrangement such that the hot metal and the products of combustion, coming out of the furnace, would give up their heat upon the cold ore going into the furnace, so that comparatively little of the heat energy would be lost in the smelting. I calculated that probably forty thousand pounds of iron could be produced per horse-power per annum by this method. Liberal allowances were made for those losses which are unavoidable, the above quantity being about half of that theoretically obtainable. Relying on this estimate and on practical data with reference to a certain kind of sand ore existing in abundance in the region of the Great Lakes, including cost of transportation and labor, I found that in some localities iron could be manufactured in this manner cheaper than by any of the adopted methods. This result would be obtained all the more surely if the oxygen obtained from the water, instead of being used for smelting of ore, as assumed, should be more profitably employed. Any new demand for this gas would secure a higher revenue from the plant, thus cheapening the iron. This project was advanced merely in the interest of industry. Some day, I hope, a beautiful industrial butterfly will come out of the dusty and shriveled chrysalis.

The production of iron from sand ores by a process of magnetic separation is highly commendable in principle, since it involves no waste of coal; but the usefulness of this method is largely reduced by the necessity of melting the iron afterward. As to the crushing of iron ore, I would consider it rational only if done by water-power, or by energy otherwise obtained without consumption of fuel. An electrolytic cold process, which would make it possible to extract iron cheaply, and also to mold it into the required forms without any fuel consumption, would, in my opinion, be a very great advance in iron manufacture. In common with some other metals, iron has so far resisted electrolytic treatment, but there can be no doubt that such a cold process will ultimately replace in metallurgy the present crude method of casting, and thus obviating the enormous waste of fuel necessitated by the repeated heating of metal in the foundries.

Up to a few decades ago the usefulness of iron was based almost wholly on its remarkable mechanical properties, but since the advent of the commercial dynamo and electric motor its value to mankind has been greatly increased by its unique magnetic qualities. As regards the latter, iron has been greatly improved of late. The signal progress began about thirteen years ago, when I discovered that in using soft Bessemer steel instead of wrought iron, as then customary, in an alternating motor, the performance of the machine was doubled. I brought this fact to the attention of Mr. Albert Schmid, to whose untiring efforts and ability is largely due the supremacy of American electrical machinery, and who was then superintendent of an industrial corporation engaged in this field. Following my suggestion, he constructed transformers of steel, and they showed the same marked improvement. The investigation was then systematically continued under Mr. Schmid's guidance, the impurities being gradually eliminated from the "steel" (which was only such in name, for in reality it was pure soft iron), and soon a product resulted which admitted of little further improvement.

> The coming of age of aluminium — Doom of the copper industry — The great civilizing potency of the new metal

With the advances made in iron of late years we have arrived virtually at the limits of improvement. We cannot hope to increase very materially its tensile strength, elasticity, hardness, or malleability, nor can we expect to make it much better as regards its magnetic qualities. More recently a notable gain was secured by the mixture of a small percentage of nickel with the iron, but there is not much room for further advance in this direction. New discoveries may be expected, but they cannot greatly add to the valuable properties of the metal, though they may considerably reduce the cost of manufacture. The immediate future of iron is assured by its cheapness and its unrivaled mechanical and magnetic qualities. These are such that no other product can compete with it now. But there can be no doubt that, at a time not very distant, iron, in many of its now uncontested domains, will have to pass the scepter to another: the coming age will be the age of aluminium. It is only seventy years since this wonderful metal was discovered by Woehler, and the aluminium industry, scarcely forty years old, commands already the attention of the entire world. Such rapid growth has not been recorded in the history of civilization before. Not long ago aluminium was sold at the fanciful price of thirty or forty dollars per pound; today it can be had in any desired amount for as many cents. What is more, the time is not far off when this price, too, will be considered fanciful, for great improvements are possible in the methods of its manufacture. Most of the metal is now produced in the electric furnace by a process combining fusion and electrolysis, which offers a number of advantageous features, but involves naturally a great waste of the electrical energy of the current. My estimates show that the price of aluminium could be considerably reduced by adopting in its manufacture a method similar to that proposed by me for the production of iron. A pound of aluminium requires for fusion only about 70% of the heat needed for melting a pound of iron, and inasmuch as its weight is only about one third of that of the latter, a volume of aluminium four times that of iron could be obtained from a given amount of heat-energy. But a cold electrolytic process of manufacture is the ideal solution, and on this I have placed my hope.

The absolutely unavoidable consequence of the advancement of the aluminium industry will be the annihilation of the copper industry. They cannot exist and prosper together, and the latter is doomed beyond any hope of recovery. Even now it is cheaper to convey an electric current through aluminium wires than through copper wires; aluminium castings cost less, and in many domestic and other uses copper has no chance of successfully competing. A further material reduction of the price of aluminium cannot but be fatal to copper. But the progress of the former will not go on unchecked, for, as it ever happens in such cases, the larger industry will absorb the smaller one: the giant copper interests will control the pygmy aluminium interests, and the slow-pacing copper will reduce the lively gait of aluminium. This will only delay, not avoid the impending catastrophe.

Aluminium, however, will not stop at downing copper. Before many years have passed it will be engaged in a fierce struggle with iron, and in the latter it will find an adversary not easy to conquer. The issue of the contest will largely depend on whether iron shall be indispensable in electric machinery. This the future alone can decide. The magnetism as exhibited in iron is an isolated phenomenon in nature. What it is that makes this metal behave so radically different from all other materials in this respect has not yet been ascertained, though many theories have been suggested. As regards magnetism, the molecules of the various bodies behave like hollow beams partly filled with a heavy fluid and balanced in the middle in the manner of a seesaw. Evidently some disturbing influence exists in nature which causes each molecule, like such a beam, to tilt either one or the other way. If the molecules are tilted one way, the body is magnetic; if they are tilted the other way, the body is non-magnetic; but both positions are stable, as they would be in the case of the hollow beam, owing to the rush of the fluid to the lower end. Now, the wonderful thing is that the molecules of all known bodies went one way, while those of iron went the other way. This metal, it would seem, has an origin entirely different from that of the rest of the globe. It is highly improbable that we shall discover some other and cheaper material which will equal or surpass iron in magnetic qualities.

Unless we should make a radical departure in the character of the electric currents employed, iron will be indispensable. Yet the advantages it offers are only apparent. So long as we use feeble magnetic forces it is by far superior to any other material; but if we find ways of producing great magnetic forces, than better results will be obtainable without it. In fact, I have already produced electric transformers in which no iron is employed, and which are capable of performing ten times as much work per pound of weight as those of iron. This result is attained by using electric currents of a very high rate of vibration, produced in novel ways, instead of the ordinary currents now employed in the industries. I have also succeeded in operating electric motors without iron by such rapidly vibrating currents, but the results, so far, have been inferior to those obtained with ordinary motors constructed of iron, although theoretically the former should be capable of performing incomparably more work per unit of weight than the latter. But the seemingly insuperable difficulties which are now in the way may be overcome in the end, and then iron will be done away with, and all electric machinery will be manufactured of aluminium, in all probability, at prices ridiculously low. This would be a severe, if not fatal, blow to iron. In many other branches of industry, as ship-building, or wherever lightness of structure is required, the progress of the new metal will be much quicker. For such uses it is eminently suitable, and is sure to supersede iron sooner or later. It is highly probable that in the course of time we shall be able to give it many of those qualities which make iron so valuable.

While it is impossible to tell when this industrial revolution will be consummated, there can be no doubt that the future belongs to aluminium, and that in times to come it will be the chief means of increasing human performance. It has in this respect capacities greater by far than those of any other metal. I should estimate its civilizing potency at fully one hundred times that of iron. This estimate, though it may astonish, is not at all exaggerated. First of all, we must remember that there is thirty times as much aluminium as iron in bulk, available for the uses of man. This in itself offers great possibilities. Then, again, the new metal is much more easily workable, which adds to its value. In many of its properties it partakes of the character of a precious metal, which gives it additional worth. Its electric conductivity, which, for a given weight, is greater than that of any other metal, would be alone sufficient to make it one of the most important factors in future human progress. Its extreme lightness makes it far more easy to transport the objects manufactured. By virtue of this property it will revolutionize naval construction, and in facilitating transport and travel it will add enormously to the useful performance of mankind. But its greatest civilizing property will be, I believe, in aerial travel, which is sure to be brought about by means of it. Telegraphic instruments will slowly enlighten the barbarian. Electric motors and lamps will do it more quickly, but quicker than anything else the flying-machine will do it. By rendering travel ideally easy it will be the best means for unifying the heterogeneous elements of humanity. As the first step toward this realization we should produce a lighter storage-battery or get more energy from coal.

> Efforts toward obtaining more energy from coal — The electric transmission — The gas-engine — The cold-coal battery

I remember that at one time I considered the production of electricity by burning coal in a battery as the greatest achievement toward the advancing civilization, and I am surprised to find how much the continuous study of these subjects has modified my views. It now seems to me that to burn coal, however efficiently, in a battery would be a mere makeshift, a phase in the evolution toward something much more perfect. After all, in generating electricity in this manner, we should be destroying material, and this would be a barbarous process. We ought to be able to obtain the energy we need without consumption of material. But I am far from underrating the value of such an efficient method of burning fuel. At the present time most motive power comes from coal, and, either directly or by its products, it adds vastly to human energy. Unfortunately, in all the process now adopted, the larger portion of the energy of the coal is uselessly dissipated. The best steam-engines utilize only a small part of the total energy. Even in gas-engines, in which, particularly of late, better results are obtainable, there is still a barbarous waste going on. In our electric-lighting systems we scarcely utilize one third of 1%, and in lighting by gas a much smaller fraction, of the total energy of the coal. Considering the various uses of coal throughout the world, we certainly do not utilize more than 2% of its energy theoretically available. The man who should stop this senseless waste would be a great benefactor of humanity, though the solution he would offer could not be a permanent one, since it would ultimately lead to the exhaustion of the store of material. Efforts toward obtaining more energy from coal are now being made chiefly in two directions—by generating electricity and by producing gas for motive-power purposes. In both of these lines notable success has already been achieved.

The advent of the alternating-current system of electric power-transmission marks an epoch in the economy of energy available to man from coal.

Evidently all electrical energy obtained from a waterfall, saving so much fuel, is a net gain to mankind, which is all the more effective as it is secured with little expenditure of human effort, and as this most perfect of all known methods of deriving energy from the sun contributes in many ways to the advancement of civilization. But electricity enables us also to get from coal much more energy than was practicable in the old ways. Instead of transporting the coal to distant places of consumption, we burn it near the mine, develop electricity in the dynamos, and transmit the current to remote localities, thus effecting a considerable saving. Instead of driving the machinery in a factory in the old wasteful way of belts and shafting, we generate electricity by steam-power and operate electric motors. In this manner it is not uncommon to obtain two or three times as much effective motive power from the fuel, besides securing many other important advantages. It is in this field as much as in the transmission of energy to great distance that the alternating system, with its ideally simple machinery, is bringing about an industrial revolution. But in many lines this progress has not been yet fully felt. For example, steamers and trains are still being propelled by the direct application of steam-power to shafts or axles. A much greater percentage of the heat-energy of the fuel could be transformed into motive energy by using, in place of the adopted marine engines and locomotives, dynamos driven by specially designed high-pressure steam- or gas-engines and by utilizing the electricity generated for the propulsion. A gain of 50 to 100% in the effective energy derived from the coal could be secured in this manner. It is difficulty to understand why a fact so plain and obvious is not receiving more attention from engineers. In ocean steamers such an improvement would be particularly desirable, as it would do away with noise and increase materially the speed and the carrying capacity of the liners.

Still more energy is now being obtained from coal by the latest improved gas-engine, the economy of which is, on the average, probably twice that of the best steam-engine. The introduction of the gas-engine is very much facilitated by the importance of the gas industry. With the increasing use of the electric light more and more of the gas is utilized for heating and motive-power purposes. In many instances gas is manufactured close to the coal-mine and conveyed to distant places of consumption, a considerable saving both in cost of transportation and in utilization of the energy of the fuel being thus effected. In the present state of the mechanical and electrical arts the most rational way of deriving energy from coal is evidently to manufacture gas close to the coal store, and to utilize it, either on the spot or elsewhere, to generate electricity for industrial uses in dynamos driven by gas engines. The commercial success of such a plant is largely dependent upon the production of gas-engines of great nominal horse-power, which, judging from the keen activity in this field will soon be forthcoming. Instead of consuming coal directly, as usual, gas should be manufactured from it and burned to economize energy.

But all such improvements cannot be more than passing phases in the evolution toward something far more perfect, for ultimately we must succeed in obtaining electricity from coal in a more direct way, involving no great loss of heat-energy. Whether coal can be oxidized by a cold process is still a question. Its combination with oxygen always involves heat, and whether the energy of the combination of the carbon with another element can be turned directly into electrical energy has not yet been determined. Under certain conditions nitric acid will burn the carbon, generating an electric current, but the solution does not remain cold. Other means of oxidizing coal have been proposed, but they have offered no promise of leading to an efficient process. My own lack of success has been complete, though perhaps not quite so complete as that of some who have "perfected" the cold-coal battery. This problem is essentially one for the chemist to solve. It is not for the physicist, who determines all his results in advance, so that, when the experiment is tried, it cannot fail. Chemistry, though a positive science, does not yet admit of a solution by such positive methods as those which are available in the treatment of many physical problems. The result, if possible, will be arrived at through patent trying rather than through deduction or calculation. The time will soon come, however, when the chemist will be able to follow a course clearly mapped out beforehand, and when the process of his arriving at a desired result will be purely constructive. The cold-coal battery would give a great impetus to electrical development; it would lead very shortly to a practical flying-machine, and would enormously enhance the introduction of the automobile. But these and many other problems will be better solved, and in a more scientific manner, by a light storage battery.

> Energy from the medium — The windmill and the solar engine — Motive power from terrestrial heat — Electricity from natural sources

Besides fuel, there is abundant material from which we might eventually derive power. An immense amount of energy is locked up in limestone, for instance, and machines can be driven by liberating the carbonic acid through sulfuric acid or otherwise. I once constructed such an engine, and it operated satisfactorily.

But, whatever our resources of primary energy may be in the future, we must, to be rational, obtain it without consumption of any material. Long ago I came to this conclusion, and to arrive at this result only two ways, as before indicated, appeared possible — either to turn to use the energy of the sun stored in the ambient medium, or to transmit, through the medium, the sun's energy to distant places from some locality where it was obtainable without consumption of material. At that time I at once rejected the latter method as entirely impracticable, and turned to examine the possibilities of the former.

It is difficult to believe, but it is, nevertheless, a fact, that since time immemorial man has had at his disposal a fairly good machine which has enabled him to utilize the energy of the ambient medium. This machine is the windmill. Contrary to popular belief, the power obtainable from wind is very considerable. Many a deluded inventor has spent years of his life in endeavoring to "harness the tides," and some have even proposed to compress air by tide- or wave-power for supplying energy, never understanding the signs of the old windmill on the hill, as it sorrowfully waved its arms about and bade them stop. The fact is that a wave- or tide-motor would have, as a rule, but a small chance of competing commercially with the windmill, which is by far the better machine, allowing a much greater amount of energy to be obtained in a simpler way. Wind-power has been, in old times, of inestimable value to man, if for nothing else but for enabling him, to cross the seas, and it is even now a very important factor in travel and transportation. But there are great limitations in this ideally simple method of utilizing the sun's energy. The machines are large for a given output, and the power is intermittent, thus necessitating the storage of energy and increasing the cost of the plant.

A far better way, however, to obtain power would be to avail ourselves of the sun's rays, which beat the earth incessantly and supply energy at a maximum rate of over four million horsepower per square mile. Although the average energy received per square mile in any locality during the year is only a small fraction of that amount, yet an inexhaustible source of power would be opened up by the discovery of some efficient method of utilizing the energy of the rays. The only rational way known to me at the time when I began the study of this subject was to employ some kind of heat- or thermodynamic-engine, driven by a volatile fluid evaporate in a boiler by the heat of the rays. But closer investigation of this method, and calculation, showed that, notwithstanding the apparently vast amount of energy received from the sun's rays, only a small fraction of that energy could be actually utilized in this manner. Furthermore, the energy supplied through the sun's radiations is periodical, and the same limitations as in the use of the windmill I found to exist here also. After a long study of this mode of obtaining motive power from the sun, taking into account the necessarily large bulk of the boiler, the low efficiency of the heat-engine, the additional cost of storing the energy and other drawbacks, I came to the conclusion that the "solar engine," a few instances excepted, could not be industrially exploited with success.

Another way of getting motive power from the medium without consuming any material would be to utilize the heat contained in the earth, the water, or the air for driving an engine. It is a well-known fact that the interior portions of the globe are very hot, the temperature rising, as observations show, with the approach to the center at the rate of approximately $1°$ C for every hundred feet of depth. The difficulties of sinking shafts and placing boilers at depths of, say, twelve thousand feet, corresponding to an increase in temperature of about $120°$ C., are not insuperable, and we could certainly avail ourselves in this way of the internal heat of the globe. In fact, it would not be necessary to go to any depth at all in order to derive energy from the stored terrestrial heat. The superficial layers of the earth and the air strata close to the same are at a temperature sufficiently high to evaporate some extremely volatile substances, which we might use in our boilers instead of water. There is no doubt that a vessel might be propelled on the ocean by an engine driven by such a volatile fluid, no other energy being used but the heat abstracted from the water. But the amount of power which could be obtained in this manner would be, without further provision, very small.

Electricity produced by natural causes is another source of energy which might be rendered available. Lightning discharges involve great amounts of electrical energy, which we could utilize by transforming and storing it. Some years ago I made known a method of electrical transformation which renders the first part of this task easy, but the storing of the energy of lightning discharges will be difficult to accomplish. It is well known, furthermore, that electric currents circulate constantly through the earth, and that there exists between the earth and any air stratum a difference of electrical pressure, which varies in proportion to the height.

In recent experiments I have discovered two novel facts of importance in this connection. One of these facts is that an electric current is generated in a wire extending from the ground to a great height by the axial, and probably also by the translatory, movement of the earth. No appreciable current, however, will flow continuously in the wire unless the electricity is allowed to leak out into the air. Its escape is greatly facilitated by providing at the elevated end of the wire a conducting terminal of great surface, with many sharp edges or points. We are thus enabled to get a continuous supply of electrical energy by merely supporting a wire at a height, but, unfortunately, the amount of electricity which can be so obtained is small.

The second fact which I have ascertained is that the upper air strata are permanently charged with electricity opposite to that of the earth. So, at least, I have interpreted my observations, from which it appears that the earth, with its adjacent insulating and outer conducting envelope, constitutes a highly charged electrical condenser containing, in all probability, a great amount of electrical energy which might be turned to the uses of man, if it were possible to reach with a wire to great altitudes.

It is possible, and even probable, that there will be, in time, other resources of energy opened up, of which we have no knowledge now. We may even find ways of applying forces such as magnetism or gravity for driving machinery without using any other means. Such realizations, though highly improbable, are not impossible. An example will best convey an idea of what we can hope to attain and what we can never attain. Imagine a disk of some homogeneous material turned perfectly true and arranged to turn in frictionless bearings on a horizontal shaft above the ground. This disk, being under the above conditions perfectly balanced, would rest in any position. Now, it is possible that we may learn how to make such a disk rotate continuously and perform work by the force of gravity without any further effort on our part; but it is perfectly impossible for the disk to turn and to do work without any force from the outside. If it could do so, it would be what is designated scientifically as a "perpetuum mobile," a machine creating its own motive power. To make the disk rotate by the force of gravity we have only to invent a screen against this force. By such a screen we could prevent this force from acting on one half of the disk, and the rotation of the latter would follow. At least, we cannot deny such a possibility until we know exactly the nature of the force of gravity. Suppose that this force were due to a movement comparable to that of a stream of air passing from above toward the center of the earth. The effect of such a stream upon both halves of the disk would be equal, and the latter would not rotate ordinarily; but if one half should be guarded by a plate arresting the movement, then it would turn.

> A departure from known methods — Possibility of a "self-acting" engine or machine, inanimate, yet capable, like a living being, of deriving energy from the medium — The ideal way of obtaining motive power

When I began the investigation of the subject under consideration, and when the preceding or similar ideas presented themselves to me for the first time, though I was then unacquainted with a number of the facts mentioned, a survey of the various ways of utilizing the energy of the medium convinced me, nevertheless, that to arrive at a thoroughly satisfactory practical solution a radical departure from the methods then known had to be made. The windmill, the solar engine, the engine driven by terrestrial heat, had their limitations in the amount of power obtainable. Some new way had to be discovered which would enable us to get more energy. There was enough heat-energy in the medium, but only a small part of it was available for the operation of an engine in the ways then known. Besides, the energy was obtainable only at a very slow rate. Clearly, then, the problem was to discover some new method which would make it possible both to utilize more of the heat-energy of the medium and also to draw it away from the same at a more rapid rate.

I was vainly endeavoring to form an idea of how this might be accomplished, when I read some statements from Carnot and Lord Kelvin (then Sir William Thomson) which meant virtually that it is impossible for an inanimate mechanism or self-acting machine to cool a portion of the medium below the temperature of the surrounding, and operate by the heat abstracted. These statements interested me intensely. Evidently a living being could do this very thing, and since the experiences of my early life which I have related had convinced me that a living being is only an automaton, or, otherwise stated, a "self-acting-engine," I came to the conclusion that it was possible to construct a machine which would do the same. As the first step toward this realization I conceived the following mechanism. Imagine a thermopile consisting of a number of bars of metal extending from the earth to the outer space beyond the atmosphere. The heat from below, conducted upward along these metal bars, would cool the earth or the sea or the air, according to the location of the lower parts of the bars, and the result, as is well known, would be an electric current circulating in these bars. The two terminals of the thermopile could now be joined through an electric motor, and, theoretically, this motor would run on and on, until the media below would be cooled down to the temperature of the outer space. This would be an inanimate engine which, to all evidence, would be cooling a portion of the medium below the temperature of the surrounding, and operating by the heat abstracted.

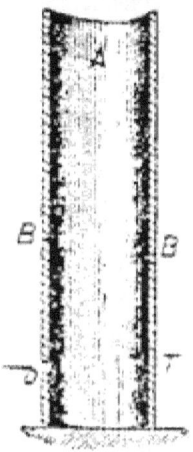

Diagram b. Obtaining energy from the ambient medium

Note: A, medium with little energy; B, B, ambient medium with much energy; O, path of the energy.

But was it not possible to realize a similar condition without necessarily going to a height? Conceive, for the sake of illustration, [a cylindrical] enclosure T, as illustrated in diagram b, such that energy could not be transferred across it except through a channel or path O, and that, by some means or other, in this enclosure a medium were maintained which would have little energy, and that on the outer side of the same there would be the ordinary ambient medium with much energy. Under these assumptions the energy would flow through the path O, as indicated by the arrow, and might then be converted on its passage into some other form of energy. The question was, Could such a condition be attained? Could we produce artificially such a "sink" for the energy of the ambient medium to flow in? Suppose that an extremely low temperature could be maintained by some process in a given space; the surrounding medium would then be compelled to give off heat, which could be converted into mechanical or other form of energy, and utilized. By realizing such a plan, we should be enabled to get at any point of the globe a continuous supply of energy, day and night. More than this, reasoning in the abstract, it would seem possible to cause a quick circulation of the medium, and thus draw the energy at a very rapid rate.

Here, then, was an idea which, if realizable, afforded a happy solution of the problem of getting energy from the medium. But was it realizable? I convinced myself that it was so in a number of ways, of which one is the following. As regards heat, we are at a high level, which may be represented by the surface of a mountain lake considerably above the sea, the level of which may mark the absolute zero of temperature existing in the interstellar space. Heat, like water, flows from high to low level, and, consequently, just as we can let the water of the lake run down to the sea, so we are able to let heat from the earth's surface travel up into the cold region above. Heat, like water, can perform work in flowing down, and if we had any doubt as to whether we could derive energy from the medium by means of a thermopile, as before described, it would be dispelled by this analogue. But can we produce cold in a given portion of the space and cause the heat to flow in continually? To create such a "sink," or "cold hole," as we might say, in the medium, would be equivalent to producing in the lake a space either empty or filled with something much lighter than water. This we could do by placing in the lake a tank, and pumping all the water out of the latter. We know, then, that the water, if allowed to flow back into the tank, would, theoretically, be able to perform exactly the same amount of work which was used in pumping it out, but not a bit more. Consequently nothing could be gained in this double operation of first raising the water and then letting it fall down. This would mean that it is impossible to create such a sink in the medium. But let us reflect a moment. Heat, though following certain general laws of mechanics, like a fluid, is not such; it is energy which may be converted into other forms of energy as it passes from a high to a low level. To make our mechanical analogy complete and true, we must, therefore, assume that the water, in its passage into the tank, is converted into something else, which may be taken out of it without using any, or by using very little, power. For example, if heat be represented in this analogue by the water of the lake, the oxygen and hydrogen composing the water may illustrate other forms of energy into which the heat is transformed in passing from hot to cold. If the process of heat transformation were absolutely perfect, no heat at all would arrive at the low level, since all of it would be converted into other forms of energy. Corresponding to this ideal case, all the water flowing into the tank would be decomposed into oxygen and hydrogen before reaching the bottom, and the result would be that water would continually flow in, and yet the tank would remain entirely empty, the gases formed escaping. We would thus produce, by expending initially a certain amount of work to create a sink for the heat or, respectively, the water to flow in, a condition enabling us to get any amount of energy without further effort. This would be an ideal way of obtaining motive power. We do not know of any such absolutely perfect process of heat-conversion, and consequently some heat will generally reach the low level, which means to say, in our mechanical analogue, that some water will arrive at the bottom of the tank, and a gradual and slow filling of the latter will take place, necessitating continuous pumping out. But evidently there will be less to pump out than flows in, or, in other words, less energy will be needed to maintain the initial condition than is developed by the fall, and this is to say that some energy will be gained from the medium. What is not converted in flowing down can just be raised up with its own energy, and what is converted is clear gain. Thus the virtue of the principle I have discovered resides wholly in the conversion of the energy on the downward flow.

First efforts to produce the self-acting engine — The mechanical oscillator — Work of dewar and linde — Liquid air

Having recognized this truth, I began to devise means for carrying out my idea, and, after long thought, I finally conceived a combination of apparatus which should make possible the obtaining of power from the medium by a process of continuous cooling of atmospheric air. This apparatus, by continually transforming heat into mechanical work, tended to become colder and colder, and if it only were practicable to reach a very low temperature in this manner, then a sink for the heat could be produced, and energy could be derived from the medium. This seemed to be contrary to the statements of Carnot and Lord Kelvin before referred to, but I concluded from the theory of the process that such a result could be attained. This conclusion I reached, I think, in the latter part of 1883, when I was in Paris, and it was at a time when my mind was being more and more dominated by an invention which I had evolved during the preceding year, and which has since become known under the name of the "rotating magnetic field." During the few years which followed I elaborated further the plan I had imagined, and studied the working conditions, but made little headway. The commercial introduction in this country of the invention before referred to required most of my energies until 1889, when I again took up the idea of the self-acting machine. A closer investigation of the principles involved, and calculation, now showed that the result I aimed at could not be reached in a practical manner by ordinary machinery, as I had in the beginning expected. This led me, as a next step, to the study of a type of engine generally designated as "turbine," which at first seemed to offer better chances for a realization of the idea. Soon I found, however, that the turbine, too, was unsuitable. But my conclusions showed that if an engine of a peculiar kind could be brought to a high degree of perfection, the plan I had conceived was realizable, and I resolved to proceed with the development of such an engine, the primary object of which was to secure the greatest economy of transformation of heat into mechanical energy. A characteristic feature of the engine was that the work-performing piston was not connected with anything else, but was perfectly free to vibrate at an enormous rate. The mechanical difficulties encountered in the construction of this engine were greater than I had anticipated, and I made slow progress. This work was continued until early in 1892, when I went to London, where I saw Professor Dewar's admirable experiments with liquefied gases. Others had liquefied gases before, and notably Ozlewski and Pictet had performed creditable early experiments in this line, but there was such a vigor about the work of Dewar that even the old appeared new. His experiments showed, though in a way different from that I had imagined, that it was possible to reach a very low temperature by transforming heat into mechanical work, and I returned, deeply impressed with what I had seen, and more than ever convinced that my plan was practicable. The work temporarily interrupted was taken up anew, and soon I had in a fair state of perfection the engine which I have named "the mechanical oscillator." In this machine I succeeded in doing away with all packings, valves, and lubrication, and in producing so rapid a vibration of the piston that shafts of tough steel, fastened to the same and vibrated longitudinally, were torn asunder. By combining this engine with a dynamo of special design I produced a highly efficient electrical generator, invaluable in measurements and determinations of physical quantities on account of the unvarying rate of oscillation obtainable by its means. I exhibited several types of this machine, named "mechanical and electrical oscillator," before the Electrical Congress at the World's Fair in Chicago during the summer of 1893, in a lecture which, on account of other pressing work, I was unable to prepare for publication. On that occasion I exposed the principles of the mechanical oscillator, but the original purpose of this machine is explained here for the first time.

In the process, as I had primarily conceived it, for the utilization of the energy of the ambient medium, there were five essential elements in combination, and each of these had to be newly designed and perfected, as no such machines existed. The mechanical oscillator was the first element of this combination, and having perfected this, I turned to the next, which was an air-compressor of a design in certain respects resembling that of the mechanical oscillator.

Similar difficulties in the construction were again encountered, but the work was pushed vigorously, and at the close of 1894 I had completed these two elements of the combination, and thus produced an apparatus for compressing air, virtually to any desired pressure, incomparably simpler, smaller, and more efficient than the ordinary. I was just beginning work on the third element, which together with the first two would give a refrigerating machine of exceptional efficiency and simplicity, when a misfortune befell me in the burning of my laboratory, which crippled my labors and delayed me. Shortly afterward Dr. Carl Linde announced the liquefaction of air by a self-cooling process, demonstrating that it was practicable to proceed with the cooling until liquefaction of the air took place. This was the only experimental proof which I was still wanting that energy was obtainable from the medium in the manner contemplated by me.

The liquefaction of air by a self-cooling process was not, as popularly believed, an accidental discovery, but a scientific result which could not have been delayed much longer, and which, in all probability, could not have escaped Dewar. This fascinating advance, I believe, is largely due to the powerful work of this great Scotchman. Nevertheless, Linde's is an immortal achievement. The manufacture of liquid air has been carried on for four years in Germany, on a scale much larger than in any other country, and this strange product has been applied for a variety of purposes. Much was expected of it in the beginning, but so far it has been an industrial ignis fatuus. By the use of such machinery as I am perfecting, its cost will probably be greatly lessened, but even then its commercial success will be questionable. When, used as a refrigerant it is uneconomical, as its temperature is unnecessarily low. It is as expensive to maintain a body at a very low temperature as it is to keep it very hot; it takes coal to keep air cold. In oxygen manufacture it cannot yet compete with the electrolytic method. For use as an explosive it is unsuitable, because its low temperature again condemns it to a small efficiency, and for motive-power purposes its cost is still by far too high. It is of interest to note, however, that in driving an engine by liquid air a certain amount of energy may be gained from the engine, or, stated otherwise, from the ambient medium which keeps the engine warm, each two hundred pounds of iron-casting of the latter contributing energy at the rate of about one effective horsepower during one hour. But this gain of the consumer is offset by an equal loss of the producer.

Much of this task on which I have labored so long remains to be done. A number of mechanical details are still to be perfected and some difficulties of a different nature to be mastered, and I cannot hope to produce a self-acting machine deriving energy from the ambient medium for a long time yet, even if all my expectations should materialize. Many circumstances have occurred which have retarded my work of late, but for several reasons the delay was beneficial.

One of these reasons was that I had ample time to consider what the ultimate possibilities of this development might be. I worked for a long time fully convinced that the practical realization of this method of obtaining energy from the sun would be of incalculable industrial value, but the continued study of the subject revealed the fact that while it will be commercially profitable if my expectations are well founded, it will not be so to an extraordinary degree.

Discovery of unexpected properties of the atmosphere — Strange experiments — Transmission of electrical energy through one wire without return — Transmission through the earth without any wire

Another of these reasons was that I was led to recognize the transmission of electrical energy to any distance through the media as by far the best solution of the great problem of harnessing the sun's energy for the uses of man. For a long time I was convinced that such a transmission on an industrial scale, could never be realized, but a discovery which I made changed my view. I observed that under certain conditions the atmosphere, which is normally a high insulator, assumes conducting properties, and so becomes capable of conveying any amount of electrical energy. But the difficulties in the way of a practical utilization of this discovery for the purpose of transmitting electrical energy without wires were seemingly insuperable. Electrical pressures of many millions of volts had to be produced and handled; generating apparatus of a novel kind, capable of withstanding the immense electrical stresses, had to be invented and perfected, and a complete safety against the dangers of the high-tension currents had to be attained in the system before its practical introduction could be even thought of. All this could not be done in a few weeks or months, or even years. The work required patience and constant application, but the improvements came, though slowly. Other valuable results were, however, arrived at in the course of this long-continued work, of which I shall endeavor to give a brief account, enumerating the chief advances as they were successively effected.

The discovery of the conducting properties of the air, though unexpected, was only a natural result of experiments in a special field which I had carried on for some years before. It was, I believe, during 1889 that certain possibilities offered by extremely rapid electrical oscillations determined me to design a number of special machines adapted for their investigation. Owing to the peculiar requirements, the construction of these machines was very difficult, and consumed much time and effort; but my work on them was generously rewarded, for I reached by their means several novel and important results. One of the earliest observations I made with these new machines was that electrical oscillations of an extremely high rate act in an extraordinary manner upon the human organism. Thus, for instance, I demonstrated that powerful electrical discharges of several hundred thousand volts, which at that time were considered absolutely deadly, could be passed through the body without inconvenience or hurtful consequences. These oscillations produced other specific physiological effects, which, upon my announcement, were eagerly taken up by skilled physicians and further investigated. This new field has proved itself fruitful beyond expectation, and in the few years which have passed since, it has been developed to such an extent that it now forms a legitimate and important department of medical science. Many results, thought impossible at that time, are now readily obtainable with these oscillations, and many experiments undreamed of then can now be readily performed by their means. I still remember with pleasure how, nine years ago, I passed the discharge of a powerful induction-coil through my body to demonstrate before a scientific society the comparative harmlessness of very rapidly vibrating electric currents, and I can still recall the astonishment of my audience. I would now undertake, with much less apprehension that I had in that experiment, to transmit through my body with such currents the entire electrical energy of the dynamos now working at Niagara—forty or fifty thousand horse-power. I have produced electrical oscillations which were of such intensity that when circulating through my arms and chest they have melted wires which joined my hands, and still I felt no inconvenience. I have energized with such oscillations a loop of heavy copper wire so powerfully that masses of metal, and even objects of an electrical resistance specifically greater than that of human tissue brought close to or placed within the loop, were heated to a high temperature and melted, often with the violence of an explosion, and yet into this very space in which this terribly-destructive turmoil was going on I have repeatedly thrust my head without feeling anything or experiencing injurious after-effects.

Another observation was that by means of such oscillations light could be produced in a novel and more economical manner, which promised to lead to an ideal system of electric illumination by vacuum-tubes, dispensing with the necessity of renewal of lamps or incandescent filaments, and possibly also with the use of wires in the interior of buildings. The efficiency of this light increases in proportion to the rate of the oscillations, and its commercial success is, therefore, dependent on the economical production of electrical vibrations of transcending rates. In this direction I have met with gratifying success of late, and the practical introduction of this new system of illumination is not far off.

The investigations led to many other valuable observations and results, one of the more important of which was the demonstration of the practicability of supplying electrical energy through one wire without return. At first I was able to transmit in this novel manner only very small amounts of electrical energy, but in this line also my efforts have been rewarded with similar success.

Fig. 3. Experiment to illustrate the supplying of electrical energy through a single wire without return

An ordinary incandescent lamp, connected with one or both of its terminals to the wire forming the upper free end of the coil shown in the photograph, is lighted by electrical vibrations conveyed to it through the coil from an electrical oscillator, which is worked only to one fifth of 1% of its full capacity.

The photograph shown in Fig. 3 illustrates, as its title explains, an actual transmission of this kind effected with apparatus used in other experiments here described. To what a degree the appliances have been perfected since my first demonstrations early in 1891 before a scientific society, when my apparatus was barely capable of lighting one lamp (which result was considered wonderful), will appear when I state that I have now no difficulty in lighting in this manner four or five hundred lamps, and could light many more. In fact, there is no limit to the amount of energy which may in this way be supplied to operate any kind of electrical device.

Fig. 4. Experiment to illustrate the transmission of electrical energy through the earth without wire.

The coil shown in the photograph has its lower end or terminal connected to the ground, and is exactly attuned to the vibrations of a distant electrical oscillator. The lamp lighted is in an independent wire loop, energized by induction from the coil excited by the electrical vibrations transmitted to it through the ground from the oscillator, which is worked only to 5% of its full capacity.

After demonstrating the practicability of this method of transmission, the thought naturally occurred to me to use the earth as a conductor, thus dispensing with all wires. Whatever electricity may be, it is a fact that it behaves like an incompressible fluid, and the earth may be looked upon as an immense reservoir of electricity, which, I thought, could be disturbed effectively by a properly designed electrical machine. Accordingly, my next efforts were directed toward perfecting a special apparatus which would be highly effective in creating a disturbance of electricity in the earth. The progress in this new direction was necessarily very slow and the work discouraging, until I finally succeeded in perfecting a novel kind of transformer or induction-coil, particularly suited for this special purpose. That it is practicable, in this manner, not only to transmit minute amounts of electrical energy for operating delicate electrical devices, as I contemplated at first, but also electrical energy in appreciable quantities, will appear from an inspection of Fig. 4, which illustrates an actual experiment of this kind performed with the same apparatus. The result obtained was all the more remarkable as the top end of the coil was not connected to a wire or plate for magnifying the effect.

"Wireless" telegraphy — The secret of tuning — Errors in the hertzian investigations — A receiver of wonderful sensitiveness

As the first valuable result of my experiments in this latter line a system of telegraphy without wires resulted, which I described in two scientific lectures in February and March, 1893. It is mechanically illustrated in diagram c, the upper part of which shows the electrical arrangement as I described it then, while the lower part illustrates its mechanical analogue. The system is extremely simple in principle. Imagine two tuning-forks F, F_1, one at the sending- and the other at the receiving-station respectively, each having attached to its lower prong a minute piston p, fitting in a cylinder. Both the cylinders communicate with a large reservoir R, with elastic walls, which is supposed to be closed and filled with a light and incompressible fluid. By striking repeatedly one of the prongs of the tuning-fork F, the small piston p below would be vibrated, and its vibrations, transmitted through the fluid, would reach the distant fork F_1, which is "tuned" to the fork F, or, stated otherwise, of exactly the same note as the latter. The fork F_1 would now be set vibrating, and its vibration would be intensified by the continued action of the distant fork F until its upper prong, swinging far out, would make an electrical connection with a stationary contact c", starting in this manner some electrical or other appliances which may be used for recording the signals. In this simple way messages could be exchanged between the two stations, a similar contact c' being provided for this purpose, close to the upper prong of the fork F, so that the apparatus at each station could be employed in turn as receiver and transmitter.

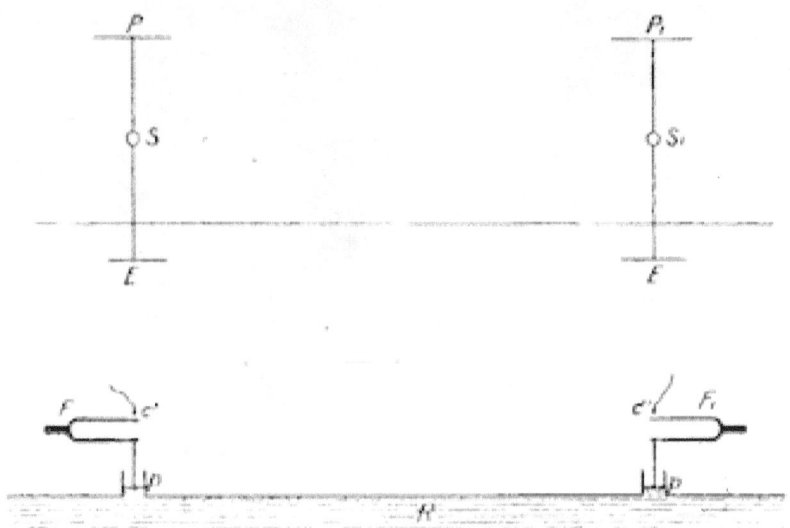

Diagram C. "Wireless" telegraphy mechanically illustrated

The electrical system illustrated in the upper figure of diagram c is exactly the same in principle, the two wires or circuits ESP and $E_1S_1P_1$, which extend vertically to a height, representing the two tuning-forks with the pistons attached to them. These circuits are connected with the ground by plates E, E_1, and to two elevated metal sheets P, P_1, which store electricity and thus magnify considerably the effect. The closed reservoir R, with elastic walls, is in this case replaced by the earth, and the fluid by electricity. Both of these circuits are "tuned" and operate just like the two tuning-forks. Instead of striking the fork F at the sending-station, electrical oscillations are produced in the vertical sending- or transmitting-wire ESP, as by the action of a source S, included in this wire, which spread through the ground and reach the distant vertical receiving-wire $E_1S_1P_1$, exciting corresponding electrical oscillations in the same. In the latter wire or circuit is included a sensitive device or receiver S_1, which is thus set in action and made to operate a relay or other appliance. Each station is, of course, provided both with a source of electrical oscillations S and a sensitive receiver S_1, and a simple provision is made for using each of the two wires alternately to send and to receive the messages.

Fig. 5. Photographic view of the coils responding to electrical oscillations

The picture shows a number of coils, differently attuned and responding to the vibrations transmitted to them through the earth from an electrical oscillator. The large coil on the right, discharging strongly, is tuned to the fundamental vibration, which is fifty thousand per second; the two larger vertical coils to twice that number; the smaller white wire coil to four times that number, and the remaining small coils to higher tones. The vibrations produced by the oscillator were so intense that they affected perceptibly a small coil tuned to the twenty-sixth higher tone.

The exact attunement of the two circuits secures great advantages, and, in fact, it is essential in the practical use of the system. In this respect many popular errors exist, and, as a rule, in the technical reports on this subject circuits and appliances are described as affording these advantages when from their very nature it is evident that this is impossible. In order to attain the best results it is essential that the length of each wire or circuit, from the ground connection to the top, should be equal to one quarter of the wave-length of the electrical vibration in the wire, or else equal to that length multiplied by an odd number. Without the observation of this rule it is virtually impossible to prevent the interference and insure the privacy of messages. Therein lies the secret of tuning. To obtain the most satisfactory results it is, however, necessary to resort to electrical vibrations of low pitch. The Hertzian spark apparatus, used generally by experimenters, which produces oscillations of a very high rate, permits no effective tuning, and slight disturbances are sufficient to render an exchange of messages impracticable. But scientifically designed, efficient appliances allow nearly perfect adjustment. An experiment performed with the improved apparatus repeatedly referred to, and intended to convey an idea of this feature, is illustrated in Fig. 5, which is sufficiently explained by its note.

Since I described these simple principles of telegraphy without wires I have had frequent occasion to note that the identical features and elements have been used, in the evident belief that the signals are being transmitted to considerable distance by "Hertzian" radiations. This is only one of many misapprehensions to which the investigations of the lamented physicist have given rise. About thirty-three years ago Maxwell, following up a suggestive experiment made by Faraday in 1845, evolved an ideally simple theory which intimately connected light, radiant heat, and electrical phenomena, interpreting them as being all due to vibrations of a hypothetical fluid of inconceivable tenuity, called the ether. No experimental verification was arrived at until Hertz, at the suggestion of Helmholtz, undertook a series of experiments to this effect. Hertz proceeded with extraordinary ingenuity and insight, but devoted little energy to the perfection of his old-fashioned apparatus. The consequence was that he failed to observe the important function which the air played in his experiments, and which I subsequently discovered. Repeating his experiments and reaching different results, I ventured to point out this oversight. The strength of the proofs brought forward by Hertz in support of Maxwell's theory resided in the correct estimate of the rates of vibration of the circuits he used. But I ascertained that he could not have obtained the rates he thought he was getting. The vibrations with identical apparatus he employed are, as a rule, much slower, this being due to the presence of air, which produces a dampening effect upon a rapidly vibrating electric circuit of high pressure, as a fluid does upon a vibrating tuning-fork. I have, however, discovered since that time other causes of error, and I have long ago ceased to look upon his results as being an experimental verification of the poetical conceptions of Maxwell. The work of the great German physicist has acted as an immense stimulus to contemporary electrical research, but it has likewise, in a measure, by its fascination, paralyzed the scientific mind, and thus hampered independent inquiry. Every new phenomenon which was discovered was made to fit the theory, and so very often the truth has been unconsciously distorted.

When I advanced this system of telegraphy, my mind was dominated by the idea of effecting communication to any distance through the earth or environing medium, the practical consummation of which I considered of transcendent importance, chiefly on account of the moral effect which it could not fail to produce universally. As the first effort to this end I proposed at that time, to employ relay-stations with tuned circuits, in the hope of making thus practicable signaling over vast distances, even with apparatus of very moderate power then at my command. I was confident, however, that with properly designed machinery signals could be transmitted to any point of the globe, no matter what the distance, without the necessity of using such intermediate stations. I gained this conviction through the discovery of a singular electrical phenomenon, which I described early in 1892, in lectures I delivered before some scientific societies abroad, and which I have called a "rotating brush." This is a bundle of light which is formed, under certain conditions, in a vacuum-bulb, and which is of a sensitiveness to magnetic and electric influences bordering, so to speak, on the supernatural. This light-bundle is rapidly rotated by the earth's magnetism as many as twenty thousand times pre second, the rotation in these parts being opposite to what it would be in the southern hemisphere, while in the region of the magnetic equator it should not rotate at all. In its most sensitive state, which is difficult to obtain, it is responsive to electric or magnetic influences to an incredible degree. The mere stiffening of the muscles of the arm and consequent slight electrical change in the body of an observer standing at some distance from it, will perceptibly affect it. When in this highly sensitive state it is capable of indicating the slightest magnetic and electric changes taking place in the earth. The observation of this wonderful phenomenon impressed me strongly that communication at any distance could be easily effected by its means, provided that apparatus could be perfected capable of producing an electric or magnetic change of state, however small, in the terrestrial globe or environing medium.

> Development of a new principle — The electrical oscillator — Production of immense electrical movements — The earth responds to man — Interplanetary communication now probable

I resolved to concentrate my efforts upon this venturesome task, though it involved great sacrifice, for the difficulties to be mastered were such that I could hope to consummate it only after years of labor. It meant delay of other work to which I would have preferred to devote myself, but I gained the conviction that my energies could not be more usefully employed; for I recognized that an efficient apparatus for, the production of powerful electrical oscillations, as was needed for that specific purpose, was the key to the solution of other most important electrical and, in fact, human problems. Not only was communication, to any distance, without wires possible by its means, but, likewise, the transmission of energy in great amounts, the burning of the atmospheric nitrogen, the production of an efficient illuminant, and many other results of inestimable scientific and industrial value. Finally, however, I had the satisfaction of accomplishing the task undertaken by the use of a new principle, the virtue of which is based on the marvelous properties of the electrical condenser. One of these is that it can discharge or explode its stored energy in an inconceivably short time. Owing to this it is unequaled in explosive violence. The explosion of dynamite is only the breath of a consumptive compared with its discharge. It is the means of producing the strongest current, the highest electrical pressure, the greatest commotion in the medium. Another of its properties, equally valuable, is that its discharge may vibrate at any rate desired up to many millions per second.

Fig. 6. Photographic view of the essential parts of the electrical oscillator used in the experiments described

I had arrived at the limit of rates obtainable in other ways when the happy idea presented itself to me to resort to the condenser. I arranged such an instrument so as to be charged and discharged alternately in rapid succession through a coil with a few turns of stout wire, forming the primary of a transformer or induction-coil. Each time the condenser was discharged the current would quiver in the primary wire and induce corresponding oscillations in the secondary. Thus a transformer or induction-coil on new principles was evolved, which I have called "the electrical oscillator," partaking of those unique qualities which characterize the condenser, and enabling results to be attained impossible by other means. Electrical effects of any desired character and of intensities undreamed of before are now easily producible by perfected apparatus of this kind, to which frequent reference has been made, and the essential parts of which are shown in Fig. 6. For certain purposes a strong inductive effect is required; for others the greatest possible suddenness; for others again, an exceptionally high rate of vibration or extreme pressure; while for certain other objects immense electrical movements are necessary. The photographs in Figs. 7, 8, 9, and 10, of experiments performed with such an oscillator, may serve to illustrate some of these features and convey an idea of the magnitude of the effects actually produced. The completeness of the titles of the figures referred to makes a further description of them unnecessary.

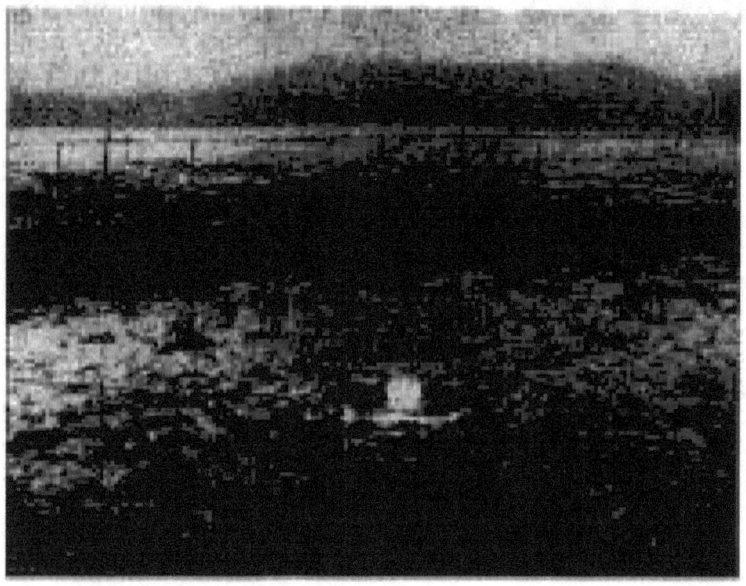

Fig. 7. Experiment to illustrate an inductive effect of an electrical oscillator of great power

The photograph shows three ordinary incandescent lamps lighted to full candle-power by currents induced in a local loop consisting of a single wire forming a square of fifty feet each side, which includes the lamps, and which is at a distance of one hundred feet from the primary circuit energized by the oscillator. The loop likewise includes an electrical condenser, and is exactly attuned to the vibrations of the oscillator, which is worked at less than 5% of its total capacity.

Fig. 8. Experiment to illustrate the capacity of the oscillator for producing electrical explosions of great power

Note to Fig. 8: The coil, partly shown in the photograph, creates an alternative movement of electricity from the earth into a large reservoir and back at a rate of one hundred thousand alternations per second. The adjustments are such that the reservoir is filled full and bursts at each alternation just at the moment when the electrical pressure reaches the maximum. The discharge escapes with a deafening noise, striking an unconnected coil twenty-two feet away, and creating such a commotion of electricity in the earth that sparks an inch long can be drawn from a water main at a distance of three hundred feet from the laboratory.

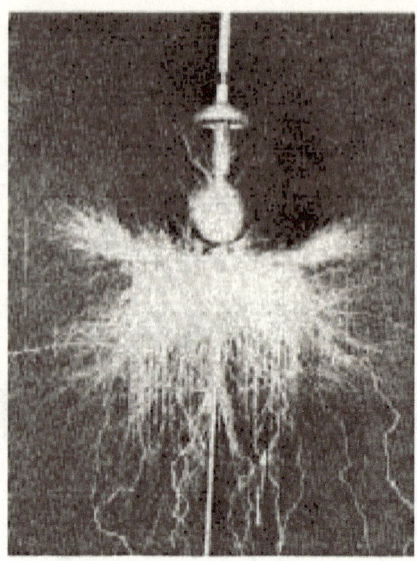

Fig. 9. Experiment to illustrate the capacity on the oscillator for creating a great electrical movement

The ball shown in the photograph, covered with a polished metallic coating of twenty square feet of surface, represents a large reservoir of electricity, and the inverted tin pan underneath, with a sharp rim, a big opening through which the electricity can escape before filling the reservoir. The quantity of electricity set in movement is so great that, although most of it escapes through the rim of the pan or opening provided, the ball or reservoir is nevertheless alternately emptied and filled to over-flowing (as is evident from the discharge escaping on the top of the ball) one hundred and fifty thousand times per second.

The discharge, creating a strong draft owing to the heating of the air, is carried upward through the open roof of the building. The greatest width across is nearly seventy feet. The pressure is over twelve million volts, and the current alternates one hundred and thirty thousand times per second.

However extraordinary the results shown may appear, they are but trifling compared with those which are attainable by apparatus designed on these same principles. I have produced electrical discharges the actual path of which, from end to end, was probably more than one hundred feet long; but it would not be difficult to reach lengths one hundred times as great. I have produced electrical movements occurring at the rate of approximately one hundred thousand horse-power, but rates of one, five, or ten million horse-power are easily practicable. In these experiments effects were developed incomparably greater than any ever produced by human agencies, and yet these results are but an embryo of what is to be.

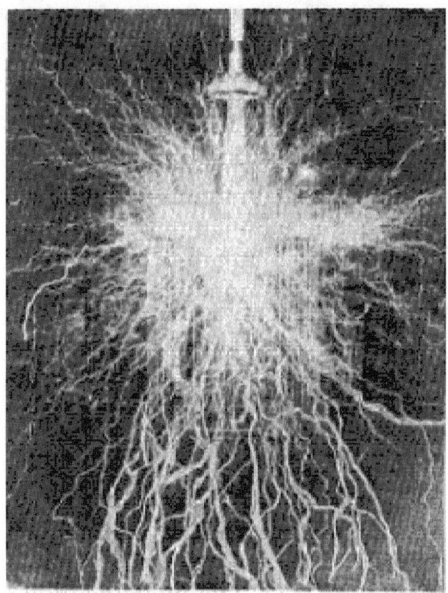

Fig. 10. Photographic view of an experiment to illustrate an effect of an electrical oscillator delivering energy at a rate of seventy-five thousand horse-power

That communication without wires to any point of the globe is practicable with such apparatus would need no demonstration, but through a discovery which I made I obtained absolute certitude. Popularly explained, it is exactly this: When we raise the voice and hear an echo in reply, we know that the sound of the voice must have reached a distant wall, or boundary, and must have been reflected from the same. Exactly as the sound, so an electrical wave is reflected, and the same evidence which is afforded by an echo is offered by an electrical phenomenon known as a "stationary" wave — that is, a wave with fixed nodal and ventral regions. Instead of sending sound-vibrations toward a distant wall, I have sent electrical vibrations toward the remote boundaries of the earth, and instead of the wall the earth has replied. In place of an echo I have obtained a stationary electrical wave, a wave reflected from afar.

Stationary waves in the earth mean something more than mere telegraphy without wires to any distance. They will enable us to attain many important specific results impossible otherwise. For instance, by their use we may produce at will, from a sending-station, an electrical effect in any particular region of the globe; we may determine the relative position or course of a moving object, such as a vessel at sea, the distance traversed by the same, or its speed; or we may send over the earth a wave of electricity traveling at any rate we desire, from the pace of a turtle up to lightning speed.

With these developments we have every reason to anticipate that in a time not very distant most telegraphic messages across the oceans will be transmitted without cables. For short distances we need a "wireless" telephone, which requires no expert operators. The greater the spaces to be bridged, the more rational becomes communication without wires. The cable is not only an easily damaged and costly instrument, but it limits us in the speed of transmission by reason of a certain electrical property inseparable from its construction. A properly designed plant for effecting communication without wires ought to have many times the working capacity of a cable, while it will involve incomparably less expense. Not a long time will pass, I believe, before communication by cable will become obsolete, for not only will signaling by this new method be quicker and cheaper, but also much safer. By using some new means for isolating the messages which I have contrived, an almost perfect privacy can be secured.

I have observed the above effects so far only up to a limited distance of about six hundred miles, but inasmuch as there is virtually no limit to the power of the vibrations producible with such an oscillator, I feel quite confident of the success of such a plant for effecting transoceanic communication. Nor is this all. My measurements and calculations have shown that it is perfectly practicable to produce on our globe, by the use of these principles, an electrical movement of such magnitude that, without the slightest doubt, its effect will be perceptible on some of our nearer planets, as Venus and Mars. Thus from mere possibility interplanetary communication has entered the stage of probability. In fact, that we can produce a distinct effect on one of these planets in this novel manner, namely, by disturbing the electrical condition of the earth, is beyond any doubt. This way of effecting such communication is, however, essentially different from all others which have so far been proposed by scientific men. In all the previous instances only a minute fraction of the total energy reaching the planet — as much as it would be possible to concentrate in a reflector — could be utilized by the supposed observer in his instrument. But by the means I have developed he would be enabled to concentrate the larger portion of the entire energy transmitted to the planet in his instrument, and the chances of affecting the latter are thereby increased many millionfold.

Besides machinery for producing vibrations of the required power, we must have delicate means capable of revealing the effects of feeble influences exerted upon the earth. For such purposes, too, I have perfected new methods. By their use we shall likewise be able, among other things, to detect at considerable distance the presence of an iceberg or other object at sea. By their use, also,

I have discovered some terrestrial phenomena still unexplained. That we can send a message to a planet is certain, that we can get an answer is probable: man is not the only being in the Infinite gifted with a mind.

> Transmission of electrical energy to any distance without wires — Now practicable — The best means of increasing the force accelerating the human mass

The most valuable observation made in the course of these investigations was the extraordinary behavior of the atmosphere toward electric impulses of excessive electromotive force. The experiments showed that the air at the ordinary pressure became distinctly conducting, and this opened up the wonderful prospect of transmitting large amounts of electrical energy for industrial purposes to great distances without wires, a possibility which, up to that time, was thought of only as a scientific dream. Further investigation revealed the important fact that the conductivity imparted to the air by these electrical impulses of many millions of volts increased very rapidly with the degree of rarefaction, so that air strata at very moderate altitudes, which are easily accessible, offer, to all experimental evidence, a perfect conducting path, better than a copper wire, for currents of this character.

Thus the discovery of these new properties of the atmosphere not only opened up the possibility of transmitting, without wires, energy in large amounts, but, what was still more significant, it afforded the certitude that energy could be transmitted in this manner economically. In this new system it matters little — in fact, almost nothing — whether the transmission is effected at a distance of a few miles or of a few thousand miles.

While I have not, as yet, actually effected a transmission of a considerable amount of energy, such as would be of industrial importance, to a great distance by this new method, I have operated several model plants under exactly the same conditions which will exist in a large plant of this kind, and the practicability of the system is thoroughly demonstrated. The experiments have shown conclusively that, with two terminals maintained at an elevation of not more than thirty thousand to thirty-five thousand feet above sea-level, and with an electrical pressure of fifteen to twenty million volts, the energy of thousands of horse-power can be transmitted over distances which may be hundreds and, if necessary, thousands of miles. I am hopeful, however, that I may be able to reduce very considerably the elevation of the terminals now required, and with this object I am following up an idea which promises such a realization. There is, of course, a popular prejudice against using an electrical pressure of millions of volts, which may cause sparks to fly at distances of hundreds of feet, but, paradoxical as it may seem, the system, as I have described it in a technical publication, offers greater personal safety than most of the ordinary distribution circuits now used in the cities. This is, in a measure, borne out by the fact that, although I have carried on such experiments for a number of years, no injury has been sustained either by me or any of my assistants.

But to enable a practical introduction of the system, a number of essential requirements are still to be fulfilled. It is not enough to develop appliances by means of which such a transmission can be effected. The machinery must be such as to allow the transformation and transmission, of electrical energy under highly economic and practical conditions. Furthermore, an inducement must be offered to those who are engaged in the industrial exploitation of natural sources of power, as waterfalls, by guaranteeing greater returns on the capital invested than they can secure by local development of the property.

From that moment when it was observed that, contrary to the established opinion, low and easily accessible strata of the atmosphere are capable of conducting electricity, the transmission of electrical energy without wires has become a rational task of the engineer, and one surpassing all others in importance. Its practical consummation would mean that energy would be available for the uses of man at any point of the globe, not in small amounts such as might be derived from the ambient medium by suitable machinery, but in quantities virtually unlimited, from waterfalls. Export of power would then become the chief source of income for many happily situated countries, as the United States, Canada, Central and South America, Switzerland, and Sweden. Men could settle down everywhere, fertilize and irrigate the soil with little effort, and convert barren deserts into gardens, and thus the entire globe could be transformed and made a fitter abode for mankind. It is highly probable that if there are intelligent beings on Mars they have long ago realized this very idea, which would explain the changes on its surface noted by astronomers. The atmosphere on that planet, being of considerably smaller density than that of the earth, would make the task much more easy.

It is probable that we shall soon have a self-acting heat-engine capable of deriving moderate amounts of energy from the ambient medium. There is also a possibility—though a small one—that we may obtain electrical energy direct from the sun. This might be the case if the Maxwellian theory is true, according to which electrical vibrations of all rates should emanate from the sun. I am still investigating this subject. Sir William Crookes has shown in his beautiful invention known as the "radiometer" that rays may produce by impact a mechanical effect, and this may lead to some important revelation as to the utilization of the sun's rays in novel ways. Other sources of energy may be opened up, and new methods of deriving energy from the sun discovered, but none of these or similar achievements would equal in importance the transmission of power to any distance through the medium. I can conceive of no technical advance which would tend to unite the various elements of humanity more effectively than this one, or of one which would more add to and more economize human energy. It would be the best means of increasing the force accelerating the human mass. The mere moral influence of such a radical departure would be incalculable. On the other hand if at any point of the globe energy can be obtained in limited quantities from the ambient medium by means of a self-acting heat-engine or otherwise, the conditions will remain the same as before. Human performance will be increased, but men will remain strangers as they were.

I anticipate that any, unprepared for these results, which, through long familiarity, appear to me simple and obvious, will consider them still far from practical application. Such reserve, and even opposition, of some is as useful a quality and as necessary an element in human progress as the quick receptivity and enthusiasm of others. Thus, a mass which resists the force at first, once set in movement, adds to the energy. The scientific man does not aim at an immediate result. He does not expect that his advanced ideas will be readily taken up. His work is like that of the planter—for the future.

June 1900

TESLA'S NEW DISCOVERY

—The Sun, New York, January 30, 1901. Capacity of Electrical Conductors is Variable, Not Constant, and Formulas Will Have to Be Rewritten—Capacity Varies With Absolute Height Above Sea Level, Relative Height From Earth and Distance From the Sun

Nikola Tesla announced yesterday another new discovery in electricity. This time it is a new law and by reason of it, Mr. Tesla asserts, a large part of technical literature will have to be rewritten. Ever since anything has been known about electricity, scientific men have taken for granted that the capacity of an electrical conductor is constant. When Tesla was experimenting in Colorado he found out that this capacity is not constant—but variable. Then he determined to find out the law governing this phenomenon. He did so, and all this he explained to The Sun yesterday. Here is what he said:

"Since many years scientific men engaged in the study of physics and electrical research have taken it for granted that certain quantities, entering continuously in their estimates and calculations, are fixed and unalterable. The exact determination of these quantities being of particular importance in electrical vibrations, which are engrossing more and more the attention of experimenters all over the world, it seems to be important to acquaint others with some of my observations, which have finally led me to the results now attracting universal attention. These observations, with which I have long been familiar, show that some of the quantities referred to are variable and that, owing to this, a large portion of the technical literature is defective. I shall endeavor to convey the knowledge of the facts I have discovered in plain language, devoid as much as possible of technicalities.

"It is well known that an electric circuit compacts itself like a spring with a weight attached to it. Such a spring vibrates at a definite rate, which is determined by two quantities, the pliability of the spring and the mass of the weight. Similarly an electric circuit vibrates, and its vibration, too, is dependent on two quantities, designated as electrostatic capacity and inductance. The capacity of the electric circuit corresponds to the pliability of the spring and the inductance to the mass of the weight.

"Exactly as mechanics and engineers have taken it for granted that the pliability of the spring remains the same, no matter how it be placed or used, so electricians and physicists have assumed that the electrostatic capacity of a conducting body, say of a metallic sphere, which is frequently used in experiments, remains a fixed and unalterable quantity, and many scientific results of the greatest importance are dependent on this assumption. Now, I have discovered that this capacity is not fixed and unalterable at all. On the contrary, it is susceptible to great changes, so that under certain conditions it may amount to many times its theoretical value, or may eventually be smaller. Inasmuch as every electrical conductor, besides possessing an inductance, has also a certain amount of capacity, owing to the variations of the latter, the inductance, too, is seemingly modified by the same causes that tend to modify the capacity. These facts I discovered some time before I gave a technical description of my system of energy transmission and telegraphy without wires, which, I believe, became first known through my Belgian and British patents.

"In this system, I then explained, that, in estimating the wave-length of the electrical vibration in the transmitting and receiving circuits, due regard must be had to the velocity with which the vibration is propagated through each of the circuits, this velocity being given by the product of the wave-length and the number of vibrations per second. The rate of vibration being, however, as before stated, dependent on the capacity and inductance in each case, I obtained discordant values.

"Continuing the investigation of this astonishing phenomenon I observed that the capacity varied with the elevation of the conducting surface above the ground and I soon ascertained the law of this variation. The capacity increased as the conducting surface was elevated, in open space, from 0.5 to 0.75 of 1% per foot of elevation. In buildings, however, or near large structures, this increase often amounted to 50% per foot of elevation, and this alone will show to what extent many of the scientific experiments recorded in technical literature are erroneous. In determining the length of the coils or conductors such as I employ in my system of wireless telegraphy, for instance, the rule which I have given is, in view of the above, important to observe.

"Far more interesting, however, for men of science is the fact I observed later, that the capacity undergoes an annual variation with a maximum in summer, and a minimum in winter. In Colorado, where I continued with improved methods of investigations begun in New York, and where I found the rate of increase slightly greater, I furthermore observed that there was a diurnal variation with a maximum during the night. Further, I found that sunlight causes a slight increase in capacity. The moon also produces an effect, but I do not attribute it to its light.

"The importance of these observations will be better appreciated when it is stated that owing to these changes of a quantity supposed to be constant an electrical circuit does not vibrate at a uniform rate, but its rate is modified in accordance with the modifications of the capacity. Thus a circuit vibrates a little slower at an elevation than when at a lower level. An oscillating system, as used in telegraphy without wires, vibrates a little quicker when the ship gets into the harbor than when on open sea. Such a circuit oscillates quicker in the winter than in the summer, though it be at the same temperature, and a trifle quicker at night than in daytime, particularly if the sun is shining.

"Taking together the results of my investigations I find that this variation of the capacity and consequently of the vibration period is evidently dependent, first, on the absolute height above sea level, though in a smaller degree; second, on the relative height of the conducting surface or capacity with respect to the bodies surrounding it; third, on the distance of the earth from the sun, and fourth, on the relative change of the circuit with respect to the sun, caused by the diurnal rotation of the earth. These facts may be of particular interest to meteorologists and astronomers, inasmuch as practical methods of inquiry may result from these observations, which may be useful in their respective fields. It is probable that we shall perfect instruments for indicating the altitude of a place by means of a circuit, properly constructed and arranged, and I have thought of a number of other uses to which this principle may be put.

"It was in the course of investigations of this kind in Colorado that I first noted certain variations in electrical systems arranged in peculiar ways. These variations I first discovered by calculating over the results I had previously noted, and it was only subsequently that I actually perceived them. It will thus be clear that some who have ventured to attribute the phenomena I have observed to ordinary atmospheric disturbances have made a hasty conclusion."

Colorado Springs
Oct. 23, 1899

The coil referred to on a previous occasion was finished with exactly 689 turns on a drum of eight feet in length and 14" diam. The wire used was cord No. 20 as before stated so that the approximate estimate of self-induction and other particulars holds good. The coil was set up upright outside of the building at some distance to reduce any errors due to the influence of the woodwork. From the building extended a structure of dry pine to a height of about sixty feet from the ground. This framework supported, on a projecting crossbeam, a pulley (wood) with cord for pulling up a ball or other object to any desired height within the limits permitted and this beam also carried on its extreme end and close to the pulley a strong glass bottle within which was fastened a bare wire No. 10, which extended vertically downward to the top of the coil. The bottle was an ordinary Champagne bottle, from which the wine had been poured out! and the bottom broken in. It was forced neck downward into a hole bored into the beam and fastened besides with a cord. A tapering plug of hard wood was wedged into the neck and into this plug was fastened the wire. The bottle was finally filled with melted wax.

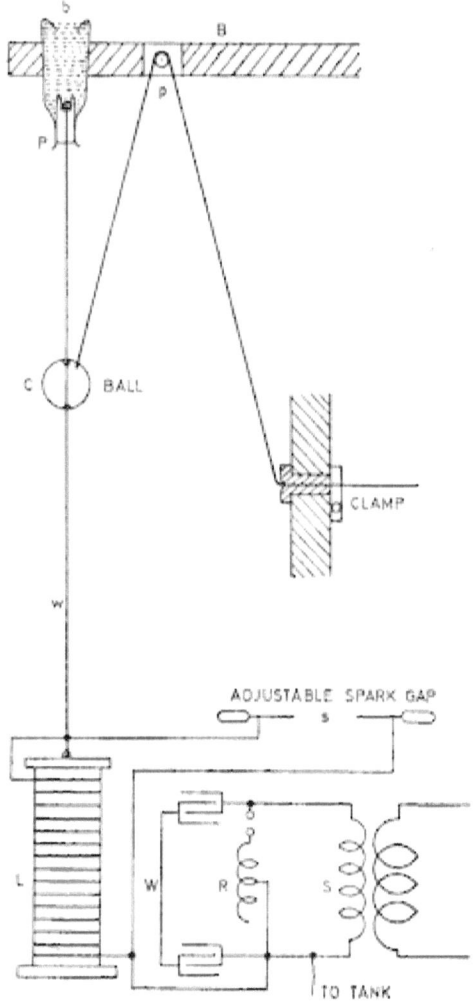

Photo from RADIOTECHNICA, *Muzej Nikole TesleExperiments* to further ascertain the influence of elevation upon capacity.

The whole arrangement is illustrated in the sketch shown in which b is the bottle with wooden plug p supported on beam B also carrying pulley p, over which passes the cord for pulling up the object, which in this case is shown as the sphere C. The spheres used were of wood and hollow and covered very smoothly with tin foil and any points of the foil were pressed in so as to be below the surface of the sphere...

Colorado Springs Notes, pp. 235, 236
January 30, 1901

TALKING WITH THE PLANETS

—Collier's Weekly, February 19, 1901, page 4-5

Editor's note: *Mr. Nikola Tesla has accomplished some marvelous results in electrical discoveries. Now, with the dawn of the new century, he announces an achievement which will amaze the entire universe, and which eclipses the wildest dream of the most visionary scientist. He has received communication, he asserts, from out the great void of space: a call from the inhabitants of Mars, or Venus, or some other sister planet! And, furthermore, noted scientists like Sir Norman Lockyer are disposed to agree with Mr. Tesla in his startling deductions.*

Mr. Tesla has not only discovered many important principles, but most of his inventions are in practical use: notably in the harnessing of the Titanic forces of Niagara Falls, and the discovery of a new light by means of a vacuum tube. He has, he declares, solved the problem of telegraphing without wires or artificial conductors of any sort, using the earth as his medium. By means of this principle he expects to be able to send messages under the ocean, and to any distance on the earth's surface. Interplanetary communication has interested him for years, and he sees no reason why we should not soon be within talking distance of Mars or of all worlds in the solar system that may be tenanted by intelligent beings.

At the request of Collier's Weekly Mr. Tesla presents herewith a frank statement of what he expects to accomplish and how he hopes to establish communication with the planets.

The idea of communicating with the inhabitants of other worlds is an old one. But for ages it has been regarded merely as a poet's dream, forever unrealizable. And with the invention and perfection of the telescope and the ever-widening knowledge of the heavens, its hold upon our imaginations has been increased, and the scientific achievements during the latter part of the nineteenth century, together with the development of the tendency toward the nature ideal of Goethe, have intensified it to such a degree that it seems as if it were destined to become the dominating idea of the century that has just begun. The desire to know something of our neighbors in the immense depths of space does not spring from idle curiosity nor from thirst for knowledge, but from a deeper cause, and it is a feeling firmly rooted in the heart of every human being capable of thinking at all.

Whence, then, does it come? Who knows? Who can assign limits to the subtlety of nature's influences? Perhaps, if we could clearly perceive all the intricate mechanism of the glorious spectacle that is continually unfolding before us, and could, also, trace this desire to its distant origin, we might find it in the sorrowful vibrations of the earth which began when it parted from its celestial parent.

But in this age of reason it is not astonishing to find persons who scoff at the very thought of effecting communication with a planet. First of all, the argument is made that there is only a small probability of other planets being inhabited at all. This argument has never appealed to me. In the solar system, there seem to be only two planets—Venus and Mars—capable of sustaining life such as ours: but this does not mean that there might not be on all of them some other forms of life.

MR. TESLA'S EXPERIMENT TO ILLUSTRATE THE CAPACITY OF HIS OSCILLATOR FOR PRODUCING ELECTRICAL EXPLOSIONS OF GREAT POWER

Chemical processes may be maintained without the aid of oxygen, and it is still a question whether chemical processes are absolutely necessary for the sustenance of organized beings. My idea is that the development of life must lead to forms of existence that will be possible without nourishment and which will not be shackled by consequent limitations. Why should a living being not be able to obtain all the energy it needs for the performance of its life functions from the environment, instead of through consumption of food, and transforming, by a complicated process, the energy of chemical combinations into life-sustaining energy?

If there were such beings on one of the planets we should know next to nothing about them. Nor is it necessary to go so far in our assumptions, for we can readily conceive that, in the same degree as the atmosphere diminishes in density, moisture disappears and the planet freezes up, organic life might also undergo corresponding modifications, leading finally to forms which, according to our present ideas of life, are impossible. I will readily admit, of course, that if there should be a sudden catastrophe of any kind all life processes might be arrested; but if the change, no matter how great, should be gradual, and occupied ages, so that the ultimate results could be intelligently foreseen, I cannot but think that reasoning beings would still find means of existence. They would adapt themselves to their constantly changing environment. So I think it quite possible that in a frozen planet, such as our moon is supposed to be, intelligent beings may still dwell, in its interior, if not on its surface.

Signalling at 100,000,000 miles!

Then it is contended that it is beyond human power and ingenuity to convey signals to the almost inconceivable distances of fifty million or one hundred million miles. This might have been a valid argument formerly. It is not so now. Most of those who are enthusiastic upon the subject of interplanetary communication have reposed their faith in the light-ray as the best possible medium of such communication. True, waves of light, owing to their immense rapidity of succession, can penetrate space more readily than waves less rapid, but a simple consideration will show that by their means an exchange of signals between this earth and its companions in the solar system is, at least now, impossible. By way of illustration, let us suppose that a square mile of the earth's surface — the smallest area that might possibly be within reach of the best telescopic vision of other worlds — were covered with incandescent lamps, packed closely together so as to form, when illuminated, a continuous sheet of light. It would require not less than one hundred million horse-power to light this area of lamps, and this is many times the amount of motive power now in the service of man throughout the world.

But with the novel means, proposed by myself, I can readily demonstrate that, with an expenditure not exceeding two thousand horse-power, signals can be transmitted to a planet such as Mars with as much exactness and certitude as we now send messages by wire from New York to Philadelphia. These means are the result of long-continued experiment and gradual improvement.

Some ten years ago, I recognized the fact that to convey electric currents to a distance it was not at all necessary to employ a return wire, but that any amount of energy might be transmitted by using a single wire. I illustrated this principle by numerous experiments, which, at that time, excited considerable attention among scientific men.

This being practically demonstrated, my next step was to use the earth itself as the medium for conducting the currents, thus dispensing with wires and all other artificial conductors. So I was led to the development of a system of energy transmission and of telegraphy without the use of wires, which I described in 1893. The difficulties I encountered at first in the transmission of currents through the earth were very great. At that time I had at hand only ordinary apparatus, which I found to be ineffective, and I concentrated my attention immediately upon perfecting machines for this special purpose. This work consumed a number of years, but I finally vanquished all difficulties and succeeded in producing a machine which, to explain its operation in plain language, resembled a pump in its action, drawing electricity from the earth and driving it back into the same at an enormous rate, thus creating ripples or disturbances which, spreading through the earth as through a wire, could be detected at great distances by carefully attuned receiving circuits. In this manner I was able to transmit to a distance, not only feeble effects for the purposes of signalling, but considerable amounts of energy, and later discoveries I made convinced me that I shall ultimately succeed in conveying power without wires, for industrial purposes, with high economy, and to any distance, however great.

SUPPLYING ELECTRICAL ENERGY THROUGH A SINGLE WIRE WITHOUT RETURN   TRANSMITTING ELECTRICAL ENERGY THROUGH THE EARTH WITHOUT WIRE   AN ELECTRICAL OSCILLATOR DELIVERING ENERGY AT A RATE OF 75,000 HORSE-POWER

Experiments in Colorado

To develop these inventions further, I went to Colorado in where I continued my investigations along these and other lines, one of which in particular I now consider of even greater importance than the transmission of power without wires. I constructed a laboratory in the neighborhood of Pike's Peak. The conditions in the pure air of the Colorado Mountains proved extremely favorable for my experiments, and the results were most gratifying to me. I found that I could not only accomplish more work, physically and mentally, than I could in New York, but that electrical effects and changes were more readily and distinctly perceived. A few years ago it was virtually impossible to produce electrical sparks twenty or thirty foot long; but I produced some more than one hundred feet in length, and this without difficulty. The rates of electrical movement involved in strong induction apparatus had measured but a few hundred horse-power, and I produced electrical movements of rates of one hundred and ten thousand horse-power. Prior to this, only insignificant electrical pressures were obtained, while I have reached fifty million volts.

The accompanying illustrations, with their descriptive titles, taken from an article I wrote for the *Century Magazine*, may serve to convey an idea of the results I obtained in the directions indicated.

Many persons in my own profession have wondered at them and have asked what I am trying to do. But the time is not far away now when the practical results of my labors will be placed before the world and their influence felt everywhere. One of the immediate consequences will be the transmission of messages without wires, over sea or land, to an immense distance. I have already demonstrated, by crucial tests, the practicability of signalling by my system from one to any other point of the globe, no matter how remote, and I shall soon convert the disbelievers.

I have every reason for congratulating myself that throughout these experiments, many of which were exceedingly delicate and hazardous, neither myself nor any of my assistants received any injury. When working with these powerful electrical oscillations the most extraordinary phenomena take place at times. Owing to some interference of the oscillations, veritable balls of fire are apt to leap out to a great distance, and if any one were within or near their paths, he would be instantly destroyed. A machine such as I have used could easily kill, in an instant, three hundred thousand persons. I observed that the strain upon my assistants was telling, and some of them could not endure the extreme tension of the nerves. But these perils are now entirely overcome, and the operation of such apparatus, however powerful, involves no risk whatever.

As I was improving my machines for the production of intense electrical actions, I was also perfecting the means for observing feeble effects. One of the most interesting results, and also one of great practical importance, was the development of certain contrivances for indicating at a distance of many hundred miles an approaching storm, its direction, speed and distance travelled. These appliances are likely to be valuable in future meteorological observations and surveying, and will lend themselves particularly to many naval uses.

It was in carrying on this work that for the first time I discovered those mysterious effects which have elicited such unusual interest. I had perfected the apparatus referred to so far that from my laboratory in the Colorado mountains I could feel the pulse of the globe, as it were, noting every electrical change that occurred within a radius of eleven hundred miles.

Terrified by success

I can never forget the first sensations I experienced when it dawned upon me that I had observed something possibly of incalculable consequences to mankind. I felt as though I were present at the birth of a new knowledge or the revelation of a great truth. Even now, at times, I can vividly recall the incident, and see my apparatus as though it were actually before me. My first observations positively terrified me, as there was present in them something mysterious, not to say supernatural, and I was alone in my laboratory at night; but at that time the idea of these disturbances being intelligently controlled signals did not yet present itself to me.

The changes I noted were taking place periodically, and with such a clear suggestion of number and order that they were not traceable to any cause then known to me. I was familiar, of course, with such electrical disturbances as are produced by the sun, Aurora Borealis and earth currents, and I was as sure as I could be of any fact that these variations were due to none of these causes. The nature of my experiments precluded the possibility of the changes being produced by atmospheric disturbances, as has been rashly asserted by some. It was some time afterward when the thought flashed upon my mind that the disturbances I had observed might be due to an intelligent control. Although I could not decipher their meaning, it was impossible for me to think of them as having been entirely accidental. The feeling is constantly growing on me that I had been the first to hear the greeting of one planet to another. A purpose was behind these electrical signals; and it was with this conviction that I announced to the Red Cross Society, when it asked me to indicate one of the great possible achievements of the next hundred years, that it would probably be the confirmation and interpretation of this planetary challenge to us.

Since my return to New York more urgent work has consumed all my attention; but I have never ceased to think of those experiences and of the observations made in Colorado. I am constantly endeavoring to improve and perfect my apparatus, and just as soon as practicable I shall again take up the thread of my investigations at the point where I have been forced to lay it down for a time.

Communicating with the Martians

At the present stage of progress, there would be no insurmountable obstacle in constructing a machine capable of conveying a message to Mars, nor would there be any great difficulty in recording signals transmitted to us by the inhabitants of that planet, if they be skilled electricians. Communication once established, even in the simplest way, as by a mere interchange of numbers, the progress toward more intelligible communication would be rapid. Absolute certitude as to the receipt and interchange of messages would be reached as soon as we could respond with the number "four," say, in reply to the signal "one, two, three." The Martians, or the inhabitants of whatever planet had signalled to us, would understand at once that we had caught their message across the gulf of space and had sent back a response. To convey a knowledge of form by such means is, while very difficult, not impossible, and I have already found a way of doing it.

PHOTOGRAPHIC VIEW OF THE ESSENTIAL PARTS OF THE ELECTRICAL OSCILLATOR USED IN MR. TESLA'S EXPERIMENTS

What a tremendous stir this would make in the world! How soon will it come? For that it will some time be accomplished must be clear to every thoughtful being.

Something, at least, science has gained. But I hope that it will also be demonstrated soon that in my experiments in the West I was not merely beholding a vision, but had caught sight of a great and profound truth.

February 9, 1901

INVENTOR TESLA'S PLANT NEARING COMPLETION

—Brooklyn Eagle, February 8, 1902. Buildings at Wardencliff to be Used in Developing His Electrical Discoveries

Sinking a Very Deep Well

A Big Tower 100 Feet in Diameter and Over 200 Feet High—Power House and Dynamos (Special to the Eagle)

Wardenclyffe, L. I., February 8—Work on the buildings at Wardenclyffe, L. I. to be used by Electrician Tesla in the development of his electrical discoveries, is progressing rapidly. The power house is completed and the foundations of the big tower have been laid.

The working room, or tower, which will be the foundation for Mr. Tesla's across-the-world flashes, will be octagonal in shape and will be 210 feet high, 100 feet In diameter at the base, narrowing down to 80 feet in diameter at the top. It will be constructed chiefly of wood, though the builders say that fifty tons of iron and steel and 50,000 bolts of various sizes are used in Its construction.

Inside the big tower a well 120 feet deep has been sunk, the well being 12 feet square, cased Its entire depth with 8 inch timbers, which will be finished off with brick and cement. A staircase, which will lead down into the well, is nearly completed.

Transversely across the bottom of the well will be a series of four tunnels, each to be 100 feet long, and a force of workmen has begun work on these subterranean passages.

The power house Is constructed of pressed brick, and is 100 feet square. In it are a boiler room, engine and dynamo room, machine shop and laboratory. The Westinghouse Company furnished the electrical equipment, boilers, engines and dynamos. Only the finishing touches are needed on the electrical equipment to make it available and Mr. Tesla promises to begin active operations very soon.

February 8, 1902

THE TRANSMISSION OF ELECTRICAL ENERGY WITHOUT WIRES

—Electrical World and Engineer, March 5, 1904

It is impossible to resist your courteous request extended on an occasion of such moment in the life of your journal. Your letter has vivified the memory of our beginning friendship, of the first imperfect attempts and undeserved successes, of kindnesses and misunderstandings. It has brought painfully to my mind the greatness of early expectations, the quick flight of time, and alas! the smallness of realizations. The following lines which, but for your initiative, might not have been given to the world for a long time yet, are an offering in the friendly spirit of old, and my best wishes for your future success accompany them.

Towards the close of 1898 a systematic research, carried on for a number of years with the object of perfecting a method of transmission of electrical energy through the natural medium, led me to recognize three important necessities: First, to develop a transmitter of great power; second, to perfect means for individualizing and isolating the energy transmitted; and, third, to ascertain the laws of propagation of currents through the earth and the atmosphere. Various reasons, not the least of which was the help proffered by my friend Leonard E. Curtis and the Colorado Springs Electric Company, determined me to select for my experimental investigations the large plateau, two thousand meters above sea-level, in the vicinity of that delightful resort, which I reached late in May, 1899. I had not been there but a few days when I congratulated myself on the happy choice and I began the task, for which I had long trained myself, with a grateful sense and full of inspiring hope. The perfect purity of the air, the unequaled beauty of the sky, the imposing sight of a high mountain range, the quiet and restfulness of the place—all around contributed to make the conditions for scientific observations ideal. To this was added the exhilarating influence of a glorious climate and a singular sharpening of the senses. In those regions the organs undergo perceptible physical changes. The eyes assume an extraordinary limpidity, improving vision; the ears dry out and become more susceptible to sound. Objects can be clearly distinguished there at distances such that I prefer to have them told by someone else, and I have heard—this I can venture to vouch for—the claps of thunder seven and eight hundred kilometers away. I might have done better still, had it not been tedious to wait for the sounds to arrive, in definite intervals, as heralded precisely by an electrical indicating apparatus — nearly an hour before.

In the middle of June, while preparations for other work were going on, I arranged one of my receiving transformers with the view of determining in a novel manner, experimentally, the electric potential of the globe and studying its periodic and casual fluctuations. This formed part of a plan carefully mapped out in advance. A highly sensitive, self-restorative device, controlling a recording instrument, was included in the secondary circuit, while the primary was connected to the ground and an elevated terminal of adjustable capacity. The variations of potential gave rise to electric surgings in the primary; these generated secondary currents, which in turn affected the sensitive device and recorder in proportion to their intensity. The earth was found to be, literally, alive with electrical vibrations, and soon I was deeply absorbed in the interesting investigation. No better opportunities for such observations as I intended to make could be found anywhere. Colorado is a country famous for the natural displays of electric force. In that dry and rarefied atmosphere the sun's rays beat the objects with fierce intensity. I raised steam, to a dangerous pressure, in barrels filled with concentrated salt solution, and the tin-foil coatings of some of my elevated terminals shriveled up in the fiery blaze. An experimental high-tension transformer, carelessly exposed to the rays of the setting sun, had most of its insulating compound melted out and was rendered useless. Aided by the dryness and rarefaction of the air, the water evaporates as in a boiler, and static electricity is developed in abundance. Lightning discharges are, accordingly, very frequent and sometimes of inconceivable violence. On one occasion approximately twelve thousand discharges occurred in two hours, and all in a radius of certainly less than fifty kilometers from the laboratory. Many of them resembled gigantic trees of fire with the trunks up or down. I never saw fire balls, but as compensation for my disappointment I succeeded later in determining the mode of their formation and producing them artificially.

In the latter part of the same month I noticed several times that my instruments were affected stronger by discharges taking place at great distances than by those near by. This puzzled me very much. What was the cause? A number of observations proved that it could not be due to the differences in the intensity of the individual discharges, and I readily ascertained that the phenomenon was not the result of a varying relation between the periods of my receiving circuits and those of the terrestrial disturbances. One night, as I was walking home with an assistant, meditating over these experiences, I was suddenly staggered by a thought. Years ago, when I wrote a chapter of my lecture before the *Franklin Institute* and the *National Electric Light Association*, it had presented itself to me, but I dismissed it as absurd and impossible. I banished it again. Nevertheless, my instinct was aroused and somehow I felt that I was nearing a great revelation.

It was on the third of July—the date I shall never forget—when I obtained the first decisive experimental evidence of a truth of overwhelming importance for the advancement of humanity. A dense mass of strongly charged clouds gathered in the west and towards the evening a violent storm broke loose which, after spending much of its fury in the mountains, was driven away with great velocity over the plains. Heavy and long persisting arcs formed almost in regular time intervals. My observations were now greatly facilitated and rendered more accurate by the experiences already gained. I was able to handle my instruments quickly and I was prepared. The recording apparatus being properly adjusted, its indications became fainter and fainter with the increasing distance of the storm, until they ceased altogether. I was watching in eager expectation. Surely enough, in a little while the indications again began, grew stronger and stronger and, after passing through a maximum, gradually decreased and ceased once more. Many times, in regularly recurring intervals, the same actions were repeated until the storm which, as evident from simple computations, was moving with nearly constant speed, had retreated to a distance of about three hundred kilometers. Nor did these strange actions stop then, but continued to manifest themselves with undiminished force. Subsequently, similar observations were also made by my assistant, Mr. Fritz Lowenstein, and shortly afterward several admirable opportunities presented themselves which brought out, still more forcibly, and unmistakably, the true nature of the wonderful phenomenon. No doubt, whatever remained: I was observing stationary waves.

As the source of disturbances moved away the receiving circuit came successively upon their nodes and loops. Impossible as it seemed, this planet, despite its vast extent, behaved like a conductor of limited dimensions. The tremendous significance of this fact in the transmission of energy by my system had already become quite clear to me. Not only was it practicable to send telegraphic messages to any distance without wires, as I recognized long ago, but also to impress upon the entire globe the faint modulations of the human voice, far more still, to transmit power, in unlimited amounts, to any terrestrial distance and almost without loss.

With these stupendous possibilities in sight, and the experimental evidence before me that their realization was henceforth merely a question of expert knowledge, patience and skill, I attacked vigorously the development of my magnifying transmitter, now, however, not so much with the original intention of producing one of great power, as with the object of learning how to construct the best one. This is, essentially, a circuit of very high self-induction and small resistance which in its arrangement, mode of excitation and action, may be said to be the diametrical opposite of a transmitting circuit typical of telegraphy by Hertzian or electromagnetic radiations. It is difficult to form an adequate idea of the marvelous power of this unique appliance, by the aid of which the globe will be transformed. The electromagnetic radiations being reduced to an insignificant quantity, and proper conditions of resonance maintained, the circuit acts like an immense pendulum, storing indefinitely the energy of the primary exciting impulses and impressions upon the earth of the primary exciting impulses and impressions upon the earth and its conducting atmosphere uniform harmonic oscillations of intensities which, as actual tests have shown, may be pushed so far as to surpass those attained in the natural displays of static electricity.

Simultaneously with these endeavors, the means of individualization and isolation were gradually improved. Great importance was attached to this, for it was found that simple tuning was not sufficient to meet the vigorous practical requirements. The fundamental idea of employing a number of distinctive elements, cooperatively associated, for the purpose of isolating energy transmitted, I trace directly to my perusal of Spencer's clear and suggestive exposition of the human nerve mechanism. The influence of this principle on the transmission of intelligence, and electrical energy in general, cannot as yet be estimated, for the art is still in the embryonic stage; but many thousands of simultaneous telegraphic and telephonic messages, through one single conducting channel, natural or artificial, and without serious mutual interference, are certainly practicable, while millions are possible. On the other hand, any desired degree of individualization may be secured by the use of a great number of cooperative elements and arbitrary variation of their distinctive features and order of succession. For obvious reasons, the principle will also be valuable in the extension of the distance of transmission.

Progress though of necessity slow was steady and sure, for the objects aimed at were in a direction of my constant study and exercise. It is, therefore, not astonishing that before the end of 1899 I completed the task undertaken and reached the results which I have announced in my article in the Century Magazine of June, 1900, every word of which was carefully weighed.

Much has already been done towards making my system commercially available, in the transmission of energy in small amounts for specific purposes, as well as on an industrial scale. The results attained by me have made my scheme of intelligence transmission, for which the name of "World

Telegraphy" has been suggested, easily realizable. It constitutes, I believe, in its principle of operation, means employed and capacities of application, a radical and fruitful departure from what has been done heretofore. I have no doubt that it will prove very efficient in enlightening the masses, particularly in still uncivilized countries and less accessible regions, and that it will add materially to general safety, comfort and convenience, and maintenance of peaceful relations. It involves the employment of a number of plants, all of which are capable of transmitting individualized signals to the uttermost confines of the earth. Each of them will be preferably located near some important center of civilization and the news it receives through any channel will be flashed to all points of the globe. A cheap and simple device, which might be carried in one's pocket, may then be set up somewhere on sea or land, and it will record the world's news or such special messages as may be intended for it. Thus the entire earth will be converted into a huge brain, as it were, capable of response in every one of its parts. Since a single plant of but one hundred horse-power can operate hundreds of millions of instruments, the system will have a virtually infinite working capacity, and it must needs immensely facilitate and cheapen the transmission of intelligence.

The first of these central plants would have been already completed had it not been for unforeseen delays which, fortunately, have nothing to do with its purely technical features. But this loss of time, while vexatious, may, after all, prove to be a blessing in disguise. The best design of which I know has been adopted, and the transmitter will emit a wave complex of total maximum activity of ten million horse-power, 1% of which is amply sufficient to "girdle the globe." This enormous rate of energy delivery, approximately twice that of the combined falls of Niagara, is obtainable only by the use of certain artifices, which I shall make known in due course.

For a large part of the work which I have done so far I am indebted to the noble generosity of Mr. J. Pierpont Morgan, which was all the more welcome and stimulating, as it was extended at a time when those, who have since promised most, were the greatest of doubters. I have also to thank my friend, Stanford White, for much unselfish and valuable assistance. This work is now far advanced, and though the results may be tardy, they are sure to come.

Meanwhile, the transmission of energy on an industrial scale is not being neglected. The Canadian Niagara Power Company have offered me a splendid inducement, and next to achieving success for the sake of the art, it will give me the greatest satisfaction to make their concession financially profitable to them. In this first power plant, which I have been designing for a long time, I propose to distribute ten thousand horse-power under a tension of one hundred million volts, which I am now able to produce and handle with safety.

This energy will be collected all over the globe preferably in small amounts, ranging from a fraction of one to a few horse-power. One its chief uses will be the illumination of isolated homes. I takes very little power to light a dwelling with vacuum tubes operated by high-frequency currents and in each instance a terminal a little above the roof will be sufficient. Another valuable application will be the driving of clocks and other such apparatus. These clocks will be exceedingly simple, will require absolutely no attention and will indicate rigorously correct time. The idea of impressing upon the earth American time is fascinating and very likely to become popular. There are innumerable devices of all kinds which are either now employed or can be supplied, and by operating them in this manner I may be able to offer a great convenience to whole world with a plant of no more than ten thousand horse-power. The introduction of this system will give opportunities for invention and manufacture such as have never presented themselves before.

Knowing the far-reaching importance of this first attempt and its effect upon future development, I shall proceed slowly and carefully. Experience has taught me not to assign a term to enterprises the consummation of which is not wholly dependent on my own abilities and exertions. But I am hopeful that these great realizations are not far off, and I know that when this first work is completed they will follow with mathematical certitude.

When the great truth accidentally revealed and experimentally confirmed is fully recognized, that this planet, with all its appalling immensity, is to electric currents virtually no more than a small metal ball and that by this fact many possibilities, each baffling imagination and of incalculable consequence, are rendered absolutely sure of accomplishment; when the first plant is inaugurated and it is shown that a telegraphic message, almost as secret and non-interferable as a thought, can be transmitted to any terrestrial distance, the sound of the human voice, with all its intonations and inflections, faithfully and instantly reproduced at any other point of the globe, the energy of a waterfall made available for supplying light, heat or motive power, anywhere-on sea, or land, or high in the air-humanity will be like an ant heap stirred up with a stick: See the excitement coming! March 5, 1904

ELECTRIC AUTOS

—*Special Correspondence, Manufacturers' Record. December 29, 1904*

Nicola Tesla's View of the Future in Motive Power

New York, December 27

In view of the great interest which is being taken in the articles published by the Manufacturers' Record and some of the magazines on the development of new power-producers, through the internal-combustion engine, for use for transportation purposes both by land and sea, the following signed statement, made by Mr. Nicola Tesla after a discussion of a new type of autobus designed by Mr. Charles A. Lieb, mechanical engineer of the Manhattan Transit Co., will doubtless be read with much general interest:

New York, December 17

Mr. Albert Phenis, Special Correspondent Manufacturers' Record, New York:

Dear Sir—Replying to your inquiry of yesterday, the application of electricity to the propulsion of automobiles is certainly a rational idea. I am glad to know that Mr. Lieb has undertaken to put it into practice. His long experience with the General Electric Co. and other concerns must have excellently fitted him for the task.

There is no doubt that a highly-successful machine can be produced on these lines. The field is inexhaustible, and this new type of automobile, introducing electricity between the prime mover and the wheels, has, in my opinion, a great future.

I have myself for many years advocated this principle. Your will find in numerous technical publications statements made by me to this effect. In my article in the Century, June, 1900, I said, in dealing with the subject: 'Steamers and trains are still being propelled by the direct application of steam power to shafts or axles. A much greater percentage of the heat energy of the fuel could be transformed in motive energy by using, in place of the adopted marine engines and locomotives, dynamos driven by specially designed high-pressure steam or gas engines, by utilizing the electricity generated for the propulsion. A gain of 50 to 100%, in the effective energy derived from the fuel could be secured in this manner. It is difficult to understand why a fact so plain and obvious is not receiving more attention from engineers.

At first glance it may appear that to generate electricity by an engine and then apply the current to turn a wheel, instead of turning it by means of some mechanical connection with the engine, is a complicated and more or less wasteful process. But it is not so; on the contrary, the use of electricity in this manner secures great practical advantages. It is but a question of time when this idea will be extensively applied to railways and also to ocean liners, though in the latter case the conditions are not quite so favorable. How the railroad companies can persist in using the ordinary locomotive is a mystery. By providing an engine generating electricity and operating with the current motors under the cars a train can be propelled with greater speed and more economically. In France this has already been done by Heilman, and although his machinery was not the best, the results he obtained were creditable and encouraging. I have calculated that a notable gain in speed and economy can also be secured in ocean liners, on which the improvement is particularly desirable for many reasons. It is very likely that in the near future oil will be adopted as fuel, and that will make the new method of propulsion all the more commendable. The electric manufacturing companies will scarcely be able to meet this new demand for generators and motors.

In automobiles practically nothing has been done in this direction, and yet it would seem they offer the greatest opportunities for application of this principle. The question, however, is which motor to employ the direct-current or my induction motor. The former has certain preferences as regards the starting and regulation, but the commutators and brushes are very objectionable on an automobile. In view of this I would advocate the use of the induction motor as an ideally simple machine which can never get out of order. The conditions are excellent, inasmuch as a very low frequency is practicable and more than three phases can be used. The regulation should offer little difficulty, and once an automobile on this novel plan is produced its advantages will be readily appreciated.

Yours very truly,
N. Tesla.

THE TRANSMISSION OF ELECTRICAL ENERGY WITHOUT WIRES AS A MEANS FOR FURTHERING PEACE

—*Electrical World and Engineer, January 7, 1905, pp. 21-24*

Universal peace, assuming it to be in the fullest sense realizable, might not require eons for its accomplishment, however probable this may appear, judging from the imperceptibly slow growth of all great reformatory ideas of the past. Man, as a mass in movement, is inseparable from sluggishness and persistence in his life manifestations, but it does not follow from this that any passing phase, or any permanent state of his existence, must necessarily be attained through a *stataclitic* process of development.

Our accepted estimates of the duration of natural metamorphoses, or changes in general, have been thrown in doubt of late. The very foundations of science have been shaken. We can no longer believe in the Maxwellian hypothesis of transversal ether-undulations of electrical vibrations, this most important field of human endeavor, particularly in the advancement of philanthropy and peace, was in no small measure retarded by that fascinating illusion, which I since long hoped to dispel. I have noted with satisfaction the first signs of a change of scientific opinion. The brilliant discovery of the exceptionally "radio-active" substances, radium and polonium, by Mrs. Sklodowska Curie, has likewise afforded me much personal gratification, being an *eclatante* confirmation of my early experimental demonstrations, of electrified radian streams of primary matter or corpuscular emanations (Electrical Review, New York, 1896-1897), which were then received with incredulity. They have awakened us from the poetical dream of an intangible conveyor of energy, weightless, structureless ether, to the plain, palpable reality of a ponderous medium of coarse particles, or bodily carriers of force. They have led us to a radically new interpretation of the changes and transformations we perceive. Enlightened by this recognition, we cannot say the sun is hot, the moon is cold, the star is bright, for all these might be purely electrical phenomena. If this be the case, then even our conceptions of time and space may have to be modified.

So, too, as regards the organic world, a similar revolution of thought is distinctly observable. In biological and zoological research the bold ideas of Haensel have found support in recent discoveries. A heretic belief in such possibilities as the artificial production of simple living material aggregates, the spontaneous natural creation of complex organisms and willful sex control, is gaining ground. We still brush it aside, but not with pedantic disdain as before. The fact is our faith in the orthodox theory of slow evolution is being destroyed!

Thus a state of human life vaguely defined by the term "Universal Peace," while a result of cumulative effort through centuries past, might come into existence quickly, not unlike a crystal suddenly forms in a solution which has been slowly prepared. But just as no effect can precede its cause, so this state can never be brought on by any pact between nations, however solemn. Experience is made before the law is formulated, both are related like cause and effect. So long as we are clearly conscious of the expectation, that peace is to result from such a parliamentary decision, so long have we a conclusive evidence that we are not fit for peace. Only then when we shall feel that such international meetings are mere formal procedures, unnecessary except in so far as they might serve to give definite expression to a common desire, will peace be assured.

To judge from current events we must be, as yet, very distant from that blissful goal. It is true that we are proceeding towards it rapidly. There are abundant signs of this progress everywhere. The race enmities and prejudices are decidedly waning. A recent act of His Excellency, the President of the United States, is significant in this respect. We begin to think cosmically. Our sympathetic feelers reach out into the dim distance. The bacteria of the "Weltschmerz," are upon us. So far, however, universal harmony has been attained only in a single sphere of international relationship. That is the postal service. Its mechanism is working satisfactorily, but—how remote are we still from that scrupulous respect of the sanctity of the mail bag! And how much farther again is the next milestone on the road to peace—an international judicial service equally reliable as the postal!

The coming meeting at the Hague, now indefinitely postponed, can only consider temporary expedients. General disarmament being for the present entirely out of question, a proportionate reduction might be recommended. The safety of any country and of the world's commerce depending not on the absolute, but relative amount of war material, this would be evidently the first reasonable step to take towards universal economy and peace. But it would be a hopeless task to establish an equitable basis of adjustment. Population, naval strength, force of army, commercial importance, water-power, or any other natural resource, actual or prospective, are equally unsatisfactory standards to consider.

In view of this difficulty a measure suggested by Carnegie might be adopted by a few strong countries to scare all the weaker ones into peace. But while for the time being such a course may seem advisable, the beneficial effects of this homeopathic treatment of the martial disease could hardly be lasting. In the first place, a coalition of the leading powers could not fail to create an organized opposition, which might result in a disaster all the greater as it was long deferred. The ultimate falling out of the virtuous, peace-dictating nations, as certain as the law of gravitation, should be all the more reckoned with, as it would be extremely demoralizing. Again, it is by no means sufficient authority.

To conquer by sheer force is becoming harder and harder every day. Defensive is getting continuously the advantage of offensive, as we progress in the satanic science of destruction. The new art of controlling electrically the movements and operations of individualized automata at a distance without wires, will soon enable any country to render its coasts impregnable against all naval attacks. It is to be regretted, in this connection, that my proposal to the United States Navy four years ago, to introduce this invention, did not receive the least encouragement. Also that my offer to Secretary Long to establish telegraphic communication across the Pacific Ocean by my wireless system was thrown in the naval waste basket in Washington, quite *sans facon*. At that time I had already announced in The Century Magazine of June, 1900, my successful "girdling" of the globe with electrical impulses (stationary waves), and my "telautomata" had been publicly exhibited. But that was not the fault of the naval officials, for then these inventions of mine were decried as bold, visionary schemes, loudest indeed by those who have since become Croesuses of Promise—in "light" storage batteries, "Ocean" telephony and "transatlantic" wireless telegraphy, yet remained to this day—Sisyphuses of Attainment. Had only a few "telautomatic" torpedoes been constructed and adopted by our navy, the mere moral influence of this would have been powerfully and most beneficially felt in the present Eastern Complication. Not to speak of the advantages which might have been secured through the direct and instantaneous transmission of messages to our distant colonies and scenes of the present barbarous conflicts. Since advancing that principle, I have invented a number of improvements, making it possible to direct such a torpedo, submersible at will, from a distance much greater than the range of the largest gun, with unerring precision, upon the object to be destroyed. What is still more surprising, the operator will not need to see the infernal engine or even know its location, and the enemy will be unable to interfere, in the slightest, with its movements by any electrical means. One of these devil-telautomata will soon be constructed, and I shall bring it to the attention of governments. The development of this art must unavoidably arrest the construction of expensive battleships as well as land fortifications, and revolutionize the means and methods of warfare. The distance at which it can strike, and the destructive power of such a quasi-intelligent machine being for all practical purposes unlimited, the gun, the armor of the battleship and the wall of the fortress, lose their import and significance. One can prophesy with a Daniel's confidence that skilled electricians will settle the battles of the near future. But this is the least. In its effect upon war and peace, electricity offers still much greater and more wonderful possibilities. To stop war by the perfection of engines of destruction alone, might consume centuries and centuries. Other means must be employed to hasten the end. What are these to be? Let us consider.

Fights between individuals, as well as governments and nations, invariably result from misunderstandings in the broadest interpretation of this term. Misunderstandings are always caused by the inability of appreciating one another's point of view. This again is due to the ignorance of those concerned, not so much in their own, as in their mutual fields. The peril of a clash is aggravated by a more or less predominant sense of combativeness, posed by every human being. To resist this inherent fighting tendency the best was is to dispel ignorance of the doings of others by a systematic spread of general knowledge. With this object in view, it is most important to aid exchange of thought and intercourse.

Mutual understanding would be immensely facilitated by the use of one universal tongue. But which shall it be, is the great question. At present it looks as if the English might be adopted as such, though it must be admitted that it is not the most suitable. Each language, of course, excels in some feature. The English lends itself to a terse, forceful expression of facts. The French is precise and finely distinctive. The Italian is probably the most melodious and easiest to learn. The Slavic tongues are very rich in sound but extremely difficult to master. The German is unequaled in the facility it offers for coining and combining words. A practical answer to that momentous question must perforce be found in times to come, for it is manifest that by adopting one common language the onward march of man would be prodigiously quickened. I do not believe that an artificial concoction, like Volapuk, will ever find universal acceptance, however time-saving it might be. That would be contrary to human nature. Languages have grown into our hearts. I rather look to the possibility of a reversion to the old Latin or Greek mother tongues, basing myself in this conclusion on the Spencerian law of rhythm [see Spencer's First Principles]. It seems unfortunate that the English-speaking nations, who are now fittest to rule the world, while endowed with extraordinary energy and practical intelligence, are singularly wanting in linguistic talent.

Next to speech we must consider permanent records of all kinds as a means for disseminating general information, or that knowledge of mutual endeavor which is chiefly conducive to harmony. Here the newspapers play by far the most important part. They are undoubtedly more effective than institutions of learning, libraries, museums and individual correspondence, all combined. The knowledge they convey is, on the whole, superficial and sometimes defective, but it is poured out in a mighty stream that reaches far and wide. Disregarding the force of electrical invention, that of journalism is the greatest in urging peace. Our schools are instrumental, mainly, in the furtherance of special thorough knowledge in our own fields, which is destructive of concordance. A world composed of crass specialists only would be perpetually at war. The diffusion of general knowledge through libraries and similar sources of information is very slow. As to individual correspondence, it is principally useful as an indispensable ingredient of the cement of commercial interest, that most powerful binding material between heterogeneous masses of humanity. It would be hard to overestimate the beneficial influence of the marvelous and precise art of photography, nor can that of other arts or means of recording be ignored. But a simple reflection will show that the peace-making force of all permanent, printed, printed or other records, resides not in themselves. It must be sought elsewhere. This is also true of speech.

Our senses enable us to perceive only a minute portion of the outside world. Our hearing extends to a small distance. Our sight is impeded by intervening bodies and shadows. To know each other we must reach beyond the sphere of our sense perceptions. We must transmit our intelligence, travel, transport the materials and transfer the energies necessary for our existence. Following this thought we now realize, forcibly enough to dispense with argument, that of all other conquests of man, without exception, that which is most desirable, which would be most helpful in the establishment of universal peaceful relations is — the complete annihilation of distance.

To achieve this wonder, electricity is the one and only means. Inestimable good has already been done by the use of this all powerful agent, the nature of which is still a mystery. Our astonishment at what has been accomplished would be uncontrollable were it not held in check by the expectation of greater miracles to come. That one, the greatest of all, can be viewed in three aspects: Dissemination of intelligence, transportation, and transmission of power.

Referring to the first, the present systems of telegraphic and telephonic communication are very limited in scope. The conducting channels are costly and of small working capacity. There is serious inductive disturbance, and storms render the service unsafe which, moreover, is too expensive: A vast improvement will be effected by placing the wires underground and insulating them artificially, by refrigeration. Their working capacity also could by indefinitely augmented by resorting to the new principle of "individualization," which I have more recently announced, permitting the simultaneous transmission of thousands of telegraphic and telephonic messages, without interference, over a single wire. The public would be already profiting from these great advances were it not for the stolid indifference of the leading companies engaged in the transmission of intelligence. But new concerns are springing into existence and the near future will witness a great transformation along these two lines of invention. The submarine cables are subject to still greater limitations. Some obstacles to rapid signaling, through them, seem insuperable. The attempts to overcome these have been numerous, but so far all have proved futile. The celebrated mathematician, O. Heaviside, and several able electricians following in his footsteps, have fallen into the singular error that rapid telegraphy and even telephony through ocean cables would be made practicable by the use of induction coils. Inductances might be to some extent helpful on comparatively short lines with thick paper insulation; on long lines insulated with rubber or gutta-percha they would be positively detrimental. Improvements will, undoubtedly, be made, but great electrostatic capacity and unavoidable loss of energy in the insulation and surrounding conductors will always restrict the usefulness of the transmission through artificial conductors is necessarily confined to a small number of stations.

It is therefore evident that the abolishment of all these drawbacks by the conveyance of signals or messages without wires, as I have undertaken in my "world" telegraphy and telephony, will be of the greatest moment in the furtherance of peace. The unifying influence of this advance will be felt all the more, as it will not only completely annihilate distance, but also make it possible to operate from a singe "world" telegraphy plant, an unlimited number of receiving stations distributed all over the globe, and with equal facility, irrespective of location. Within a few years a simple and inexpensive device, readily carried about, will enable one to receive on land or sea the principal news, to hear a speech, a lecture, a song or play of a musical instrument, conveyed from any other region of the globe. The invention will also meet the crying need for cheap transmission to great distances, more especially over the oceans. The small working capacity of the cables and the excessive cost of messages are now fatal impediments in the dissemination of intelligence which can only be removed by transmission without wires.

The deficiencies of Hertzian telegraphy have created in the public mind the impression that exclusive or private messages without the use of artificial channels are impracticable. As a matter of fact, nothing could be more erroneous. Ever since its first appearance in 1891, I have denied the commercial possibilities of the system of signaling by Hertzian or electromagnetic waves, and my forecasts have been fully confirmed. It lends itself little to tuning, still less to the higher artifices of "individualization," and transmission to considerable distances is wholly out of the question. Portentous claims for this method of communication were made three years ago, but they have been unable to stand the hard, cruel test of time. Moreover, I have recently learned through the leading British electrical journal (*Electrician*, London, February 27, 1903), that some experimenters have abandoned all their own and have been "converted" to my methods and appliances, without my approval and officiation. I was both astonished and pained—astonished at the nonchalance and lack of appreciation of these men, pained at the inability exhibited in the construction and use of my apparatus. My high hopes raised by that excellent journal, however, are still to be realized, for I have ascertained that His Majesty the King of England, His excellency the President of the United States, and other persons of exalted positions have, after all, not conferred upon me the imperishable honor of graciously condescending to the use for my coils, transformers and high-potential methods of transmission, but have exchanged their august greetings through the medium of a cable in the old-fashioned way. What has been actually achieved by Hertzian telegraphy can only be conjectured.

Quite different conditions exist in my system in which the electromagnetic waves or radiations are designedly minimized, the connection of one of the terminals of the transmitting circuit to the ground having, itself, the effect of reducing the energy of these radiations to about one-half. Under observance of proper rules and artifices the distance is of little or no consequence, and by skillful application of the principle of "individualization," repeatedly referred to the messages may be rendered both non-interfering and non-interferable. This invention, which I have described in technical publications, attempts to imitate, in a very crude way, the nervous system in the human body. It was the outcome of long-continued tests demonstrating the impossibility of satisfying rigorous commercial requirements by my earlier system, based on simple tuning, in which the selective quality is dependent on a single characteristic feature. In this later improvement the exclusiveness and non-interferability of impulses transmitted through a common channel result from cooperative association for a *number* of distinctive elements, and can be pushed as far as desired. In actual practice it is found that by combining only two vibrations or tones, a degree of privacy sufficient for most purposes is attained. When three vibrations are combined, it is extremely difficult, even for a skilled expert, to read or to disturb signals not intended for him, with four it is a vain undertaking. The probability of his getting the secret combinations at the right moments and in proper order, is much smaller than that of drawing an ambo, terno or quaterno, respectively, in a lottery. From experimental facts, I conclude that the invention will permit the simultaneous transmission of several millions of separately distinguishable messages through the earth, which, strangely enough, is in this respect much superior to an artificial conductor. This number ought to be sufficient to meet all the pressing necessities of intelligence transmission for at least one century to come. It is important to observe that but one "world" telegraphy plant, such as I am now completing, will have a greater working capacity than all the ocean cables combined. Once these facts are recognized this new art, which I am inaugurating, will sweep the world with the force of a hurricane.

In transportation a great change is now going on. The trolley lines are being extended, the steam locomotive is making place for the electric motor. The ocean liners are adopting the turbine. Land travel is being improved by the automobile. The waterfalls are being harnessed and the energy used in the propulsion of cars. The advantages of first generating electricity by a prime mover, are being more and more appreciated. To the majority, this may appear a roundabout way of doing, but in reality it is as direct as the driving of a pulley from another by a belt. The idea is already being applied to railroads, and automobiles of this new type are making their appearance. The ocean vessels are bound to follow. An immense and virgin field will be thus opened up to the manufacturers of electric machinery. Effort towards saving time and money is characteristic of all modern methods of transportation. In many of these new developments, the artificial insulation of the high-tension mains by refrigeration will be very useful. However paradoxical, it is true, that by the use of this invention, power for all industrial purposes can be transmitted to distances of many hundreds of miles, not only without any loss, but with appreciable gain of energy. This is due to the fact that the conductor is much colder than the surrounding medium. The operativeness of this method is restricted to the use of a gaseous refrigerant, no known liquid permitting the attainment of a sufficiently low temperature of the transmission line. Hydrogen is by far the best cooling agent to employ. By its use electric railways can be extended to any desired distance. Owing to the smallness of ohmic loss, the objections to the multiphase system disappear and induction motors with closed coil armatures can be adopted. I find that even transmission through a submarine cable, as from Sweden to England, of great amounts of power is perfectly practicable. But the ideal solution of the problem of transportation will be arrived at only when the complete annihilation of distance in the transmission of power in large amounts shall have become a commercial reality. That day we shall invade the domain of the bird. When the vexing problem of aerial navigation, which has defied his attempts for ages, is solved, man will advance with giant strides.

That electrical energy can be economically transmitted without wires to any terrestrial distance, I have unmistakably established in numerous observations, experiments and measurements, qualitative and quantitative. These have demonstrated that is practicable to distribute power from a central plant in unlimited amounts, with a loss not exceeding a small fraction of 1% in the transmission, even to the greatest distance, twelve thousand miles — to the opposite end of the globe. This seemingly impossible feat can now be readily performed by any electrician familiar with the design and construction of my "high-potential magnifying transmitter," the most marvelous electrical apparatus of which I have knowledge, enabling the production of effects of unlimited intensities in the earth and its ambient atmosphere. It is, essentially, a freely vibrating secondary circuit of definite length, very high self-induction and small resistance, which has one of its terminals in intimate direct or inductive connection with the ground and the other with an elevated conductor, and upon which the electrical oscillations of a primary or exciting circuit are impressed under conditions of resonance. To give an idea of the capabilities of this wonderful appliance, I may state that I have obtained, by its means, spark discharges extending through more than one hundred feet and carrying currents of one thousand amperes, electromotive forces approximating twenty million volts, chemically active streamers covering areas of several thousand square feet, and electrical disturbances in the natural media surpassing those caused by lightning, in intensity.

Whatever the future may bring, the universal application of these great principles is fully assured, though it may be long in coming. With the opening of the first power plant, incredulity will give way to wonderment, and this to ingratitude, as ever before. The time is not distant when the energy of falling water will be man's life energy. So far only about million horse-power have been harnessed by my system of alternating-current transmission. This is little, but corresponds, nevertheless to the adding of sixty million indefatigable laborers, working virtually without food and pay, to the world's population. The projects which have come to my own attention, however, contemplate the exploitation of water powers aggregating something like one hundred and fifty million horse-power. Should they be carried out in a quarter of a century, as seems probable from present indications, there will be, on the average two such untiring laborers for every individual. Long before this consummation, coal and oil must cease to be important factors in the sustenance of human life on this planet. It should be borne in mind that electrical energy obtained by harnessing a waterfall is probably fifty times more effective than fuel energy. Since this is the most perfect way of rendering the sun's energy available, the direction of the future material development of man is clearly indicated. He will live on "white coal." Like a babe to the mothers's breast will he cling to his waterfall. "Give us our daily waterfall," will be the prayer of the coming generations. *Deus futurus est deus aquae deiectus!*

But the fact that stationary waves are producible in the earth is of special and, in many ways, still greater significance in the intellectual development of humanity. Popularly explained, such a wave is a phenomenon generically akin to an echo — a result of reflection. It affords a positive and uncontrovertible experimental evidence that the electric current, after passing into the earth travels to the diametrically opposite region of the same and rebounding from there, returns to its point of departure with virtually undiminished force. The outgoing and returning currents clash and form nodes and loops similar to those observable on a vibrating cord. To traverse the entire distance of about twenty-five thousand miles, equal to the circumference of the globe, the current requires a certain time interval, which I have approximately ascertained. In yielding this knowledge, nature has revealed one of its most precious secrets, of inestimable consequence to man. So astounding are the facts in this connection, that it would seem as though the Creator, himself, had electrically designed this planet just for the purpose of enabling us to achieve wonders which, before my discovery, could not have been conceived by the wildest imagination. A full account of my discoveries and improvements will be given to the world in a special work which I am preparing. In so far, however, as they relate to industrial and commercial uses, they will be disclosed in patent specifications most carefully drawn.

As stated in a recent article (*Electrical World and Engineer*, March 5, 1904), I have been since some time at work on designs of a power plant which is to transmit ten thousand horse-power without wires. The energy is to be collected all over the earth at many places and in varying amounts. It should not be understood that the practical realization of this undertaking is necessarily far off. The plans could be easily finished this winter, and if some preliminary work on the foundations could be done in the meantime the plant might be ready for operation before the close of next fall. We would then have at our disposal a unique and invaluable machine. Just this one oscillator would advance the world a century. Its civilizing influence would be felt even by the humblest dweller in the wilderness. Millions of instruments of all kinds, for all imaginable purposes, could be operated from that one machine. Universal time could be distributed by simple inexpensive clocks requiring no attention and running with nearly mathematical precision. Stock-tickers, synchronous movements and innumerable devices of this character could be worked in unison all over the earth. Instruments might be provided for indicating the course of a vessel at sea, the distance traversed, the speed, the hour at any particular place, the latitude and longitude. Incalculable commercial advantages could be thus secured and countless accidents and disasters avoided. Here and there a house might be lighted or some other work requiring a few horse-power performed. What is far more important than this, flying machines might be driven in any part of the world. They could be made to travel swiftly because of their small weight and great motive power. My intention would be to utilize this first plant rather as means of enlightenment, to collect its power in very small amounts, and at as many places as possible. The knowledge that there is throbbing through the earth energy readily available everywhere, would exert a strong stimulus on students, mechanics and inventors of all countries. This would be productive of infinite good. Manufacture would receive a fresh and powerful incentive. Conditions, such as never existed before in commerce, would be brought about. Supply would be ever inadequate to demand. The industries of iron, copper, aluminum, insulated wire and many others, could not fail to derive great and lasting benefits from this development.

The economic transmission of power without wires is of all-surpassing importance to man. By its means he will gain complete mastery of the air, the sea and the desert. It will enable him to dispense with the necessity of mining, pumping, transporting and burning fuel, and so do away with innumerable causes of sinful waste. By its means, he will obtain at any place and in any desired amount, the energy of remote waterfalls — to drive his machinery, to construct his canals, tunnels and highways, to manufacture the materials of his want, his clothing and food, to heat and light his home — year in, year out, ever and ever, by day and by night. It will make the living glorious sun his obedient, toiling slave. It will bring peace and harmony on earth.

Over five years have elapsed since that providential lightning storm on the 3d of July, 1899, of which I told in the article before mentioned, and through which I discovered the terrestrial stationary waves; nearly five years since I performed the great experiment which on that unforgettable day, the dark God of Thunder mercifully showed me in his vast, awe-sounding laboratory.

I thought then that it would take a year to establish commercially my wireless girdle around the world. Alas! my first "world telegraphy" plant is not yet completed, its construction has progressed but slowly during the past two years. And this machine I am building is but a plaything, an oscillator of a maximum activity of only ten million horse-power, just enough to throw this planet into feeble tremors, by sign and word—to telegraph and to telephone. When shall I see completed that first power plant, that big oscillator which I am designing! From which a current stronger than that of a welding machine, under a tension of one hundred million volts, is to rush through the earth! Which will deliver energy at the rate of one thousand million horse-power—one hundred Falls of Niagara combined in one, striking the universe with blows—blows that will wake from their slumber the sleepiest electricians, if there be any, on Venus or Mars! ... It is not a dream, it is a simple feat of scientific electrical engineering, only expensive—blind, faint-hearted, doubting world! ... Humanity is not yet sufficiently advanced to be willingly led by the discover's keen searching sense. But who knows? Perhaps it is better in this present world of ours that a revolutionary idea or invention instead of being helped and patted, be hampered and ill-treated in its adolescence—by want of means, by selfish interest, pedantry, stupidity and ignorance; that it be attacked and stifled; that it pass through bitter trials and tribulations, through the heartless strife of commercial existence. So do we get our light. So all that was great in the past was ridiculed, condemned, combated, suppressed—only to emerge all the more powerfully, all the more triumphantly from the struggle.

January 7, 1905

TUNED LIGHTNING

—English Mechanic and World of Science, March 8, 1907

I read with interest an article in the Sunday World of Jan. 20 on "Tuned Lightning," described as a mysterious new energy, which is to turn every wheel on earth, and is supposed to have been recently discovered by the Danish inventors Waldemar Poulsen and P. O. Pederson.

From other reports I have gathered that these gentlemen have so far confined themselves to the peaceful production of miniature bolts not many inches long, and I am wondering what an account of their prospective achievements would read like if they had succeeded in obtaining, like myself, electrical discharges of 100 ft., far surpassing lightning in some features of intensity and power.

In view of their limited Jovian experience, the programme outlined by the Danish engineers is rather extensive, Lord Armstrong's vast resources notwithstanding. Naturally enough, I shall look with interest to their telephoning across the Atlantic, supplying light and propelling airships without wires. *Anch in suito pittore* (I, too, am a painter.) In the mean time it may not be amiss to state here incidentally that all the essential processes of and appliances for the generation, transmission, transformation, distribution, storage, regulation, control, and economic utilisation of "tuned lightning" have been patented by me, and that I have long since undertaken, and am sparing no effort to render these advances instrumental in insuring the welfare, comfort, and convenience, primarily, of my fellow citizens.

There is nothing remarkable in the demonstration reported to have been made before Sir William Preece and Prof. Sylvanus P. Thompson, nor is there any novelty in the electrical devices employed. The lighting of arc lamps through the human body, the fusing of a piece of copper in mid-air, as described, are simple experiments which by the use of my high-frequency transformers any student of electricity can readily perform. They teach nothing new, and have no bearing on wireless transmission, for the actions virtually cease at a distance of a few feet from the source of vibratory energy. Years ago I gave exhibitions of similar and other much more striking experiments with the same kind of apparatus, many of which have been illustrated and explained in technical journals. The published records are open to inspection.

Regardless of all that, the Danish inventors have not as yet offered the slightest proof that their expectations are realisable, and before advancing seriously the claim that an efficient wireless distribution of light and power to great distances is possible, they should, at least, repeat those of my experiments which have furnished this evidence.

A scientific audience cannot help being impressed by a display of interesting phenomena, but the originality and significance of a demonstration such as that referred to can only be judged by an expert possessed of full knowledge and capable of drawing correct conclusions. A novel effect, spectacular and surprising, might be quite unimportant, while another, seemingly trifling, is of the greatest consequence.

To illustrate, let me mention here two widely different experiments of mine. In one the body of a person was subjected to the rapidly-alternating pressure of an electrical oscillator of two and a half million volts; in the other a small incandescent lamp was lighted by means of a resonant circuit grounded on one end, all the energy being drawn through the earth electrified from a distant transmitter.

The first presents a sight marvellous and unforgettable. One sees the experimenter standing on a big sheet of fierce, blinding flame, his whole body enveloped in a mass of phosphorescent wriggling streamers like the tentacles of an octopus. Bundles of light stick out from his spine. As he stretches out the arms, thus forcing the electric fluid outwardly, roaring tongues of fire leap from his fingertips. Objects in his vicinity bristle with rays, emit musical notes, glow, grow hot. He is the centre of still more curious actions, which are invisible. At each throb of the electric force myriads of minute projectiles are shot off from him with such velocities as to pass through the adjoining walls. He is in turn being violently bombarded by the surrounding air and dust. He experiences sensations which are indescribable.

A layman, after witnessing this stupendous and incredible spectacle, will think little of the second modest exhibit. But the expert will not be deceived. He realizes at once that the second experiment is ever so much more difficult to perform and immensely more consequential. He knows that to make the little filament glow, the entire surface of the planet, two hundred million square miles, must be strongly electrified. This calls for peculiar electrical activities, hundreds of times greater than those involved in the lighting of an arc lamp through the human body. What impresses him most, however, is the knowledge that the little lamp will spring into the same brilliancy anywhere on the globe, there being no appreciable diminution of the effect with the increase of distance from the transmitter.

This is a fact of overwhelming importance, pointing with certitude to the final and lasting solution of all the great social, industrial, financial, philanthropic, international, and other problems confronting humanity, a solution of which will be brought about by the complete annihilation of distance in the conveyance of intelligence, transport of bodies and materials, and the transmission of the energy necessary to man's existence. More light has been thrown on this scientific truth lately through Prof. Slaby's splendid and path-breaking experiment in establishing perfect wireless telephone connection between Naum and Berlin, Germany, a distance of twenty miles. With apparatus properly organised such telephonic communication can be effected with the same facility and precision at the greatest terrestrial distance.

The discovery of the stationary terrestrial waves, showing that, despite its vast extent, the entire planet can be thrown into resonant vibration like a little tuning fork; that electrical oscillations suited to its physical properties and dimensions pass through it unimpeded, in strict obedience to a simple mathematical law, has proved beyond the shadow of a doubt that the earth, considered as a channel for conveying electrical energy, even in such delicate and complex transmissions as human speech or musical composition, is infinitely superior to a wire or cable, however well designed.

Very soon it will be possible to talk across an ocean as clearly and distinctly as across a table. The first practical success, already forecast by Slaby's convincing demonstration, will be the signal for revolutionary improvements which will take the world by storm.

However great the success of the telephone, it is just beginning its evidence of usefulness. Wireless transmission of speech will not only provide new but also enormously extend existing facilities. This will be merely the forerunner of ever so much more important development, which will proceed at a furious pace until, by the application of these same great principles, the power of waterfalls can be focused whenever desired; until the air is conquered, the soil fructified and embellished; until, in all departments of human life distance has lost its meaning, and even the immense gulf separating us from other worlds is bridged.

March 8, 1907

TESLA'S WIRELESS TORPEDO

—*New York Times, March 19, 1907*

Inventor Says He did Show that it Worked Perfectly

To the Editor of The New York Times:

"A report in the Times of this morning says that I have attained no practical results with my dirigible wireless torpedo. I have constructed such machines, and shown them in operation on frequent occasions. They have worked perfectly and everybody who saw them was amazed at their performance.

"It is true that my efforts to have this novel means for attack and defense adopted by our Government have been unsuccessful, but this is no discredit to my invention. I have spent years in fruitless endeavor before the world recognized the value of my rotating field discoveries which are now universally applied. The time is not yet ripe for the telautomatic art. If its possibilities were appreciated the nations would not be building large battleships. Such a floating fortress may be safe against an ordinary torpedo, but would be helpless in a battle with a machine which carries twenty tons of explosive, moves swiftly underwater, and is controlled with precision by an operator beyond the range of the largest gun.

"As to projecting wave-energy to any particular region of the globe, I have given a clear description of the means in technical publications. Not only can this be done by the means of my devices, but the spot at which the desired effect is to be produced can be calculated very closely, assuming the accepted terrestrial measurements to be correct. This, of course, is not the case. Up to this day we do not know a diameter of the globe within one thousand feet. My wireless plant will enable me to determine it within fifty feet or less, when it will be possible to rectify many geodetical data and make such calculations as those referred to with greater accuracy."

Nikola Tesla.
New York, March 19, 1907

POSSIBILITIES OF WIRELESS

—*New York Times, October 22, 1907, p. 8, col. 6.*

Nikola Tesla says distance forms no obstacle to transmission of energy

To the Editor of The New York Times:

"In your issue of the 19th inst. Edison makes statements which cannot fail to create erroneous impressions.

"There is a vast difference between primitive Hertzwave signalling, practicable to but a few miles, and the great art of wireless transmission of energy, which enables an expert to transmit, to any distance, not only signals, but power in unlimited amounts, and of which the experiments across the Atlantic are a crude application. The plants are quite inefficient, unsuitable for finer work, and totally doomed to an effect less than 1% of that I attained in my test in 1899.

"Edison thinks that Sir Hiram Maxim is blowing hot air. The fact is my Long Island plant will transmit almost its entire energy to the antipodes, if desired. As to [Martian communications] I can only say, that I shall be able to attain a wave activity of 800,000,000 horse power and a simple calculation will show, that the inhabitants of that planet, if there be any, need not have a Lord Raleigh to detect the disturbance.

"Referring to your editorial comment of even date, the question of wireless interference is puzzling only because of its novelty. The underlying principle is old, and it has presented itself for consideration in numerous forms. Just now it appears in the novel aspects of aerial navigation and wireless transmission. Every human effort must of necessity create a disturbance. What difference is there in essence, between the commotion produced by any revolutionary idea or improvement and that of a wireless transmitter? The spectre of interference has been conjured by Hertzwave or radio telegraphy in which attunement is absolutely impossible, simply because the effect diminishes rapidly with distance. But to my system of energy transmission, based on the use of impulses not sensibly diminishing with distance, perfect attunement and the higher artifice of individualization are practicable. As ever, the ghost will vanish with the wireless dawn."

Nikola Tesla. New York, Oct. 21, 1907

THE FUTURE OF THE WIRELESS ART

Wireless telegraphy & telephony

By Walter W. Massie & Charles R. Underhill, 1908, pp. 67-71.

Mr. Nikola Tesla, in a recent interview by the authors, as to the future of the Wireless Art, volunteered the following statement which is herewith produced in his own words.

"A mass in movement resists change of direction. So does the world oppose a new idea. It takes time to make up the minds to its value and importance. Ignorance, prejudice and inertia of the old retard its early progress. It is discredited by insincere exponents and selfish exploiters. It is attacked and condemned by its enemies. Eventually, though, all barriers are thrown down, and it spreads like fire. This will also prove true of the wireless art.

"The practical applications of this revolutionary principle have only begun. So far they have been confined to the use of oscillations which are quickly damped out in their passage through the medium. Still, even this has commanded universal attention. What will be achieved by waves which do not diminish with distance, baffles comprehension.

"It is difficult for a layman to grasp how an electric current can be propagated to distances of thousands of miles without diminution of intention. But it is simple after all. Distance is only a relative conception, a reflection in the mind of physical limitation. A view of electrical phenomena must be free of this delusive impression. However surprising, it is a fact that a sphere of the size of a little marble offers a greater impediment to the passage of a current than the whole earth. Every experiment, then, which can be performed with such a small sphere can likewise be carried out, and much more perfectly, with the immense globe on which we live. This is not merely a theory, but a truth established in numerous and carefully conducted experiments. When the earth is struck mechanically, as is the case in some powerful terrestrial upheaval, it vibrates like a bell, its period being measured in hours. When it is struck electrically, the charge oscillates, approximately, twelve times a second. By impressing upon it current waves of certain lengths, definitely related to its diameter, the globe is thrown into resonant vibration like a wire, stationary waves forming, the nodal and ventral regions of which can be located with mathematical precision. Owing to this fact and the spheroidal shape of the earth, numerous geodetical and other data, very accurate and of the greatest scientific and practical value, can be readily secured. Through the observation of these astonishing phenomena we shall soon be able to determine the exact diameter of the planet, its configuration and volume, the extent of its elevations and depressions, and to measure, with great precision and with nothing more than an electrical device, all terrestrial distances. In the densest fog or darkness of night, without a compass or other instruments of orientation, or a timepiece, it will be possible to guide a vessel along the shortest or orthodromic path, to instantly read the latitude and longitude, the hour, the distance from any point, and the true speed and direction of movement. By proper use of such disturbances a wave may be made to travel over the earth's surface with any velocity desired, and an electrical effect produced at any spot which can be selected at will and the geographical position of which can be closely ascertained from simple rules of trigonometry.

"This mode of conveying electrical energy to a distance is not 'wireless' in the popular sense, but a transmission through a conductor, and one which is incomparably more perfect than any artificial one. All impediments of conduction arise from confinement of the electric and magnetic fluxes to narrow channels. The globe is free of such cramping and hinderment. It is an ideal conductor because of its immensity, isolation in space, and geometrical form. Its singleness is only an apparent limitation, for by impressing upon it numerous non-interfering vibrations, the flow of energy may be directed through any number of paths which, though bodily connected, are yet perfectly distinct and separate like ever so many cables. Any apparatus, then, which can be operated through one or more wires, at distances obviously limited, can likewise be worked without artificial conductors, and with the same facility and precision, at distances without limit other than that imposed by the physical dimensions of the globe.

"It is intended to give practical demonstrations of these principles with the plant illustrated. As soon as completed, it will be possible for a business man in New York to dictate instructions, and have them instantly appear in type at his office in London or elsewhere. He will be able to call up, from his desk, and talk to any telephone subscriber on the globe, without any change whatever in the existing equipment. An inexpensive instrument, not bigger than a watch, will enable its bearer to hear anywhere, on sea or land, music or song, the speech of a political leader, the address of an eminent man of science, or the sermon of an eloquent clergyman, delivered in some other place, however distant. In the same manner any picture, character, drawing, or print can be transferred from one to another place. Millions of such instruments can be operated from but one plant of this kind. More important than all of this, however, will be the transmission of power, without wires, which will be shown on a scale large enough to carry conviction. These few indications will be sufficient to show that the wireless art offers greater possibilities than any invention or discovery heretofore made, and if the conditions are favorable, we can expect with certitude that in the next few years wonders will be wrought by its application."

1908

MR. TESLA'S VISION

—New York Times, April 21, 1908

How the electrician's lamp of Aladdin may construct New Worlds

To the Editor of the New York Times:

"From a report in your issue of March 11, which escaped my attention, I notice that some remarks I made on the occasion referred to have been misunderstood. Allow me to make a correction.

"When I spoke of future warfare I meant that it should be conducted by direct application of electrical waves without the use of aerial engines or other implements of destruction. This means, as I pointed out, would be ideal, for not only would the energy of war require no effort for the maintenance of its potentiality, but it would be productive in times of peace. This is not a dream. Even now wireless power plants could be constructed by which any region of the globe might be rendered uninhabitable without subjecting the population of other parts to serious danger or inconvenience.

"What I said in regard to the greatest achievement of the man of science whose mind is bent upon the mastery of the physical universe, was nothing more than what I stated in one of my unpublished addresses, from which I quote: "According to an adopted theory, every ponderable atom is differentiated from a tenuous fluid, filling all space merely by spinning motion, as a whirl of water in a calm lake. By being set in movement this fluid, the ether, becomes gross matter. Its movement arrested, the primary substance reverts to its normal state. It appears, then, possible for man through harnessed energy of the medium and suitable agencies for starting and stopping ether whirls to cause matter to form and disappear. At his command, almost without effort on his part, old worlds would vanish and new ones would spring into being. He could alter the size of this planet, control its seasons, adjust its distance from the sun, guide it on its eternal journey along any path he might choose, through the depths of the universe. He could make planets collide and produce his suns and stars, his heat and light; he could originate life in all its infinite forms. To cause at will the birth and death of matter would be man's grandest deed, which would give him the mastery of physical creation, make him fulfill his ultimate destiny."

"Nothing could be further from my thought than to call wireless telephony around the world "the greatest achievement of humanity" as reported. This is a feat which, however stupefying, can be readily performed by any expert. I have myself constructed a plant for this very purpose. The wireless wonders are only seeming, not results of exceptional skill, as popularly believed. The truth is the electrician has been put in possession of a veritable lamp of Aladdin. All he has to do is to rub it. Now, to rub the lamp of Aladdin is no achievement.

"If you are desirous of hastening the accomplishment of still greater and further reaching wonders you can do no better than by emphatically opposing any measure tending to interfere with the free commercial exploitation of water power and the wireless art. So absolutely does human progress depend on the development of these that the smallest impediment, particularly through the legislative bodies of this country, may set back civilization and the cause of peace for centuries."

Nikola Tesla
New York, April 19, 1908

NIKOLA TESLA'S NEW WIRELESS

—The Electrical Engineer—London, Dec. 24, 1909

Mr. Nikola Tesla has announced that as the result of experiments conducted at Shoreham, Long Island, he has perfected a new system of wireless telegraphy and telephony in which the principles of transmission are the direct opposite of Hertzian wave transmission. In the latter, he says, the transmission is effected by rays akin to light, which pass through the air and cannot be transmitted through the ground, while in the former the Hertz waves are practically suppressed and the entire energy of the current is transmitted through the ground exactly as though a big wire. Mr. Tesla adds that in his experiments in Colorado it was shown that a very powerful current developed by the transmitter traversed the entire globe and returned to its origin in an interval of 84 one-thousandths of a second, this journey of 24,000 miles being effected almost without loss of energy.

Dec. 24, 1909

DR. TESLA TALKS OF GAS TURBINES

—*Motor World, September 18, 1911*

*So Confident, He Offers to Build Them for Motor Cars
Considers New Power His Greatest Invention.*

Gas turbines of practical and efficient construction, light, flexible and in every way suitable for automobile propulsion, are not a dream of the future only but a probability of only a very few years hence. At least such is the conviction of Dr. Nikola Tesla, whose newly developed method of fluid propulsion, as he calls it, and which was illustrated and described in last week's Motor World, is attracting so much attention in scientific circles. Dr. Tesla himself considers it the greatest of all his inventions. By his own statement the scientist already has built, run and carefully tested internal combustion engines operating on the new turbine principle and so confident is he of the thorough practicability of the idea that on Friday of last week he informed a Motor World man that he would even be willing today to sign a contract to build and install turbines for automobiles. He readily admits, however, that he would like to have more time, considerable more time probably, in which to develop a method of combustion entirely suited to the turbine.

Automobile motors, as a matter of fact, play a distinct part in the inventor's plans for the future. So do airship motors, pumps of various sorts, steam engines in every conceivable size, shape and capacity, and apparatus of other and varied uses. If in steam engines and pumps wonderful results already have been obtained, it is his expectation later to accomplish equally wonderful results with internal combustion engines.

Bearing in mind that a 110 horsepower steam engine already has been built so diminutive that its rotor or active part would drop into an ordinary water bucket, it would seem that the part of the program which might be supposed to concern the automobile industry would be well worth investigating. It was the tempting prospect of a pocket-size motor, therefore, which led a representative of the Motor World to seek the famous scientist in his offices high up in the Metropolitan Tower in New York.

Contrary to popular impression, not all great and famous men are inaccessible, and Tesla proved not only thoroughly approachable but extremely ready to discuss the new turbine principle in many of its bearings. His easy predictions of the future developments of the system and his confident bearing when he declared it to be the greatest of his accomplishments, might have been merely the vaporings of an over enthusiastic inventor; but the broad bearing of the man, his record and the depth of perception revealed in some of his conclusions would have laid at rest such doubts. The new principle unquestionably is a great contribution to science and engineering, great in its simplicity and breadth of application. Just when its fullest realization will be given to the markets of the world is another question; one that Tesla himself cannot answer, though he explained that he is "under great pressure from all sides" to complete the development of certain kinds of apparatus, steam engines and pumps having received a great deal of attention up to the present, and that he expects to have some of them ready for production "before very long."

Tall, erect, almost angular, with the broad brow of the philosopher and the sharply chiseled features of the habitual student, Dr. Tesla bears few of the earmarks of the traditional genius. He wears his iron-gray hair a little low in the back, to be sure, but not for an inventor, and when he walks there is just a bare suggestion of histrionic attainments utterly at variance with the hurried preoccupation of the conventional type of man whose brains are stored in the archives of the Patent Office.

"We understand that you are doing remarkable things with steam, but how far has your confidence been extended to the gas turbine?" the scientist was asked.

He laughed.

"Why, I am working with them all the time," he answered.

"You mean to say that you already have built and operated internal combustion turbines employing your principle of fluid propulsion?"

"Yes. But I am not satisfied—not yet. You see there are many things to be considered. The turbine, that is one thing; it is complete in itself and there is no question of its applicability. But when you come to the combustion of the gas you have a new difficulty. I am not satisfied with the present methods of gasification. I have tried one of my turbines discharging the gas into a chamber and then spraying water into it. You see in that way you get an intermittent flow through the nozzle, but you also have better thermal action because you get your adiabatic expansion, [meaning that in which theoretically there is no loss of heat]. And then, I have tried with gasoline using a constant jet, in which you get less efficient thermal action but better action for the turbine. But I am not yet satisfied. I think that some day we shall get better processes of combustion that will enable me to work more advantage with my turbine.

"You see, that is one great trouble," continued Dr. Tesla. "The human mind thinks but to complicate. As soon as one problem is solved, that solution introduces new complications, other problems that perhaps did not exist before. That was one of my great troubles when I was younger, I invented many things that were very fine, but always I was getting into complications. I have had to work very hard to overcome that. But here you see what I have done. Do you see how very simple it is? You take, for instance, the ordinary turbine, a bucket turbine. Here you have around the outside of the wheel a row of little jets, and within, on the periphery of a wheel, a row of buckets — many of them and very small, even on a large wheel. But don't you see that in that entire wheel you have only a narrow strip, a ring perhaps three or four inches wide, that is really useful — that is really active?

"In my invention practically the whole surface is active. In the bucket turbine the action does not even extend all the way around; you must have a series of jets. But in my turbine you have the gas traveling all the way around in free spirals — always seeking the path of least resistance — and expending its full energy."

Here he laid aside the pencil with which he had been illustrating the point, and reverted to the beginning of what he evidently considers his "big idea."

"I have been working at this a long time. Many years ago I invented a pump for pumping mercury. Just a plain disk, like this, and it would work very well. ' All right,' I said, 'that is friction.' But one day I thought it out, and I thought, 'No, that is not friction, it is something else. The particles are not always sliding by the disk, but some of them at least are carried along with it. Therefore it cannot be friction. It must be adhesion. ' And that, you see, was the real beginning.

"For if you can imagine a wheel rotating in a medium, whether the fluid is receiving or imparting energy, and moving at nearly the same velocity as the fluid, then you have a minimum of friction, you get little or no 'slip.' Then you are getting something very different from friction; you are making use of adhesion alone. It's all so simple, so very simple.

"This is the greatest of my inventions," Tesla went on with great enthusiasm. "Now take my 'rotating field'— do you know my rotating field — are you familiar at all with electricity? There are millions invested in it already. Well, that is a very useful thing, but the field is limited to dynamos and motors. But here you have a new power for pumps, steam engines, gasoline motors, for automobiles, for airships, for many other uses, and all so simple."

"But is it really true that you have produced 110 horsepower from a wheel only 9 3/4 inches in diameter and two inches wide, as has been reported?" asked the interviewer incredulously.

"Oh, yes!" was the reply. "And more. We could get more power. We had

125 pounds steam pressure and no vacuum. We ran it that way for hours."

"Was it sustained power?"

"Yes, sustained power. And we could only use part of the drop in pressure; we would have twisted off the shaft, it was so light, if we had been able to use all the energy of the steam. I had to put in a smaller nozzle on that account."

"And they are very light, these steam engines?"

"I can build a steam engine that will develop one horsepower for every one tenth of a pound of weight," was the instant and amazing response. "I am now building a double turbine, one with two wheels which must revolve in opposite directions. It is for a special purpose, and I cannot talk much about that, but each wheel develops 200 horsepower, that is 400 horsepower, and it weighs 88 pounds."

A no less amazing claim made at another time was that the steam turbine could be made to return in power at the shaft no less than 97% of the energy of the steam. There seems to be no limit to what the inventor thinks the new system will accomplish, though, of course, a waiting world may be pardoned for withholding a full verdict of confidence until it has had opportunity to witness some of the promised marvels.

As far as demonstration of the basic principle is concerned, however, the success of the idea is unquestionable. A small pump, originally put together for purposes of exhibition before a body of scientists, to whom Dr. Tesla first disclosed his invention, was operated for the benefit of the Motor World man; the inventor himself obligingly switching on and off the current from the little electric motor which drove it, and operating the valve by means of which the discharge could be regulated to increase the flow and decrease the pressure, or vice versa.

The rotor, mounted in a casing of volute form hardly more than six inches in diameter, contained five disks of three inches diameter. From a small tank, which was part of the model, water was drawn into the casing and forced through a pipe with a lift of 18 inches or so to a long strainer in a horizontal pipe, whence, after passing a baffle plate, to break up the flow and prevent splashing, it fell back into the tank over a miniature weir in a beautiful clear sheet. The hand of a pressure gauge indicated four pounds when the valve was closed, but fell to a little under two pounds when the full discharge was permitted.

With the valve closed, the action of the disks was shown to good advantage. Rapidly snapping on and off the switch, the inventor gleefully pointed out how the hand of the gauge jumped up and fell back again so closely in response to the speed of the motor, as judged by the hum of its commutator, that eye and ear failed to detect the difference. "And so you really believe that a practical form of gas turbine can be developed on this principle and in such shape that it could be profitably adapted to automobile use?" asked the Motor World man.

"I am so sure that I would make a contract today to build gas turbines and equip automobiles with them."

September 18, 1911

TESLA'S NEW MONARCH OF MACHINES

—New York Herald, Oct. 15, 1911

Suppose some one should discover a new mechanical principle—something as fundamental as James Watt's discovery of the expansive power of steam—by the use of which it became possible to build a motor that would give ten horse power for every pound of the engine's weight, a motor so simple that the veriest novice in mechanics could construct it and so elemental that it could not possibly get out of repair. Then suppose that this motor could be run forward or backward at will, that it could be used as either an engine or a pump, that it cost almost nothing to build as compared with any other known form of engine, that it utilized a larger percentage of the available power than any existing machine, and, finally, that it would operate with gas, steam, compressed air or water, any one of them, as its driving power.

It does not take a mechanical expert to imagine the limitless possibilities of such an engine. It takes very little effort to conjure up a picture of a new world of industry and transportation made possible by the invention of such a device. Revolutionary—seems a mild term to apply to it. That, however, is the word the inventor uses in describing it—Nikola Tesla, the scientist whose electrical discoveries underlie all modern electrical power development, whose experiments and deductions made the wireless telegraph possible, and who now, in the mechanical field, has achieved a triumph even more far reaching than anything he accomplished in electricity.

There is something of the romantic in this discovery of the famous explorer of the hidden realms of knowledge. The pursuit of an ideal is always romantic, and it was in the pursuit of an ideal which he has been seeking twenty years that Dr. Tesla made his great discovery. That ideal is the power to fly—to fly with certainty and absolute safety—not merely to go up in an aeroplane and take chances on weather conditions,—holes in the air,—tornadoes, lightning and the thousand other perils the aviator of today faces, but to fly with the speed and certainty of a cannon ball, with power to overcome any of nature's aerial forces, to start when one pleases, go whither one pleases and alight where one pleases. That has been the aim of Dr. Tesla's life for nearly a quarter of a century. He believes that with the discovery of the principle of his new motor he has solved this problem and that incidentally he has laid the foundations for the most startling new achievements in other mechanical lines.

There was a time when men of science were skeptical—a time when they ridiculed the announcement of revolutionary discoveries. Those were the days when Nikola Tesla, the young scientist from the Balkans, was laughed at when he urged his theories on the engineering world. Times have changed since then, and the—practical—engineer is not so incredulous about—scientific—discoveries. The change came about when young Tesla showed the way by which the power of Niagara Falls could be utilized. The right to divert a portion of the waters of Niagara had been granted; then arose the question of how best to utilize the tremendous power thus made available—how to transmit it to the points where it could be commercially utilized. An international commission sat in London and listened to theories and practical plans for months.

Up to that time the only means of utilizing electric power was the direct current motor, and direct current dynamos big enough to be of practical utility for such a gigantic power development were not feasible.

Then came the announcement of young Tesla's discovery of the principle of the alternating current motor. Practical tests showed that it could be built—that it would work.

That discovery, at that opportune time, decided the commission. Electricity was determined upon as the means for the transmission of Niagara's power to industry and commerce. Today a million horse power is developed on the brink of the great cataract, turning the wheels of Buffalo, Rochester, Syracuse and the intervening cities and villages operating close at hand the great new electro-chemical industries that the existence of this immense source of power has made possible, while all around the world a thousand waterfalls are working in the service of mankind, sending the power of their—white coal—into remote and almost inaccessible corners of the globe, all because of Nikola Tesla's first great epoch making discovery.

Today the engineering world listens respectfully when Dr. Tesla speaks. The first announcement of the discovery of his new mechanical principle was made in a technical periodical in mid-September, 1911. Immediately it became the principal topic of discussions wherever engineers met.

"It is the greatest invention in a century," wrote one of the foremost American engineers, a man whose name stands close to the top of the list of those who have achieved scientific fame and greatness.

"No invention of such importance in the automobile trade has yet been made," declared the editor of one of the leading engineering publications. Experts in other engineering lines pointed out other applications of the new principle and letters asking for further information poured in on Dr. Tesla from the four quarters of the globe.

"Oh, I've had too much publicity," he said, when I telephoned to him to ask for an interview in order to explain his new discovery to the non-technical public. It took a good deal of persuasion before he reluctantly fixed an hour when he would see me, and a good bit more after that before he talked at all freely. When he did speak, however, he opened up vistas of possible applications of the new engine that staggered the imagination of the interviewer.

Looking out over the city from the windows of his office, on the twentieth floor of the Metropolitan Tower, his face lit up as he told of his life dream and its approaching realization, and the listener's fancy could almost see the air full of strange flying craft, while huge steamships propelled at unheard of speeds ploughed the waters of the North River, automobiles climbed the very face of the Palisades, locomotives of incredible power whisked wheeled palaces many miles a minute and all the discomforts of summer heat vanished as marvelous refrigerating plants reduced the temperature of the whole city to a comfortable maximum—for these were only a few of the suggestions of the limitless possibilities of the latest Tesla discovery.

"Just what is your new invention?" I asked.

"I have accomplished what mechanical engineers have been dreaming about ever since the invention of steam power," replied Dr. Tesla. "That is the perfect rotary engine. It happens that I have also produced an engine which will give at least twenty-five times as much power to a pound of weight as the lightest weight engine of any kind that has yet been produced.

"In doing this I have made use of two properties which have always been known to be possessed by all fluids, but which have not heretofore been utilized. These properties are adhesion and viscosity.

"Put a drop of water on a metal plate. The drop will roll off, but a certain amount of the water will remain on the plate until it evaporates or is removed by some absorptive means. The metal does not absorb any of the water, but the water adheres to it.

"The drop of water may change its shape, but until its particles are separated by some external power it remains intact. This tendency of all fluids to resist molecular separation is viscosity. It is especially noticeable in the heavier oils.

"It is these properties of adhesion and viscosity that cause the skin friction that impedes a ship in its progress through the water or an aeroplane in going through the air. All fluids have these qualities—and you must keep in mind that air is a fluid, all gases are fluids, steam is fluid. Every known means of transmitting or developing mechanical power is through a fluid medium.

"Now, suppose we make this metal plate that I have spoken of circular in shape and mount it at its centre on a shaft so that it can be revolved. Apply power to rotate the shaft and what happens? Why, whatever fluid the disk happens to be revolving in is agitated and dragged along in the direction of rotation, because the fluid tends to adhere to the disk and the viscosity causes the motion given to the adhering particles of the fluid to be transmitted to the whole mass. Here, I can show you better than tell you"

Dr. Tesla led the way into an adjoining room. On a desk was a small electric motor and mounted on the shaft were half a dozen flat disks, separated by perhaps a sixteenth of an inch from one another, each disk being less than that in thickness. He turned a switch and the motor began to buzz. A wave of cool air was immediately felt.

"There we have a disk, or rather a series of disks, revolving in a fluid—the air," said the inventor. "You need no proof to tell you that the air is being agitated and propelled violently. If you will hold your hand over the centre of these disks—you see the centres have been cut away—you will feel the suction as air is drawn in to be expelled from the peripheries of the disks.

"Now, suppose these revolving disks were enclosed in an air tight case, so constructed that the air could enter only at one point and be expelled only at another—what would we have?"

"You'd have an air pump," I suggested.

"Exactly—an air pump or blower," said Dr. Tesla.

"There is one now in operation delivering ten thousand cubic feet of air a minute.—Now, come over here."

He stepped across the hall and into another room, where three or four draughtsmen were at work and various mechanical and electrical contrivances were scattered about. At one side of the room was what appeared to be a zinc or aluminum tank, divided into two sections, one above the other, while a pipe that ran along the wall above the upper division of the tank was connected with a little aluminum case about the size and shape of a small alarm clock. A tiny electric motor was attached to a shaft that protruded from one side of the aluminum case. The lower division of the tank was filled with water.

"Inside of this aluminum case are several disks mounted on a shaft and immersed in a fluid, water," said Dr. Tesla. "From this lower tank the water has free access to the case enclosing the disks. This pipe leads from the periphery of the case. I turn the current on, the motor turns the disks and as I open this valve in the pipe the water flows."

He turned the valve and the water certainly did flow. Instantly a stream that would have filled a barrel in a very few minutes began to run out of the pipe into the upper part of the tank and thence into the lower tank.

"This is only a toy," said Dr. Tesla. "There are only half a dozen disks — runners, — I call them — each less than three inches in diameter, inside of that case. They are just like the disks you saw on the first motor — no vanes, blades or attachments of any kind. Just perfectly smooth, flat disks revolving in their own planes and pumping water because of the viscosity and adhesion of the fluid. One such pump now in operation, with eight disks, eighteen inches in diameter, pumps four thousand gallons a minute to a height of 360 feet."

We went back into the big, well lighted office. I was beginning to grasp the new Tesla principle.

"Suppose now we reversed the operation," continued the inventor. "You have seen the disks acting as a pump. Suppose we had water, or air under pressure, or steam under pressure, or gas under pressure, and let it run into the case in which the disks are contained — what would happen?"

"The disks would revolve and any machinery attached to the shaft would be operated — you would convert the pump into an engine," I suggested.

"That is exactly what would happen — what does happen," replied Dr. Tesla. "It is an engine that does all that engineers have ever dreamed of an engine doing, and more. Down at the Waterside power station of the New York Edison Company, through their courtesy, I have had a number of such engines in operation. In one of them the disks are only nine inches in diameter and the whole working part is two inches thick. With steam as the propulsive fluid it develops 110-horse power, and could do twice as much."

"You have got what Professor Langley was trying to evolve for his flying machine — an engine that will give a horse power for a pound of weight," I suggested.

Ten Horse Power to the Pound.

"I have got more than that," replied Dr. Tesla. "I have an engine that will give ten horse power to the pound of weight. That is twenty-five times as powerful as the lightest weight engine in use today. The lightest gas engine used on aeroplanes weighs two and one-half pounds to the horse power. With two and one-half pounds of weight I can develop twenty-five horse power."

"That means the solution of the problem of flying," I suggested.

"Yes, and many more," was the reply. "The applications of this principle, both for imparting power to fluids, as in pumps, and for deriving power from fluids, as in turbine, are boundless. It costs almost nothing to make, there is nothing about it to get out of order, it is reversible — simply have two ports for the gas or steam, to enter by, one on each side, and let it into one side or other. There are no blades or vanes to get out of order — the steam turbine is a delicate thing."

I remembered the bushels of broken blades that were gathered out of the turbine casings of the first turbine equipped steamship to cross the ocean, and realized the importance of this phase of the new engine.

"Then, too," Dr. Tesla went on, "there are no delicate adjustments to be made. The distance between the disks is not a matter of microscopic accuracy and there is no necessity for minute clearances between the disks and the case. All one needs is some disks mounted on a shaft, spaced a little distance apart and cased so that a fluid can enter at one point and go out at another. If the fluid enters at the centre and goes out at the periphery it is a pump. If it enters at the periphery and goes out at the center it is a motor.

"Coupling these engines in series, one can do away with gearing in machinery. Factories can be equipped without shafting. The motor is especially adapted to automobiles, for it will run on gas explosions as well as on steam. The gas or steam can be let into a dozen ports all around the rim of the case if desired. It is possible to run it as a gas engine with a continuous flow of gas, gasoline and air being mixed and the continuous combustion causing expansion and pressure to operate the motor. The expansive power of steam, as well as its propulsive power, can be utilized as in a turbine or a reciprocating engine. By permitting the propelling fluid to move along the lines of least resistance a considerably larger proportion of the available power is utilized.

"As an air compressor it is highly efficient. There is a large engine of this type now in practical operation as an air compressor and giving remarkable service. Refrigeration on a scale hitherto never attempted will be practical, through the use of this engine in compressing air, and the manufacture of liquid air commercially is now entirely feasible.

"With a thousand horse power engine, weighing only one hundred pounds, imagine the possibilities in automobiles, locomotives and steamships. In the space now occupied by the engines of the Lusitania twenty-five times her 80,000 horse power could be developed, were it possible to provide boiler capacity sufficient to furnish the necessary steam."

"And it makes the aeroplane practical," I suggested.

"Not the aeroplane, the flying machine," responded Dr. Tesla. "Now you have struck the point in which I am most deeply interested — the object toward which I have been devoting my energies for more than twenty years — the dream of my life. It was in seeking the means of making the perfect flying machine that I developed this engine.

"Twenty years ago I believed that I would be the first man to fly; that I was on the track of accomplishing what no one else was anywhere near reaching. I was working entirely in electricity then and did not realize that the gasoline engine was approaching a perfection that was going to make the aeroplane feasible. There is nothing new about the aeroplane but its engine, you know.

"What I was working on twenty years ago was the wireless transmission of electric power. My idea was a flying machine propelled by an electric motor, with power supplied from stations on the earth. I have not accomplished this as yet, but am confident that I will in time.

"When I found that I had been anticipated as to the flying machine, by men working in a different field, I began to study the problem from other angles, to regard it as a mechanical rather than an electrical problem. I felt certain there must be some means of obtaining power that was better than any now in use. And by vigorous use of my gray matter for a number of years, I grasped the possibilities of the principle of the viscosity and adhesion of fluids and conceived the mechanism of my engine. Now that I have it, my next step will be the perfect flying machine."

"An aeroplane driven by your engine?" I asked.

"Not at all," said Dr. Tesla. "The aeroplane is fatally defective. It is merely a toy—a sporting play-thing. It can never become commercially practical. It has fatal defects. One is the fact that when it encounters a downward current of air it is helpless. The—hole in the air—of which aviators speak is simply a downward current, and unless the aeroplane is high enough above the earth to move laterally but can do nothing but fall.

"There is no way of detecting these downward currents, no way of avoiding them, and therefore the aeroplane must always be subject to chance and its operator to the risk of fatal accident. Sportsmen will always take these chances, but as a business proposition the risk is too great.

"The flying machine of the future—my flying machine—will be heavier than air, but it will not be an aeroplane. It will have no wings. It will be substantial, solid, stable. You cannot have a stable airplane. The gyroscope can never be successfully applied to the airplane, for it would give a stability that would result in the machine being torn to pieces by the wind, just as the unprotected aeroplane on the ground is torn to pieces by a high wind.

"My flying machine will have neither wings nor propellers. You might see it on the ground and you would never guess that it was a flying machine. Yet it will be able to move at will through the air in any direction with perfect safety, higher speeds than have yet been reached, regardless of weather and oblivious of—holes in the air—or downward currents. It will ascend in such currents if desired. It can remain absolutely stationary in the air, even in a wind, for great length of time. Its lifting power will not depend upon any such delicate devices as the bird has to employ, but upon positive mechanical action."

"You will get stability through gyroscopes?" I asked.

"Through gyroscopic action of my engine, assisted by some devices I am not yet prepared to talk about," he replied.

"Powerful air currents that may be deflected at will, if produced by engines and compressors sufficiently light and powerful, might lift a heavy body off the ground and propel it through the air," I ventured, wondering if I had grasped the inventor's secret.

Dr. Tesla smiled an inscrutable smile.

"All I have to say on that point is that my airship will have neither gas bag, wings nor propellers," he said. "It is the child of my dreams, the product of years of intense and painful toil and research. I am not going to talk about it any further. But whatever my airship may be, here at least is an engine that will do things that no other engine ever has done, and that is something tangible."

Oct. 15, 1911

THE DISTURBING INFLUENCE OF SOLAR RADIATION ON THE WIRELESS TRANSMISSION OF ENERGY

—*Electrical Review and Western Electrician, July 6, 1912*

When Heinrich Hertz announced the results of his famous experiments in confirmation of the Maxwellian electromagnetic theory of light, the scientific mind at once leaped to the conclusion that the newly discovered dark rays might be used as a means for transmitting intelligible messages through space. It was an obvious inference, for heliography, or signaling by beams of light, was a well recognized wireless art. There was no departure in principle, but the actual demonstration of a cherished scientific idea surrounded the novel suggestion with a nimbus of originality and atmosphere of potent achievement. I also caught the fire of enthusiasm but was not long deceived in regard to the practical possibilities of this method of conveying intelligence.

Granted even that all difficulties were successfully overcome, the field of application was manifestly circumscribed. Heliographic signals had been flashed to a distance of 200 miles, but to produce Hertzian rays of such penetrating power as those of light appeared next to impossible, the frequencies obtainable through electrical discharges being necessarily of a much lower order. The rectilinear propagation would limit the action on the receiver to the extent of the horizon and entail interference of obstacles in a straight line joining the stations. The transmission would be subject to the caprices of the air and, chief of all drawbacks, the intensity of disturbances of this character would rapidly diminish with distance.

But a few tests with apparatus, far ahead of the art of that time, satisfied me that the solution lay in a different direction, and after a careful study of the problem I evolved a new plan which was fully described in my addresses before the Franklin Institute and National Electric Light Association in February and March, 1893. It was an extension of the transmission through a single wire without return, the practicability of which I had already demonstrated. If my ideas were rational, distance was of no consequence and energy could be conveyed from one to any point of the globe, and in any desired amount. The task was begun under the inspiration of these great possibilities.

While scientific investigation had laid bare all the essential facts relating to Hertz-wave telegraphy, little knowledge was available bearing on the system proposed by me. The very first requirement, of course, was the production of powerful electrical vibrations. To impart these to the earth in an efficient manner, to construct proper receiving apparatus, and develop other technical details could be confidently undertaken. But the all important question was, how would the planet be affected by the oscillations impressed upon it? Would not the capacity of the terrestrial system, composed of the earth and its conducting envelope, be too great? As to this, the theoretical prospect was for a long time discouraging. I found that currents of high frequency and potential, such as had to be necessarily employed for the purpose, passed freely through air moderately rarefied. Judging from these experiences, the dielectric stratum separating the two conducting spherical surfaces could be scarcely more than 20 kilometers thick and, consequently, the capacity would be over 220,000 microfarads, altogether too great to permit economic transmission of power to distances of commercial importance. Another observation was that these currents cause considerable loss of energy in the air around the wire. That such waste might also occur in the earth's atmosphere was but a logical inference.

A number of years passed in efforts to improve the apparatus and to study the electrical phenomena produced. Finally my labors were rewarded and the truth was positively established; the globe did not act like a conductor of immense capacity and the loss of energy, due to absorption in the air, was insignificant. The exact mode of propagation of the currents from the source and the laws governing the electrical movement had still to be ascertained. Until this was accomplished the new art could not be placed on the plane of scientific engineering. One could bridge the greatest distance by sheer force, there being virtually no limit to the intensity of the vibrations developed by such a transmitter, but the installment of economic plants and the predetermination of the effects, as required in most practical applications, would be impossible.

Such was the state of things in 1899 when I discovered a new difficulty which I had never thought of before. It was an obstacle which could not be overcome by any improvement devised by man and of such nature as to fill me with apprehension that transmission of power without wires might never be quite practicable. I think it useful, in the present phase of development, to acquaint the profession with my investigations.

It is a well know fact that the action on a wireless receiver is appreciably weaker during the day than at night and this is attributed to the effect of sunlight on the elevated aerials, an explanation naturally suggested through an early observation of Heinrich Hertz. Another theory, ingenious but rather fine-spun, is that some of the energy of the waves is absorbed by ions or electrons, freed in sunlight and caused to move in the direction of propagation. The Electrical Review and Western Electrician of June 1, 1912, contains a report of a test, during the recent solar eclipse, between the station of the Royal Dock Yard in Copenhagen and the Blaavandshuk station on the coast of Jutland, in which it was demonstrated that the signals in that region became more distinct and reliable when the sunlight was partially cut off by the moon. The object of this communication is to show that in all the instances reported the weakening of the impulses was due to an entirely different cause.

It is indispensable to first dispel a few errors under which electricians have labored for years, owing to the tremendous momentum imparted to the scientific mind through the work of Hertz which has hampered independent thought and experiment. To facilitate understanding, attention is called to the annexed diagrams in which Fig. 1 and Fig. 2 represent, respectively, the well known arrangements of circuits in the Hertz-wave system and my own. In the former the transmitting and receiving conductors are separated from the ground through spark gaps, choking coils, and high resistances. This is necessary, as a ground connection greatly reduces the intensity of the radiation by cutting off half of the oscillator and also by increasing the length of the waves from 40 to 100 percent, according to the distribution of capacity and inductance. In the system devised by me a connection to earth, either directly or through a condenser is essential. The receiver, in the first case, is affected only by rays transmitted through the air, conduction being excluded; in the latter instance there is no appreciable radiation and the receiver is energized through the earth while an equivalent electric displacement occurs in the atmosphere.

Now, an error which should be the focus of investigation for experts is, that in the arrangement shown in Fig. 1 the Hertzian effect has been gradually reduced through the lowering of frequency, so as to be negligible when the usual wavelengths are employed. That the energy is transmitted chiefly, if not wholly, by conduction can be demonstrated in a number of ways. One is to replace the vertical transmitting wire by a horizontal one of the same effective capacity, when it will be found that the action on the receiver is as before. Another evidence is afforded by quantitative measurement which proves that the energy received does not diminish with the square of the distance, as it should, since the Hertzian radiation propagates in a hemisphere. One more experiment in support of this view may be suggested. When transmission through the ground is prevented or impeded, as by severing the connection or otherwise, the receiver fails to respond, at least when the distance is considerable. The plain fact is that the Hertz waves emitted from the aerial are just as much of a loss of power as the short radiations of heat due to frictional waste in the wire. It has been contended that radiation and conduction might both be utilized in actuating the receiver, but this view is untenable in the light of my discovery of the wonderful law governing the movement of electricity through the globe, which may be conveniently expressed by the statement that the projection of the wave-lengths (measured along the surface) on the earth's diameter or axis of symmetry of movement are all equal. Since the surfaces of the zones so defined are the same the law can also be expressed by stating that the current sweeps in equal times over equal terrestrial areas. (See among others "Handbook of Wireless Telegraphy" by James Erskine-Murray.) Thus the velocity propagation through the superficial layers is variable, dependent on the distance from the transmitter, the mean value being n/2 times the velocity of light, while the ideal flow along the axis of propagation takes place with a speed of approximately 300,000 kilometers per second.

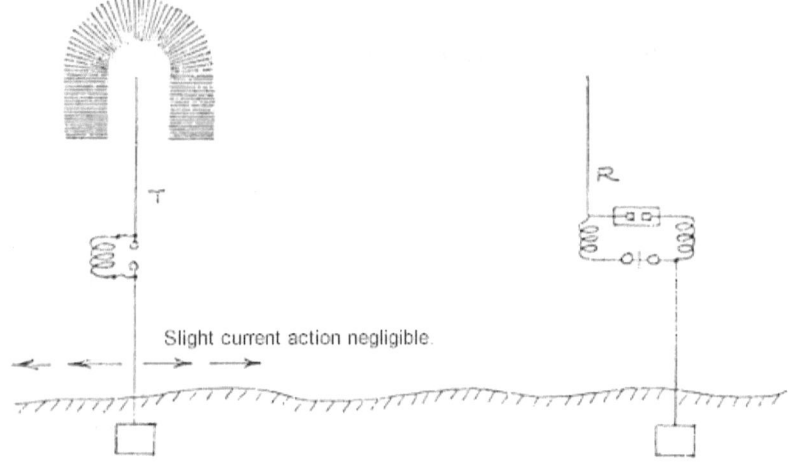

Fig. 1—Hertz Wave System

To illustrate, the current from a transmitter situated at the Atlantic Coast will traverse that ocean—a distance of 4,800 kilometers—in less than 0.006 second with an average speed of 800,000 kilometers. If the signaling were done by Hertz waves the time required would be 0.016 second.

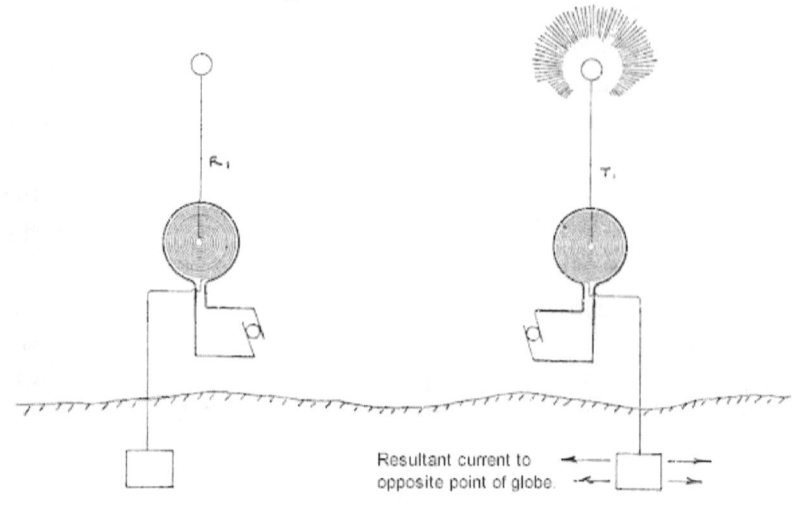

Fig. 2—System Devised by Tesla

Bearing, then, in mind, that the receiver is operated simply by currents conducted along the earth as through a wire, energy radiated playing no part, it will be at once evident that the weakening of the impulses could not be due to any changes in the air, making it turbid or conductive, but should be traced to an effect interfering with the transmission of the current through the superficial layers of the globe. The solar radiations are the primary cause, that is true, not those of light, but of heat. The loss of energy, I have found, is due to the evaporation of the water on that side of the earth which is turned toward the sun, the conducting particles carrying off more or less of the electrical charges imparted to the ground. This subject has been investigated by me for a number of years and on some future occasion I propose to dwell on it more extensively. At present it may be sufficient, for the guidance of experts, to state that the waste of energy is proportional to the product of the square of the electric density induced by the transmitter at the earth's surface and the frequency of the currents. Expressed in this manner it may not appear of very great practical significance. But remembering that the surface density increases with the frequency it may also be stated that the loss is proportional to the cube of the frequency. With waves 300 meters [1 MHz] in length economic transmission of energy is out of the question, the loss being too great. When using wave-lengths of 6,000 meters [50 kHz] it is still noticeable though not a serious drawback. With wave-lengths of 12,000 meters [25 kHz] it becomes quite insignificant and on this fortunate fact rests the future of wireless transmission of energy.

To assist investigation of this interesting and important subject, Fig. 3 has been added, showing the earth in the position of summer solstice with the transmitter just emerging from the shadow. Observation will bring out the fact that the weakening is not noticeable until the aerials have reached a position, with reference to the sun, in which the evaporation of the water is distinctly more rapid. The maximum will not be exactly when the angle of incidence of the suns rays is greatest, but some time after. It is noteworthy that the experimenters who watched the effect of the recent eclipse, above referred to, have observed the delay.

Fig. 3—Illustrating Disturbing Effect of the Sun on Wireless Transmission.

July 6, 1912

HOW COSMIC FORCES SHAPE OUR DESTINIES
("DID THE WAR CAUSE THE ITALIAN EARTHQUAKE")

—New York American, February 7, 1915

Every living being is an engine geared to the wheelwork of the universe. Though seemingly affected only by its immediate surrounding, the sphere of external influence extends to infinite distance. There is no constellation or nebula, no sun or planet, in all the depths of limitless space, no passing wanderer of the starry heavens, that does not exercise some control over its destiny—not in the vague and delusive sense of astrology, but in the rigid and positive meaning of physical science.

More than this can be said. There is no thing endowed with life—from man, who is enslaving the elements, to the humblest creature—in all this world that does not sway it in turn. Whenever action is born from force, though it be infinitesimal, the cosmic balance is upset and universal motion result.

Herbert Spencer has interpreted life as a continuous adjustment to the environment, a definition of this inconceivably complex manifestation quite in accord with advanced scientific thought, but, perhaps, not broad enough to express our present views. With each step forward in the investigation of its laws and mysteries our conceptions of nature and its phases have been gaining in depth and breadth.

In the early stages of intellectual development man was conscious of but a small part of the macrocosm. He knew nothing of the wonders of the microscopic world, of the molecules composing it, of the atoms making up the molecules and of the dwindlingly small world of electrons within the atoms. To him life was synonymous with voluntary motion and action. A plant did not suggest to him what it does to us—that it lives and feels, fights for its existence, that it suffers and enjoys. Not only have we found this to be true, but we have ascertained that even matter called inorganic, believed to be dead, responds to irritants and gives unmistakable evidence of the presence of a living principle within.

Thus, everything that exists, organic or inorganic, animated or inert, is susceptible to stimulus from the outside. There is no gap between, no break of continuity, no special and distinguishing vital agent. The same law governs all matter, all the universe is alive. The momentous question of Spencer, "What is it that causes inorganic matter to run into organic forms!" has been answered. It is the sun's heat and light. Wherever they are there is life. Only in the boundless wastes of interstellar space, in the eternal darkness and cold, is animation suspended, and, possibly, at the temperature of absolute zero all matter may die.

Man as a machine

This realistic aspect of the perceptible universe, as a clockwork wound up and running down, dispensing with the necessity of a hypermechanical vital principle, need not be in discord with our religious and artistic aspirations — those undefinable and beautiful efforts through which the human mind endeavors to free itself from material bonds. On the contrary, the better understanding of nature, the consciousness that our knowledge is true, can only be all the more elevating and inspiring.

It was Descartes, the great French philosopher, who in the seventeenth century, laid the first foundation to the mechanistic theory of life, not a little assisted by Harvey's epochal discovery of blood circulation. He held that animals were simply automata without consciousness and recognized that man, though possessed of a higher and distinctive quality, is incapable of action other than those characteristic of a machine. He also made the first attempt to explain the physical mechanism of memory. But in this time many functions of the human body were not as yet understood, and in this respect some of his assumptions were erroneous.

Great strides have since been made in the art of anatomy, physiology and all branches of science, and the workings of the man-machine are now perfectly clear. Yet the very fewest among us are able to trace their actions to primary external causes. It is indispensable to the arguments I shall advance to keep in mind the main facts which I have myself established in years of close reasoning and observation and which may be summed up as follows:

1. The human being is a self-propelled automaton entirely under the control of external influences. Willful and predetermined though they appear, his actions are governed not from within, but from without. He is like a float tossed about by the waves of a turbulent sea.
2. There is no memory or retentive faculty based on lasting impression. What we designate as memory is but increased responsiveness to repeated stimuli.
3. It is not true, as Descartes taught, that the brain is an accumulator. There is no permanent record in the brain, there is no stored knowl-

edge. Knowledge is something akin to an echo that needs a disturbance to be called into being.

4. All knowledge or form conception is evoked through the medium of the eye, either in response to disturbances directly received on the retina or to their fainter secondary effects and reverberations. Other sense organs can only call forth feelings which have no reality of existence and of which no conception can be formed

5. Contrary to the most important tenet of Cartesian philosophy that the perceptions of the mind are illusionary, the eye transmits to it the true and accurate likeness of external things. This is because light propagates in straight lines and the image cast on the retina is an exact reproduction of the external form and one which, owing to the mechanism of the optic nerve, can not be distorted in the transmission to the brain. What is more, the process must be reversible, that in to say, a form brought to consciousness can, by reflex action, reproduce the original image on the retina just as an echo can reproduce the original disturbance If this view is borne out by experiment an immense revolution in all human relations and departments of activity will be the consequence.

Natural forces influence us

Accepting all this as true let us consider some of the forces and influences which act on such a wonderfully complex automatic engine with organs inconceivably sensitive and delicate, as it is carried by the spinning terrestrial globe in lightning flight through space. For the sake of simplicity we may assume that the earth's axis is perpendicular to the ecliptic and that the human automaton is at the equator. Let his weight be one hundred and sixty pounds then, at the rotational velocity of about 1,520 feet per second with which he is whirled around, the mechanical energy stored in his body will be nearly 5,780,000 foot pounds, which is about the energy of a hundred-pound cannon ball.

This momentum is constant as well as upward centrifugal push, amounting to about fifty-five hundredth of a pound, and both will probably be without marked influence on his life functions. The sun, having a mass 332,000 times that of the earth, but being 23,000 times farther, will attract the automaton with a force of about one-tenth of one pound, alternately increasing and diminishing his normal weight by that amount

Though not conscious of these periodic changes, he is surely affected by them. The earth in its rotation around the sun carries him with the prodigious speed of nineteen miles per second and the mechanical energy imparted to him is over 25,160,000,000 foot pounds. The largest gun ever made in Germany hurls a projectile weighing one ton with a muzzle velocity of 3,700 feet per second, the energy being 429,000,000 foot pounds. Hence the momentum of the automaton's body is nearly sixty times greater. It would be sufficient to develop 762,400 horse-power for one minute, and if the motion were suddenly arrested the body would be instantly exploded with a force sufficient to carry a projectile weighing over sixty tons to a distance of twenty-eight miles.

This enormous energy is, however, not constant, but varies with the position of the automaton in relation to the sun. The circumference of the earth has a speed of 1,520 feet per second, which is either added to or subtracted from the translatory velocity of nineteen miles through space. Owing to this the energy will vary from twelve to twelve hours by an amount approximately equal to 1,533,000,000 foot pounds, which means that energy streams in some unknown way into and out of the body of the automaton at the rate of about sixty-four horse-power.

But this is not all. The whole' solar system is urged towards the remote constellation Hercules at a speed which some estimate at some twenty miles per second and owing to this there should be similar annual changes in the flux of energy, which may reach the appalling figure of over one hundred billion foot pounds. All these varying and purely mechanical effects are rendered more complex through the inclination of the orbital planes and many other permanent or casual mass actions.

This automaton, is, however subjected to other forces and influences. His body is at the electric potential of two billion volts, which fluctuates violently and incessantly. The whole earth is alive with electrical vibrations in which he takes part. The atmosphere crushes him with a pressure of from sixteen to twenty tons, according, to barometric condition. He receives the energy of the sun's rays in varying intervals at a mean rate of about forty foot pounds per second, and is subjected to periodic bombardment of the sun's particles, which pass through his body as if it were tissue paper. The air is rent with sounds which beat on his eardrums, and he is shaken by the unceasing tremors of the earth's crust. He is exposed to great temperature changes, to rain and wind.

What wonder then that in such a terrible turmoil, in which cast iron existence would seem impossible, this delicate human engine should act in an exceptional manner? If all automata were in every respect alike they would react in exactly the same way, but this is not the case. There is concordance in response to those disturbances only which are most frequently repeated, not to all. It is quite easy to provide two electrical systems which, when subjected to the same influence, will behave in just the opposite way.

So also two human beings, and what is true of individuals also holds good for their large aggregations. We all sleep periodically. This is not an indispensable physiological necessity any more than stoppage at intervals is a requirement for an engine. It is merely a condition gradually imposed upon us by the diurnal revolution of the globe, and this is one of the many evidences of the truth of the mechanistic theory. We note a rhythm or ebb and tide, in ideas and opinions, in financial and political movements, in every department of our intellectual activity.

How wars are started

It only shows that in all this a physical system of mass inertia is involved which affords a further striking proof. If we accept the theory as a fundamental truth and, furthermore, extend the limits of our sense perceptions beyond those within which we become conscious of the external impressions, then all the states in human life, however unusual, can be plausibly explained. A few examples may be given in illustration.

The eye responds only to light vibrations through a certain rather narrow range, but the limits are not sharply defined. It is also affected by vibrations beyond, only in lesser degree. A person may thus become aware of the presence of another in darkness, or through intervening obstacles, and people laboring under illusions ascribe this to telepathy. Such transmission of thought is absurdly impossible.

The trained observer notes without difficulty that these phenomena are due to suggestion or coincidence. The same may be said of oral impressions, to which musical and imitative people are especially susceptible. A person possessing these qualities will often respond to mechanical shocks or vibrations which are inaudible.

To mention another instance of momentary interest reference may be made to dancing, which comprises certain harmonious muscular contractions and contortions of the body in response to a rhythm. How they come to be in vogue just now, can be satisfactorily explained by supposing the existence of some new periodic disturbances in the environment, which are transmitted through the air or the ground and may be of mechanical, electrical or other character.

Exactly so it is with wars, revolutions and similar exceptional states of society.

Though it may seem so, a war can never be caused by arbitrary acts of man. It is invariably the more or less direct result of cosmic disturbance in which the sun is chiefly concerned.

In many international conflicts of historical record which were precipitated by famine, pestilence or terrestrial catastrophes the direct dependence of the sun is unmistakable. But in most cases the underlying primary causes are numerous and hard to trace.

In the present war it would be particularly difficult to show that the apparently willful acts of a few individuals were not causative. Be it so, the mechanistic theory, being founded on truth demonstrated in everyday experience, absolutely precludes the possibility of such a state being anything but the inevitable consequence of cosmic disturbance.

The question naturally presents itself as to whether there is some intimate relation between wars and terrestrial upheavals. The latter are of decided influence on temperament and disposition, and might at times be instrumental in accelerating the clash but aside from this there seems to be no mutual dependence, though both may be due to the same primary cause.

What can be asserted with perfect confidence is that the earth may be thrown into convulsions through mechanical effects such as are produced in modern warfare. This statement may be startling, but it admits of a simple explanation.

Earthquakes are principally due to two causes—subterranean explosions or structural adjustments. The former are called volcanic, involve immense energy and are hard to start. The latter are named tectonic; their energy is comparatively insignificant and they can be caused by the slightest shock or tremor. The frequent slides in the Culebra are displacements of this kind.

War and the earthquake

Theoretically, it may be said that one might think of a tectonic earthquake and cause it to occur as a result of the thought, for just preceding the release the mass may be in the most delicate balance. There is a popular error in regard to the energy of such displacements. In a case recently reported as quite extraordinary, extending as it did over a vast territory, the energy was estimated at 65,000,000,000,000 foot tons. Assuming even that the whole work was performed in one minute it would only be equivalent to that of 7,500,000 horse-power during one year, which seems much, but is little for a terrestrial upheaval. The energy of the sun's rays falling on the same area is a thousand times greater.

The explosions of mines, torpedoes, mortars and guns develop reactive forces on the ground which are measured in hundreds or even thousands of tons and make themselves felt all over the globe. Their effect, however, may be enormously magnified by resonance. The earth is a sphere of a rigidity slightly greater than that of steel and vibrates once in about one hour and forty-nine minutes.

If, as might well be possible, the concussions happen to be properly timed their combined action could start tectonic adjustments in any part of the earth, and the Italian calamity may thus have been the result of explosions in France. That man can produce such terrestrial convulsions is beyond any doubt, and the time may be near when it will be done for purposes good or apt.

February 7, 1915

SOME PERSONAL RECOLLECTIONS

—Scientific American, June 5, 1915

I am glad to be accorded this opportunity for two reasons. In the first place I have long since desired to express my great appreciation of the Scientific American and to acknowledge my indebtedness for the timely and useful information which its columns are pouring out in a steady stream. It is a publication remarkable for the high quality of special articles as well as for the accurate review of technical advances. The knowledge it conveys is always reliable and rendered still more valuable through the scrupulous observance of literary courtesy in the quotation of the sources. The services it has rendered in helping invention and spreading enlightenment are inestimable. The Scientific American is a periodical ably and conscientiously conducted, measured and dignified in tone to the point of serving as a model, and in these features, as much as in the wealth and excellence of its contributions, it reflects great credit, not only on its staff and publishers, but on the whole country. This is not an idle compliment, but a genuine and well-deserved tribute to which I add my best wishes for continued success on this memorable occasion.

The second reason is one that concerns me personally. Many erroneous statements have appeared in print relative to my discovery of the rotating magnetic field and invention of the induction motor which I was compelled to pass in silence. Great interests have waged a long and bitter contest for my patent rights; commercial animosities and professional jealousies were aroused, and I was made to suffer in more than one way. But despite of all pressure and efforts of ingenious lawyers and experts, the rulings of the courts were in support of my claims for priority in every instance without exception. The battles have been fought and forgotten, the thirty or forty patents granted to me on the alternating system have expired, I have been released of burdensome obligations and am free to speak.

Every experience which I have lived through bearing on that early discovery is vividly present in my memory. I see the faces of the persons, the scenes and objects of my attention, with a sharpness and distinction and in a fullness of light which is astonishing, and is a measure of the intensity and depth of the original impressions. I have always been fortunate in ideas, but no other invention, however great, could be as dear to me as that first one. This will be understood if I dwell briefly on the circumstances surrounding it and some of the phases and incidents of my young life.

From my childhood I had been intended for the clergy. This prospect hung like a dark cloud on my mind. After passing eleven years at a public school and a higher institution, I obtained my certificate of maturity and found myself at the critical point of my career. Should I disobey my father, ignore the fondest wishes of my mother, or should I resign myself to fate. The thought oppressed me, and I looked to the future with dread.

Just at that time a terrible epidemic of cholera broke out in my native land. People knew nothing of the character of the disease and the means of sanitation were of the poorest kind. They burned huge piles of odorous shrubbery to purify the air, but drank freely of the infected water and died in crowds like sheep. Contrary to peremptory orders from my father I rushed home and was stricken down. Nine months in bed with scarcely the ability to move seemed to exhaust all my vitality, and I was given up by the physicians. It was an agonizing experience, not so much because of physical suffering as on account of my intense desire to live. On the occasion of one of the fainting spells my father cheered me by a promise to let me study engineering; but it would have remained unfulfilled had it nor been for a marvelous cure brought about by an old lady. There was no force of suggestion or mysterious influence about it. Such means would have had no effect whatever on me, for I was a firm believer in natural laws. The remedy was purely medicinal, heroic if not desperate; but it worked and in one year of mountain climbing and forest life I was fit for the most arduous bodily exertion. My father kept his word, and in 1877 I entered the Joanneum in Gratz, Styria, one of the oldest technical institutions of Europe. I proposed to show results which would repay my parents for their bitter disappointment due to my change of vocation. It was not a passing determination of a light-hearted youth; it was iron resolve. As some young reader of the Scientific American might draw profit from my example I will explain.

Three rotors used with the early induction motor shown below

When I was a boy of seven or eight I read a novel untitled "Abafi"—The Son of Aba—a Servian translation from the Hungarian of Josika, a writer of renown. The lessons it teaches are much like those of "Ben Hur," and in this respect it might be viewed as anticipatory of the work of Wallace. The possibilities of will-power and self-control appealed tremendously to my vivid imagination, and I began to discipline myself. Had I a sweet cake or a juicy apple which I was dying to eat I would give it to another boy and go through the tortures of Tantalus, pained but satisfied. Had I some difficult task before me which was exhausting I would attack it again and again until it was done. So I practiced day by day from morning till night. At first it called for a vigorous mental effort directed against disposition and desire, but as years went by the conflict lessened and finally my will and wish became identical. They are so today, and in this lies the secret of whatever success I have achieved. These experiences are as intimately linked with my discovery of the rotating magnetic field as if they formed an essential part of it; but for them I would never have invented the induction motor.

In the first year of my studies at the Joanneum I rose regularly at three o'clock in the morning and worked till eleven at night; no Sundays or holidays excepted. My success was unusual and excited the interest of the professors. Among these was Dr. All—, who lectured on differential equations and other branches of higher mathematics and whose addresses were unforgettable intellectual treats, and Prof. Poeschl, who held the chair of Physics, theoretical and experimental. These men I always remember with a sense of gratitude. Prof. Poeschl was peculiar; it was said of him that he wore the same coat for twenty years. But what he lacked in personal magnetism he made up in the perfection of his exposition. I never saw him miss a word or gesture, and his demonstrations and experiments always went off with clocklike precision. Some time in the winter of 1878 a new apparatus was installed in the lecture room. It was a dynamo with a laminated permanent magnet and a Gramme armature. Prof. Poeschl had wound some wire around the field to show the principle of self-excitation, and provided a battery for running the machine as a motor. As he was illustrating this latter feature there was lively sparking at the commutator and brushes, and I ventured to remark that these devices might be eliminated. He said that it was quite impossible and likened my proposal to a perpetual motion scheme, which amused my fellow students and embarrassed me greatly. For a time I hesitated, impressed by his authority, but my conviction grew stronger and I decided to work out the solution. At that time my resolve meant more to me than the most solemn vow.

One of the earliest of induction motors. Although it weighed only a little over 20 pounds, it developed 1/4 horse-power at a speed of 1,800 revolutions, a performance considered remarkable at the time.

I undertook the task with all the fire and boundless confidence of youth. To my mind it was simply a test of will-power. I knew nothing of the technical difficulties. All my remaining term in Gratz was passed in intense but fruitless effort, and I almost convinced myself that the problem was unsolvable. Indeed, I thought, was it possible to transform the steady pull of gravitation into a whirling force! The answer was an emphatic no. And was this not also true of magnetic attraction? The two propositions appeared very much the same.

In 1880 I went to Prague, Bohemia, carrying out my father's wish to complete my school education at a university. The atmosphere of that old and interesting city was favorable to invention. Hungry artists were plentiful and intelligent company could be found everywhere. Here I made the first distinct step in advance, by detaching the commutators from the machines and placing them on distant arbors. Every day I imagined arrangements on this plan without result, but feeling that I was nearing the solution. In the following year there was a sudden change in my views of life. I realized that my parents were making too great sacrifices for me and resolved to relieve them of the burden. The American telephone wave had reached the European continent, and the system was to be installed in Budapest. It appeared an ideal opportunity, and I took the train for that city. By an irony of fate my first employment was as a draughtsman. I hated drawing; it was for me the very worst of annoyances. Fortunately it was not long before I secured the position I sought, that of chief electrician to the telephone company. My duties brought me in contact with a number of young men in whom I became interested. One of these was Mr. Szigety, who was a remarkable specimen of humanity. A big head with an awful lump on one side and a sallow complexion made him distinctly ugly, but from the neck own his body might have served for a statue of Apollo. His strength was phenomenal. At that time I had exhausted myself through hard work and incessant thinking. He impressed me with the necessity of systematic physical development, and I accepted his offer to train me in athletics. We exercised every day and I gained rapidly in strength. My mind also seemed to grow more vigorous and as my thoughts turned to the subject which absorbed me I was surprised at my confidence of success. On one occasion, ever present in my recollection, we were enjoying ourselves in the Varos-liget or City Park. I was reciting poetry, of which I was passionately fond. At that age I knew entire books by heart and could read them from memory word by word. One of these was Faust. It was late in the afternoon, the sun was setting, and I was reminded of the passage:

> "The glow retreats, done is the day of toil
> It yonder hastes, new fields of life exploring;
> Ah, that no wing can lift me from the soil
> Upon its track to follow, follow soaring!
> "Alas! the wings that lift the mind no aid
> Of wings to lift the body can bequeath me."

As I spoke the last words, plunged in thought and marveling at the power of the poet, the idea came like a lightning flash. In an instant I saw it all, and I drew with a stick on the sand the diagrams which were illustrated in my fundamental patents of May, 1888, and which Szigety understood perfectly.

It is extremely difficult for me to put this experience before the reader in its true light and significance for it is so altogether extraordinary. When an idea presents itself it is, as a rule, crude and imperfect. Birth, growth and development are phases normal and natural. It was different with my invention. In the very moment I became conscious of it. I saw it fully developed and perfected. Then again, a theory, however plausible, must usually be confirmed by experiment. Not so the one I had formulated. It was being daily demonstrated every dynamo and motor was absolute proof of its soundness. The effect on me was indescribable. My imaginings were equivalent to realities. I had carried out what I had undertaken and pictured myself achieving wealth and fame. But more than all this was to me the revelation that I was an inventor. This was the one thing I wanted to be. Archimedes was my ideal. I admired the works of artists, but to my mind, they were only shadows and semblances. The inventor, I thought, gives to the world creations which are palpable, which live and work.

The telephone installation was now completed and in the spring of 1882 an offer was made me to go to Paris, which I accepted eagerly. Here I met a number of Americans whom I befriended and to whom I talked of my invention, and one of them, Mr. D. Cunningham, proposed to form a company for exploitation. This might have been done had not my duties called me to Strasburg, Alsace. It was in this city that I constructed my first motor. I had brought some material from Paris, and a disk of iron with bearings was made for me in a mechanical shop close to the railroad station in which I was installing the light and power plant. It was a crude apparatus, but afforded me the supreme satisfaction of seeing for the first time, rotation affected by alternating currents without commutator. I repeated the experiment with my assistant twice in the summer of 1883. My intercourse with Americans had directed my attention to the practical introduction and I endeavored to secure capital, but was unsuccessful in this attempt and returned to Paris early in 1884. Here, too, I made several ineffectual efforts, and finally resolved to go to America, where I arrived in the summer of 1884. By a previous understanding I entered the Edison Machine Works, where I undertook the design of dynamos and motors. For nine months my regular hours were from 10:30 A. M. till 5 A. M. the next day. All this time I was getting more and more anxious about the invention and was making up my mind to place it before Edison. I still remember an odd incident in this connection. One day in the latter part of 1884 Mr. Bachelor, the manager of the works, took me to Coney Island, where we met Edison in company with his former wife. The moment that I was waiting for was propitious, I was just about to speak when a horrible looking tramp took hold of Edison and drew him away, preventing me from carrying out my intention. Early in 1885 people approached me with a proposition to develope an arc light system and to form a company under my name. I signed the contract, and a year and a half later I was free and in a position to devote myself to the practical development of my discovery. I found financial support, and in April, 1887, a company was organized for the purpose, and what has followed since is well known.

A few words should be said in regard to the various claims for anticipation which were made upon the issuance of my patents in 1888, and in numerous suits conducted subsequently. There were three contestants for the honor, Ferraris, Schallenberger and Cabanellas. All three succumbed to grief. The opponents of my patents advanced the Ferraris claim very strongly, but any one who will peruse his little Italian pamphlet, which appeared in the spring of 1888, and compare it with the patent record filed by me seven months before, and with my paper before the *American Institute of Electrical Engineers*, will have no difficulty in reaching a conclusion. Irrespective of being behind me in time, Prof. Ferraris's publication concerned only my split-phase motor, and in an application for a patent by him priority was awarded to me. He never suggested any of the essential practical features which constitute my system, and in regard to the split-phase motor he was very decided in his opinion that it was of no value. Both Ferraris and Schallenberger discovered the rotation accidentally while working with a Gullard and Gibbs transformer, and had difficulty in explaining the actions. Neither of them produced a rotating field motor like mine, nor were their theories the same as my own. As to Cabanellas, the only reason for his claim is an abandoned and defective technical document. Some over-zealous friends have interpreted a United States patent granted to Bradley as a contemporary record, but there is no foundation whatever for such a claim. The original application only described a generator with two circuits which were provided for the sole purpose of increasing the output. There was not much novelty in the idea, since a number of such machines existed at that time. To say that these machines were anticipations of my rotary transformer is wholly unjustified. They might have served as one of the elements in my system of transformation, but were nothing more than dynamos with two circuits constructed with other ends in view and in utter ignorance of the new and wonderful phenomena revealed through my discovery.

June 5, 1915

THE WONDER WORLD TO BE CREATED BY ELECTRICITY

—Manufacturer's Record, September 9, 1915

Whoever wishes to get a true appreciation of the greatness of our age should study the history of electrical development. There he will find a story more wonderful than any tale from Arabian Nights. It begins long before the Christian era when Thales, Theophrastus and Pliny tell of the magic properties of electron—the precious substance we call amber—that came from the pure tears of the Heliades, sisters of Phaeton, the unfortunate youth who attempted to run the blazing chariot of Phoebus and nearly burned up the earth. It was but natural for the vivid imagination of the Greeks to ascribe the mysterious manifestations to a hyperphysical cause, to endow the amber with life and with a soul.

Whether this was actual belief or merely poetic interpretation is still a question. When at this very day many of the most enlightened people think that the pearl is alive, that it grows more lustrous and beautiful in the warm contact of the human body. So too, it is the opinion of men of science that a crystal is a living being and this view is being extended to embrace the entire physical universe since Prof. Jagadis Chunder Bose has demonstrated, in a series of remarkable experiments, that inanimate matter responds to stimuli as plant fiber and animal tissue.

The superstitious belief of the ancients, if it existed at all, can therefore not be taken as a reliable proof of their ignorance, but just how much they knew about electricity can only be conjectured. A curious fact is that the ray or torpedo fish, was used by them in electro-therapy. Some old coins show twin stars, or sparks, such as might be produced by a galvanic battery. The records, though scanty, are of a nature to fill us with conviction that a few initiated, at least, had a deeper knowledge of amber-phenomena. To mention one, Moses was undoubtedly a practical and skillful electrician far in advance of his time. The Bible describes precisely and minutely arrangements constituting a machine in which electricity was generated by friction of air against silk curtains and stored in a box constructed like a condenser. It is very plausible to assume that the sons of Aaron were killed by a high tension discharge and that the vestal fires of the Romans were electrical. The belt drive must have been known to engineers of that epoch and it is difficult to see how the abundant evolution of static electricity could have escaped their notice. Under favorable atmospheric conditions a belt may be transformed into a dynamic generator capable of producing many striking actions. I have lighted incandescent lamps, operated motors and performed numerous other equally interesting experiments with electricity drawn from belts and stored in tin cans.

That many facts in regard to the subtle force were known to the philosophers of old can be safely concluded, the wonder is, why two thousand years elapsed before [William Gilbert] in 1600 published his famous work, the first scientific treatise on electricity and magnetism. To an extent this long period of unproductiveness can be explained. Learning was the privilege of a few and all information was jealously guarded. Communication was difficult and slow and a mutual understanding between widely separated investigators hard to reach. Then again, men of those times had no thought of the practical, they lived and fought for abstract principles, creeds, traditions and ideals. Humanity did not change much in Gilbert's time but his clear teachings had a telling effect on the minds of the learned. Friction machines were produced in rapid succession and experiments and observations multiplied. Gradually fear and superstition gave way to scientific in-sight and in 1745 the world was thrilled with the news that Kleist and Leyden had succeeded in imprisoning the uncanny agent in a phial from which it escaped with an angry snap and destructive force. This was the birth of the condenser, perhaps the most marvelous electrical device ever invented.

Two tremendous leaps were made in the succeeding forty years. One was when Franklin demonstrated the identity between the gentle soul of amber and the awe-inspiring belt of Jupiter; the other when Galvany and Volta brought out the contact and chemical battery, from which the magic fluid could be drawn in unlimited quantities. The succeeding forty years bore still greater fruit. Oersted made a significant advance in deflecting a magnetic needle by an electric current, Arago produced the electro-magnet, Seebeck the thermo-pile and in 1831, as the crowning achievement of all, Faraday announced that he had obtained electricity from a magnet, thus discovering the principle of that wonderful engine — the dynamo, and inaugurating a new era both in scientific research and practical application.

From that time on inventions of inestimable value have followed one another at a bewildering rate. The telegraph, telephone, phonograph and incandescent lamp, the induction motor, oscillatory transformer, Roentgen ray, Radium, wireless and numerous other revolutionary advances have been made and all conditions of existence eighty-four years which have since elapsed, the subtle agents dwelling in the living amber and loadstone have been transformed into cyclopean forces turning the wheels of human progress with ever increasing speed. This, in brief, is the fairy tale of electricity from Thales to the present day. The impossible has happened, the wildest dreams have been surpassed and the astounded world is asking: What is coming next?

Electrical possibilities in coal and iron

Many a would-be discoverer, failing in his efforts, has felt the regret to have been born at a time when everything has been already accomplished and nothing is left to be done. This erroneous impression that, as we are advancing, the possibilities of invention are being exhausted, is not uncommon. In reality it is just the opposite. Spencer has conveyed the right idea when he likened civilization to the sphere of light which a lamp throws out in darkness. The brighter the lamp and the larger the sphere the greater is its dark boundary. It is paradoxical, yet true, to say that the more we know the more ignorant we become in the absolute sense, for it is only through enlightenment that we become conscious of our limitations. Precisely one of the most gratifying results of intellectual evolution is the continuous opening up of new and greater prospects. We are progressing at an amazing pace, but the truth is that, even in fields most successfully exploited, the ground has only been broken. What has been so far done by electricity is nothing compared with what the future has in store. Not only this, but there are now innumerable things done in old-fashioned ways which are much inferior in economy, convenience and many other respects to the new method. So great are the advantages of the latter that whenever an opportunity presents itself the engineer advises his client to "do it electrically."

Consider, in illustration, one of the largest industries, that of coal. From this valuable mineral we chiefly draw the sun's stored energy which is required to meet our industrial and commercial needs. According to statistical records, the output in the United States during the past year was 480,000,000 tons. In perfect engines this fuel would have been sufficient to develop 500,000,000 horse-power steadily for one year, but the squandering is so reckless that we do not get more than 5% of its heating value on the average. There is an appalling waste in mining, handling, transportation, store and use of coal, which could be very much reduced through the adoption of a comprehensive electrical plan in all these operations. The market value of the yearly product would be easily doubled and an immense sum added to the revenues of the country. What is more, inferior grades, billions of tons of which are being thrown away, might be turned to profitable use.

Similar considerations apply to natural gas and mineral oil, the annual loss of which amounts to hundreds of millions of dollars. In the very near future such waste will be looked upon as criminal and the introduction of the new methods will be forced upon the owners of such properties. Here, then, is an immense field for the use of electricity in many ways, vast industries which are bound to be revolutionized through its extensive application.

To give another example, I may refer to the manufacture of iron and steel, which is carried on in this country on a scale truly colossal. During the last year, notwithstanding unfavorable business conditions, 31,000,000 tons of steel have been produced. It would lead too far to dwell on the possibilities of electrical improvements in the manufacturing processes themselves, and I will only indicate what is likely to be accomplished in using the waste gases from the coke ovens and blast furnaces to generate electricity for industrial purposes.

Since in the production of pig-iron for every ton about one ton of coke is employed, the yearly consumption of coke may be put at 31,000,000 tons. The combustion in the blast furnaces yields, per minute, 7,000,000 cubic feet of gas of a heating value of 110 B. T. units per cubic foot. Of this total, without making special provision, 4,000,000 cubic feet may be made available for power purposes. If all the heat energy of this gas could be transformed into mechanical effort, it would develop 10,389,000 horse-power. This result is impossible, but it is perfectly practicable to obtain 2,500,000 horse-power electrical energy at the terminals of the dynamos.

In the manufacture of coke approximately 9400 cubic feet of gas are evolved per ton of coal. This gas is excellent for power purposes, having an average heating value of 600 B. T. units, but very little is now used in engines, largely because of their great cost and other imperfections. A ton of coke requires about 1.32 tons of American coal; hence the total coal consumption per annum on the above basis is nearly 41,000,000 tons, which give, per minute, 733,000 cubic feet of gas. Assuming the yield of surplus or rich gas to be 333,000 cubic feet, the balance of 400,000 cubic feet could be used in gas engines. The heat contents would be, theoretically, sufficient to develop 5,660,000 horse-power, of which 1,500,000 horse-power could be obtained in the form of electric energy.

I have devoted much thought to this industrial proposition, and find that with new, efficient, extremely cheap and simple thermo-dynamic transformers not less than 4,000,000 horse-power could be developed in electric generators by utilizing the heat of these gases, which, if not entirely wasted, are only in part and inefficiently employed.

With systematic improvements and refinements much better results could be secured and an annual revenue of $50,000,000 or more derived. The electrical energy could be advantageously used in the fixation of atmospheric nitrogen and production of fertilizers, for which there is an unlimited demand and the manufacture of which is restricted here on account of the high cost of power. I expect confidently the practical realization of the project in the very near future, and look to exceptionally rapid electrical development in this direction.

Hydro-electric development

Water-power offers great opportunities for novel electrical applications, particularly in the department of electro-chemistry. The harness of waterfalls is the most economical method known for drawing energy from the sun. This is due to the fact that both water and electricity are incompressible. The net efficiency of the hydro-electric process can be as high as 85%. The initial outlay is generally great, but the cost of maintenance is small and the convenience offered ideal. My alternating system is invariably employed , and so far about 7,000,000 horse-power have been developed. As generally used we do not get more than six-hundredths of a horse-power per ton of coal per year. This water energy is therefore equivalent to that obtainable from an annual supply of 120,000,000 tons of coal, which is about 25% of the total output in the United States. The estimate is conservative, and in view of the immense waste of coal 50% may be a closer guess.

We get better appreciation of the tremendous value of this power in our economic development when we remember that, unlike fuel, which demands a terrible sacrifice of human energy and is consumed, it is supplied without effort and destruction of material and equals the mechanical performance of 150,000,000 men—one and one-half times the entire population of this country. These figures are imposing; nevertheless, we have only begun the exploitation of this vast national resource.

There are two chief limitations at present—one in the availability of the energy, the other in its transmission to distance. The theoretical power of the falling water is enormous. If we assume for the rain clouds an average height of 15,000 feet and annual precipitation of 33 inches, the 24 horse-power per square mile is over 4000, and for the whole area in the United States more than 12,000,000,000 horse-power. As a matter of fact, the larger portion of the potential energy is used up in air friction. This, while disappointing to the economist, is a fortunate circumstance, for otherwise the drops would reach the ground with a speed of 800 feet per second—sufficient to raise blisters on our bodies, while hail would be positively deadly. Most of the water, which is available for power purposes comes from a height of about 2000 feet and represents over one and one-half billion horse-power, but we are only able to use an average fall of, say, 100 feet, which means that if all the water-power in this country were harnessed under the existing conditions only 80,000,000 horse-power could be obtained.

The next great achievement – Electrical control of atmospheric moisture

But the time is very near when we shall have the precipitation of the moisture of the atmosphere under complete control, and then it will be possible to draw unlimited quantities of water from the oceans, develop any desired amount of energy, and completely transform the globe by irrigation and intensive farming. A greater achievement of man through the medium of electricity can hardly be imagined.

The present limitations in the transmission of power to distance will be overcome in two ways — through the adoption of underground conductors insulated by power, and through the introduction of the wireless art. The first plan I have advanced years ago. The underlying principle is to convey through a tubular conductor hydrogen at a very low temperature, freeze the surrounding material and thus secure a perfect insulation by indirect use of electric energy. In this manner the power derived from falls can be transmitted to distances of hundreds of miles with the highest economy and at a small cost. This innovation is sure to greatly extend the fields of electrical application. As to the wireless method, we have now the means for economic transmission of energy in any desired amount and to distances only limited by the dimensions of this planet. In view of assertions of some misinformed experts to the effect that in the wireless system I have perfected the power of the transmitter is dissipated in all directions, I wish to be emphatic in my statement that such is not the case. The energy goes only to the place where it is needed and to no other.

When these advanced ideas are practically realized we shall get the full benefit of water-power, and it will become our chief dependence in the supply of electricity for domestic, public and other uses in the arts of peace and war.

Economy in light and power – Electric propulsion

In the great departments of electric light and power immense opportunities are offered through the introduction of all kinds of novel devices which can be attached to the circuits at convenient hours for the purpose of equalizing the loads and increasing the revenues from the plants. I have myself knowledge of a number of new appliances of this kind. The most important among them is probably an electrical ice machine which obviates entirely the use of dangerous and otherwise objectionable chemicals. The new machine will also require absolutely no attention and will be extremely economical in operation, so that the refrigeration will be effected very cheaply and conveniently in every household.

An interesting fountain, electrically operated, has been brought out which is likely to be extensively introduced, and will afford an unusual and pleasing sight in squares, parks, hotels and residences.

Cooking devices for all domestic purposed are being provides, and there is great demand for practical designs and suggestions in this field. The same may be stated of electric signs and other attractive means of advertising which can be electrically operated. Some of the effects which it is possible to produce by electric currents are wonderful and lend themselves to exhibitions, and there is no doubt that much can be done in that direction. Theaters, public halls and private dwellings are in need of a great many devices and instruments for convenience and offer ample opportunities to an ingenious and practical inventor.

A vast and absolutely untouched field is the use of electricity for the propulsion of ships. The leading electrical company in this country has just equipped a large vessel with high-speed turbines and electric motors and has achieved a signal success. Applications of this kind will multiply at a rapid rate, for the advantages of the electrical drive are now patent to everybody. In this connection gyroscopic apparatus will probably play an important part, as its general adoption on vessels is sure to come. Very little has yet been done in the introduction of electrical drive in the various branches of industry and manufacture, and the prospects are unlimited.

A few of the wonders to come

Books have already been written on the agricultural uses of electricity, but the fact is that hardly anything has been practically done. The beneficial effects of electricity of high tension have been unmistakably established, and a revolution will be brought about through the extensive adoption of agricultural electrical apparatus. The safeguarding of forests against fires, the destruction of microbes, insects and rodents will, in due course, be accomplished by electrical means.

In the near future we shall see a great many new uses of electricity aiming at safety, particularly vessels at sea. We shall have electrical instruments for preventing collisions, and we shall even be able to disperse fogs by electric force and powerful and penetrative rays. I am hopeful that within the next few years wireless plants will be installed for the purpose of illuminating the oceans. The project is perfectly feasible, and if carried out will contribute more than any other provision to the safety of property and human lives at sea. The same plant could also produce stationary electrical waves and enable vessels to get at any time accurate bearings and other valuable practical data without resorting to the present means. It could also be used for time signaling and many other purposes of similar nature.

Electrotherapy is another great field in which there are unlimited possibilities for electrical applications. High-frequency currents especially have a great future. The time will come when this form of electrical energy will be available in every private residence. I consider it quite possible that through their surface actions we may do away with the customary bath, as the cleaning of the body can be instantaneously effected simply by connecting it to a source of currents or electric energy of very high potential, which results in the throwing off of dust or any small particles adhering to the skin. Such a dry bath, besides being convenient and time-saving, would also be of beneficial therapeutic influence. New electric devices for use of the deaf and blind are coming and will be a blessing to the afflicted.

In the prevention of crime electrical instruments will soon become an important factor. In court proceedings electric evidence will often be decisive. In a time not distant it will be possible to flash any image formed in thought on a screen and render it visible at any place desired. The perfection of this means of reading thought will create a revolution for the better in all our social relations. Unfortunately, it is true, that cunning lawbreakers will avail themselves of such advantages to further their nefarious business.

Telegraphic photography and other advances

Great improvements are still possible in telegraphy and telephony. The use of a new receiving device which will be shortly described, and the sensitiveness of which can be increased almost without limit, will enable telephoning through aerial lines or cables however long by reducing the necessary working current to an infinitesimal value. This invention will dispense with the necessity of resorting to expensive constructions, which, however, are of circumscribed usefulness. It will also enormously extend the wireless transmission of intelligence in all its departments.

The next art to be inaugurated is that of picture transmission by ordinary telegraphic methods and existing apparatus. This idea of telegraphing or telephoning pictures is old, but practical difficulties have hampered commercial realizations. A number of improvements of great promise have been made, and there is every reason to expect that success will soon be achieved.

Another valuable novelty will be a typewriter electrically operated by the human voice. This advance will fill a long-felt want, as it will do away with the operator and save a great deal of labor and time in offices.

A new and extremely simple electric tachometer is being prepared for the market, and it is expected that it will prove useful in power plants and central stations, on boats, locomotives and automobiles.

Many municipal improvements based on the use of electricity are about to be introduced. We have soon to have everywhere smoke annihilators, dust absorbers, ozonizers, sterilizers of water, air, food and clothing, and accident preventers on streets, elevated roads and in subways. It will become next to impossible to contract disease germs or get hurt in the city, and country folk will got to town to rest and get well.

Electric inventions in war

The present international conflict is a powerful stimulus to invention of devices and implements of warfare. An electric gun will soon be brought out. The wonder is that it was not produced long ago. Dirigibles and aeroplanes will be equipped with small electric generators of high tension, from which the deadly currents will be conveyed through the wires to the ground. Battleships and submarines will be provided with electric and magnetic feelers so delicate that the approach of any body underwater or in darkness will be detected. Torpedoes and floating mines are almost in sight which will direct themselves automatically and without fail get in fatal contact with the object to be destroyed. The art of telautomatics, or wireless control of automatic machines at a distance, will play a very important part in future wars and, possibly, in the next phases of the present one. Such contrivances which act as if endowed with intelligence will be used in innumerable ways for attack as well as defense. They may take the shape of aeroplanes, balloons, automobiles, surface or under-water boats, or any other form according to the requirement in each special case, and will be of greater range and destructiveness than the implements now employed. I believe that the telautomatic aerial torpedo will make the large siege gun, on which so much dependence is placed at present, obsolete.

A volume might be filled with such suggestions without exhausting the possibilities. The advance even under the conditions existing is rapid enough, but when the wireless transmission of energy for general use becomes a practical fact the human progress will assume the character of a hurricane. So all-surpassing is the importance of this marvelous art to the future existence and welfare of the human race that every enlightened person should have a clear idea of the chief factors bearing on its development.

The power of the future

We have at our disposal three main sources of life-sustaining energy—fuel, water-power and the heat of the sun's rays. Engineers often speak of harnessing the tides, but the discouraging truth is that the tidewater over one acre of ground will, on the average, develop only one horse-power. Thousands of mechanics and inventors have spent their best efforts in trying to perfect wave motors, not realizing that the power so obtained could never compete with that derived from other sources. The force of wind offers much better chances and is valuable in special instances, but is by far inadequate. Moreover, the tides, waves and winds furnish only periodic and often uncertain power and necessitate the employment of large and expensive storage plants. Of course, there are other possibilities, but they are remote, and we must depend on the first of three resources. If we use fuel to get our power, we are living on our capital and exhausting it rapidly. This method is barbarous and wantonly wasteful, and will have to be stopped in the interest of coming generations. The heat of the sun's rays represents an immense amount of energy vastly in excess of water-power. The earth receives an equivalent of 83 foot-pounds per second for each square foot on which the rays fall perpendicularly. From simple geometrical rules applying to a spherical body it follows that the mean rate per square foot of the earth's surface is one-quarter of that, or 20 3/4 foot-pounds. This is to say over one million horse-power per square mile, or 250 times the water-power for the same area. But that is only true in theory; the practical facts put this in a different aspect. For instance, considering the United States, and taking into account the mean latitude, the daily variation, the diurnal changes, the seasonal variations and casual changes, this power of the sun's rays reduces to about one-tenth, or 100,00 horse-power per square mile, of which we might be able to recover in high-speed low-pressure turbines 10,000 horse-power. To do this would mean the installment of apparatus and storage plants so large and expensive that such a project is beyond the pale of the practical. The inevitable conclusion is that water-power is by far our most valuable resource. On this humanity must build its hopes for the future. With its full development and a perfect system of wireless transmission of the energy to any distance man will be able to solve all the problems of material existence. Distance, which is the chief impediment to human progress, will be completely annihilated in thought, word and action. Humanity will be united, wars will be made impossible and peace will reign supreme.

September 9, 1915

NIKOLA TESLA SEES A WIRELESS VISION

Thinks his "World System" will allow hundreds to talk at once through the earth. Ends static disturbance. Inventor hopes also to transmit pictures by the same medium which carries the voice.

Nikola Tesla announced to The Times last night that he had received a patent on an invention which would not only eliminate static interference, the present bugaboo of wireless telephony, but would enable thousands of persons to talk at once between wireless stations and make it possible for those talking to see one another by wireless, regardless of the distance separating them. He said also that with his wireless station now in the process of construction on Long Island he hoped to make New York one of the central exchanges in a world system of wireless telephony. Mr. Tesla has been working on wireless problems for many years. Yesterday he exhibited an article published in *The Electrical World* eleven years ago, in which he predicted not only wireless telephony on a commercial basis but that it would be possible to identify the voice of an acquaintance over any distance. That its operator in Hawaii was able to distinguish the voice of an engineer friend at Arlington, Va., was announced by the American Telephone and Telegraph Company as the most marked triumph of its communication by wireless telephone from the naval radio station at Arlington to Pearl Harbor, Hawaii, a distance of 4,000 miles.

The inventor, who has won fame by his electrical inventions, dictated this statement yesterday.

"The experts carrying out this brilliant experiment are naturally deserving of great credit for the skill they have shown in perfecting the devices. These are of two kinds: First, those serving to control transmission, and, second, those magnifying the received impulse. That the control of transmission is perfect is plain to experts from the fact the Arlington, Mare Island, and Pearl Harbor plants are all inefficient and that the distance of telephonic transmission is equal to that of telegraphic transmission. It is also perfectly apparent that the chief merit of the application lies in the magnification of the microphonic impulse. It must not be imagined that we deal here with new discoveries. The improvement simply concerns the control of the transmitted and the magnification of the received impulse, but the wireless 'system is the same. This can never be changed.

"That it is practicable to project the human voice not only to a distance of

5,000 miles, but clear across the globe, I demonstrated by experiments in Colorado in 1899. Among my publications I would refer to an article in *The Electrical World* of March 5, 1904, but describing really tests I made in 1899. The facts which I pointed out in the article were of much greater significance than that of the experiments reported, although this should be taken in a scientific sense, as the experiments were simply scientific demonstrations. I pointed out then that the modulations of the human voice can be reproduced more clearly through the earth than through wire. It is difficult for the layman to understand, but it is an absolute fact that transmission through the earth with the proper apparatus is not more difficult than the sending of a message on a wire strung across a room. This wonderful property of the planet, that, electrically speaking, is through its very bigness, small, is of incalculable significance for the future of mankind.

"These tests made between Washington and Honolulu will act as an immense stimulus to wireless telephony and would be of much more value to the world if the principles of the transmission were understood. But they are not. Even now, fifteen years after the fundamental principles have been demonstrated and the possibilities shown, there are many experts in the dark.

"For instance, it is claimed that static disturbances will fatally interfere with the transmission, while, as a matter of fact, there is no static disturbance possible in properly designed transmission and receiving circuits. Quite recently I have described in a patent, circuits which are absolutely immune to static and other interferences *so much so that when a telephone is attached, there is absolute silence, even lightning in the immediate vicinity not producing a click of the diaphragm, while 'in the ordinary telephonic conversation there are all kinds of noises. Transmission without static interference has many wonderful properties, besides, first of which is that unlimited amounts of power can be transmitted with very small loss.

"Another contention is that there can be no secrecy in wireless telephone conversation. I say it is absurd to raise this contention when it is positively demonstrated by experiments that the earth is more suitable for transmission than any wire could ever be. A wireless telephone conversation can be made as secret as thought.

"I have myself erected a plant for the purpose of connecting by wireless telephone the chief centers of the world, and from this plant as many as a hundred will be able to talk absolutely without interference and with absolute secrecy. This plant would simply be connected with the telephone central exchange of New York City, and any subscriber will be able to talk to any other telephone subscriber in the world, and all this without any change in his apparatus. This plan has been called my "world system." By the same means I propose also to transmit pictures and project images, so that the subscriber will not only hear the voice, but see the person to whom he is talking. Pictures transmitted over wires is a perfectly simple art practiced today. Many inventors have labored on it, but the chief credit is due Professor Korn of Munich.

"His apparatus can be attached to a wireless plant and at any other wireless plant can be reproduced. I have undertaken this in the hope of establishing a service which would greatly facilitate the work of the press. A picture could be sent from a battlefield in Europe to New York in five minutes if the proper instruments were available.

"A further advantage would be that transmission is instant and free of the unavoidable delay experienced with the use of wires and cables. As I have already made known, the current passes through the earth, starting from the transmission station with infinite speed, slowing down to the speed of light at a distance of 6,000 miles, then increasing in speed from that region and reaching the receiving station again with infinite velocity.

"It's all a wonderful thing. Wireless is coming to mankind in its full meaning like a hurricane some of these days. Some day there will be, say, six great wireless telephone stations in a world system connecting all the inhabitants on this earth to one another, not only by voice, but by sight. It's surely coming.

New York Times, Oct. 4, 1915 p. 4, col. 3

CORRECTION BY MR. TESLA

—Of Wireless Apparatus he meant to say "Ineffective," not "Inefficient."

The Times received last night from Nikola Tesla a letter saying the inventor wished to correct a statement in his forecast of the possibilities of wireless, published in the Times of yesterday morning, when he was quoted as saying that the apparatus used by the American Telephone and Telegraph Co. to talk from Arlington to Hawaii was "inefficient." The inventor wrote that he wished to say that the apparatus was "ineffective."

"Although I can guess the character of the apparatus which was employed in projecting the human voice through 4,600 miles of space," Mr. Tesla wrote, "I am unable to judge of its efficiency, but from the technical particulars available I know that the plants are ineffective, inasmuch as they would have furnished currents of much greater volume and tension had they been differently designed. Incidentally, they would then have been immune against static disturbances, unfailing in their operation and adapted to secure secrecy of messages.

"In calling attention to this fact I have meant to give testimony to the excellence of the means of control and magnification resorted to by the experimenters. Had the same devices been used in connection with plants designed for maximum effect the results would have been such as to cause a most profound sensation and to stir great commercial interests all the world over, perhaps to the point of powerfully affecting and hastening the finish of the awful struggle in which nations of the earth are now engaged."

TESLA'S NEW DEVICE LIKE BOLTS OF THOR

He Seeks to Patent Wireless Engine for Destroying Navies by Pulling a Lever. To Shatter Armies Also. "Impractical," He says of Westerner's Plan to Circle Country with Electric Fire.

Nikola Tesla, the inventor, winner of the 1915 Nobel Physics Prize, has filed patent applications on the essential parts of a machine the possibilities of which test a layman's imagination and promise a parallel of Thor's shouting thunderbolts from the sky to punish those who angered the gods. Dr. Tesla insists there is nothing sensational about it, that it is but the fruition of many years of work and study. He is not yet ready to give the details of the engine which he says will render fruitless any military expedition against a country which possesses it. Suffice to say that the destructive invention will go through space with a speed of 300 miles a second, and manless airship without propelling engine or wings, sent by electricity to any desired point on the globe on its errand of destruction, if destruction its manipulator wishes to effect.

Ten miles or a thousand miles, it will be all the same to the machine, the inventor says. Straight to the point, on land or on sea, it will be able to go with precision, delivering a blow that will paralyze or kill, as is desired. A man in a tower on Long Island could shield New York against ships or army by working a lever, if the inventor's anticipations become realizations.

"It is not the time," said Dr. Tesla yesterday, "to go into the details of this thing. It is founded on a principle that means great things in peace, it can be used for great things in war. But I repeat, this is no time to talk of such things."

"It is perfectly practicable to transmit electrical energy without wires and produce destructive effects at a distance. I have already constructed a wireless transmitter which makes this possible, and have described it in my technical publications, among which I may refer to my patent 1,119,732 [1] recently granted. With transmitters of this kind we are enabled to project electrical energy in any amount to any distance and apply it for innumerable purposes, both in peace and war. Through the universal adoption of this system, ideal conditions for the maintenance of law and order will be realized, for then the energy necessary to the enforcement of right and justice will be normally productive, yet potential, and in any moment available, for attack and defense. The power transmitted need not be necessarily destructive, for, if existence is made to depend upon it, its withdrawal or supply will bring about the same results as those now accomplished by force of arms.

"But when unavoidable, the same agent may be used to destroy property and life. The art is already so far developed that great destructive effects can be produced at any point on the globe, determined beforehand and with great accuracy. In view of this I have not thought it hazardous to predict a few years ago that the wars of the future will not be waged with explosives but with electrical means."

Dr. Tesla then said that it would be possible with his wireless mechanism to direct an ordinary aeroplane, manless, to any point over a ship or an army, and to discharge explosives of great strength from the base of operations.

Asked to express an opinion upon the announcement last Sunday of Charles H. Harris, and electrical engineer of Los Angeles, that he would be able to surround this country with an electrical wall of fire in time of war, Dr. Tesla gave it as his opinion that Mr. Harris was not practical.

"It is hard to stamp as impossible such results as those described in the press dispatches to which you refer. Granted, however, that the project is feasible, it would take more than all the motive power obtainable in the United States to throw a wall of fire around the country. In fact, even the passage of small currents at considerable distances through air consumes a great deal of energy on account of the immense pressure required. So, for instance, in lightening discharges, energy may be delivered at the rated of billions of horsepower, though the currents are of smaller volume than those developed by electrical generators in our power houses."

(1) Apparatus For Transmitting Electrical Energy, No. 1,119,732, Dec. 1, 1914.

New York Times, December 8, 1915

WONDERS OF THE FUTURE

—Collier's Weekly, December 2, 1916

Nikola Tesla is an inventor, electrical wizard, and seer. He is the discoverer of alternating-current power transmission, the system of electrical conversion and distribution by oscillatory discharges, transmission of energy through a single wire without return, a system of wireless transmission of intelligence, transformer, etc. His laboratory is at Shoreham, L. I.

Many a would-be discoverer, failing in his efforts, has felt regret at having been born at a time when, as he thinks, everything has been already accomplished and nothing is left to be done. This erroneous impression that, as we are advancing, the possibilities of invention are being exhausted is not uncommon. In reality it is just the opposite. What has been so far done by electricity is nothing as compared with what the future has in store. Not only this, but there are now innumerable things done in old-fashioned ways which are much inferior in economy, convenience, and many other respects to the new method. So great are the advantages of the latter that whenever an opportunity presents itself the engineer advises his client to "do it electrically."

Water power offers great opportunities for novel electrical applications, particularly in the department of electrochemistry. The harnessing of waterfalls is the most economical method known for drawing energy from the sun. This is due to the fact that both water and electricity are incompressible. The net efficiency of the hydroelectric process can be as high as 85%. The initial outlay is generally great, but the cost of maintenance is small and the convenience offered ideal. My alternating system is invariably employed, and so far about 7,000,000 horsepower as been developed. As generally used, we do not get more than six-hundredths of a horsepower per ton of coal per year. This water energy is therefore equivalent to that obtainable from an annual supply of 120,000,000 tons of coal, which is from 25 to 50% of the total output of the United States.

Great possibilities also lie in the use of coal. From this valuable mineral we chiefly draw the sun's stored energy, which is required to meet our industrial and commercial needs. According to statistical records the output in the United States during an average year is 480,000,000 tons. In perfect engines this fuel would be sufficient to develop 500,000,000 horsepower steadily for one year, but the squandering is so reckless that we do not get more than

5% of its heating value on the average. A comprehensive electrical plan for mining, transporting, and using coal could much reduce this appalling waste. What is more, inferior grades, billions of tons of which are being thrown away, might be turned to profitable use.

Similar considerations apply to natural gas and mineral oil, the annual loss of which amounts to hundreds of millions of dollars. In the very near future such waste will be looked upon as criminal and the introduction of the new methods will be forced upon the owners of such properties. Here, then, is an immense field for the use of electricity in many ways. The manufacture of iron and steel offers another large opportunity for the effective application of electricity.

In the production of pig iron about one ton of coke is employed for every ton. Thus 31,000,000 tons of coke are used a year. There are 4,000,000 cubic feet of gases from the blast furnaces which may be used for power purposes. It is practicable to obtain 2,500,000 horsepower electrical energy in this way.

In the manufacture of coke some 41,000,000 tons of coal are employed in this country. From the gases produced in this process some 1,500,000 horsepower could be produced in the form of electrical energy.

I have devoted much thought to this industrial proposition, and find that with new, efficient, extremely cheap, and simple thermodynamic transformers not less than 4,000,000 horsepower could be developed in electric generators by utilizing the heat of these gases, which, if not entirely wasted, are only in part and inefficiently employed.

With systematic improvements and refinements much better results could be secured and an annual revenue of $50,000,000 or more derived. The electrical energy could be advantageously used in the fixation of atmospheric nitrogen and production of fertilizers, for which there is an unlimited demand and the manufacture of which is restricted here on account of the high cost of power. I expect confidently the practical realization of this project in the very near future, and look to exceptionally rapid electrical development in this direction.

But the time is very near when we shall have the precipitation of the moisture of the atmosphere under complete control, and then it will be possible to draw unlimited quantities of water from the oceans, develop any desired amount of energy, and completely transform the globe by irrigation and intensive farming. A greater achievement of man through the medium of electricity can hardly be imagined.

The present limitations in the transmission of power to distance will be overcome in two ways: through the adoption of underground conductors insulated by power, and through the introduction of the wireless art.

When these advanced ideas are practically realized we shall get the full benefit of water power, and it will become our chief dependence in the supply of electricity for domestic, public, and other uses in the arts of peace and war.

A vast and absolutely untouched field is the use of electricity for the propulsion of ships. The leading electrical company in this country equipped a large vessel with high-speed turbines and electric motors. The new equipment was a signal success. Applications of this kind will multiply at a rapid rate, for the advantages of the electrical drive are not patent to everybody. Gyroscopic apparatus will probably play an important part, as its general adoption on vessels is sure to come, Very little has yet been done in the introduction of electrical drive in the various branches of industry and manufacture, but the prospects here are unlimited.

Books have already, been written on the uses of electricity in agriculture, but the fact is that very little has been practically done. The beneficial effects of electricity of high tension have been unmistakably established, so that we are warranted in believing that a revolution will be brought about through the extensive adoption of agricultural electrical apparatus. The safeguarding of forests against fires, the destruction of microbes, insects, and rodents will, in due course, be accomplished by electricity.

In the not far distant future we shall see a great many new uses of electricity that will aim at safety. The safety of vessels at sea will be particularly affected. We shall have electrical instruments which will prevent collisions, and we shall even be able to disperse fogs by electric force and powerful and penetrative rays. I am hopeful that within the next few years wireless plants will be installed for the purpose of illuminating the oceans. The project is perfectly feasible; if carried out it will contribute more than any other provision to the safety of property and human lives at sea. The same plant could also produce stationary electrical waves and enable ships to .get any time accurate bearings and other valuable practical data, thus making the present means unnecessary. It could also be used for time signaling and many other such purposes.

In the great departments of electric light and power great opportunities are offered through the introduction of many kinds of novel devices which can be attached to the circuits at convenient hours to equalize the loads and increase the revenues from the plants. I myself have knowledge of a number of new appliances of this kind. The most important of them is probably an electrical ice machine which obviates entirely the use of dangerous and otherwise objectionable chemicals. The new machine will also require no attention and will be very economical in operation. In this way refrigeration will be effected very cheaply and conveniently in every household.

An interesting fountain, electrically operated, has already been brought out. It will very likely be extensively introduced, and will afford an unusual and pleasing sight in squares, parks, and hotels.

Cooking devices for all domestic purposes are now being made, and there is a large demand for practical designs and suggestions in this field, and for electric signs and other attractive means of advertising which can be electrically operated. Some of the effects which it is possible to produce by electric currents are wonderful and lend themselves to exhibitions. There is no doubt that much can be done in this direction. Theatres, public halls, and private dwellings are in need of a great many devices and instruments for convenience, and offer ample opportunities to ingenious and practical inventors.

Great improvements are also still possible in telegraphy and telephony. The use of a new receiving device, the sensitiveness of which can be increased almost without limit, will enable us to telephone through aerial lines or cables of any length by reducing the necessary working current to an infinitesimal value. This invention will enormously extend the wireless transmission of intelligence in all its departments.

The next art to be introduced is that of picture transmission telegraphically. Existing apparatus will be used. This idea of telegraphing or telephoning pictures was arrived at long ago, but practical difficulties have hampered commercial realization. There have been promising experiments, and there is every reason to believe that success will soon be achieved. Another valuable invention will be a typewriter electrically operated by the human voice. This advance will be of the utmost value, as it will do away with the operator and save a great deal of labor and time in business offices.

Many municipal improvements based on the use of electricity are soon to be introduced. There will be smoke annihilators, dust absorbers, ozonizers, sterilizers of water, air, food, and clothing, and accident preventers on streets, elevated roads, and in subways. It will become next to impossible to contract diseases from germs or get hurt in the city. Country folk will go to town to rest and get well.

Electrotherapy is another great field in which there are unlimited possibilities for the application of electricity. High-frequency currents especially have a great future. The time is bound to come when this form of electrical energy will be on tap in every private residence. It is possible that we may be able to do away with the customary bath. The cleaning of the body can be instantaneously effected simply by connecting it to a source of electric energy of very high potential, which will result in the throwing off of dust or any small particles adhering to the skin. Such a bath, besides being dry and time-saving would also be of beneficial therapeutic influence. New electric devices that will be a blessing to the deaf and blind are coming.

Electrical instruments will soon become an important factor in the prevention of crime. In court proceedings electric evidence can be made decisive. It will, no doubt, be possible before very long to flash any image formed in the mind on a screen and make it visible to a spectator at any place desired. The perfection of this sort of reading thought will create a revolution for the better in all our social relations. It is true that cunning lawbreakers will avail themselves of the same means to further their nefarious business.

The present international conflict is a powerful stimulus to invention of destructive devices and implements. An electric gun will soon be brought out. The wonder is that it was not invented long ago. Dirigibles and aeroplanes will be furnished with small electric generators of high tension, from which the deadly currents will be conveyed through thin wires to the ground. Battleships and submarines will be provided with electric and magnetic feelers so delicate that the approach of any body under water or in darkness may be easily detected. Torpedoes and floating mines will direct themselves automatically and without fail get in fatal contact with the object to be destroyed—in fact, these are almost in sight. The art of *telautomatics*, or wireless control of automatic machines at a distance, will play a very important role in future wars and, possibly, in the later phases of the present one. Such contrivances, which act as if endowed with intelligence, may take the shape of aeroplanes, balloons, automobiles, surface, or underwater boats, or any other form according to the requirement in each special case. They will have far greater ranges and will be much more destructive than the implements now employed. I believe that the *telautomatic* aerial torpedo will make the large siege gun, on which so much dependence is now placed, utterly obsolete.

December 2, 1916

ELECTRIC DRIVE FOR BATTLE SHIPS

—*New York Herald, February 25, 1917*

The ideal simplicity of the induction motor, its perfect reversibility and other unique qualities render it eminently suitable for ship propulsion, and ever since I brought my system of power transmission to the attention of the profession through the *American Institute of Electrical Engineers* I have vigorously insisted on its application for that purpose. During many years the scheme was declared to be impracticable and I was assailed in a manner as vicious as incompetent. In 1900, when an article from me advocating the electric drive appeared in the Century Magazine, Marine Engineering pronounced the plan to be the "climax of asininity," and such was the fury aroused by my proposals that the editor of another technical periodical resigned and severed his connection rather than to allow the publication of some attacks.

A similar reception was accorded to my wireless boat repeatedly described in the Herald of 1898. The patents on these inventions have since expired and they are now common property. Meanwhile insane antagonism and ignorance have been replaced by helpful interest and appreciation of their value. Recently the Navy Department has let contracts aggregating $100,000,000 for the construction of seven war vessels with the induction motor drive, and an equal sum is appropriated to cover the cost of four huge battle cruisers which are to be fitted out in the same way. This latter project is resisted by some shipbuilders, turbine makers, electrical manufacturers and engineers who, in fear of a fatal mistake by the government and under the sway of patriotic motives, urged upon the authorities the employment of the geared turbine.

Controversial correspondence

Numerous letters of protest have been written to C. A. Swanson, of the Senate Naval Committee, but what has so far come out of this correspondence is purely controversial and of no profit whatever to those who seek information. It is regrettable that the question should have been raised at this critical moment, when speedy preparation against threatening national perils is recognized as imperative, and in view of this no doubt should be permitted to remain in the public mind as to the superiority of the equipment recommended by the naval experts. In the following I shall endeavor to make this clear to the general reader.

The most efficient means of propulsion is a jet of water expelled astern from the body of the vessel. Though the theoretical laws governing its action were precisely expressed fifty years ago by Rankine, a singular and inexplicable prejudice against this device still prevails among engineers and writers of text books on hydraulics. But the far sighted are keenly alive to its possibilities. While our present motive resources do not admit of an advantageous use of the jet, it can be confidently predicted that it will soon be instrumental in a more complete conquest of the deep. I firmly believe that at this writing it is being applied to the submarines devastating the oceans, for their silence alone can explain why they escape so easily detection by microphonic instruments. The sound emitted is the Achilles talon of the undersea boat. Its suppression materially increases the destructiveness of the new weapon.

The helical propeller

However, under the existing conditions the best results are obtained in all kinds of surface craft with a helical propeller, which is operated in four different ways. First, straight from the shaft of the prime mover; second, by means of a gear wheel; third, through a hydraulic transformer, and, fourth, by an electric transmitter of power. As the screw in order to save energy must be revolved at a moderate rate, the first mentioned, or "direct drive," lends itself best to a reciprocating or rotary engine. The former is clumsy, the latter impossible, and competition has forced the turbine on the market. But, excessive speed being indispensable to its good performance, it had to be adapted to the propeller. This was, in a measure, accomplished by "staging"—that is, passing the steam through a number of turbines in succession—a plan obviously entailing great drawbacks, financial and otherwise. Need of reducing the bulk and cost of the machinery and insuring better working then compelled the adoption of the second arrangement—"geared turbine drive"—in which a peculiar pinioned wheel, first introduced by De Laval, transmits motion to the screw. Next, efforts to do away with certain limitations of this combination resulted in the third, or "hydraulic drive," the turbine actuating the propeller through a centrifugal pump and water motor. Finally, as the furthest step toward perfection, the last named disposition—"electric drive"—was resorted to. In this case the turbine imparts rotation to a dynamo, which in turn runs a motor carrying the screw on its shaft.

Advantages of types

Each of these forms has its supporters and champions. In principle the first would be the preferable were it not handicapped in many respects. The second type is cheap, but the gear is a serious objection. Though less economical, the third commends itself by a number of practical and valuable features. As to the last, it is not only very efficient but obtains results impossible with other forms. The law of survival of the fittest is asserting itself, and the struggle for supremacy is now on between the geared turbine and the electric drive.

Through gradual improvement of the cutting tools, scientific design, metallurgical advances and refinement of lubricants, the so-called herringbone gear has been brought to great perfection. De Laval attained an efficiency of 97% and MacAlpine, Melville and Westinghouse 98.5% in the transmission from the driving to the driven shaft. On the other hand, 93.75% may be considered as the maximum with electrical apparatus. This means that with the gear the same turbine would impart 5% more power to the propeller, which should increase the speed of the cruiser from thirty-five to a little more than thirty-five and one-half knots. As it also appears at first sight that the electric drive requires additional space, is heavier and more costly, it is only natural for those who have not made a thorough study of all its phases to decide in favor of the gear.

Some fatal mistakes

But a careful inquiry into the subject would induce them to reverse their opinion. In estimating the relative merits of these essentially different propelling means they make two fatal mistakes. The first is to take the power transmitted under abnormal conditions as a criterion; the second to draw a parallel between installments entirely unlike, one primitive, the other elaborate, the former being incapable of fulfilling important functions of the latter. When premises are erroneous deductions therefrom must needs be faulty. Thus the opponents of the electric drive have been led to the conclusion that it is less efficient than the gear, of greater weight, more expensive and uncertain of success. How much truth there is in these contentions will be apparent from an examination of well established facts.

The electric drive is of complex influence on results in ship operation. For the sake of brevity it will be viewed only in the following principal aspects: (1) turbine performance, (2) power transmitted to the propeller, (3) efficiency of the screw, (4) low power cruising, (5) high power action, (6) fuel consumption by auxiliaries and apparatus for ship use, (7) general economy and (8) promptness and precision of control of all effects, internal and external.

The present turbines are extremely unsuitable for ship propulsion. They offer a striking example of an antiquated invention of small value elevated to a position of extraordinary commercial utility through profound research and astonishing mechanical skill. With hundreds of thousands of thin blades easily destroyed, buckets that through corrosion and erosion soon become wasteful and small clearances between surfaces rotating at terrific rates, they are a cause of constant danger and hazard.

Irreversible turbines

But their cardinal defect is that they are irreversible, which necessitates the employment of separate turbines for backing. These, besides involving great expense and considerable friction loss, impose narrow limits on the temperature of the working medium. Very high superheat, so desirable in thermodynamic conversion, is out of the question in such, perishable structures, but from 200 to 300 degrees F. are permissible.

To that extent, then, the turbine is at an advantage when driving a dynamo. Two hundred degrees superheat will usually effect a saving of about 23% of steam and 10% of fuel. This, however, is not the only gain. The turbine, freed from all the impediments of the gear drive, is capable of being safely run at a higher peripheral speed with a correspondingly increased efficiency and output. Thus, by moderate superheat and other simple and allowable expedients, it becomes practicable to develop 25% more power from the same fuel, and this alone would make the electric drive decidedly superior to its competitor.

A mechanical tour de force

As regards the power transmitted from the turbine to the propeller, it will seem in the light of the preceding that the gear is better by 5%. That may be the case in exceptional tests, but it is quite different in actual service. To this is to be traced the error of those who are taking results obtained at constant load as standard of comparison. The perfection of the modern high speed gear was a veritable tour de force of the scientific machinist. It is a wonderful device, but it also has its inseparable weaknesses and shortcomings. Since the friction loss in it is sensibly constant through a wide range of performance, a relatively great amount of energy is absorbed at small load. The gear is likewise very sensitive to shocks and vibrations, which break down the capillary oil film, vital to smooth running. In consequence, there is great waste of power when the resisting force is subject to frequent and sudden fluctuations, Measurements I made with turbine gears have shown that while the efficiency with steady and normal effort was 96%, not more than 90% was realized with a rapidly varying load. This is what might be expected in practice. Any one who has listened to the tortured engines of a steamship in heavy sea could not have failed to observe how the turning effort varies as the vessel rolls, pitches and ploughs through billows and conflicting undercurrents. A similar state of things is apt to confront a war ship in action, as was evidenced in recent naval engagements, when mountains of water were raised by the exploding shells. Under such circumstances the gear is at a great disadvantage, as the electric drive is susceptible to these drawbacks in a much smaller degree. The idea that the gear transmits more of the primary power to the propeller than the combination of dynamo and motor is, therefore, largely illusionary. There is ample evidence, experimental and inferential, that rather the opposite is true.

Superiority of electric drive

Considering the efficiency of the screw as distinct from that attained in the transmission of power, it is admittedly better with the electric drive, this conclusion being entirely based on the superior adaptability and flexibility of the system. But there are deeper causes which should be taken into account. The interposition of electromagnetic means between the turbine and propeller materially reduces the loss due to shocks, vibrations, racing and other disturbances owing to inherent elastic resilience and equalizing tendency. The saving of energy thus effected at high speed and in a heavy sea is considerable.

Economy in cruising is one of the most desirable qualities of a war vessel. This is its ordinary use, for the chance of ever being engaged in battle is remote. The bitterest opponents of the electric drive do not deny that it excels in this feature, upon which the manufacturer chiefly relies in guaranteeing a fuel consumption from 10 to 12% smaller than with the gear. The latter is hopelessly condemned by inability of adjustment to varying speed and wasteful in cruising operations, while the former is readily adaptable and economical under all conditions.

Another quality of the electric drive, which may prove especially valuable in action, is its capacity of carrying great overload without danger, owing to the nature of the connection between the turbine and propeller, as explained. The gear is rigid and unyielding and any increase of effort, particularly if sudden, may cause a breakdown.

Saving of power

Referring to the auxiliaries proper and other apparatus for ship use, to which are chargeable approximately 20% of the fuel consumed, a very substantial saving of power will be achieved through the introduction of the electrical method.

Quite apart from this central station supply will be operative in reducing other waste, many accessories will be dispensed with and general economy materially increased.

But from the military point of view the quickness, ease and precision of control will perhaps be the most significant of the advantages gained. Everything may be made to respond instantly to the pressure of a button. By reversing the motors the vessel may be brought from full speed to a stop within its length. It will be possible to make it go through all evolutions with extraordinary rapidity and a perfection of manoeuvre, undreamed of before, will be attained.

A curious mistake is made by the advocates of the turbine gear in estimating relative weight. It hardly needs be stated that it is unfair, if not absurd, to compare arrangements of widely different character and scope. Only such as are capable of accomplishing the same results should be considered. Now, a gear drive corresponding to the electric would consist of four main turbines with gears, four reversing turbines of the same capacity, and eight smaller driving and reversing turbines for cruising. This agglomeration of complex and not all too rugged machinery, with its network of water, air and oil pipes, valves, pumps and attachments, would by far exceed in weight the proposed electric drive, and would also require better structural protection, not to speak of other defects and shortcomings.

Question of weight

It should be observed, however, that the weights must be taken in their relation to that of the ship. One equipment may be heavier than another, but if it is more efficient and thereby reduces the weight of fuel and other cargo, it is for all purposes the lighter of the two.

The same is true of the cost. Comparative figures mean nothing. The question is whether the investment of capital is justified by what is to be accomplished. But enough has been said to show that for results in all respects equivalent, assuming them to be possible, the gear type, notwithstanding all claims to the contrary, would be more expensive.

That the electric drive is experimental and uncertain of performance is the least tenable of adverse assertions. In the first place, it has been successfully employed on a number of vessels and a great many more are being built. It also was found to be capable of an efficiency higher than that of any other form. But this is quite immaterial. The confidence that in the present instance all expectations will be realized is not based on a few demonstrations, but on years of experience with power plants ever since my system was commercially inaugurated. Tens of millions of horse power of induction motors are now in use the world over and no failure is recorded.

New cruisers' requirements

The new cruisers will require 180,000 horse power each, which, if necessary can be developed by four units of 45,000 horse power. Turbines of that capacity have been constructed and are in operation today. Dynamos of corresponding output have been installed at several places and are supplying light and power to large cities and districts. Induction motors of 15,000 horse power are being turned out by the manufacturers and can be produced in any size that may be desired, for of all kinds of motors this is the simplest and most dependable. Long since the whole system has been worked out and perfected to the least detail. The project is colossal, but it can be readily carried out by any of the few concerns who are possessed of the proper facilities. Not even a new tool has to be made. There is nothing whatever untried or hazardous about the electric drive.

Much stress is laid on reports, still to be verified, that it was rejected by England and Germany. But this is of no consequence. It has been rejected here more than once. Besides, there was war in the air of Europe and the time for radical innovations unpropitious. Moreover, the Diesel engine was looming up with large possibilities and Dr. Fautinger's hydraulic drive was being tried. The beginning must be made somewhere and it would be deplorable indeed if the United States, where the invention was first announced and introduced on a gigantic scale, were the last to recognize it. Such mistakes have happened only too often. The foreign navies are not in the habit of keeping the press informed of their doings and it is safe to predict that if progress in this country is much retarded there will be a repetition of previous disappointments.

It is unnecessary to dwell on other objections which are of minor importance and of no bearing on the principle. Without going into tedious technical discussion, it may be stated that the electric drive, if judiciously designed, will save not less than 25% of fuel and, with due regard to this and certain specific and invaluable advantages, will be lighter, cheaper and in every respect more dependable than the gear. In fact, I believe that a scheme can be devised permitting the placing of all vital parts below the water line. In view of this, it is to be hoped that the Secretary of the Navy will not pay attention to the protests of rivals, however patriotic, but will cause the good work to be pushed to completion with all the power at his command.

These statements are to be understood as reflecting the present state of the art. The advent of a reversible turbine will profoundly affect the situation in favor of the gear. Such a machine has been perfected and was described in the Herald of October 15, 1911. It is the lightest prime mover ever produced and can be operated without trouble at red heat, thereby obtaining a very high economy in the transformation of heat energy.. I anticipate its speedy and extensive application to ship propulsion. But although an ideally simple and very inexpensive drive will be thus provided, there will still be weighty reasons for adopting the electric method on war ships. In order to dissipate all doubt created in the mind by diversified engineering opinion, I will make known but One of them, which in itself is sufficiently consequential and convincing to dispense with further arguments.

Disarmament impossible

It is idle to dream of disarmament and universal peace in the face of the terrible events which are now unfolding. They prove conclusively that no country will be allowed to control all others by any means. Before all peoples can feel secure of their national existence and worldwide harmony established certain obstacles will have to be removed, the chief of which are German militarism, British domination of the seas, the rising tide of Russian millions, the yellow peril and the money power of America. These adjustments will be slow and *pennible* in conformity with natural laws. International friction and armed conflict will not be banished from the earth for a long period to come. The drag on human progress would not be so great if war energy could be maintained in purely potential form. This can and will be done through the universal introduction of wireless power. Then all destructive energy will be obtained without effort merely by controlling the life sustaining forces of peace.

The maintenance of war ships and other military implements involves an appalling waste. A vessel costing twenty millions of dollars is rendered virtually worthless in the short span of ten years, deteriorating at the mean rate of two million dollars a year, interest not considered. Hardly more than one out of fifty serves its real purpose. To lessen this ruinous loss and exploit certain inventions I elaborated a scheme some years ago. It was recognized as rational bur financially and in other ways difficult of realization. Now, when national economy and preparedness have become burning questions, it assumes special import and significance.

To use war ships in peace

The underlying idea is to make war ships available for purposes of peace in a profitable manner, at the same time improving them in a number of features. I am aware of the proposal lately advanced to employ them as carriers of commerce, but this is not feasible and would be an impediment to further perfection. My project primarily contemplated the installment of the electric drive and the use of the turbo-dynamos for light and power supply and manufacture of various valuable products and articles aboard or on land. This would be a step in the direction of present development meeting the objects of both military and industrial preparedness. I further intended the creation of a type of vessel on radically different principles, which would be a precious asset in peace and ever so much more destructive in war. The new cruisers, if equipped as planned by the Navy Department, will constitute four floating central stations of 180,000 horse power each. The turbines and dynamos are designed for highest efficiency and operate under most favorable conditions. The power they are capable of developing represents a market value of several million dollars a year and could be advantageously utilized at places where fuel can be readily obtained and transport is convenient. The plants would also prove of great value in cases of emergency. They could be quickly sent to any point along the coast of the United States or elsewhere and would enable the government to lend speedy assistance whenever necessary.

But this is not all. There is another and still more potent reason for adopting the electric method. It is founded on the knowledge that at a time not distant the present means and methods of warfare will be revolutionized through novel applications of electric force.

February 25, 1917

PRESENTATION OF THE EDISON MEDAL TO NIKOLA TESLA

Minutes of the annual meeting of the *American Institute of Electrical Engineers*, held at the Engineering Societies Building, New York City, Friday evening, May 18, 1917

President Buck called the meeting to order at 8:30 o'clock.

The President: As you know, gentlemen, this is the Annual Meeting of the Institute, and the first thing on the program will be the presentation of the Report of the Board of Directors by our Secretary, Mr. Hutchinson.

Secretary Hutchinson: The annual report of the Institute for the year has been printed and distributed, and it is not my intention to take the time to read it. It consists of a brief resume of the activities of the institute for the entire year, and includes abstracts of the reports of the various committees.

The President: Gentlemen, the next order of business of the evening will be the announcement of the election of officers and managers for the coming year. The report of the Tellers will be presented by the Secretary, Mr. Hutchinson.

Secretary Hutchinson then presented the report of the Tellers, which showed elections as follows:

President:
W. W. Rice, Jr.
Vice-Presidents:
Frederich Bedell,
John H. Finney,
A. S. McAllister
Managers:
Walter A. Hall,
William A. DelMar,
Wilfred Sykes
Treasurer:
George A. Hamilton

The President: It is our privilege from time to time to honor those in the electrical profession who have rendered conspicuous service towards this advance. We have the pleasure this evening of so honoring Mr. Nikola Tesla. Dr. Kennelly, who is Chairman of the Edison Medal Committee, will tell us what the Edison Medal is and what it stands for. I take pleasure in introducing Dr. A. E. Kennelly.

Dr. A. E. Kennelly: Mr. President, Ladies and Gentlemen: It is my privilege to say a few words to you upon the origin and purpose of the Edison Medal. First of all, many people suppose that the Edison Medal is a medal presented by Mr. Edison. That is a mistake. Mr. Edison has been so busy during his life receiving medals that he has not time for the delivery of any. The Edison Medal owes its existence to the action of a group of his admirers who in a very remarkable Deed of Gift, a printed copy of which I have here, have set apart a fund for the purpose of the annual award of a medal for meritorious achievement in the electrical science and art. This deed of gift originally recited, in 1904, that the medal should be annually awarded for the best graduating thesis by the students of electrical engineering in the United States and Canada, but in the years that elapsed between 1904 and 1908, I think I am correct in saying that there were no successful candidates, at least for the medal under those terms, although there may have been many aspirants. It is supposed that the dignity of the medal and the junior character of the tyros restrained them in their modesty from making proper application.

Be that as it may, finding that the applicants held back under the original terms of the deed of gift, the matter was taken up further and the original body of men redrafted the deed and placed it in the hands of the *American Institute of Electrical Engineers* to award the medal, under the choice of a Committee, annually, for meritorious achievement, as indicated, to any resident of the United States, its dependencies, or Canada, during each administration year. The monument which they raised to Mr. Edison by their act is, I think you will admit, one of the most wonderful that has ever been raised to any scientist.

The Deed of Gift says that there shall be twenty-four members appointed by the *American Institute of Electrical Engineers*, sixteen from the membership at large, three ex officio members, the President, Secretary and Treasurer, and the balance from the Board of Directors.

Every year the medal is due to be awarded. There have been already six medals awarded, not counting the medal which is to be awarded to-night, and the recipients of these medals have been Elihu Thomson, Frank J. Sprague, George Westinghouse, William Stanley, Charles F. Brush, Alexander Graham Bell. I think you will say that is a fitting selection for the galaxy of names that we look forward to in the future, all of them, in honoring Mr. Edison's achievements, which have been so noteworthy, that every household in the land holds his name as a cherished household word. We may look forward to a time say a thousand years hence, when, like this evening, the *American*

Institute of Electrical Engineers, or its successors or assigns, shall be convoked, and at which the medal of the year will be awarded to its One Thousand and Seventh recipient, and all that long galaxy of names will represent those individuals who have contributed to the recognition of the achievements of Mr. Edison and his gift to humanity.

In addition to what this deed of gift shows in honor of Mr. Edison himself, there is, of course, the very great honor that it bestows upon the recipient. The Deed of Gift says there shall be twenty-four jurors, which you see is twice the number of jurors that is allowed in the palladium of our liberties, but whereas the jurors of ordinary life convict by unanimous vote, the twenty-four jurors of the Edison Medal convict, at least, by a two-thirds vote, so I think I am correct in saying that their convictions have hitherto been entirely unanimous, and in this particular case I can certainly declare that it has been unanimous.

The galaxy of names that will be produced and has already been produced under this deed of gift will be great and noteworthy. It will not be necessary to look into a "Who's Who" to see who has been great and notorious and worthy of merit in electrical science and art. The historian of the future will simply say "Give me the list of the Edison Medallists."

This deed of gift is also wonderful in other respects. It has marvelous flexibility and marvelous rigidity in certain directions. It provides for the possibility of a change of personnel, a change of procedure and a change of administration as time and things may change. It only makes one rigid restriction, and that is that the name "Edison Medal" shall never be changed. Times may change and persons and institutions, the Institute itself may go out of existence, and there is provided machinery whereby if the Institute should say it is tired, or it has gone out of existence, or can no longer administer the medal, that the five oldest universities of the country, maintaining a course in electrical engineering, shall be able to place the administration of the medal by their vote in the hands of some new institution, so you see that this is a very wonderful Deed of Gift that I have the honor of bringing to your notice here this evening in connection with the bestowal of this medal. Another great advantage that the medal presents is that its recipient shall be alive, that is to say he must not only have been convicted of great merit and meritorious achievement, but he must also have escaped being run over by automobiles up to the time of the presentation. That represents a great advance over those methods of awarding distinction which depend upon the demise of the individual. You know somebody has said that a great statesman is a successful politician who is dead, but we may say that the Edison Medallist is a great electrician who is alive, and you know it is wonderful how little is known sometimes about a man's demise, however much may be known about his work. The other day I met a *negro* in the South, and I happened to mention Washington, and what was done by George Washington who died so many years ago, and he said, "For the Lord's sake, I didn't even heard the man was sick." So you see that even George Washington, no matter how meritorious he might have been in electrical matters, could not possibly be the recipient of an Edison medal.

We have recently received the sad news in this country of the demise of the great English electrical engineer Silvanus P. Thomson, a man who had many admirers and many friends in this country, many students here, a man whose name and work is dear to so many of us, and efforts are now being made to contribute to a fitting memorial for him by the purchase of his library as an appendix to the great library of the *British Institution of Electrical Engineers*, and a notice is given on page 126 of the *May Proceedings of the Institute* regarding that movement, and you will find it a very worthy movement. Subscription lists are open to the members of this Institute, as a matter of courtesy, and a matter of recognition, that so many of his friends in this country could be allowed to give some contribution to this great Thomson Memorial. It is a fact, as I dare say many of you know, that the funds for Lord Kelvin's Memorial Window in Westminster Abbey were largely raised in America, more largely, I believe, than they were in England itself. In this case I am led to believe that they do not want the funds so much, as they want the names of sympathizers with the project, the support of those who recognize the work and merit of Silvanus P. Thomson. But how much better it would be if we were presenting a memorial to Silvanus P. Thomson living, as we are able to do in the case of the Edison Medal, than presenting a memorial to Silvanus P. Thomson passed away.

Then one thing more: This deed of gift between its lines suggests a third and by no means least important purpose, and that is a safeguard, lest we forget. We in this time and of this continent, particularly we of the electrical profession, with our faces ever turned to the rising sun, are so apt to forget that there has been a preceding night of trouble, difficulty and dismay, and that the tools of our trade which lie to our hand were only secured by hard work and toil against all sorts of distress and discouragements. The Edison Medal is our means for reviving your memories of the past and pointing out that the things we look upon as the sunshine of heaven now have been arrived at by the hard work, the inspiration, or, as Edison himself would say, the perspiration of those who have worked in the past.

We remember that beautiful book, "The Twins", where Budge and Toddy the children always insisted at all times of the day and night to see the wheels go round and have their father's watch opened for them. The medallist tonight was a man who saw in his mind wheels going around when there was no means of getting alternating current motors to rotate, when the alternating current would do everything but make wheels go round, and he devised the rotating magnetic field so prophetically in his mind's eye that the rotating magnetic wheel would set wheels going 'round all over the land and all over the world, and the vision is carried out, and we recognize that vision here, and the Medal is partly as a reminder that we should not forget the fact, that the medallist also made the phenomenon of high frequency known to us all practically for the first time, and that what he showed was a revelation to science and art unto all time.

For this third purpose the Edison Medal has been created, and we may look far forward into the future and see it given year after year for, let us hope, a thousand years from now, in the year 2917, to witness the ceremony which we may well expect will be furnished at that time. (Applause)

The President: Dr. Kennelly has referred to the struggles of the past, and we are very fortunate in having with us to-night one who was associated with Mr. Tesla in his struggles of the past. Gentlemen, I want to introduce to you Mr. Charles A. Terry, who will tell us something about these struggles and the early work of Mr. Tesla, for which we assign to him the Medal tonight.

Charles A. Terry: Mr. Kennelly spoke of the thousandth award of the Medal. I think there is a peculiar significance in the fact that Mr. Tesla is to receive the seventh medal — the seventh in most calculations is considered a most excellent number to have.

The convolutions of the brain of one man impel him to paint upon canvas the visions of his soul; another conceives beauty of form which he must express in plastic art or in architectural structure; others are driven by an inner force to devote their lives to the discovery of the secrets of unexplored regions of the earth, or to search out the mysteries of the stars; some find themselves compelled by an irresistible desire to learn through archeological research the forgotten achievements of ancient races; still others seek to ascertain and formulate the physical laws which govern the processes of nature, and men with other talents find themselves urged by a like persistent force to devise and disclose new means whereby those laws may be utilized for the further benefit of mankind.

It is this God-given desire to accomplish and to give, that has produced the Michelangelos, the Galileos, the Sir Christopher Wrens, the Livingstons, Newtons, Franklins, Westinghouses, Edisons and scores of other makers of history; men whose names we retain in affectionate remembrance, because they earnestly responded to the call from within and by patient toil conceived thoughts and discovered things of value which they promulgated for the benefit of their fellow men.

Although hope of reward may and properly should exist as an added impulse to such endeavors, the chiefly effective force compelling to the long hours of hard work and personal sacrifices of such men is the—I must—which speaks from within the soul, and with our truly great men the desire for reward is better satisfied by a consciousness of achieving their aims and by the just commendation of their fellows than by material gain, except insofar as the latter may aid in the further advancement of their tasks.

Fortunately, men generally are not jealous nor envious of the doers of great deeds and the givers of large benefits, but from the depths of their hearts are grateful and they are satisfied only when evidence of their gratitude can be brought home to the giver.

It is because of this desire to show gratitude to, and appreciation of, one of our fellow members, whose name history will rightly record in the same distinguished class with those we have mentioned that we are gathered to-night.

Twenty-nine years ago this month, there was presented before this Institute, a paper of unusual import. It is entitled "A New System of Alternate Current Motors and Transformers". The author, Nikola Tesla, was then only 31 years of age, and but four years a resident of this country. His early life was spent near his birthplace not far from the Eastern Adriatic Coast. His father a Greek Clergyman and his mother, herself of an inventive mind, secured for their young son a comprehensive training in mathematics, physics and philosophy. At the age of 22 he had completed his studies in engineering at the Polytechnic School in Gratz and also a course in the University of Prague; and in 1881 began his practical work at Budapest. In 1883 he was located in Strasbourg, engaged in completing the lighting of a newly erected railway station. Shortly after finishing this task he came to the United States. Mr. Tesla's first work in this country was upon new designs of direct current arc and incandescent lighting systems for the Edison Company.

Throughout all these years his desire had been to find an opportunity to demonstrate the truth of a conviction which became fixed in his mind while studying direct current motors in school at Gratz in 1878; the conviction was that it should be possible to create a rotating magnetic field without the use of commutators. While at Strasbourg, Tesla had succeeded in producing the rotation of a pivoted iron disc placed in a coil traversed by alternating currents, a steel bar being projected into the coil in the neighborhood of the disc. His conception of the reason for this rotation at that time was that a lag occurred in the subsidence of the magnetism of both the disc and the steel bar between successive current waves, and that the mutual repulsions caused the disc to revolve. By some fortunate process of reasoning he conceived while in Budapest (in 1882) that by using two or more out-of-phase alternating currents respectively passing through geometrically displaced coils it would be possible to develop his long sought progressively shifting magnetic field.

Lack of funds and facilities for working out his theory compelled still further postponement, but in 1885 Tesla had the good fortune to interest men of means in a direct current arc light which he had devised, and subsequently a laboratory was equipped for him in Liberty Street, New York, and here at last he found opportunity to demonstrate the correctness of his long cherished theory. In 1887 he was able to exhibit to his business associates and to Professor William A. Anthony, whose expert opinion they sought, motors having such progressively shifting fields without the use of commutators, as he had foreseen nine years before.

Having thus demonstrated the correctness of his theory and the feasibility of its application, it remained for Tesla to work out various practical methods of applying the principle, and the rapidity and wonderful way in which he surrounded the entire field of constant speed, synchronous, induction and split-phase motors is beautifully set forth in his paper of May 18th, and in the numerous patents issued May 1st, 1888, and succeeding years, covering the forms of electric motors which have since become the almost universal means for transforming the energy of alternating currents into mechanical energy.

It is somewhat difficult to eliminate from our minds the developments of the past thirty years which have now become every day features of the electrical industry, and to realize the meagreness of the then existing knowledge of alternating current phenomena. The commercial use of alternating current systems of distributions was then scarcely two years old. The Gaulard & Gibbs system of series transformers had been used abroad in a limited way for a slightly longer period but the multiple arc system based upon the so-called — Stanley Rule — which initiated the great development of the present system, was not put in practical operation in the pioneer Great Barrington plant until March 1886. It was then recognized that while the alternating -current possessed wonderful possibilities for electrical distribution for lighting purposes, two almost necessary devices were lacking to render it a complete success, one a meter, the other a power motor. Professor Elihu Thomson promptly devised a successful form of meter, the motive portion of which comprised a laminated field and armature, the coils of the latter being periodically close-circuited during revolution by a commutator. To fill the demand for a power motor, however, the most promising device then suggested was a series commutator motor with laminated field and armature cores, but no satisfactory results had been obtained. Such was the situation when Tesla's achievement was announced in the Institute paper to which reference has been made.

His Honor Judge Townsend of the United States Circuit Court, in an opinion rendered in August, 1900, as the outgrowth of some patent litigation on the Tesla inventions, concisely defines the underlying characteristic of the Tesla motor as follows:

"Tesla's invention, considered in it essence, was the production of a continuously rotating or whirling field of magnetic forces for power purposes by generating two or more displaced or differing phases of the alternating current, transmitting such phases, with their independence preserved, to the motor, and utilizing the displaced phases as such in the motor."

Among the first to recognize the immense importance of Mr. Tesla's motors were Mr. Westinghouse and his advisors, Mr. T. B. Kerr, Mr. Byllesby, Mr. Shallenberger and Mr. Schmid, and in June Mr. Westinghouse secured an option which shortly resulted in the purchase of the patents, thus bringing under one ownership the alternating current transformer system of distribution, and the Tesla motor. It is interesting to here note that Mr. Shallenberger had about two weeks before the publication of the Tesla patents independently devised an alternating current meter, the principle of operation of which was that of the Tesla motor, and whatever might have been Mr. Shallenberger's natural disappointment upon finding himself thus anticipated, he at once recognized that to Mr. Tesla belonged the honor of being the first to solve the great fundamental problem of an alternating current motor. A warm friendship between these two men began at once and continued throughout Mr. Shallenberger's life, and Mr. Tesla rejoiced to accord to Mr. Shallenberger full credit for the latter's brilliant work in producing what is now the standard meter for alternating currents.

As illustrating the generous gentleness of Tesla's character, I wish to here quote from testimony given by him in 1903. Referring to Shallenberger, Tesla said:

"I clearly remember that in the first days when I came to Pittsburgh he took me to lunch at the Duquesne Hotel, and when I told him that I was sorry that I had anticipated him, I saw tears in his eyes. That incident I remember vividly; but what has preceded it I cannot remember now. Perhaps it is because this impression was so vivid that it has destroyed the preceding ones, which were weaker."

It is characteristic of Tesla that he should so deeply regret the disappointments of another.

Owing in a measure to the circumstance that the then prevailing rate of alternation of the alternating current system was 16,000, the commercial introduction of Tesla motors was somewhat retarded during the first few years, that rate being found less adapted to the motor work than a lower rate. Today, however, wherever alternating current systems are used Tesla motors abound. Without such motors the alternating current system would have remained seriously restricted in its use.

Before passing to a consideration of other of Tesla's activities, it will be appropriate to refer again to the opinion of Judge Townsend, from which I quote the following:

> "The Tesla discovery for which these patents were granted revolutionized the art of electrical power transmission, as well demonstrated in the record from both judicial and scientific standpoints."

In the closing passage of the opinion, Judge Townsend pays further tribute to Tesla in the following words:

> "It remained to the genius of Tesla to capture the unruly, unrestrained, and hitherto opposing elements in the field of nature and art and to harness them to draw the machines of man. It was he who first showed how to transform the toy of Arago into an engine of power, the—Laboratory experiment—of Baily into a practically successful motor, the indicator into a driver. He first conceived the idea that the very impediments of reversal in direction, the contradictions of alternations, might be transformed into power-producing rotation, a whirling field of force.
>
> "What others looked upon as only invincible barriers, impassable currents, and contradictory forces, he brought under control and by harmonizing their directions taught how to utilize in practical motors in distant cities the power of Niagara."

Imagination developed to a high degree is a marked characteristic of all great inventors, so it is of our great poets, artists, philosophers, generals, and, in fact, of all great originators of thought and motion. The power to picture in the mind things not yet existent is an underlying characteristic of most great men. But imagination to be effective must be combined with a just sense of proportion, a logical appreciation of limitations, and a capacity for unremitting application. Mr. Tesla combines these qualities in a marked degree, and particularly does he possess the faculty of projecting his thought far into unexplored regions, not only of science but of philosophy. His passion for searching out the ultimate is charmingly evidenced by the following extract from his lecture before this Institute at Columbia College, May 20th, 1891:

"In how far we can understand the world around us is the ultimate thought of every student of nature. The coarseness of our senses prevents us from recognizing the ulterior construction of matter, and astronomy, this grandest and most positive of natural sciences, can only teach us something that happens, as it were, in our immediate neighborhood; of the remoter portions of the boundless universe, with its numberless stars and suns, we know nothing. But far beyond the limit of perception of our senses the spirit still can guide us, and so we may hope that even these unknown worlds—infinitely small and great—may in a measure become known to us. Still, even if this knowledge should reach us, the searching mind will find a barrier, perhaps forever unsurpassable, to the true recognition of that which seems to be, the mere appearance of which is the only and slender basis of all our philosophy.

"Of all the forms of nature's immeasurable, all-pervading energy, which, ever and ever changing and moving, like a soul animates the inert universe, those of electricity and magnetism are perhaps the most fascinating."

The impress made upon the world by the deeds of a great inventor cannot be measured by the number of patents which he has received nor by the monetary reward secured nor by the mere exploitation of his name. Often his greatest gifts are in the form of inspiring contributions to the literature, filled with suggestions of lines of thought which lead others to work in untried fields. This is especially true of a series of lectures delivered by Mr. Tesla upon the subject of high frequency, high potential currents. The first of the series was given at Columbia College in 1891, before this Institute. During 1892 and 1893 this lecture with additional data and experiments was repeated in London, Paris, Philadelphia and St. Louis.

Referring to an interesting interview with Mr. Tesla appearing in a New York daily in 1893 regarding the St. Louis lecture the Editor of *The Electrical World* says:

"Mr. Tesla, in his own graceful way, tells the story of his life and the history of some of his more important inventions. Perhaps there is no living scientist in whose life and work the general public takes a deeper interest, especially in this country."

Tesla's fundamental purpose was to publish the results of an extended research and of a series of experiments patiently conducted at his laboratory and elsewhere through many years. During these lectures he exhibited to the audience numerous experiments displaying striking and instructive phenomena. He also described many novel pieces of apparatus such, for instance, as his high-frequency generator and induction coils and his magnetically quenched arc. Mr. Erskine Murray in his treatise upon Wireless Telegraphy, referring to certain of these early inventions of Tesla says:

"Among many other inventions, made as early as 1893, perhaps the most important to wireless telegraphists is his method of producing long trains of waves of high frequency, and of transforming them to higher voltage. After several unsuccessful attempts he completed an alternator which could be run at 30,000 periods per second, and designed a form of transformer capable of transforming these currents to very high voltage. He also showed that his transformer, or "Tesla coil" as it is usually called nowadays, could transform currents of much higher frequencies than were obtainable from his alternator, even currents of 100,000 or 1,000,000 periods per second, such as are produced by the oscillatory discharge of a Leyden jar."

The London lecture was given under the auspices of the British Institution of Electrical Engineers and because of the intense public interest manifested after its announcement the ample capacity of the Theatre of the Royal Institution was required to accommodate the audience.

At the completion of the lecture Prof. Aytron spoke as follows:

"It is my most pleasing duty to propose a very hearty vote of thanks to our lecturer, who has entertained us, it is true, for two hours, but we would willingly wait for another hour's similar entertainment."

Mr. Fleming in his authoritative book on wireless telegraphy and telephony pays the following tribute:

"In 1892 Nikola Tesla captured the attention of the whole scientific world by his fascinating experiments on high frequency electric currents. He stimulated the scientific imagination of others as well as displayed his own, and created a widespread interest in his brilliant demonstrations.

"Amongst those who witnessed these things no one was more able to appreciate their inner meaning than Sir William Crookes."

An article by E. Raverot appearing in *The Electrical World* of March 26, 1892, closes a review of the Tesla Paris lecture with the following appreciative comment:

"One sees from this lecture the deep interest which the works and discoveries of Mr. Tesla have inspired among physicists since the first appearance of his publication, and it is with great satisfaction that we are able to express the feeling of admiration which his experiments have inspired in us."

In his London lecture delivered in February, 1892, Tesla had occasion to describe a special construction of insulated cable designed to guard against electro-static disturbances, but immediately added the following significant prediction:

"But such cables will not be constructed, for before long intelligence — transmitted without wires — will throb through the earth like a pulse through a living organism. The wonder is that, with the present state of knowledge and experiences gained, no attempt is being made to disturb the electrostatic or magnetic condition of the earth and transmit, if nothing else, intelligence."

This was Tesla's prophecy twenty-five years ago.

In his lecture before the National Electric Light Association at St. Louis in March, 1893, Mr. Tesla elaborated certain views regarding the importance of resonance effects in this field and stated:

"I would say a few words on a subject which constantly fills my thoughts and which concerns the welfare of all. I mean the transmission of intelligible signals or perhaps even power to any distance without the use of wires."

He then announced that his conviction had grown so strong that he no longer looked upon the plan of transmitting intelligence as a mere theoretical possibility, and referring to the existing belief of some that telephony to any distance might be accomplished — by induction through the air — , concisely set forth his theory as follows:

"I cannot stretch my imagination so far, but I do firmly believe that it is practical to disturb by means of powerful machines the electro-static condition of the earth and thus transmit intelligible signals and perhaps power."

Enlarging upon this theory, he states that, although we have no possible evidence of a charged body existing in space without other oppositely electrified bodies being near, there is a fair probability that the earth is such a body, for by whatever process it was separated from other bodies it must have retained a charge and that the upper strata of the air may be conducting and contain this opposite charge. He further expanded the theory that with proper means for producing electrical oscillations it might be possible to produce electrical disturbances sufficiently powerful to be perceptible by suitable instruments at any point on the Earth's surface. He thus forecast the theory at present accepted by leading scientists as the true basis of wireless telegraphy.

Continuing the same line of thought Mr. Tesla in an interview which appeared in the New York Herald in 1393 said:

"One result of my investigations, the possibility of which has been proven by experiment, is the transmission of energy through the air. I advanced that idea some time ago, and I am happy to say it is now receiving some attention from scientific men.

The plan I have suggested is to disturb by powerful machinery the electricity of the earth, thus setting it in vibration. Proper appliances will be constructed to take up the energy transmitted by these vibrations, transforming them into suitable form of power to be made available for the practical wants of life."

Testifying in a patent suit regarding these early predictions Mr. John Stone Stone, the well-known authority on wireless telegraphy has but recently made the following striking comment:

"I misunderstood Tesla. I think we all misunderstood Tesla. We thought he was a dreamer and visionary. He did dream and his dreams came true, he did have visions but they were of a real future, not an imaginary one. Tesla was the first man to lift his eyes high enough to see that the rarified stratum of atmosphere above our earth was destined to play an important role in the radio telegraphy of the future, a fact which had to obtrude itself on the attention of most of us before we saw it. But Tesla also perceived what many of us did not in those days, namely, the currents which flowed away from the base of the antenna over the surface of the earth and in the earth itself."

Seldom is it that an art springs into being through the efforts of one man alone, but rather as a growth to which many have contributed. This is peculiarly true of the wireless art, and without detracting in the slightest from the honor which is justly due to those who have brought the system to its present wonderful efficiency, it is just to accord to Tesla highest praise not alone for his exposition of principles as set forth in his lectures but also for the more definitive work which followed, much of which is evidenced by his many patents dealing with the wireless art.

Before leaving this branch of Tesla's work, I wish to quote again from the testimony of Mr. Stone, presenting his view of the indebtedness of the wireless art to Tesla:

"Some of those whose work or whose writings during that early period must be noted are Nikola Tesla, Prof. Elihu Thomson, Prof. M. I. Pupin, Prof. Lodge, Prof. Northrup, Prof. Pierce, Hutin & Leblanc, Mr. Marconi and myself. Among all these, the name of Nikola Tesla stands out most prominently. Tesla, with his almost preternatural insight into alternating current phenomena that had enabled him some years before to revolutionize the art of electric power transmission through the invention of the rotary field motor, knew how to make resonance serve, not merely the role of microscope to make visible the electric oscillations, as Hertz had done, but he made it serve the role of a stereopticon to render spectacular to large audiences the phenomena of electric oscillations and high frequency currents. He did more to excite interest and create an intelligent understanding of these phenomena in the years 1891-92-93 than any one else, and the more we learn about high frequency phenomena, resonance and radiation today, the nearer we find ourselves approaching what we at one time were inclined, through a species of intellectual myopia, to regard as the fascinating but fantastical speculations of a man who we are now compelled, in the light of modern experience and knowledge, to admit was a prophet. He saw to the fulfillment of his prophesies and it has been difficult to make any but unimportant improvements in the art of radio-telegraph without traveling part of the way at least, along a trail blazed by this pioneer who, though eminently ingenious, practical and successful in the apparatus he devised and constructed, was so far ahead of his time that the best of us then mistook him for a dreamer."

Another well recognized wireless authority, Professor Slaby in a personal letter to Tesla took occasion to say:

"I am devoting myself since some time to investigations in wireless telegraph, which you have first founded in such a clear and precise manner. It will interest you, as father of this telegraph, to know, etc."

Throughout Tesla's work with high potential currents he had persistently in mind the wireless transmission of power in large quantities. It was in the furtherance of this line of investigation that he expended large amounts of money and years of labor at Wardenclyffe, Long Island, and at Telluride, Colorado. Late in 1914 he secured a patent upon an application filed twelve years before upon an apparatus for transmitting electric energy with which he hopes to be able to transmit unlimited power with high economy to any distance without wires. While as yet these efforts have not resulted in commercial exploitation, the future may prove that his dream of thus transmitting energy in substantial amounts is of that class which in time come true, as in the case of his dream of wireless telegraphy.

Another use to which Tesla adapted the results of his high frequency investigations was the control of the movements of torpedoes and boats. In 1898 he patented such an apparatus and also built and successfully operated such a craft. The movements of the propelling engine, the steering and other mechanisms were controlled wirelessly from the shore or other point through a distance of two miles. Apparently this, like some of his other inventions, was ahead of its time. Tesla, however, evidenced his entire faith in the future of the apparatus in an interview which appeared in 1898 from which I quote:

"But I have no desire that my fame should rest on the invention of a merely destructive device, no matter how terrible. I prefer to be remembered as the inventor who succeeded in abolishing war. That will be my highest pride. But there are many peaceful uses to which my invention can be put, conspicuously that of rescuing the shipwrecked.

"It will be perfectly feasible to equip our lifesaving stations with life cars, or boats, directed and controlled from the shores, which will approach stranded vessels and bring off the passengers and crews without risking the lives of the brave fellows who are now forced to fight their way to the rescue through the raging surf. It may also be used for the propulsion of pilot boats, for carrying letters or provisions or instruments to inaccessible regions."

On March 12th, 1895, Mr. Tesla met with a disastrous loss by the destruction of his laboratory at 33 and 35 South 5th Avenue, New York. In the Electrical Review of March 20th, 1895, there is published an interview with Mr. Tesla regarding this fire. In it he says:

"I am congratulating myself all the time it is no worse. I begin all over again, but I have the knowledge and experience of what has gone before, and fortunately I was able to show with completed apparatus that my ideas and theories are correct. Had the fire occurred a few months ago I should have been robbed of the opportunity of many highly successful demonstrations."

In his laboratory were stored a vast quantity of old models and trial apparatus with which he would have been unwilling to part for any amount of money. He further states that he was at the time engaged upon four main lines of work and investigation: his oscillator, and improved method of electric lighting, the transmission of intelligence without wires, and, an investigation relating to the nature of electricity. Mr. Tesla deeply appreciated the expressions of sympathy received from his many friends and with unabated zeal applied himself to a continuation of the work thus unfortunately interrupted.

Another field of investigation in which Mr. Tesla has contributed valuable material is related to the Roentgen Ray. In the Electrical Review of March and April, 1896, there appeared a number of communications from Mr. Tesla which while giving full credit to Roentgen for his magnificent discovery made public much additional data derived from his own careful experiments in this line of research. From an editorial in the Electrical Review of March 18th, 1896, the following is quoted:

"The announcement of Nikola Tesla's achievements in the new art first published in the Electrical Review of March 11th, in the author's own modest language has added fresh impetus to the work in this direction. His disruptive discharge coil has been universally used where the best results in radiography have been obtained, and his two marked improvements, namely, the single electrode tube and his method of rarefaction, promise great results. Other important points about Tesla's work are the fine details he has obtained in his radiographs, the great distance at which the radiographs have been made, and brief time of exposure."

And again:

"Mr. Tesla is pursuing quietly his work and giving all credit to Roentgen; and it is significant, we think, that the first radiograph he produced in his laboratory was the name of the discoverer. We wish that such courtesies among scientists would always be practiced."

Mr. J. Mount Bleyer commenting upon these investigations said:

"The results obtained by Tesla are simply marvelous, but are just what I expected."

Among the many other inventions to which Mr. Tesla has devoted much time and energy may be mentioned a thermo-magnetic motor and a pyro-magnetic generator, anti-sparking dynamo brush and commutator, auxiliary brush regulation of direct current dynamos, uni-polar dynamos, mechanical and electrical oscillators, electro-therapeutic apparatus, the oxidation of nitrogen by high frequency currents, and an electrolytic registering meter. The last named device was based upon an exceedingly interesting theory. The current to be measured was passed through two parallel conductors arranged in series. The current established a difference of potential between these conductors proportional to the strength of the current passing. This results in a transference of the metal from one conductor to the other, thereby decreasing the resistance of one and increasing that of the other. From such variations in resistance of one or both, the current energy expended is computed.

One other line of endeavor entirely outside of electricity to which Tesla has given much attention is the development of a bladeless steam turbine in which the friction of the passing steam as distinguished from its direct impact is availed of. The steam is admitted between plain parallel rotating discs and passing spirally from the circumference toward the axial center imparts energy to the discs. Such a turbine can be run at exceedingly high temperatures, is readily reversible and having no blades is extremely simple and free from liability to accidental derangement. With great ingenuity Tesla has succeeded in producing such machines of considerable power and having exceedingly interesting characteristics. It is to be hoped that with his indefatigable zeal Tesla will soon succeed in perfecting the commercial application of this invention.

It is not possible in this brief survey even to touch upon many of the lines of Mr. Tesla's activities, but we must content ourselves with this inadequate presentation of typical evidences of the fascinating genius of this man whom we delight to welcome as a citizen of our country—the country which he twenty-five years ago adopted as his own—the country of which he once said:

"When I arrived upon your hospitable shores I eagerly applied myself to work and to learn, and I have persevered in that course. If I have made any special success in this country, I attribute it largely to a feature which is characteristic of both the English and American races; that is, their keen and generous appreciation of any work that they think is good."

Mr. Tesla, we would indeed be woefully lacking in the attributes which you so kindly ascribe to us were we not most cordially appreciative of your work; work which we know is good.

The President: Gentlemen, we are fortunate in having with us to-night another man who has been familiar with Mr. Tesla's work for many years and can tell us something further about his work. I introduce Mr. B. A. Behrend.

B. A. Behrend: Mr. Chairman: Mr. President of the *American Institute of Electrical Engineers*: Fellow Members: Ladies and Gentlemen:

By an extraordinary coincidence, it is exactly twenty-nine years ago, to the very day and hour, that there stood before this Institute Mr. Nikola Tesla, and he read the following sentences:

"To obtain a rotary effort in these motors was the subject of long thought. In order to secure this result it was necessary to make such a disposition that while the poles of one element of the motor are shifted by the alternate currents of the source, the poles produced upon the other elements should always be maintained in the proper relation to the former, irrespective of the speed of the motor. Such a condition exists in a continuous current motor; but in a synchronous motor, such as described, the condition is fulfilled only when the speed is normal.

"The object has been attained by placing within the ring properly subdivided cylindrical iron core wound with several independent coils closed upon themselves. Two coils at right angles are sufficient, but a greater number may be advantageously employed. It results from this disposition that when the poles of the ring are shifted, currents are generated in the closed armature coils. These currents are the most intense at or near the points of the greatest density of the lines of force, and their effect is to produce poles upon the armature at right angles to those of the ring, at least theoretically so; and since this action is entirely independent of the speed -that is, as far as the location of the poles is concerned—a continuous pull is exerted upon the periphery of the armature. In many respects these motors are similar to the continuous current motors. If load is put on, the speed, and also the resistance of the motor, is diminished and more current is made to pass through the energizing coils, thus increasing the effort. Upon the load being taken off, the counter-electromotive force increases and less current passes through the primary or energizing coils. Without any load the speed is very nearly equal to that of the shifting poles of the field magnet.

"It will be found that the rotary effort in these motors fully equals that of the continuous current motors. The effort seems to be greatest when both armature and field magnets are without any projections."

Not since the appearance of Faraday's Experimental Researches in Electricity has a great experimental truth been voiced so simply and so clearly as this description of Mr. Tesla's great discovery of the generation and utilization of polyphase alternating currents. He left nothing to be done for those who followed him. His paper contained the skeleton even of the mathematical theory.

Three years later, in 1891, there was given the first great demonstration, by Swiss engineers, of the transmission of power at 30,000 volts from Aauffen to Frankfort by means of Mr. Tesla's system. A few years later this was followed by the development of the Cataract Construction Company, under the presidency of our member, Mr. Edward D. Adams, and with the aid of the engineers of the Westinghouse Company. It is interesting to recall here tonight that in Lord Kelvin's report to Mr. Adams, Lord Kelvin recommended the use of direct current for the development of power at Niagara Falls and for the transmission to Buffalo.

The due appreciation or even enumeration of the results of Mr. Tesla's invention is neither practicable nor desirable at this moment. There is a time for all things. Suffice it to say that, were we to seize and to eliminate from our industrial world the results of Mr. Tesla's work, the wheels of industry would cease to turn, our electric cars and trains would stop, our towns would be dark, our mills would be dead and idle. Yea, so far reaching is this work, that it has become the warp and woof of industry.

The basis for the theory of the operating characteristics of Mr. Tesla's rotating field induction motor, so necessary to its practical development, was laid by the brilliant French savant, Prof. Andre Blondel, and by Prof. Kapp of Birmingham. It fell to my lot to complete their work and to coordinate,—by means of the simple "circle diagram,"—the somewhat mysterious and complex experimental phenomena. As this was done twenty-one years ago, it is particularly pleasing to me, upon the coming of age of this now universally accepted theory,—tried out by application to several million horse power of machines operating in our great industries,—to pay my tribute to the inventor of the motor and the system which have made possible the electric transmission of energy. HIS name marks an epoch in the advance of electrical science. From THAT work has sprung a revolution in the electrical art.

We asked Mr. Tesla to accept this medal. We did not do this for the mere sake of conferring a distinction, or of perpetuating a name; for so long as men occupy themselves with our industry, his work will be incorporated in the common thought of our art, and the name of Tesla runs no more risk of oblivion than does that of Faraday, or that of Edison.

Nor indeed does this Institute give this medal as evidence that Mr. Tesla's work has received its official sanction. His work stands in no need of such sanction.

No, Mr. Tesla, we beg you to cherish this medal as a symbol of our gratitude for the new creative thought, the powerful impetus, akin to revolution, which you have given to our art and to our science. You have lived to see the work of your genius established. What shall a man desire more than this? There rings out to us a paraphrase of Pope's lines on Newton:

Nature and Nature's laws lay hid in night God said, 'Let Tesla be,' and all was light.

The President: It is easy, I think, for engineers and scientists to take for granted things that have been done in years past. When we sit under an apple tree and see the apples fall, it is an obvious phenomenon of nature. We can understand the laws of gravitation, but to Sir Isaac Newton, many years ago, this phenomenon, which to us today is so simple, helped him to an act of creative imagination of the most extraordinary kind.

So, later on, the phenomenon of electromagnetic induction, which to us today has become a matter of second nature, to Faraday was an act of the most extraordinary creative imagination.

Thirty years ago when Mr. Tesla was doing his very great work, we sometimes forget the conditions of electrical engineering which prevailed at that time. Direct current or continuous current was universally used, and the conceptions of electrical engineers with respect to electric currents were all unidirectional, so to speak. We had not arrived at that conception of currents which went first in one direction and then in another, to say nothing of electrical currents which differed by phase relations, and the work of Nikola Tesla at that time in his great conception of the rotary field seems to me one of the greatest feats of imagination which has ever been attained by human mind. Today we take the rotary field motor, the rotary field transmission, as a matter of course, because we have become used to it, and we forget what it required of the human intellect to create it thirty or thirty-five years ago.

At the time the great Niagara Falls enterprise was instituted, we were under the direct-current regime. As Mr. Behrend says, such a great authority on electrical engineering as Lord Kelvin, and also Mr. Edison, recommended direct-current for transmission of energy from Niagara Falls to Buffalo, and as a system for universal use in their great waterpower development. I think we all realize today where we should be at the present time if direct-current had been used in the development of that enterprise. There would have been a radiating copper mine running out from Niagara Falls which would have wrecked the enterprise in the first year of its existence. Mr. Tesla came along with his great mind and at the psychological moment devised the principle which made that enterprise a success, and made hundreds of other enterprises all over the world an equal success. We owe him the greatest possible debt of gratitude for what he has done for electrical engineers.

And so again, in another field of endeavor in which he was most conspicuous, that of high voltage and high frequency alternating-current, he devised and discovered phenomena which were entirely new to electrical engineers, and he introduced to the world the conception of alternating-current as being elastic or oscillating media. The direct-current engineers at the time never thought of the electric current being something that could oscillate, and Mr. Tesla showed it could, and he also showed many of the phenomena which resulted from oscillating currents. From his work followed the great work of Roentgen, who discovered the Roentgen rays, and all that work which has been carried on throughout the world in the following years by J. J. Thompson and others, which has really led to the conception of modern physics. His work, as has been stated, antedated that of Marconi and formed the basis of wireless telegraphy, which is one of the most scientific applications of the present day, and so on throughout all branches of science and engineering we find from time to time some important evidence of what Tesla has contributed to the sciences and engineering of the present day. So, Mr. Tesla, you hear to-night the many compliments which have been paid to you, but they are not bouquets merely cast for the adornment of the occasion -they have been given with the sincere appreciation of the electrical profession, and we give this medal to you in recognition of this, with full appreciation of what you have done for us, and with great hope that you may continue to contribute to our profession in the future. (Great applause)

Nikola Tesla: Mr. President, Ladies and Gentlemen. I wish to thank you heartily for your kind sympathy and appreciation. I am not deceiving myself in the fact, of which you must be aware, that the speakers have greatly magnified my modest achievements. One should in such a situation be neither diffident nor self-assertive, and in that sense I will concede that some measure of credit may be due to me for the first steps tin certain new directions;

but the ideas I advanced have triumphed, the forces and elements have been conquered, and greatness achieved, through the cooperation of many able men some of whom, I am glad to say, are present this evening. Inventors, engineers, designers, manufacturers and financiers have done their share until, as Mr. Behrend said, a gigantic revolution has been wrought in the transmission and transformation of energy. While we are elated over the results achieved we are pressing on, inspired with the hope and conviction that this is just a beginning, a forerunner of further and still greater accomplishments.

On this occasion, you might want me to say something of a personal and more intimate character bearing on my work. One of the speakers suggested: "Tell us something about yourself, about your early struggles." If I am not mistaken in this surmise I will, with your approval, dwell briefly on this rather delicate subject.

Some of you who have been impressed by what has been said, and would be disposed to accord me more than I have deserved, might be mystified and wonder how so much as Mr. Terry has outlined could have been done by a man as manifestly young as myself. Permit me to explain this. I do not speak often in public, and wish to address just a few remarks directly to the members of my profession, so that there will be no mistake in the future. In the first place, I come from a very wiry and long-lived race. Some of my ancestors have been centenarians, and one of them lived one hundred and twenty-nine years. I am determined to keep up the record and please myself with prospects of great promise. Then again, nature has given me a vivid imagination which, through incessant exercise and training, study of scientific subjects and verification of theories through experiment, has become very accurate and precise, so that I have been able to dispense, to a large extent, with the slow, laborious, wasteful and expensive process of practical development of the ideas I conceive. It has made it possible for me to explore extended fields with great rapidity and get results with the least expenditure of vital energy. By this means I have it in my power to picture the objects of my desires in forms real and tangible and so rid myself of that morbid craving for perishable possessions to which so many succumb. I may say, also, that I am deeply religious at heart, although not in the orthodox meaning, and that I give myself to the constant enjoyment of believing that the greatest mysteries of our being are still to be fathomed and that, all the evidence of the senses and the teachings of exact and dry sciences to the contrary notwithstanding, death itself may not be the termination of the wonderful metamorphosis we witness. In this way I have managed to maintain an undisturbed peace of mind, to make myself proof against adversity, and to achieve contentment and happiness to a point of extracting some satisfaction even from the darker side of life, the trials and tribulations of existence. I have fame and untold wealth, more than this, and yet — how many articles have been written in which I was declared to be an impractical unsuccessful man, and how many poor, struggling writers, have called me a visionary. Such is the folly and shortsightedness of the world!

Now that I have explained why I have preferred my work to the attainment of worldly rewards, I will touch upon a subject which will lend me to say something of greater importance and enable me to explain how I invent and develop ideas. But first I must say a few words regarding my life which was most extraordinary and wonderful in its varied impressions and incidents. In the first place, it was charmed. You have heard that one of the provisions of the Edison Medal was that the recipient should be alive. Of course the men who have received this medal have fully deserved it, in that respect, because they were alive when it was conferred upon them, but none has deserved it in anything like the measure I do, when it comes to that feature. In my youth my ignorance and lightheartedness brought me into innumerable difficulties, dangers and scrapes, from which I extricated myself as by enchantment. That occasioned my parents great concern more, perhaps, because I was the last male than because I was of their own flesh and blood. You should know that Serbians desperately cling to the preservation of the race. I was nearly drowned a dozen times. I was almost cremated three or four times and just missed being boiled alive. I was buried, abandoned and frozen. I have had narrow escapes from mad dogs, hogs and other wild animals. I have passed through dreadful diseases — have been given up by physicians three or four times in my life for good. I have met with all sorts of odd accidents — I cannot think of anything that did not happen to me, and to realize that I am here this evening, hale and hearty, young in mind and body, with all these fruitful years behind me, is little short of a miracle.

But my life was wonderful in another respect — in my capacity of inventor. Not so much, perhaps, in concentrated mentality, or physical endurance and energy; for these are common enough. If you inquire into the career of successful men in the inventor's profession you will find, as a rule, that they are as remarkable for their physical as for their mental performance. I know that when I worked with Edison, after all of his assistants had been exhausted, he said to me: "I never saw such a thing, you take the cake." That was a characteristic way for him to express what I did. We worked from half past ten in the morning until five o'clock the next morning. I carried this on for nine months without a single day's exception; everybody else gave up. Edison stuck, but he occasionally dozed off on the table. What I wish to say particularly is that my early life was really extraordinary in certain experiences which led to everything I ever did afterwards. It is important that this should be explained to you as otherwise you would not know how I discovered the rotating field. From childhood I was afflicted in a singular way — I would see images of objects and scenes with a strong display of light and of much greater vividness than those I had observed before. They were always images of objects and scenes I had actually seen, never of such as I imagined. I have asked students of psychology, physiology and other experts about it, but none of them has been able to explain the phenomena which seems to have been unique, although I was probably predisposed, because my brother also saw images in the same way. My theory is that they were simply reflex actions from the brain on the retina, superinduced by hyper-excitation of the nerves. You might think that I had hallucinations. That is impossible. They are produced only in diseased and anguished brains. My head was always clear as a bell, and I had no fear. Do you want me to tell of my recollections bearing on this? (Turning to the gentlemen on the platform). This is traditional with me, for I was too young to remember anything of what I said. I had two old aunts, I recall, with wrinkled faces, one of them with two great protruding teeth which she used to bury into my cheek when she kissed me. One day they asked me which of the two was prettier. After looking them over I answered: "This one is not as ugly as the other one." That was evidence of good sense. Now as I told you, I had no fear. They used to ask me, "Are you afraid of robbers?" and I would reply "No". "Of wolves?" "No". Then they would ask, "Are you afraid of crazy Luka?" (A fellow who would tear through the village and nothing could stop him) "No, I am not afraid of Luka." "Are you afraid of the gander?" "Yes, I am," I would reply and cling to my mother. That was because once they put me in the court yard with nothing on, and that beast ran up and grabbed me by the soft part of the stomach tearing off a piece of flesh. I still have the mark.

These images I saw caused me considerable discomfort. I will give you and illustration: Suppose I had witnessed a funeral. In my country the rites are but intensified torture. They smother the dead body with kisses, then they bathe it, expose it for three days, and finally one hears the dull thuds of the earth, when all is over. Some of the pictures as that of the coffin, for instance, would not appear vividly but were sometimes so persistent that when I would stretch my hand out I would see it penetrate the image. As I look at it now these images were simply reflex actions through the optic nerve on the retina, producing on the same an effect identical to that of a projection through the lens, and if my view is correct, then it will be possible, (and certainly my experience has demonstrated that), to project the image of any object one conceives in thought on a screen and make it visible. If this could be done it would revolutionize all human relations. I am convinced that it can and will be accomplished.

In order to free myself of these tormenting appearances, I tried to fix my mind on some other picture or image which I had seen, and in this way I would manage to get some relief; but in order to get this relief I had to let the images come one after the other very fast. Then I found that I soon exhausted all I had at my command, my—reel—was out, as it were. I had seen little of the world, only objects around my own home, and they took me a few times to some neighbors, that was all I knew. When I did so the second or third time, in order to chase the appearance from my vision, I found that this remedy lost all the force: Then I began to make excursions beyond the limits of the little world I knew, and I saw new scenes. These were at first very blurred and indistinct, and would flit away when I tried to concentrate my attention upon them, but by and by I succeeded in fixing them; they gained in force and distinctness and finally assumed the intensity of real things. Soon I observed that my best comfort was attained if I simply went on in my vision farther and farther, getting new impressions all the time, and so I started to travel—of course, in my mind. You know that there have been great discoveries made—when Columbus found America that was one, but when I hit upon the idea of traveling it seemed to me that was the greatest discovery possible to man. Every night (and sometimes during the day), as soon as I was alone I would start on my travels. I would see new places, cities and countries, I would live there, meet people and make friendships and acquaintances, and these were just as dear to me as those in real life and not a bit less intense. That is the way I did until I reached almost manhood. When I turned my thoughts to invention, I found that I could visualize my conceptions with the greatest facility. I did not need any models, drawings or experiments, I could do it all in my mind, and I did. In this way I have unconsciously evolved what I consider a new method of materializing inventive concepts and ideas, which is exactly opposite to the purely experimental of which undoubtedly Edison is the greatest and most successful exponent. The moment you construct a device to carry into practice a crude idea you will find yourself inevitably engrossed with the details and defects of the apparatus. As you go on improving and reconstructing, your force of concentration diminishes and you lose sight of the great underlying principle. You obtain results, but at the sacrifice of quality. My method is different, I do not rush into constructive work. When I get an idea, I start right away to build it up in my mind. I change the structure, I make improvements, I experiment, I run the device in my mind. It is absolutely the same to me whether I operate my turbine in thought or test it actually in my shop. It makes no difference, the results are the same. In this way, you see, I can rapidly develop and perfect an invention, without touching anything. When I have gone so far that I have put into the device every possible improvement I can think of, that I can see no fault anywhere, I then construct this final product of my brain. Every time my device works as I conceive it should and my experiment comes out exactly as I plan it. In twenty years there has not been a single solitary experiment which did not turn out precisely as I thought it would. Why should it not? Engineering, electrical and mechanical, is positive in results. Almost any subject presented can be mathematically treated and the effects calculated; but if it is such that results cannot be had by simple methods of mathematics or short cuts, there is all the experience, and all the data on which to draw and from which to build;—why, then, should one carry out the crude idea? It is not necessary, it is a waste of energy, money and time. Now, that is just the way I produced the rotating field.

If I am to give you in a few words the history of that invention, I must begin with my birthday, and you will see the reason why. I was born exactly at midnight, I have no birthday and I never celebrate it. But something else must have happened on that date. I have learned that my heart beat on the right side and did so for many years after. As I grew up it beat on both sides, and finally settled on the left. I remember that I was surprised, when I developed into a very strong man, to find my heart on the left side. Nobody understands how it happened. I had two or three falls and on one occasion nearly all my chest bones were crushed in. Something that was quite unusual must have occurred at my birth and my parents destined me for the clergy then and there. When I was six years old I managed to have myself imprisoned in a little chapel at an inaccessible mountain, and visited only once a year. It was a place of many bloody encounters and there was a grave yard near by. I was locked in there while looking for some sparrows' nests, and had the most dreadful night I ever passed in my life, in company with the ghosts of the dead. American boys will not understand it, of course, for there are no ghosts in America—the people are too sensible; but my country was full of them, and every one from the small boy to greatest hero, who was plastered all over with medals for courage and bravery, had a fear of ghosts. Finally, as by a wonder, they rescued me, and then my parents said: "Surely he must go to the clergy, he must become a churchman." Whatever happened after that, no matter what it was, simply fortified them in that resolution. One day, to tell you a little story, I fell from the top of one of the farm buildings into a large kettle of milk, which was boiling over a roaring fire. Did I say boiling milk?—It was not boiling—not according to the thermometer—though I would have sworn it was when I fell into it, and they pulled me out. But I only got a blister on the knee where I struck the hot kettle. My parents said again: "Was not that wonderful? Did you ever hear of such a thing? He will surely be a bishop, a metropolitan, perhaps a patriarch." In my eighteenth year I came to the cross roads. I had passed through the preliminary schools and had to make up my mind either to embrace the clergy or to run away. I had a profound respect for my parents, and so I resigned myself to take up studies for the clergy. Just then one thing occurred, and if it had not been for that, I would not have had my name connected with the occasion of this evening. A tremendous epidemic of cholera broke out, which decimated the population and, of course, I got immediately. Later it developed into dropsy, pulmonary trouble, and all sorts of diseases until finally my coffin was ordered. In one of the fainting spells when they thought I was dying, my father came to my bedside and cheered me: "You are going to get well." "Perhaps," I replied, "if you will let me study engineering." "Certainly I will," he assured me, "you will go to the best polytechnic school in Europe." I recovered to the amazement of everybody. My father kept his word, and after a year of roaming through the mountains and getting myself in good physical shape, I went to the Polytechnic School at Gratz, Styria, one of the oldest institutions. Something else occurred, however, of which I must tell you as it is vitally linked with this discovery. In the preparatory schools there was no liberty in the choice of subjects, and unless a student was proficient in all of them he could not pass. I found myself in this predicament every year. I could not draw. My faculty for imagining things paralyzed whatever gift I might have had in this respect. I have made some mechanical drawings, of course; practicing so many years one must needs learn to make simple sketches, but if I draw for half an hour I am all exhausted. I never was qualified and passed only through my father's influence. Now, when I went to the polytechnic school I had free choice of subjects and proposed myself to show my parents what I could do. The first year at the polytechnic school was spent in this way—I got up at three o'clock in the morning and worked until eleven o'clock at night, for one whole year, with a single day's exception. Well, you know when a man with a reasonably healthy brain works that way he must accomplish something. Naturally, I did. I graduated nine times that year and some of the professors were not satisfied with giving me the highest distinction, because they said, that did not express their idea of what I did, and here is where I come to the rotating field. In addition to the regular graduating papers they gave me some certificates which I brought to my father believing that I had achieved a great triumph. He took the certificates and threw them into the waste basket, remarking contemptuously: "I know how these testimonials are obtained." That almost killed my ambition; but later, after my father had died, I was mortified to find a package of letters, from which I could see that there had been considerable correspondence going on between him and the professors who had written to the effect that unless he took me away from school I would kill myself with work. Then I understood why he had slighted my success, which I was told was greater than any previous one at that institution; in fact the best students had only graduated twice. My record in the first year had the result that the professors became very much interested in and attached to me, particularly three of them; Prof. Rogner who was teaching arithmetical subjects and geometry; Prof. Alle, one of the most brilliant and wonderful lecturers I have ever seen, who specialized in differential equations, about which he wrote quite a number of works in German, and Prof. Poeschl, who was my instructor in physics. These three men were simply in love with me and used to give me problems to solve. Prof. Poeschl was a curious man. I never saw such feet in my life. They were about that size. (Indicating) His hands were like paws, but when he performed experiments they were so convincing and the whole went off so beautifully that one never realized how they were done. It was all in the method. He did all with the precision of a clock work, and everything succeeded.

It was in the second year of my studies that we received a Gramme machine from Paris, having a horse-shoe form of laminated magnet, and a wound armature with a commutator. We connected it up and showed various effects of currents. During the time Prof. Poeschl was making demonstrations running the machine as a motor we had some trouble with the brushes. They sparked very badly, and I observed: "Why should not we operate without the brushes?" Prof. Poeschl declared that it could not be done, and in view of my success in the past year he did me the honor of delivering a lecture touching on the subject. He remarked: "Mr. Tesla may accomplish great things, but he certainly never will do this," and he reasoned that it would be equivalent to converting a steadily pulling force, like that of gravity, into a rotary effort, a sort of perpetual motion scheme, an impossible idea. But you know that instinct is something which transcends knowledge. We have, undoubtedly, certain finer fibers that enable us to perceive truths when logical deduction, or any other willful effort of the brain, is futile. We cannot reach beyond certain limits in our reasoning, but with instinct we can go to very great lengths. I was convinced that I was right and that it was possible. It was not a perpetual motion idea, it could be done, and I started to work at once.

I will not tire you with an extended account of this undertaking, but will only say that I began in the summer of 1877 and I proceeded as follows: I would picture first of all, a direct-current machine, run it and see how the currents changed in the armature. Then I would imagine an alternator and do the same thing. Next I would visualize systems comprising motors and generators, and so on. Whatever apparatus I imagined, I would put together and operate in my mind, and I continued this practice incessantly until 1882. In that year somehow or other, I began to feel that a revelation was near. I could not yet see just exactly how to do it, but I knew that I was approaching the solution. While on my vacation, in 1882, sure enough, the idea came to me and I will never forget the moment. I was walking with a friend of mine in the city park of Budapest reciting passages from Faust. It was nothing for me to read from memory the contents of an entire book, with every word between the covers, from the first to the last. My sister and brother, however, could do much better than myself. I would like to know whether any of you has that kind of memory. It is curious, entirely visual and retroactive. To be explicit—when I had exams, I had always to read the books three or four days if not a week before, because in that time I could reconstruct the images and visualize them; but if I had an examination the next day after reading, images were not clear and the remembrance was not quite complete. As I say, I was reciting Goethe's poem, and just as the sun was setting I felt wonderfully elated, and the idea came to me like a flash. I saw the whole machinery clearly, the generator, the motor, the connections, I saw it work as if it had been real. With a stick I drew on the sand the diagrams which were shown in my paper before the *American Institute of Electrical Engineers* and illustrated in my patents, as clearly as possible, and from that time on I carried this image in my mind. Had I been a man possessed of the practical gifts of Edison, I would have gone right away to perform an experiment and push the invention along, but I did not have to do this. I could see pictures so vividly, and what I imagined was so real and palpable, that I did not need any experimenting, nor would it have been particularly interesting to me. I went on and improved the plan continuously, inventing new types, and the day I came to America, practically every form, every kind of construction, every arrangement of apparatus I described in my thirty or forty patents was perfected, except just two or three kinds of motors which were the result of later development.

In 1883, I made some tests in Strasbourg, as Mr. Terry pointed out, and there at the railroad station obtained the first rotation. The same experiment was repeated twice.

Now I come to an interesting chapter of my life, when I arrived in America. I had made some improvements in dynamos for a French company who were getting their machinery from here. The improved forms were so much better that the manager of the works said to me: "You must go to America, and design the machines for the Edison Company." So, after ineffectual efforts on the other side to get somebody to interest himself in my plans financially, I came to this country. I wish that I could only give you an idea how what I saw here impressed me. You would be very much astonished. You have all undoubtedly read those charming Arabian Nights tales, in which the genie transports people into wonderful regions, to go through all sorts of delightful adventures. My case was just the opposite. The genie transported me from a world of dreams into one of realities. My world was beautiful, ethereal, as I could imagine it. The one I found here was a machine world; the contact was rough, but I liked it. I realized from the very moment I saw Castle Garden that I was a good American before I landed. Then came another event. I met Edison, and the effect he produced upon me was extraordinary. When I saw this wonderful man, who had had no theoretical training at all, no advantages, who did all himself, getting great results by virtue of his industry and application, I felt mortified that I had squandered my life. I had studied a dozen languages, delved in literature and art and had spent my best years in ruminating through libraries and reading all sorts of stuff that fell into my hands. I thought to myself, what a terrible thing it was to have wasted my life in those useless efforts. If I had only come to America earlier and devoted all of my brain power to inventive work, what might I have done? In later life though, I realized I would not have produced anything without the scientific training I got, and it is a question whether my surmise as to my possible accomplishment was correct. In Edison's works I passed nearly a year of the most strenuous labor, and then certain capitalists approached me with the project to form my own company. I went into the proposition, and developed the arc light. To show you how prejudiced people were against the alternating-current, as the President has indicated, when I told these friends of mine that I had a great invention relating to alternating-current transmission, they said: "No, we want the arc lamp. We do not care for this alternating-current." Finally I perfected my lighting system and the city adopted it. Then I succeeded in organizing another company, in April, 1886, and a laboratory was put up, where I rapidly developed these motors, and eventually the Westinghouse people approached us, and an arrangement was made for their introduction. You know what has happened since then. The invention has swept the world.

I should like to say just a few words regarding the Niagara Falls enterprise. We have a man here to-night to whom belongs really the credit for the early steps and for the first financiering of the project, which was difficult at that time. I refer to Mr. E. D. Adams. When I heard that such authorities as Lord Kelvin and Prof. W. C. Unwin had recommended—one the direct-current system and the other compressed air—for the transmission of power from Niagara Falls to Buffalo, I thought it was dangerous to let the matter go further, and I went to see Mr. Adams. I remember the interview perfectly. Mr. Adams was much impressed with what I told him. We had some correspondence afterwards, and whether it was in consequence of my enlightening him on the situation, or owing to some other influence, my system was adopted. Since that time, of course, new men, new interests have come in, and what has been done I do not know, except that the Niagara Falls enterprise was the real starting impulse in the great movement inaugurated for the transmission and transformation of energy on a huge scale.

Mr. Terry has referred to other inventions of mine. I will just make a few remarks relative to these as some of my work has been misunderstood. It seems to me that I ought to tell you a few words about an effort that absorbed my attention later. In 1892 I delivered a lecture at the Royal Institution and Lord Rayleigh surprised me by acknowledging my work in very generous terms, something that is not customary, and among other things he stated that I had really an extraordinary gift for invention. Up to that time, I can assure you, I had hardly realized that I was an inventor. I remembered, for instance, when I was a boy, I could go out into the forest and catch as many crows as I wanted, and nobody else could do it. Once, when I was seven years of age, I repaired a fire engine which the engineers could not make work, and they carried me in triumph through the city. I constructed turbines, clocks and such devices as no other boy in the community. I said to myself: "If I really have a gift for invention, I will bend it to some great purpose or task and not squander my efforts on small things." Then I began to ponder just what was the greatest deed to accomplish. One day as I was walking in the forest a storm gathered and I ran under a tree for shelter. The air was very heavy, and all at once there was a lightning flash, and immediately after a torrent of rain fell. That gave me the first idea. I realized that the sun was lifting the water vapor, and wind swept it over the regions where it accumulated and reached a condition when it was easily condensed and fell to earth again. This life-sustaining stream of water was entirely maintained by sun power, and lightning, or some other agency of this kind, simply came in a trigger-mechanism to release the energy at the proper moment. I started out and attacked the problem of constructing a machine which would enable us to precipitate this water whenever and wherever desired. If this was possible, then we could draw unlimited amounts of water from the ocean, create lakes, rivers and water falls, and indefinitely increase the hydroelectric power, of which there is now a limited supply. That led me to the production of very intense electrical effects. At the same time my wireless work, which I had already begun, was exactly in that direction, and I devoted myself to the perfection of that device, and in 1908, I filed an application describing an apparatus with which I thought the wonder could be achieved. The Patent Office Examiner was from Missouri, he would not believe that it could be done, and my patent was never granted. But in Colorado I had constructed a transmitter by which I produced effects in some respects at least greater than those of lightning. I do not mean in potential. The highest potential I reached was something like 20,000,000 volts, which is insignificant as compared to that of lightning, but certain effects produced by my apparatus were greater than those of lightning. For instance, I obtained in my antennae currents of from 1,000 to 1,100 amperes. That was in 1899 and you know that in the biggest wireless plants of today only 250 amperes are used. In Colorado I succeeded one day in precipitating a dense fog. There was a mist outside, but when I turned on the current the cloud in the laboratory became so dense that when the hand was held only a few inches from the face it could not be seen. I am positive in my conviction that we can erect a plant of proper design in an arid region, work it according to certain observations and rules, and by its means draw from the ocean unlimited amounts of water for irrigation and power purposes. If I do not live to carry it out, somebody else will, but I feel sure that I am right.

As to the transmission of power through space, that is a project which I considered absolutely certain of success long since. Years ago I was in the position to transmit wireless power to any distance without limit other than that imposed by the physical dimensions of the globe. In my system it makes no difference what the distance is. The efficiency of the transmission can be as high as 96 or 97%, and there are practically no losses except such as are inevitable in the running of the machinery. When there is no receiver there is no energy consumption anywhere. When the receiver is put on, it draws power. That is the exact opposite of the Hertz-wave system. In that case, if you have a plant of 1,000 horsepower, it is radiating all the time whether the energy is received or not; but in my system no power is lost. When there are no receivers the plant consumes only a few horsepower necessary to maintain the electric vibration; it runs idle, as the Edison plant when the lamps and motors are shut off.

I have made advances along this line in later years which will contribute to the practical features of the system. Recently I have obtained a patent on a transmitter with which it is practicable to transfer unlimited amount of energy to any distance.

I had a very interesting experience with Mr. Stone, whom I consider, if not the ablest, certainly one of the ablest living experts. I said to Mr. Stone: "Did you see my patent?" He replied: "Yes, I saw it, but I thought you were crazy." When I explained it to Mr. Stone he said, "Now, I see; why, that is great," and he understood how the energy is transmitted.

To conclude, gentlemen, we are coming to great results, but we must be prepared for a condition of paralysis for quite a while. We are facing a crisis such as the world has never seen before, and until the situation clears the best thing we can do is to devise some scheme for overcoming the submarines, and that is what I am doing now. (Applause)

Alfred H. Cowles: Here are some pictures you gave to me twenty years ago, relating to your experiments of 1899, I think you will be interested in seeing them. (Hands pictures to Mr. Tesla)

Nikola Tesla: I have learned how to put up a plant that will develop a tension of 100,000,000 volts and handle it with perfect safety. This plant (indicating) was in Colorado. If anybody, who had not been dabbling in these experiments as long as myself, had done such work, he would surely have been killed. In this plant I had the narrowest escape ever. It was a square building, in which there was a coil 52 feet in diameter, about nine feet high. When it was adjusted to resonance, the streamers passed from top to bottom and it was a most beautiful sight. You see, that was about fifteen hundred, perhaps two thousand square feet of streamer surface. To save money I had calculated the dimensions as closely as possible, and the streamers came within six or seven inches from the sides of the building. As boys had been looking through a single window provided in the rear, I nailed it up. For handling the heavy currents, I had a special switch. It was hard to pull, and I had a spring arranged so that I could just touch the handle and it would snap in. I sent one of my assistants down town and was experimenting alone. I threw up the switch and went behind the coil to examine something. While I was there the switch snapped in, when suddenly the whole room was filled with streamers, and I had no way of getting out. I tried to break through the window but in vain as I had no tools, and there was nothing else to do than to throw myself on my stomach and pass under. The primary carried 500,000 volts, and I had to crawl through the narrow place here (pointing) with the streamers going. The nitrous acid was so strong I could hardly breathe. These streamers rapidly oxidize nitrogen because of their enormous surface, which makes up for what they lack in intensity. When I came to the narrow space they closed on my back. I got away and barely managed to open the switch when the building began to burn. I grabbed a fire extinguisher and succeeded in smothering the fire. Then I had enough, I was all in. But now I can operate a plant without any fear of its destruction by fire. Mr. Cowles is responsible for excursion into this matter. (Applause)

The President: If there is no further business, we will consider this meeting adjourned.

May 18, 1917

TESLA'S VIEWS ON ELECTRICITY AND THE WAR

—by H. WINFIELD SECOR, The Electrical Experimenter, August, 1917

Exclusive Interview to *The Electrical Experimenter*

Nikola Tesla, one of the greatest of living electrical engineers and recipient of the seventh "Edison" medal, has evolved several unique and far-reaching ideas which if developed and practically applied should help to partially, if not totally, solve the much discussed submarine menace and to provide a means whereby the enemy's powder and shell magazines may be exploded at a distance of several miles.

There have been numerous stories bruited about by more or less irresponsible self-styled experts that certain American inventors, including Dr. Tesla, had invented among other things an electric ray to destroy or detect a submarine under water at a considerable distance. Mr. Tesla very courteously granted the writer an interview and some of his ideas on electricity's possible role in helping to end the great world-war are herein given:

The all-absorbing topic of daily conversation at the present time is of course the "U-boat." Therefore, I made that subject my opening shot.

> *Nikola tesla, the famous electric inventor, has proposed three different electrical schemes for locating submerged submarines. The reflected electric ray method is illustrated above; the high-frequency invisible electric ray, when reflected by a submarine hull, causes phosphorescent screens on another or even the same ship to glow, giving warning that the U-boats are near.*

"Well," said Dr. Tesla, "I have several distinct ideas regarding the subjugation of the submarine. But lest we forget, let us not underestimate the efficiency of the means available for carrying on submarine warfare. We may use microphones to detect the submarine, but on the other hand the submarine commander may employ microphones to locate a ship and even torpedo it by the range thus found, without ever showing his periscope above water.

"Many years ago while serving in the capacity of chief electrician for an electric plant situated on the river Seine, in France, I had occasion to require for certain testing purposes an extremely sensitive galvanometer. In those days the quartz fiber was an unknown quantity—and I, by be coming specially adept, managed to produce an extremely fine cocoon fiber for the galvanometer suspension. Further, the galvanometer proved very sensitive for the location in which it was to be used; so a special cement base was sunk in the ground and by using a lead sub-base suspended on springs all mechanical shock and vibration effects were finally gotten rid of.

"As a matter of actual personal experience," said Dr. Tesla. "it became a fact that the small iron-hull steam mail-packets (ships) plying up and down the river Seine at a distance of 3 miles would distinctly affect the galvanometer!"

"How could this be applied to the submarine problem?" I asked.

"Well, for one thing," the scientist replied, "I believe this magnetic method of locating or indicating the presence of an iron or steel mass might prove very practical in locating a hidden submarine. And it is of course of paramount importance that we do find a means of accurately locating the sub-sea fighters when they are submerged, so that we can, with this in formation, be ready to close in on them when they attempt to come to the surface. Especially is this important when several vessels are traveling in fleet formation; the location and presence of the enemy submarine can be radiographed to the other vessels by the one doing the magnetic surveying and, by means of nets in some cases or gun-fire and the use of hydro-aeroplanes sent aloft from the ships, the enemy under water stands a mighty good chance of being either 'bombed,' shelled or netted.

"However, a means would soon be found of nullifying this magnetic detector of the submerged undersea war-craft. They might make the 'U-boat' hulls of some non-magnetic metal, such as copper, brass, or aluminum. It is a good rule to always keep in mind that for practically every good invention of such a kind as this, there has always been invented an opposite, and equally efficient counteracting invention."

"How about this new electric ray method of locating submarines?" I ventured to ask.

"Yes, yes, I am coming to that," the master electrician parried. "Now suppose that we erect on a vessel, a large rectangular helice or inductance coil of insulated wire. Actual experiments in my laboratory at Houston Street (New York City), have proven that the presence of a local iron mass, such as the ship's hull, would not interfere with the action of this device. To this coil of wire, measuring perhaps 400 feet in length by 70 feet in width (the length and breadth of the ship) we connect a source of extremely high frequency and very powerful oscillating current. By this means there are radiated powerful oscillating electro-static currents, which as I have found by actual experiment in my Colorado tests some years ago, will first affect a metallic body (such as a submarine hull, even the made of brass or any other metal), and in turn cause that mass to react inductively on the exciting coil on the ship. To locate an iron mass it is not necessary to excite the coil with a high frequency current; the critical balance of the coil will be affected simply by the presence of the magnetic body. To be able to accurately determine the direction and range of the enemy submarine four exciting inductances should be used. With a single inductance, however, it would be possible to determine the location of a submarine by running the ship first in one direction and then in another, and noting whether the reactive effect caused by the presence of the submarine hull increased or decreased. The radiating inductance must be very sharply attuned to the measuring apparatus installed on the ship, when no trouble will be found in detecting the presence of such a large metallic mass as a submarine, even at a distance of 5 to 6 miles; of this I feel confident from my past experiments in the realm of ultra-high frequency currents and potentials."

"What particular experiments do you have in mind, Dr. Tesla?" I asked.

"The Colorado tests of 1898-1900. Wonderful were the results there obtained, both those anticipated as well as those unexpected. As an example of what has been done with several hundred kilowatts of high frequency energy liberated, it was found that the dynamos in a power house six miles away were repeatedly burned out, due to the powerful high frequency currents set up in them, and which caused heavy sparks to jump thru the windings and destroy the insulation! The lightning arresters in the power house showed a stream of blue-white sparks passing between the metal plates to the earth connection. I could walk on the sand (ordinarily considered a very good insulator) several hundred feet from my large high frequency oscillator, and sparks jumped from my shoes! At such distances all incandescent lamps glowed by wireless power, and banks of lamp, connected to a few turns of wire arranged in a coil on the ground, were lighted to full brilliancy. The effect on metallic objects at considerable distances was really remarkable."

I asked him about the "Ulivi ray," which was accorded considerable newspaper publicity some time ago.

"The 'Ulivi ray' really was transplanted from this country to Italy," asserted Dr. Tesla. "It was simply an adaptation of my ultra-powerful high-frequency phenomena as carried out in Colorado and cited previously. With a powerful oscillator developing thousands of horsepower it would become readily possible to detonate powder and munition magazines by means of the high frequency currents induced in every bit of metal, even when located five to six miles away and more. Even a powder can would have a potential of 6,000 to 7,000 volts induced in it at that distance.

"At the time of those tests I succeeded in producing the most powerful

X-rays ever seen. I could stand at a distance of 100 feet from the X-ray apparatus and see the bones of the hand clearly with the aid of a fluoroscope screen; and I could have easily seen them at a distance several times this by utilizing suitable power. In fact, I could not then procure X-ray generators to handle even a small fraction of the power I had available. But I now have apparatus designed whereby this tremendous energy of hundreds of kilowatts can be successfully transformed into X-rays."

"Could these ultra-powerful and unusually penetrating X-rays be used to locate or destroy a submarine with?" I interjected.

"Now we are coming to the method of locating such hidden metal masses as submarines by an electric ray," replied the electrical wizard. "That is the thing which seems to hold great promises. If we can shoot out a concentrated ray comprising a stream of minute electric charges vibrating electrically at tremendous frequency, say millions of cycles per second, and then intercept this ray, after it has been reflected by a submarine hull for example, and cause this intercepted ray to illuminate a fluorescent screen (similar to the X-ray method) on the same or another ship, then our problem of locating the hidden submarine will have been solved.

"This electric ray would necessarily have to have an oscillation wave length extremely short and here is where the great problem presents itself; i.e., to be able to develop a sufficiently short wave length and a large amount of power, say several hundred thousand or even several thousand horse-power. I have produced oscillators having a wave length of but a few millimeters.

"Suppose, for example, that a vessel is fitted with such an electric ray projector. The average ship has available from say 10,000 to 15,000 H.P. The exploring ray could be flashed out intermittently and thus it would be possible to hurl forth a very formidable beam of pulsating electric energy, involving a discharge of hundreds of thousands of horse-power. The electric energy would be taken from the ship's plant for a fraction of a minute only, being absorbed at a tremendous rate by suitable condensers and other apparatus, from which it could be liberated at any rate desired.

"Imagine that the ray has been shot out and that in sweeping thru the water it encounters the hull of a submarine. What happens? Just this: The ray would be reflected, and by an appropriate device we would intercept and translate this reflected ray, as for instance by allowing the ray to impinge on a phosphorescent screen, acting in a similar way to the X-ray screen. The ray would be invisible to the unaided eye. The reflected ray could be firstly, intercepted by the one or more ships in the fleet; or secondly, it would be possible for the ship originating the ray to intercept the refracted portion by sending out the ray intermittently and also by taking advantage of what is known as the after-glow effect, which means that the ray would affect the registering screen an appreciable time after its origination. This would be necessary to allow the ship to move forward sufficiently to get within range of the reflected ray from the submarine, as the reflection would not be in the same direction as the originating ray.

"To make this clearer, consider that a concentrated ray from a searchlight is thrown on a balloon at night. When the spot of light strikes the balloon, the latter at once becomes visible from many different angles. The same effect would be created with the electric ray if properly applied. When the ray struck the rough hull of a submarine it would be reflected, but not in a concentrated beam—it would spread out; which is just what we want. Suppose several vessels are steaming along in company; it thus becomes evident that several of them will intercept the reflected ray and accordingly be warned of the presence of the submarine or submarines. The vessels would at once lower their nets, if so equipped, order their gun crews to quarters and double the look-out watch. The important thing to know is that submarines are present. Forewarned is forearmed!

"The Teutons are clever, you know; very, very clever, but we shall beat them," said Dr. Tesla confidently. [It may be of interest to our readers to know that several important electrical war schemes will shortly be laid before the War and Navy Departments by Dr. Tesla, the details of which we naturally cannot now publish.]

by H. Winfield Secor
August, 1917

FAMOUS SCIENTIFIC ILLUSIONS

— *Electrical Experimenter, February 1919*

The singular misconception of the wireless

To the popular mind this sensational advance conveys the impression of a single invention but in reality it is an art, the successful practice of which involves the employment of a great many discoveries and improvements. I viewed it as such when I undertook to solve wireless problems and it is due to this fact that my insight into its underlying principles was clear from their very inception.

In the course of development of my induction motors it became desirable to operate them at high speeds and for this purpose I constructed alternators of relatively high frequencies. The striking behavior of the currents soon captivated my attention and in 1889 I started a systematic investigation of their properties and the possibilities of practical application. The first gratifying result of my efforts in this direction was the transmission of electrical energy thru one wire without return, of which I gave demonstrations in my lectures and addresses before several scientific bodies here and abroad in 1891 and 1892. During that period, while working with my oscillation transformers and dynamos of frequencies up to 200,000 cycles per second, the idea gradually took hold of me that the earth might be used in place of the wire, thus dispensing with artificial conductors altogether. The immensity of the globe seemed an insurmountable obstacle but after a prolonged study of the subject I became satisfied that the undertaking was rational, and in my lectures before the Franklin Institute and National Electric Light Association early in 1893 I gave the outline of the system I had conceived. In the latter part of that year, at the Chicago World's Fair, I had the good fortune of meeting Prof. Helmholtz to whom I explained my plan, illustrating it with experiments. On that occasion I asked the celebrated physicist for an expression of opinion on the feasibility of the scheme. He stated unhesitatingly that it was practicable, provided I could perfect apparatus capable of putting it into effect but this, he anticipated, would be extremely difficult to accomplish.

I resumed the work very much encouraged and from that date to 1896 advanced slowly but steadily, making a number of improvements the chief of which was my system of concatenated tuned circuits and method of regulation, now universally adopted. In the summer of 1897 Lord Kelvin happened to pass thru New York and honored me by a visit to my laboratory where I entertained him with demonstrations in support of my wireless theory. He was fairly carried away with what he saw but, nevertheless, condemned my project in emphatic terms, qualifying it as something impossible, "an illusion and a snare." I had expected his approval and was pained and surprised. But the next day he returned and gave me a better opportunity for explanation of the advances I had made and of the true principles underlying the system I had evolved. Suddenly he remarked with evident astonishment: "Then you are not making use of Hertz waves?" "Certainly not," I replied, "these are radiations". No energy could be economically transmitted to a distance by any such agency. In my system the process is one of true conduction which, theoretically, can be effected at the greatest distance without appreciable loss." I can never forget the magic change that came over the illustrious philosopher the moment he freed himself from that erroneous impression. The skeptic who would not believe was suddenly transformed into the warmest of supporters. He parted from me not only thoroughly convinced of the scientific soundness of the idea but strongly expressed his confidence in its success. In my exposition to him I resorted to the following mechanical analogues of my own and the Hertz wave system.

Imagine the earth to be a bag of rubber filled with water, a small quantity of which is periodically forced in and out of the same by means of a reciprocating pump, as illustrated. If the strokes of the latter are effected in intervals of more than one hour and forty-eight minutes, sufficient for the transmission of the impulse thru the whole mass, the entire bag will expand and contract and corresponding movements will be imparted to pressure gauges or movable pistons with the same intensity, irrespective of distance. By working the pump faster, shorter waves will be produced which, on reaching the opposite end of the bag, may be reflected and give rise to stationary nodes and loops, but in any case, the fluid being incompressible, its inclosure perfectly elastic, and the frequency of oscillations not very high, the energy will be economically transmitted and very little power consumed so long as no work is done in the receivers. This is a crude but correct representation of my wireless system in which, however, I resort to various refinements. Thus, for instance, the pump is made part of a resonant system of great inertia, enormously magnifying the force of the impressed impulses. The receiving devices are similarly conditioned and in this manner the amount of energy collected in them vastly increased.

The Hertz wave system is in many respects the very opposite of this. To explain it by analogy, the piston of the pump is assumed to vibrate to and fro at a terrific rate and the orifice thru which the fluid passes in and out of the cylinder is reduced to a small hole. There is scarcely any movement of the fluid and almost the whole work performed results in the production of radiant heat, of which an infinitesimal part is recovered in a remote locality. However incredible, it is true that the minds of some of the ablest experts have been from the beginning, and still are, obsessed by this monstrous idea, and so it comes that the true wireless art, to which I laid the foundation in 1893, has been retarded in its development for twenty years. This is the reason why the "statics" have proved unconquerable, why the wireless shares are of little value and why the Government has been compelled to interfere.

We are living on a planet of well-nigh inconceivable dimensions, surrounded by a layer of insulating air above which is a rarefied and conducting atmosphere (Fig. 5). This is providential, for if all the air were conducting the transmission of electrical energy thru the natural media would be impossible. My early experiments have shown that currents of high frequency and great tension readily pass thru an atmosphere but moderately rarefied, so that the insulating stratum is reduced to a small thickness as will be evident by inspection of Fig. 6, in which a part of the earth and its gaseous envelope is shown to scale. If the radius of the sphere is 12 ½", then the non-conducting layer is only 1/64" thick and it will be obvious that the Hertzian rays cannot traverse so thin a crack between two conducting surfaces for any considerable distance, without being absorbed. The theory has been seriously advanced that these radiations pass around the globe by successive reflections, but to show the absurdity of this suggestion reference is made to Fig. 7 in which this process is diagrammatically indicated. Assuming that there is no refraction, the rays, as shown on the right, would travel along the sides of a polygon drawn around the solid, and inscribed into the conducting gaseous boundary in which case the length of the side would be about 400 miles. As one-half the circumference of the earth is approximately 12,000 miles long there will be, roughly, thirty deviations. The efficiency of such a reflector cannot be more than 25%, so that if none of the energy of the transmitter were lost in other ways, the part recovered would be measured by the fraction (¼)" Let the transmitter radiate Hertz waves at the rate of 1,000 kilowatts. Then about one hundred and fifteen billionth part of one watt is all that would be collected in a perfect receiver. In truth, the reflections would be much more numerous as shown on the left of the figure, and owing to this and other reasons, on which it is unnecessary to dwell, the amount recovered would be a vanishing quantity.

Consider now the process taking place in the transmission by the instrumentalities and methods of my invention. For this purpose attention is called to Fig. 8, which gives an idea of the mode of propagation of the current waves and is largely self-explanatory. The drawing represents a solar eclipse with the shadow of the moon just touching the surface of the earth at a point where the transmitter is located. As the shadow moves downward it will spread over the earth's surface, first with infinite and then gradually diminishing velocity until at a distance of about 6,000 miles it will attain its true speed in space. From there on it will proceed with increasing velocity, reaching infinite value at the opposite point of the globe. It hardly need be stated that this is merely an illustration and not an accurate representation in the astronomical sense.

The exact law will be readily understood by reference to Fig. 9, in which a transmitting circuit is shown connected to earth and to an antenna. The transmitter being in action, two effects are produced: Hertz waves pass thru the air, and a current traverses the earth. The former propagate with the speed of light and their energy is unrecoverable in the circuit. The latter proceeds with the speed varying as the cosecant of the angle which a radius drawn from any point under consideration forms with the axis of symmetry of the waves. At the origin the speed is infinite but gradually diminishes until a quadrant is traversed, when the velocity is that of light. From there on it again increases, becoming infinite at the antipole. Theoretically the energy of this current is recoverable in its entirety, in properly attuned receivers.

Some experts, whom I have credited with better knowledge, have for years contended that my proposals to transmit power without wires are sheer nonsense but I note that they are growing more cautious every day. The latest objection to my system is found in the cheapness of gasoline. These men labor under the impression that the energy flows in all directions and that, therefore, only a minute amount can be recovered in any individual receiver. But this is far from being so. The power is conveyed in only one direction, from the transmitter to the receiver, and none of it is lost elsewhere. It is perfectly practicable to recover at any point of the globe energy enough for driving an airplane, or a pleasure boat or for lighting a dwelling. I am especially sanguine in regard to the lighting of isolated places and believe that a more economical and convenient method can hardly be devised. The future will show whether my foresight is as accurate now as it has proved heretofore.

February 1919

THE TRUE WIRELESS

—*Electrical Experimenter, May 1919*

In this remarkable and complete story of his discovery of the "True Wireless" and the principles upon which transmission and reception, even in the present day systems, are based, Dr. Nikola Tesla shows us that he is indeed the "Father of the Wireless." To him the Hertz wave theory is a delusion; it looks sound from certain angles, but the facts tend to prove that it is hollow and empty. He convinces us that the real Hertz waves are blotted out after they have traveled but a short distance from the sender. It follows, therefore, that the measured antenna current is no indication of the effect, because only a small part of it is effective at a distance. The limited activity of pure Hertz wave transmission and reception is here clearly explained, besides showing definitely that in spite of themselves, the radio engineers of today are employing the original Tesla tuned oscillatory system. He shows by examples with different forms of aerials that the signals picked up by the instruments must actually be induced by earth currents—not etheric space waves. Tesla also disproves the "Heaviside layer" theory from his personal observations and tests.

Ever since the announcement of Maxwell's electro-magnetic theory scientific investigators all the world over had been bent on its experimental verification. They were convinced that it would be done and lived in an atmosphere of eager expectancy, unusually favorable to the reception of any evidence to this end. No wonder then that the publication of Dr. Heinrich Hertz's results caused a thrill as had scarcely ever been experienced before. At that time I was in the midst of pressing work in connection with the commercial introduction of my system of power transmission, but, nevertheless, caught the fire of enthusiasm and fairly burned with desire to behold the miracle with my own eyes. Accordingly, as soon as I had freed myself of these imperative duties and resumed research work in my laboratory on Grand Street, New York, I began, parallel with high frequency alternators, the construction of several forms of apparatus with the object of exploring the field opened up by Dr. Hertz. Recognizing the limitations of the devices he had employed, I concentrated my attention on the production of a powerful induction coil but made no notable progress until a happy inspiration led me to the invention of the oscillation transformer. In the latter part of 1891 I was already so far advanced in the development of this new principle that I had at my disposal means vastly superior to those of the German physicist. All my previous efforts with Rhumkorf coils had left me unconvinced, and in order to settle my doubts I went over the whole ground once more, very carefully, with these improved appliances. Similar phenomena were noted, greatly magnified in intensity, but they were susceptible of a different and more plausible explanation. I considered this so important that in 1892 I went to Bonn, Germany, to confer with Dr. Hertz in regard to my observations. He seemed disappointed to such a degree that I regretted my trip and parted from him sorrowfully. During the succeeding years I made numerous experiments with the same object, but the results were uniformly negative. In 1900, however, after I had evolved a wireless transmitter which enabled me to obtain electromagnetic activities of many millions of horse-power, I made a last desperate attempt to prove that the disturbances emanating from the oscillator were ether vibrations akin to those of light, but met again with utter failure. For more than eighteen years I have been reading treatises, reports of scientific transactions, and articles on Hertz-wave telegraphy, to keep myself informed, but they have always impressed me like works of fiction.

The history of science shows that theories are perishable. With every new truth that is revealed we get a better understanding of Nature and our conceptions and views are modified. Dr. Hertz did not discover a new principle. He merely gave material support to hypothesis which had been long ago formulated. It was a perfectly well-established fact that a circuit, traversed by a periodic current, emitted some kind of space waves, but we were in ignorance as to their character. He apparently gave an experimental proof that they were transversal vibrations in the ether. Most people look upon this as his great accomplishment. To my mind it seems that his immortal merit was not so much in this as in the focusing of the investigators' attention on the processes taking place in the ambient medium. The Hertz-wave theory, by its fascinating hold on the imagination, has stifled creative effort in the wireless art and retarded it for twenty-five years. But, on the other hand, it is impossible to over-estimate the beneficial effects of the powerful stimulus it has given in many directions.

As regards signaling without wires, the application of these radiations for the purpose was quite obvious. When Dr. Hertz was asked whether such a system would be of practical value, he did not think so, and he was correct in his forecast. The best that might have been expected was a method of communication similar to the heliographic and subject to the same or even greater limitations.

In the spring of 1891 I gave my demonstrations with a high frequency machine before the *American Institute of Electrical Engineers* at Columbia College, which laid the foundation to a new and far more promising departure. Although the laws of electrical resonance were well known at that time and my lamented friend, Dr. John Hopkinson, had even indicated their specific application to an alternator in the Proceedings of the Institute of Electrical Engineers, London, Nov. 13, 1889, nothing had been done towards the practical use of this knowledge and it is probable that those experiments of mine were the first public exhibition with resonant circuits, more particularly of high frequency. While the spontaneous success of my lecture was due to spectacular features, its chief import was in showing that all kinds of devices could be operated thru a single wire without return. This was the initial step in the evolution of my wireless system. The idea presented itself to me that it might be possible, under observance of proper conditions of resonance, to transmit electric energy thru the earth, thus dispensing with all artificial conductors. Anyone who might wish to examine impartially the merit of that early suggestion must not view it in the light of present day science. I only need to say that as late as 1893, when I had prepared an elaborate chapter on my wireless system, dwelling on its various instrumentalities and future prospects, Mr. Joseph Wetzler and other friends of mine emphatically protested against its publication on the ground that such idle and far-fetched speculations would injure me in the opinion of conservative business men. So it came that only a small part of what I had intended to say was embodied in my address of that year before the Franklin Institute and National Electric Light Association under the chapter "On Electrical Resonance." This little salvage from the wreck has earned me the title of "Father of the Wireless" from many well-disposed fellow workers, rather than the invention of scores of appliances which have brought wireless transmission within the reach of every young amateur and which, in a time not distant, will lead to undertakings overshadowing in magnitude and importance all past achievements of the engineer.

The popular impression is that my wireless work was begun in 1893, but as a matter of fact I spent the two preceding years in investigations, employing forms of apparatus, some of which were almost like those of today. It was clear to me from the very start that the successful consummation could only be brought about by a number of radical improvements. Suitable high frequency generators and electrical oscillators had first to be produced. The energy of these had to be transformed in effective transmitters and collected at a distance in proper receivers. Such a system would be manifestly circumscribed in its usefulness if all extraneous interference were not prevented and exclusiveness secured. In time, however, I recognized that devices of this kind, to be most effective and efficient, should be designed with due regard to the physical properties of this planet and the electrical conditions obtaining on the same. I will briefly touch upon the salient advances as they were made in the gradual development of the system.

The high frequency alternator employed in my first demonstrations is illustrated in Fig. 1. It comprised a field ring, with 384 pole projections and a disc armature with coils wound in one single layer which were connected in various ways according to requirements. It was an excellent machine for experimental purposes, furnishing sinusoidal currents of from 10,000 to 20,000 cycles per second. The output was comparatively large, due to the fact that as much as 30 amperes per square millimeter could be past thru the coils without injury.

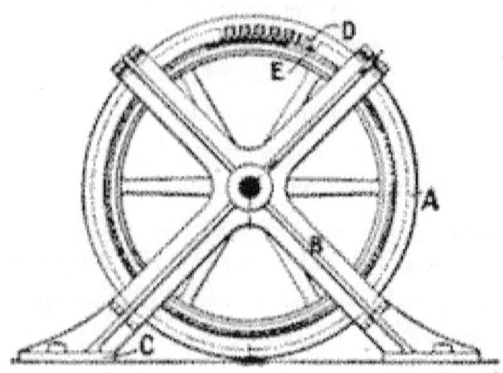

Fig. 1. Alternator of 10.000 Cycles p.s., Capacity 10KW. Which Was Employed by Tesla in His First Demonstrations of High Frequency Phenomena Before the American Institute of Electrical Engineers at Columbia College, May 20, 1891.

The diagram in Fig. 2 shows the circuit arrangements as used in my lecture. Resonant conditions were maintained by means of a condenser subdivided into small sections, the finer adjustments being effected by a movable iron core within an inductance coil. Loosely linked with the latter was a high tension secondary which was tuned to the primary.

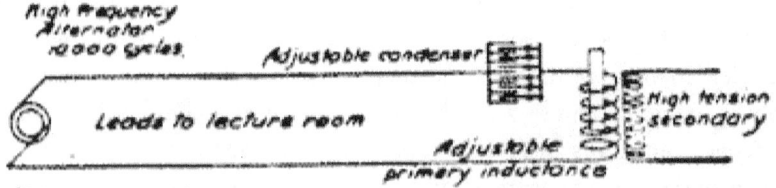

Fig. 2. Diagram Illustrating the Circuit Connections and Tuning Devices Employed by Tesla In His Experimental Demonstrations Before the American Institute of Electrical Engineers With the High Frequency Alternator Shown in Fig. 1.

The operation of devices thru a single wire without return was puzzling at first because of its novelty, but can be readily explained by suitable analogs. For this purpose reference is made to Figs. 3 and 4.

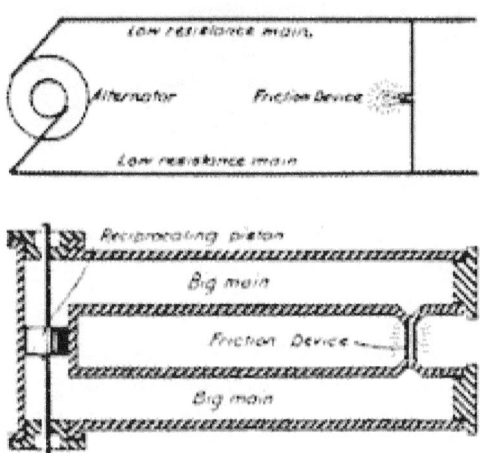

Fig. 3. Electric Transmission Thru Two Wires and Hydraulic Analog.

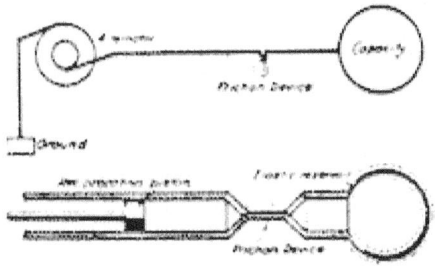

Fig. 4. Electric Transmission Thru a Single Wire Hydraulic Analog. Electric Transmission Thru Two Wires and Hydraulic Analog Electric Transmission Thru a Single Wire Hydraulic Analog.

In the former the low resistance electrical conductors are represented by pipes of large cross section, the alternator by an oscillating piston and the filament of an incandescent lamp by a minute channel connecting the pipes. It will be clear from a glance at the diagram that very slight excursions of the piston would cause the fluid to rush with high velocity thru the small channel and that virtually all the energy of movement would be transformed into heat by friction, similarly to that of the electric current in the lamp filament.

The second diagram will now be self-explanatory. Corresponding to the terminal capacity of the electric system an elastic reservoir is employed which dispenses with the necessity of a return pipe. As the piston oscillates the bag expands and contracts, and the fluid is made to surge thru the restricted passage with great speed, this resulting in the generation of heat as in the incandescent lamp. Theoretically considered, the efficiency of conversion of energy should be the same in both cases.

Granted, then, that an economic system of power transmission thru a single wire is practicable, the question arises how to collect the energy in the receivers. With this object attention is called to Fig. 5, in which a conductor is shown excited by an oscillator joined to it at one end. Evidently, as the periodic impulses pass thru the wire, differences of potential will be created along the same as well as at right angles to it in the surrounding medium and either of these may be usefully applied. Thus at a, a circuit comprising an inductance and capacity is resonantly excited in the transverse, and at b, in the longitudinal sense. At c, energy is collected in a circuit parallel to the conductor but not in contact with it, and again at d, in a circuit which is partly sunk into the conductor and may be, or not, electrically connected to the same. It is important to keep these typical dispositions in mind, for however the distant actions of the oscillator might be modified thru the immense extent of the globe the principles involved are the same.

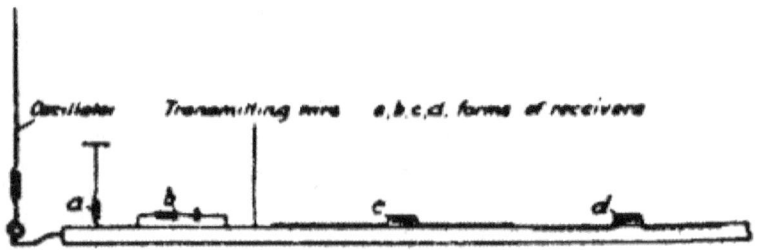

Fig. 5. Illustrating Typical Arrangements for Collecting Energy In a System of Transmission Thru a Single Wire.

Consider now the effect of such a conductor of vast dimensions on a circuit exciting it. The upper diagram of Fig. 6 illustrates a familiar oscillating system comprising a straight rod of self-inductance 2L with small terminal capacities cc and a node in the center. In the lower diagram of the figure a large capacity C is attached to the rod at one end with the result of shifting the node to the right, thru a distance corresponding to self-inductance X. As both parts of the system on either side of the node vibrate at the same rate, we have evidently, $(L+X)c = (L-X)C$ from which $X = L(C-c/C+c)$. When the capacity C becomes commensurate to that of the earth, X approximates

L, in other words, the node is close to the ground connection. The exact determination of its position is very important in the calculation of certain terrestrial electrical and geodetic data and I have devised special means with this purpose in view.

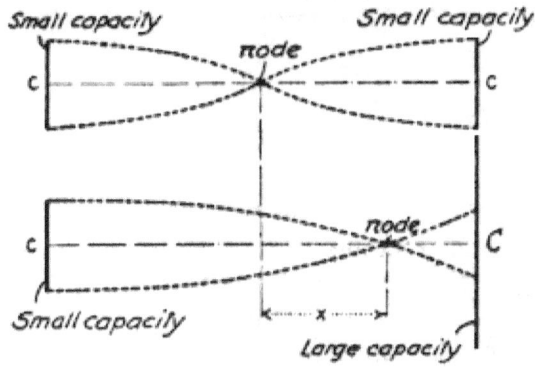

Fig. 6. Diagram Elucidating Effect of Large Capacity on One End.

My original plan of transmitting energy without wires is shown in the upper diagram of Fig. 7, while the lower one illustrates its mechanical analog, first published in my article in the Century Magazine of June, 1900. An alternator, preferably of high tension, has one of its terminals connected to the ground and the other to an elevated capacity and impresses its oscillations upon the earth. At a distant point a receiving circuit, likewise connected to ground and to an elevated capacity, collects some of the energy and actuates a suitable device. I suggested a multiplication of such units in order to intensify the effects, an idea which may yet prove valuable. In the analog two tuning forks are provided, one at the sending and the other at the receiving station, each having attached to its lower prong a piston fitting in a cylinder. The two cylinders communicate with a large elastic reservoir filled with an incompressible fluid. The vibrations transmitted to either of the tuning forks excite them by resonance and, thru electrical contacts or otherwise, bring about the desired result. This, I may say, was not a mere mechanical illustration, but a simple representation of my apparatus for submarine signaling, perfected by me in 1892, but not appreciated at that time, although more efficient than the instruments now in use.

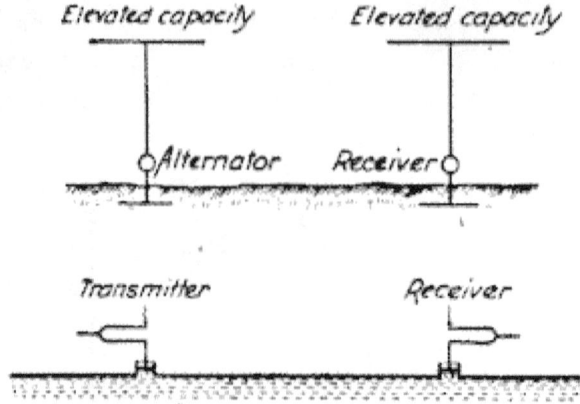

Fig. 7. Transmission of Electrical Energy Thru the Earth as Illustrated in Tesla's Lectures Before the Franklin Institute and Electric Light Association in February and March, 1893, and Mechanical Analog of the Same.

The electric diagram in Fig. 7, which was reproduced from my lecture, was meant only for the exposition of the principle. The arrangement, as I described it in detail, is shown in Fig. 8. In this case an alternator energizes the primary of a transformer, the high tension secondary of which is connected to the ground and an elevated capacity and tuned to the impressed oscillations. The receiving circuit consists of an inductance connected to the ground and to an elevated terminal without break and is resonantly responsive to the transmitted oscillations. A specific form of receiving device was not mentioned, but I had in mind to transform the received currents and thus make their volume and tension suitable for any purpose. This, in substance, is the system of today and I am not aware of a single authenticated instance of successful transmission at considerable distance by different instrumentalities. It might, perhaps, not be clear to those who have perused my first description of these improvements that, besides making known new and efficient types of apparatus, I gave to the world a wireless system of potentialities far beyond anything before conceived. I made explicit and repeated statements that I contemplated transmission, absolutely unlimited as to terrestrial distance and amount of energy. But, although I have overcome all obstacles which seemed in the beginning insurmountable and found elegant solutions of all the problems which confronted me, yet, even at this very day, the majority of experts are still blind to the possibilities which are within easy attainment.

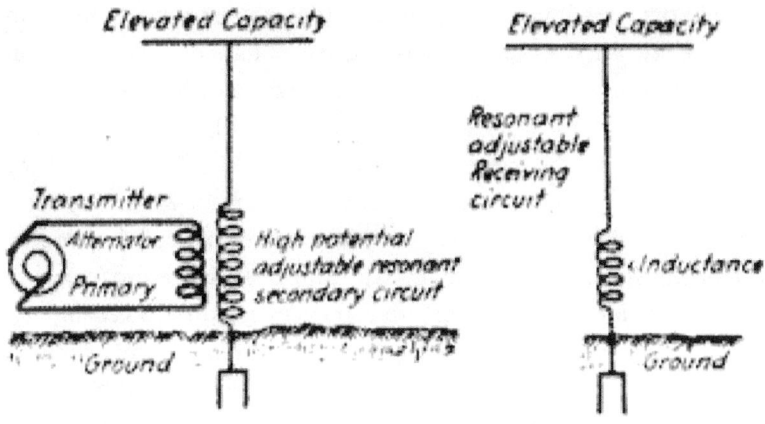

Fig. 8. Tesla's System of Wireless Transmission Thru the Earth as Actually Exposed In His Lectures Before the Franklin Institute and Electric Light Association in February and March, 1893.

My confidence that a signal could be easily flashed around the globe was strengthened thru the discovery of the "rotating brush," a wonderful phenomenon which I have fully described in my address before the Institution of Electrical Engineers, London, in 1892 Experiments with Alternate Currents of High Potential and High Frequency, and which is illustrated in Fig. 9. This is undoubtedly the most delicate wireless detector known, but for a long time it was hard to produce and to maintain in the sensitive state. These difficulties do not exist now and I am looking to valuable applications of this device, particularly in connection with the high-speed photographic method, which I suggested, in wireless, as well as in wire, transmission.

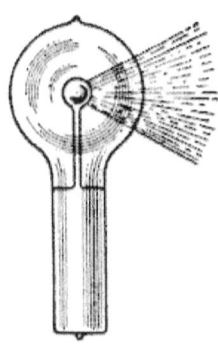

Fig. 9. The Forerunner of the Audion—the Most Sensitive Wireless Detector Known, as Described by Tesla In His Lecture Before the Institution of Electrical Engineers, London, February, 1892.

Possibly the most important advances during the following three or four years were my system of concatenated tuned circuits and methods of regulation, now universally adopted. The intimate bearing of these inventions on the development of the wireless art will appear from Fig. 10, which illustrates an arrangement described in my U.S. Patent No. 568,178 of September 22, 1896, and corresponding dispositions of wireless apparatus. The captions of the individual diagrams are thought sufficiently explicit to dispense with further comment. I will merely remark that in this early record, in addition to indicating how any number of resonant circuits may be linked and regulated, I have shown the advantage of the proper timing of primary impulses and use of harmonics. In a farcical wireless suit in London, some engineers, reckless of their reputation, have claimed that my circuits were not at all attuned; in fact they asserted that I had looked upon resonance as a sort of wild and untamable beast!

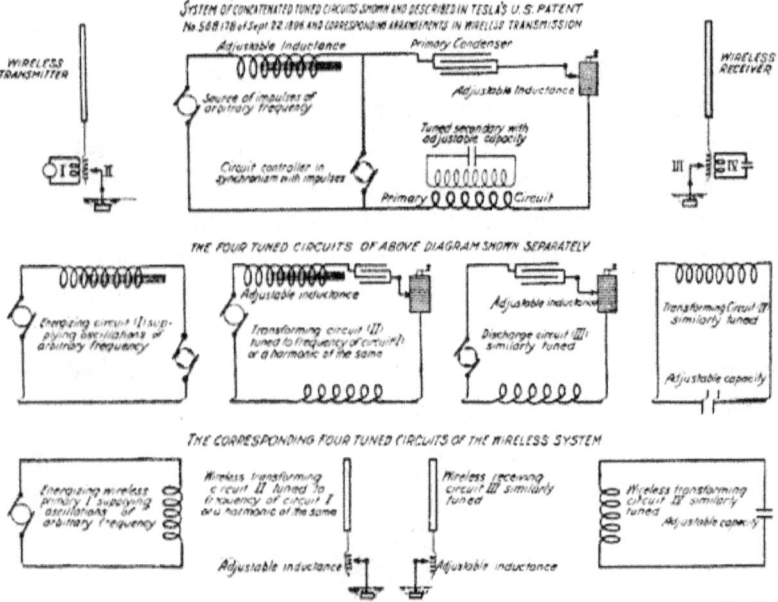

Fig. 10. Tesla's System of Concatenated Tuned Circuits Shown and Described In U. S. Patent No. 568,178 of September 22, 1896, and Corresponding Arrangements in Wireless Transmission.

It will be of interest to compare my system as first described in a Belgian patent of 1897 with the Hertz-wave system of that period. The significant differences between them will be observed at a glance. The first enables us to transmit economically energy to any distance and is of inestimable value;

the latter is capable of a radius of only a few miles and is worthless. In the first there are no spark-gaps and the actions are enormously magnified by resonance. In both transmitter and receiver the currents are transformed and rendered more effective and suitable for the operation of any desired device. Properly constructed, my system is safe against static and other interference and the amount of energy which may be transmitted is billions of times greater than with the Hertzian which has none of these virtues, has never been used successfully and of which no trace can be found at present.

A well-advertised expert gave out a statement in 1899 that my apparatus did not work and that it would take 200 years before a message would be flashed across the Atlantic and even accepted stolidly my congratulations on a supposed great feat. But subsequent examination of the records showed that my devices were secretly used all the time and ever since I learned of this I have treated these Borgia-Medici methods with the contempt in which they are held by all fair-minded men. The wholesale appropriation of my inventions was, however, not always without a diverting side. As an example to the point I may mention my oscillation transformer operating with an air gap. This was in turn replaced by a carbon arc, quenched gap, an atmosphere of hydrogen, argon or helium, by a mechanical break with oppositely rotating members, a mercury interrupter or some kind of a vacuum bulb and by such tours de force as many new "systems" have been produced. I refer to this of course, without the slightest ill-feeling, let us advance by all means. But I cannot help thinking how much better it would have been if the ingenious men, who have originated these "systems," had invented something of their own instead of depending on me altogether.

Before 1900 two most valuable improvements were made. One of these was my individualized system with transmitters emitting a wave-complex and receivers comprising separate tuned elements cooperatively associated. The underlying principle can be explained in a few words. Suppose that there are n simple vibrations suitable for use in wireless transmission, the probability that any one tune will be struck by an extraneous disturbance is $1/n$. There will then remain n-1 vibrations and the chance that one of these will be excited is $1/n-1$ hence the probability that two tunes would be struck at the same time is $1/n(n-1)$. Similarly, for a combination of three the chance will be $1/n(n-1)(n-2)$ and so on. It will be readily seen that in this manner any desired degree of safety against the statics or other kind of disturbance can be attained provided the receiving apparatus is so designed that is operation is possible only thru the joint action of all the tuned elements. This was a difficult problem which I have successfully solved so that now any desired number of simultaneous messages is practicable in the transmission thru the earth as well as thru artificial conductors.

The other invention, of still greater importance, is a peculiar oscillator enabling the transmission of energy without wires in any quantity that may ever be required for industrial use, to any distance, and with very high economy. It was the outcome of years of systematic study and investigation and wonders will be achieved by its means.

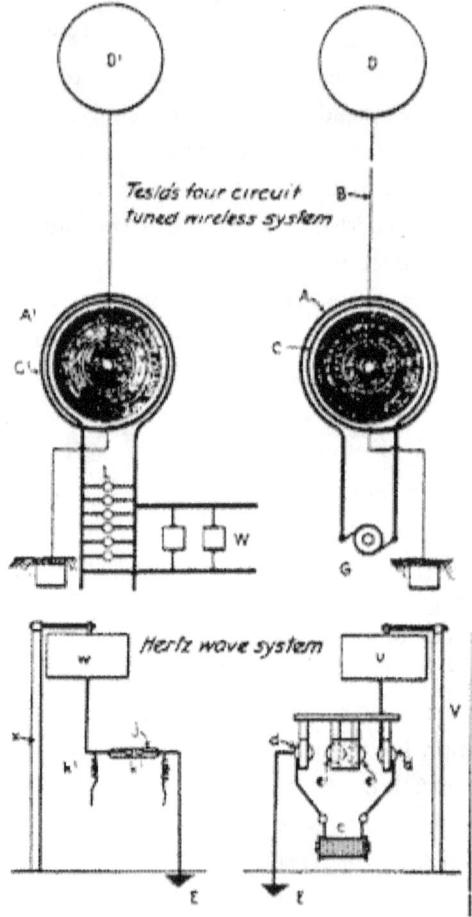

Fig. 11. Tesla's Four Circuit Tuned System Contrasted With the Contemporaneous Hertz-wave System.

The prevailing misconception of the mechanism involved in the wireless transmission has been responsible for various unwarranted announcements which have misled the public and worked harm. By keeping steadily in mind that the transmission thru the earth is in every respect identical to that thru a straight wire, one will gain a clear understanding of the phenomena and will be able to judge correctly the merits of a new scheme. Without wishing to detract from the value of any plan that has been put forward I may say that they are devoid of novelty. So for instance in Fig. 12 arrangements of transmitting and receiving circuits are illustrated, which I have described in my U.S. Patent No. 613,809 of November 8, 1898 on a Method of and Apparatus for Controlling Mechanism of Moving Vessels or Vehicles, and which have been recently dished up as original discoveries. In other patents and technical publications I have suggested conductors in the ground as one of the obvious modifications indicated in Fig. 5.

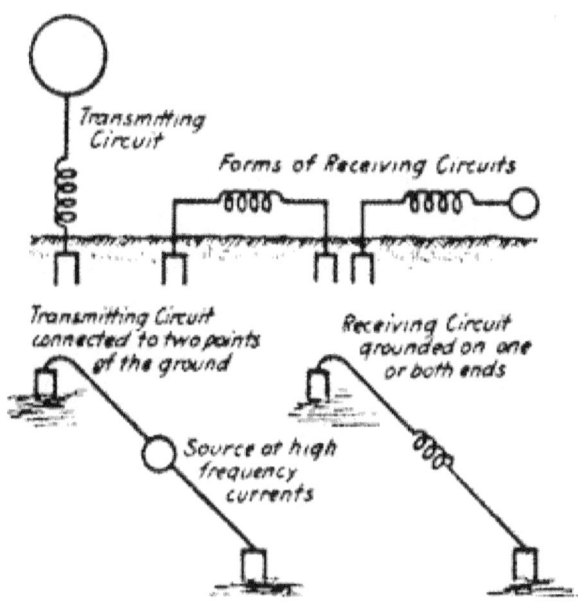

Fig. 12. Arrangements of Directive Circuits Described In Tesla's U. S. Patent No. 613,809 of November 8. 1898, on "Method of and Apparatus for Controlling Mechanism of Moving Vessels or Vehicles."

For the same reason the statics are still the bane of the wireless. There is about as much virtue in the remedies recently proposed as in hair restorers. A small and compact apparatus has been produced which does away entirely with this trouble, at least in plants suitably remodeled.

Nothing is more important in the present phase of development of the wireless art than to dispose of the dominating erroneous ideas. With this object I shall advance a few arguments based on my own observations which prove that Hertz waves have little to do with the results obtained even at small distances.

In Fig. 13 a transmitter is shown radiating space waves of considerable frequency. It is generally believed that these waves pass along the earth's surface and thus affect the receivers. I can hardly think of anything more improbable than this "gliding wave" theory and the conception of the "guided wireless" which are contrary to all laws of action and reaction. Why should these disturbances cling to a conductor where they are counteracted by induced currents, when they can propagate in all other directions unimpeded? The fact is that the radiations of the transmitter passing along the earth's surface are soon extinguished, the height of, the inactive zone indicated in the diagram, being some function of the wave length, the bulk of the waves traversing freely the atmosphere. Terrestrial phenomena which I have noted conclusively show that there is no Heaviside layer, or if it exists, it is of no effect. It certainly would be unfortunate if the human race were thus imprisoned and forever without power to reach out into the depths of space.

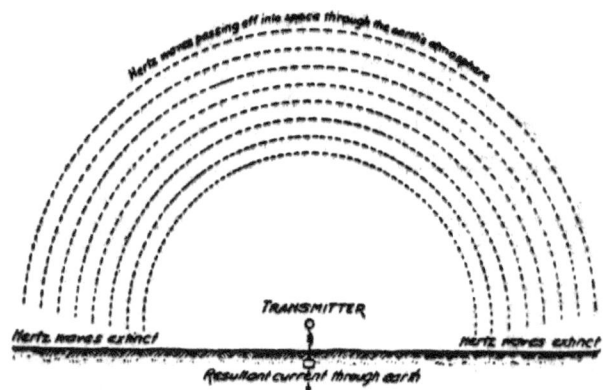

Fig. 13. Diagram Exposing the Fallacy of the Gilding Wave Theory as Propounded In Wireless Text Books.

The actions at a distance cannot be proportionate to the height of the antenna and the current in the same. I shall endeavor to make this clear by reference to diagram in Fig. 14. The elevated terminal charged to a high potential induces an equal and opposite charge in the earth and there are thus Q lines giving an average current $I = 4Qn$ which circulates locally and is useless except that it adds to the momentum. A relatively small number of lines q however, go off to great distance and to these corresponds a mean current of $ie = 4qn$ to which is due the action at a distance. The total average current in the antenna is thus $Im = 4Qn + 4qn$ and its intensity is no criterion for the performance. The electric efficiency of the antenna is $q/Q+q$ and this is often a very small fraction.

Dr. L. W. Austin and Mr. J. L. Hogan have made quantitative measurements which are valuable, but far from supporting the Hertz wave theory they are evidences in disproval of the same, as will be easily perceived by taking the above facts into consideration. Dr. Austin's researches are especially useful and instructive and I regret that I cannot agree with him on this subject. I do not think that if his receiver was affected by Hertz waves he could ever establish such relations as he has found, but he would be likely to reach these results if the Hertz waves were in a large part eliminated. At great distance the space waves and the current waves are of equal energy, the former being merely an accompanying manifestation of the latter in accordance with the fundamental teachings of Maxwell.

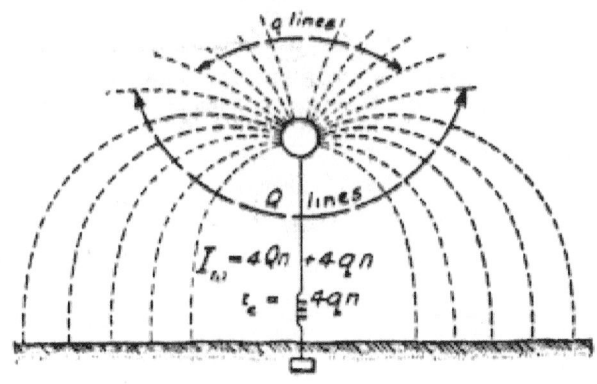

Fig. 14. Diagram Explaining the Relation Between the Effective and the Measured Current In the Antenna.

It occurs to me here to ask the question — why have the Hertz waves, been reduced from the original frequencies to those I have advocated for my system, when in so doing the activity of the transmitting apparatus has been reduced a billion fold? I can invite any expert to perform an experiment such as is illustrated in Fig. 15, which shows the classical Hertz oscillator although we may have in the Hertz oscillator an activity thousands of times greater, the effect on the receiver is not to be compared to that of the grounded circuit. This shows that in the transmission from an airplane we are merely working thru a condenser, the capacity of which is a function of a logarithmic ratio between the length of the conductor and the distance from the ground. The receiver is affected in exactly the same manner as from an ordinary transmitter, the only difference being that there is a certain modification of the action which can be predetermined from the electrical constants. It is not at all difficult to maintain communication between an airplane and a station on the ground, on the contrary, the feat is very easy.

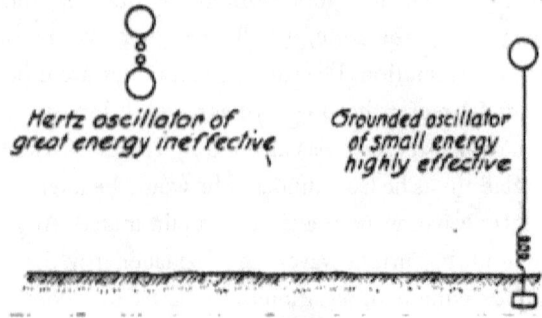

Fig. 15. Illustrating One of the General Evidences Against the Space Wave Transmission.

To mention another experiment in support of my view, I may refer to Fig. 16 in which two grounded circuits are shown excited by oscillations of the Hertzian order. It will be found that the antennas can be put out of parallelism without noticeable change in the action on the receiver, this proving that it is due to currents propagated thru the ground and not to space waves.

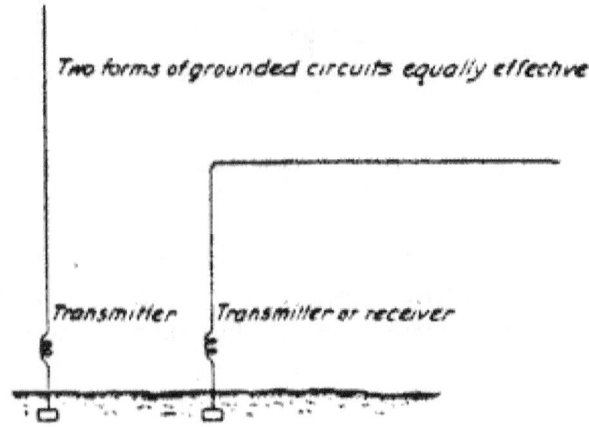

Fig. 16. Showing Unimportance of Relative Position of Transmitting and Receiving Antennae In Disproval of the Hertz-wave Theory.

Particularly significant are the results obtained in cases illustrated in Figures 17 and 18. In the former an obstacle is shown in the path of the waves but unless the receiver is within the effective electrostatic influence of the mountain range, the signals are not appreciably weakened by the presence of the latter, because the currents pass under it and excite the circuit in the same way as if it were attached to an energized wire. If, as in Fig. 18, a second range happens to be beyond the receiver, it could only strengthen the Hertz wave effect by reflection, but as a matter of fact it detracts greatly from the intensity of the received impulses because the electric level between the mountains is raised, as I have explained with my lightning protector in the *Experimenter* of February.

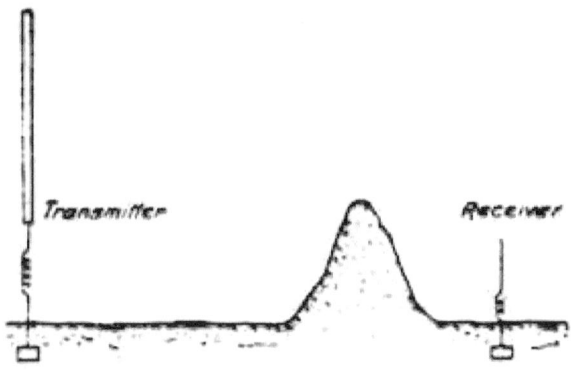

Fig. 17. Illustrating Influence of Obstacle In the Path of Transmission as Evidence Against the Hertz-wave Theory.

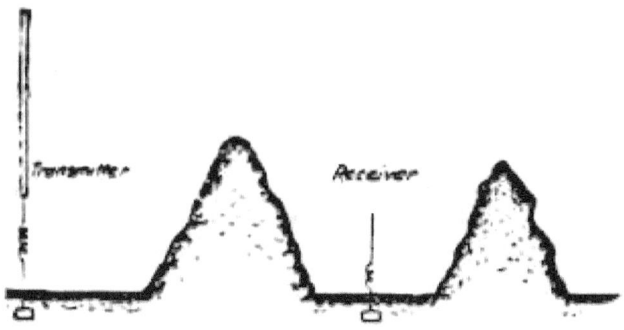

Fig. 18. Showing Effect of Two Hills as Further Proof Against the Hertz-wave Theory.

Again in Fig. 19 two transmitting circuits, one grounded directly and the other thru an air gap are shown. It is a common observation that the former is far more effective, which could not be the case with Hertz radiations. In a like manner if two grounded circuits are observed from day to day the effect is found to increase greatly with the dampness of the ground, and for the same reason also the transmission thru sea-water is more efficient.

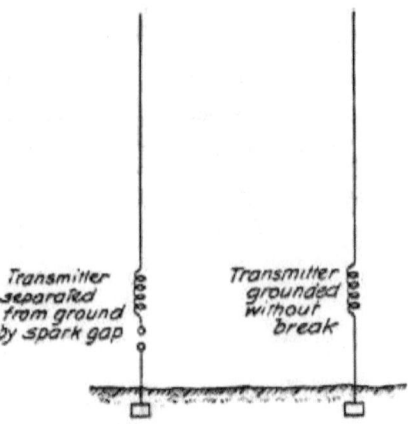

Fig. 19. Comparing the Actions of Two Forms of Transmitter as Bearing Out the Fallacy of the Hertz-wave Theory.

An illuminating experiment is indicated in Fig. 20 in which two grounded transmitters are shown, one with a large and the other with a small terminal capacity. Suppose that the latter be 1/10 of the former but that it is charged to 10 times the potential and let the frequency of the two circuits and therefore the currents in both antennas be exactly the same. The circuit with the smaller capacity will then have 10 times the energy of the other but the effects on the receiver will be in no wise proportionate.

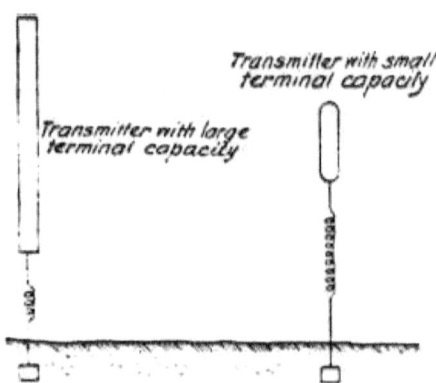

Fig. 20. Disproving the Hertz-wave Theory by Two Transmitters, One of Great and the Other of Small Energy.

The same conclusions will be reached by transmitting and receiving circuits with wires buried underground. In each case the actions carefully investigated will be found to be due to earth currents. Numerous other proofs might be cited which can be easily verified. So for example oscillations of low frequency are ever so much more effective in the transmission which is inconsistent with the prevailing idea. My observations in 1900 and the recent transmissions of signals to very great distances are another emphatic disproval.

The Hertz wave theory of wireless transmission may be kept up for a while, but I do not hesitate to say that in a short time it will be recognized as one of the most remarkable and inexplicable aberrations of the scientific mind which has ever been recorded in history.

May 1919

ELECTRICAL OSCILLATORS

—Electrical Experimenter—July 1919

Few fields have been opened up the exploration of which has proved as fruitful as that of high frequency currents. Their singular properties and the spectacular character of the phenomena they presented immediately commanded universal attention. Scientific men became interested in their investigation, engineers were attracted by their commercial possibilities, and physicians recognized in them a long-sought means for effective treatment of bodily ills. Since the publication of my first researches in 1891, hundreds of volumes have been written on the subject and many invaluable results obtained through the medium of this new agency. Yet, the art is only in its infancy and the future has incomparably bigger things in store.

From the very beginning I felt the necessity of producing efficient apparatus to meet a rapidly growing demand and during the eight years succeeding my original announcements I developed not less than fifty types of these transformers or electrical oscillators, each complete in every detail and refined to such a degree that I could not materially improve any one of them today. Had I been guided by practical considerations I might have built up an immense and profitable business, incidentally rendering important services to the world. But the force of circumstances and the ever enlarging vista of greater achievements turned my efforts in other directions. And so it comes that instruments will shortly be placed on the market which, oddly enough, were perfected twenty years ago!

These oscillators are expressly intended to operate on direct and alternating lighting circuits and to generate damped and undamped oscillations or currents of any frequency, volume and tension within the widest limits. They are compact, self-contained, require no care for long periods of time and will be found very convenient and useful for various purposes as, wireless telegraphy and telephony; conversion of electrical energy; formation of chemical compounds through fusion and combination; synthesis of gases; manufacture of ozone; lighting; welding; municipal, hospital, and domestic sanitation and sterilization, and numerous other applications in scientific laboratories and industrial institutions. While these transformers have never been described before, the general principles underlying them were fully set forth in my published articles and patents, more particularly those of September

22, 1896, and it is thought, therefore, that the appended photographs of a few types, together with a short explanation, will convey all the information that may be desired.

The essential parts of such an oscillator are: a condenser, a self-induction coil for charging the same to a high potential, a circuit controller, and a transformer which is energized by the oscillatory discharges of the condenser. There are at least three, but usually four, five or six, circuits in tune and the regulation is effected in several ways, most frequently merely by means of an adjusting screw. Under favorable conditions an efficiency as high as 85% is attainable, that is to say, that percentage of the energy supplied can be recovered in the secondary of the transformer. While the chief virtue of this kind of apparatus is obviously due to the wonderful powers of the condenser, special qualities result from concatenation of circuits under observance of accurate harmonic relations, and minimization of frictional and other losses which has been one of the principal objects of the design.

Broadly, the instruments can be divided into two classes: one in which the circuit controller comprises solid contacts, and the other in which the make and break is effected by mercury. Figures 1 to 8, inclusive, belong to the first, and the remaining ones to the second class. The former are capable of an appreciably higher efficiency on account of the fact that the losses involved in the make and break are reduced to the minimum and the resistance component of the damping factor is very small. The latter are preferable for purposes requiring larger output and a great number of breaks per second. The operation of the motor and circuit controller of course consumes a certain amount of energy which, however, is the less significant the larger the capacity of the machine.

In Fig. 1 is shown one of the earliest forms of oscillator constructed for experimental purposes. The condenser is contained in a square box of mahogany upon which is mounted the self-induction or charging coil wound, as will be noted, in two sections connected in multiple or series according to whether the tension of the supply circuit is 110 or 220 volts. From the box protrude four brass columns carrying a plate with the spring contacts and adjusting screws as well as two massive terminals for the reception of the primary of the transformer. Two of the columns serve as condenser connections while the other pair is employed to join the binding posts of the switch in front to the self-inductance and condenser. The primary coil consists of a few turns of topper ribbon to the ends of which are soldered short rods fitting into the terminals referred to. The secondary is made in two parts, wound in a manner to reduce as much as possible the distributed capacity and at the same time enable the coil to withstand a very high pressure between its terminals at the center, which are connected to binding posts on two rubber columns projecting from the primary. The circuit connections may be slightly varied but ordinarily they are as diagrammatically illustrated in the Electrical Experimenter for May on page 89, relating to my oscillation transformer photograph of which appeared on page 16 of the same number. The operation is as follows: When the switch is thrown on, the current from the supply circuit rushes through the self-induction coil, magnetizing the iron core within and separating the contacts of the controller. The high tension induced current then charges the condenser and upon closure of the contacts the accumulated energy is released through the primary, giving rise to a long series of oscillations which excite the tuned secondary circuit.

This device has proved highly serviceable in carrying on laboratory experiments of all kinds. For instance, in studying phenomena of impedance, the transformer was removed and a bent copper bar inserted in the terminals. The latter was often replaced by a large circular loop to exhibit inductive effects at a distance or to excite resonant circuits used in various investigations and measurements. A transformer suitable for any desired performance could be readily improvised and attached to the terminals and in this way much time and labor was saved. Contrary to what might be naturally expected, little trouble was experienced with the contacts, although the currents through them were heavy, namely, proper conditions of resonance existing, the great flow occurs only when the circuit is closed and no destructive arcs can develop. Originally I employed platinum and iridium tips but later replaced them by some of meteorite and finally of tungsten. The last have given the best satisfaction, permitting working for hours and days without interruption.

Fig. 2 illustrates a small oscillator designed for certain specific uses. The underlying idea was to attain great activities during minute intervals of time each succeeded by a comparatively long period of inaction. With this object a large self-induction and a quick-acting break were employed owing to which arrangement the condenser was charged to a very high potential. Sudden secondary currents and sparks of great volume were thus obtained, eminently suitable for welding thin wires, flashing lamp filaments, igniting explosive mixtures and kindred applications. The instrument was also adapted for battery use and in this form was a very effective igniter for gas engines on which a patent bearing number 609,250 was granted to me August 16, 1898.

Fig. 3 represents a large oscillator of the first class intended for wireless experiments, production of Rontgen rays and scientific research in general. It comprises a box containing two condensers of the same capacity on which are supported the charging coil and transformer. The automatic circuit controller, hand switch and connecting posts are mounted on the front plate of the inductance spool as is also one of the contact springs. The condenser box is equipped with three terminals, the two external ones serving merely for connection while the middle one carries a contact bar with a screw for regulating the interval during which the circuit is closed. The vibrating spring, itself, the sole function of which is to cause periodic interruptions, can be adjusted in its strength as well as distance from the iron core in the center of the charging coil by four screws visible on the top plate so that any desired conditions of mechanical control might be secured. The primary coil of the transformer is of copper sheet and taps are made at suitable points for the purpose of varying at will., the number of turns. As in Fig. 1 the inductance coil is wound in two sections to adapt the instrument both to 110 and 220 volt circuits and several secondaries were provided to suit the various wave lengths of the primary. The output was approximately 500 watt with damped waves of about 50,000 cycles per second. For short periods of time undamped oscillations were produced in screwing the vibrating spring tight against the iron core and separating the contacts by the adjusting screw which also performed the function of a key. With this oscillator I made a number of important observations and it was one of the machines exhibited at a lecture before the New York Academy of Sciences in 1897.

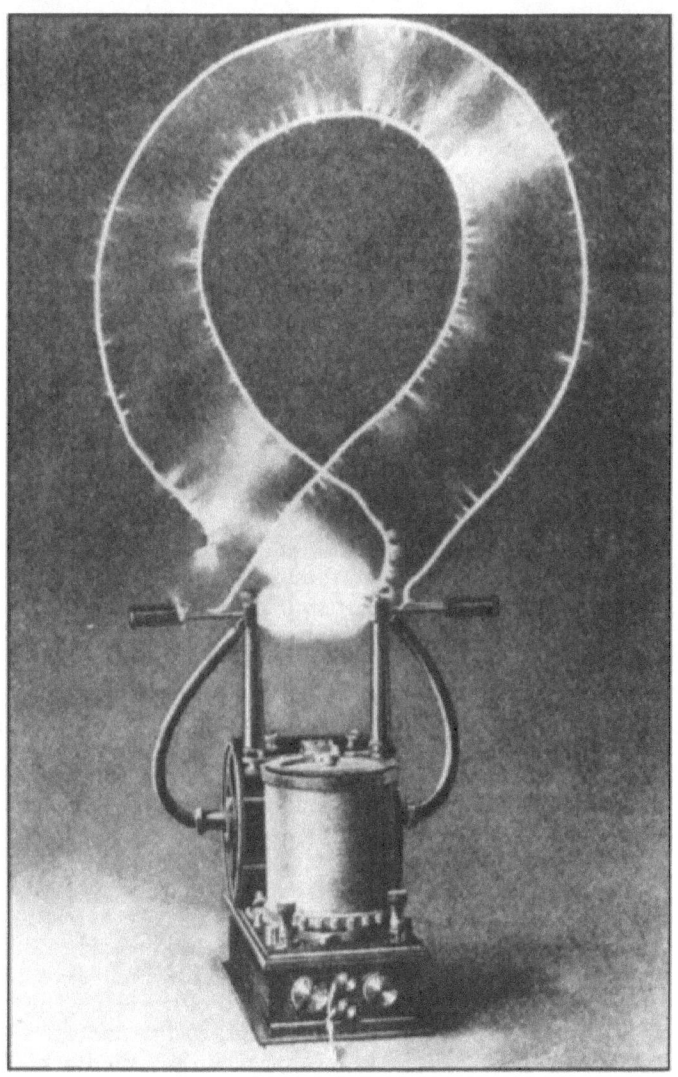

Fig. 4 is a photograph of a type of transformer in every respect similar to the one illustrated in the May, 1919, issue of the Electrical Experimenter to which reference has already been made. It contains the identical essential parts, disposed in like manner, but was specially designed for use on supply circuits of higher tension, from 220 to 500 volts or more. The usual adjustments are made in setting the contact spring and shifting the iron core within the inductance coil up and down by means of two screws. In order to prevent injury through a short-circuit, fuses are inserted in the lines. The instrument was photographed in action, generating undamped oscillations from a 220 volt lighting circuit.

Fig. 5 shows a later form of transformer principally intended to replace Rhumkorf coils. In this instance a primary is employed, having a much greater number of turns and the secondary is closely linked with the same. The currents developed in the latter, having a tension of from 10,000 to 30,000 volts, are used to charge condensers and operate an independent high frequency coil as customary. The controlling mechanism is of somewhat different construction but the core and contact spring are both adjustable as before.

Fig. 6 is a small instrument of this type, particularly intended for ozone production or sterilization. It is remarkably efficient for its size and can be connected either to a 110 or 220 volt circuit, direct or alternating, preferably the former.

In Fig. 7 is shown a photograph of a larger transformer of this kind. The construction and disposition of the parts is as before but there are two condensers in the box, one of which is connected in the circuit as in the previous cases, while the other is in shunt to the primary coil. In this manner currents of great volume are produced in the latter and the secondary effects are accordingly magnified. The introduction of an additional tuned circuit secures also other advantages but the adjustments are rendered more difficult and for this reason it is desirable to use such an instrument in the production of currents of a definite and unchanging frequency.

Fig. 8 illustrates a transformer with rotary break. There are two condensers of the same capacity in the box which can be connected in series or multiple.

The charging inductances are in the form of two long spools upon which are supported the secondary terminals. A small direct current motor, the speed of which can be varied within wide limits, is employed to drive a specially constructed make and break. In other features the oscillator is like the one illustrated in Fig. 3 and its operation will be readily understood from the foregoing. This transformer was used in my wireless experiments and frequently also for lighting the laboratory by my vacuum tubes and was likewise exhibited at my lecture before the New York Academy of Sciences above mentioned.

Coming now to machines of the second class, Fig. 9 shows an oscillatory transformer comprising a condenser and charging inductance enclosed in a box, a transformer and a mercury circuit controller, the latter being of a construction described for the first time in my patent No. 609,251 of August 16, 1898. It consists of a motor driven hollow pulley containing a small quantity of mercury which is thrown outwardly against the walls of the vessel by centrifugal force and entrains a contact wheel which periodically closes and opens the condenser circuit. By means of adjusting screws above the pulley, the depth of immersion of the vanes and consequently, also, the duration of each contact can be varied at desire and thus the intensity of the effects and their character controlled. This form of break has given thorough satisfaction, working continuously with currents of from 20 to 25 amperes. The number of interruptions is usually from 500 to 1,000 per second but higher frequencies are practicable. The space occupied is about 10" X 8" X 10" and the output approximately 1/2 kW.

In the transformer just described the break is exposed to the atmosphere and a slow oxidation of the mercury takes place. This disadvantage is overcome in the instrument shown in Fig. 10, which consists of a perforated metal box containing the condenser and charging inductance and carrying on the top a motor driving the break, and a transformer. The mercury break is of a kind to be described and operates on the principle of a jet which establishes, intermittently, contact with a rotating wheel in the interior of the pulley. The stationary parts are supported in the vessel on a bar passing through the long hollow shaft of the motor and a mercury seal is employed to effect hermetic closure of the chamber enclosing the circuit controller. The current is led into the interior of the pulley through two sliding rings on the top which are in series with the condenser and primary. The exclusion of the oxygen is a decided improvement, the deterioration of the metal and attendant trouble being eliminated and perfect working.

Fig. 11 is a photograph of a similar oscillator with hermetically inclosed mercury break. In this machine the stationary parts of the interrupter in the interior of the pulley were supported on a tube through which was led an insulated wire connecting to one terminal of the break while the other was in contact with the vessel. The sliding rings were, in this manner, avoided and the construction simplified. The instrument was designed for oscillations of lower tension and frequency requiring primary currents of comparatively smaller amperage and was used to excite other resonant circuits.

Fig. 12 shows an improved form of oscillator of the kind described in Fig.

10, in which the supporting bar through the hollow motor shaft was done away with, the device pumping the mercury being kept in position by gravity, as will be more fully explained with reference to another figure. Both the capacity of the condenser and primary turns were made variable with the view of producing oscillations of several frequencies.

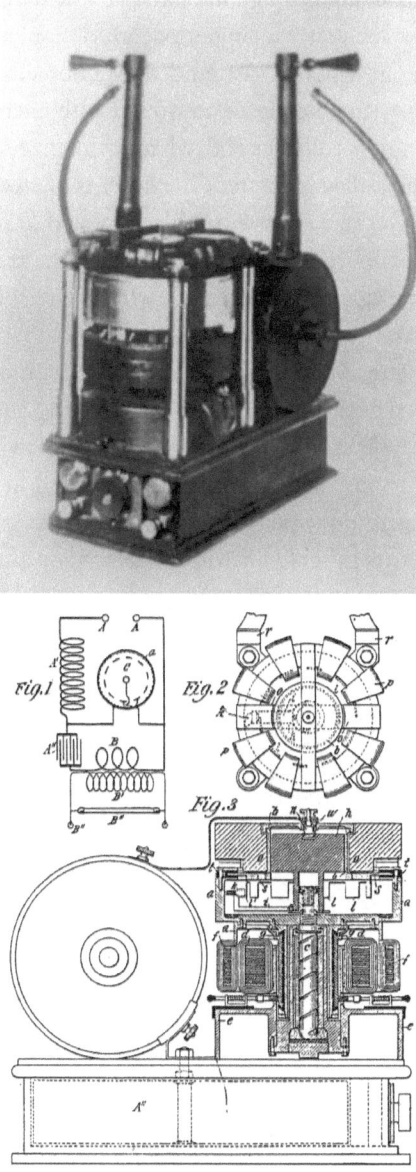

Fig. 13 is a photographic view of another form of oscillatory transformer with hermetically sealed mercury interrupter, and Fig. 14 diagrams showing the circuit connections and arrangement of parts reproduced from my patent, No. 609,245, of August 16, 1898, describing this particular device. The condenser, inductance, transformer and circuit controller are disposed as before, but the latter is of different construction, which will be clear from an inspection of Fig. 14: The hollow pulley a is secured to a shaft c which is mounted in a vertical bearing passing through the stationary field magnet d of the motor. In the interior of the vessel is supported, on frictionless bearings, a body h of magnetic material which is surrounded by a dome b in the center of a laminated iron ring, with pole pieces oo wound with energizing coils p. The ring is supported on four columns and, when magnetized, keeps the body h in position while the pulley is rotated. The latter is of steel, but the dome is preferably made of German silver burnt black by acid or nickeled. The body h carries a short tube k bent, as indicated, to catch the fluid as it is whirled around, and project it against the teeth of a wheel fastened to the pulley. The wheel is insulated and contact from it to the external circuit is established through a mercury cup. As the pulley is rapidly rotated a jet of the fluid is thrown against the wheel, thus peaking and breaking contact about 1,000 times per second. The instrument works silently and, owing to the absence of all deteriorating agents; keeps continually clean and in perfect condition. The number of interruptions per second may be much greater, however, so as to make the currents suitable for wireless telephony and like purposes.

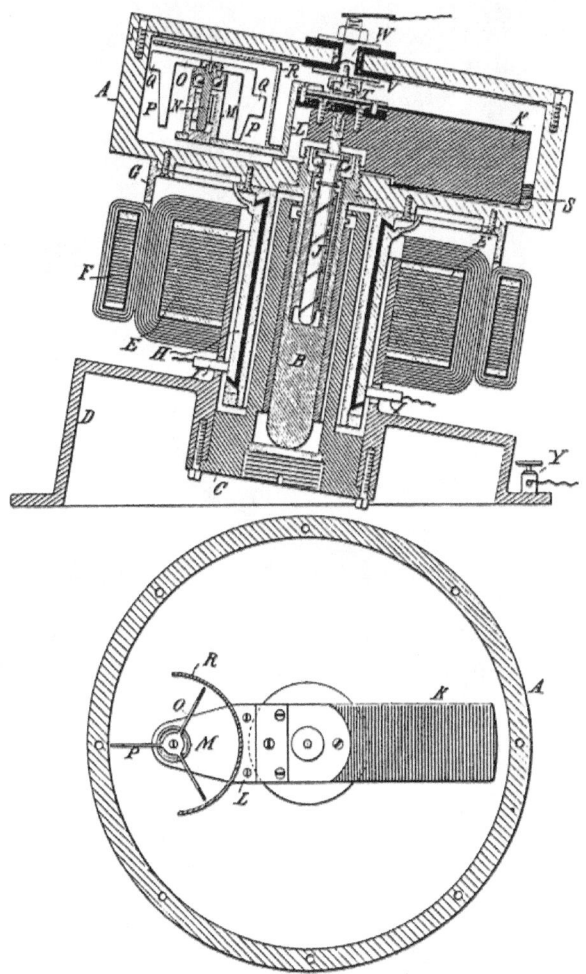

A modified form of oscillator is represented in Figs. 15 and 16, the former being a photographic view and the latter a diagrammatic illustration showing the arrangement of the interior parts of the controller. In this instance the shaft b carrying the vessel a is hollow and supports, in frictionless bearings, a spindle j to which is fastened a weight k. Insulated from the latter, but mechanically fixed to it, is a curved arm L upon which is supported, freely rotatable, a break-wheel with projections QQ. The wheel is in electrical connection with the external circuit through a mercury cup and an insulated plug supported from the top of the pulley. Owing to the inclined position of the. motor the weight k keeps the break-wheel in place by the force of gravity and as the pulley is rotated the circuit, including the condenser and primary coil of the transformer, is, rapidly made and broken.

Fig. 17 shows a similar instrument in which, however, the make and break device is a jet of mercury impinging against an insulated toothed wheel carried on an insulated stud in the center of the cover of the pulley as shown. Connection to the condenser circuit is made by brushes bearing on this plug.

Fig. 18 is a photograph of another transformer with a mercury circuit controller of the wheel type, modified in some features on which it is unnecessary to dwell.

These are but a few of the oscillatory transformers I have perfected and constitute only a small part of my high frequency apparatus of which I hope to give a full description, when I shall have freed myself of pressing duties, at some future date.

July 1919

RAIN CAN BE CONTROLLED AND HYDRAULIC FORCE PROVIDED, SAYS NIKOLA TESLA, IF WE TURN TO SUN FOR POWER

—*by Edward Marshall. Syracuse Herald, ca. February 29, 1920*

Famed inventor sees wide possibilities in new scientific theory

… "As time goes on we may find the harnessing of the sun's rays less objectionable, especially as great improvements are possible in the methods and devices so far employed"… "wind power is not to be distained"… "terrestrial heat, on an immense scale appears quite feasible and there is a strong probability that at a time not too distant projects will be seriously undertaken…"

Water power ideal

"In most of the processes of transformation we are confronted with appalling waste and definite limits exist to improvements aiming at economy. No amount of ingenuity can ever circumvent the natural laws imposing these restrictions.

"Water power is a remarkable exception in this respect. In hydraulic development the wheel can have an efficiency of 85 and the dynamo can have an efficiency of 98% so that the combined efficiency is over 83% that is to say, we are enabled in this way usefully to apply almost the entire energy furnished by the sun.

"Not only this but the apparatus is simple, well-nigh indestructible, and requires virtually no attention.

"Unfortunately this source of power supply is not adequate to meet all our needs, although the theoretical energy of falling water is, so to speak, unlimited. Assuming for the rain clouds an average height of 15,000 feet and an annual precipitation of 33 inches, the power over the whole area of the United States amounts to more than twelve billion horsepower but a large portion of the potential energy is transformed into heat by friction of the rain drops against the air so that the actual mechanical energy is much smaller.

"Most of the water comes from a height of something like 2,000 feet, and all in all represents over one-half a billion horsepower, but in the form but in the form of available waterpower we cannot obtain more than a fall of 100 feet, so that by harnessing all the falls in the United States not more than eighty million horsepower can be developed.

"So far we have harnessed approximately 8,000,000 horsepower in this country, thus effecting a saving equivalent to nearly one-third of the entire coal mined. By extensive damming the power derived can be greatly increased, possibly to several hundred million horsepower, giving us more power by far than we have now with all our coal. But this is not the limit.

On the eve of great feats

"We are on the eve of accomplishments which will be of tremendous consequence to the future advancement of the human race. One of these is the control of the precipitation of moisture.

"The water is evaporated and thus raised against the force of gravity. It is then held in suspension in the vapor which we call clouds. Air currents carry this vapor, hither and yon, often to distant regions, where it may remain for long periods at a height, in a state of delicate suspension.

"When the equilibrium is disturbed the water falls to earth [in the] form of rain and through rills and rivers flows back to the ocean.

"Thus the sun, those heat causes the evaporation, even maintains this life sustaining stream. The energy necessary to cause the precipitation of the rain, compared to that rain's potential energy when released, is like that of the spark setting off a charge of dynamite compared to the dynamite.

"If this part of the natural process were under the control of man he could transform the entire globe.

"Many schemes have been proposed to this end, none of which have knowledge offering the remotest chance of success.

"But I have ascertained that with proper apparatus this wonder can be performed.

"Any amount of power will then be at our disposal; we can make out of deserts fertile land and create lakes and rivers almost without effort on our part.

"However our triumph would not be complete if the power could not be conveyed to distances without limit. This achievement, to, is now within our reach. With my wireless system it is practicable to transmit electrical energy over a distance of 12,000 miles with a loss not exceeding 5%. I can conceive of no advances which would be more desirable at this time and be more beneficial to the further progress of mankind."

(Copyright by Edward Marshall Syndicate, Inc.)

February 29, 1920

WHEN WOMAN IS BOSS

—Colliers, January 30, 1926

The life of the bee will be the life of our race, says Nikola Tesla, world-famed scientist.

A New sex order is coming—with the female as superior. You will communicate instantly by simple vest-pocket equipment. Aircraft will travel the skies, unmanned, driven and guided by radio. Enormous power will be transmitted great distances without wires. Earthquakes will become more and more frequent. Temperate zones will turn frigid or torrid. And some of these awe-inspiring developments, says Tesla, are not so very far off.

At sixty-eight years of age Nikola Tesla sits quietly in his study, reviewing the world that he has helped to change, foreseeing other changes that must come in the onward stride of the human race. He is a tall, thin, ascetic man who wears somber clothes and looks out at life with steady, deep-set eyes. In the midst of luxury he lives meagerly, selecting his diet with a precision almost extreme. He abstains from all beverages save water and milk and has never indulged in tobacco since early manhood.

He is an engineer, an inventor and, above these as well as basic to them, a philosopher. And, despite his obsession with the practical application of what a gifted mind may learn in books, he has never removed his gaze from the drama of life.

This world, amazed many times during the last throbbing century, will rub its eyes and stand breathless before greater wonders than even the past few generations have seen; and fifty years from now the world will differ more from the present-day than our world now differs from the world of fifty years ago.

Nikola Tesla came to America in early manhood, and his inventive genius found quick recognition. When fortune was his through his revolutionary power-transmission machines he established plants, first in New York, then Colorado, later on Long Island, where his innumerable experiments resulted in all manner of important and minor advances in electrical science. Lord Kelvin said of him (before he was forty) that he had contributed more than any other man to the study of electricity.

"From the inception of the wireless system," he says, "I saw that this new art of applied electricity would be of greater benefit to the human race than any other scientific discovery, for it virtually eliminates distance. The majority of the ills from which humanity suffers are due to the immense extent of the terrestrial globe and the inability of individuals and nations to come into close contact.

"Wireless will achieve the closer contact through transmission of intelligence, transport of our bodies and materials and conveyance of energy.

"When wireless is perfectly applied the whole earth will be converted into a huge brain, which in fact it is, all things being particles of a real and rhythmic whole. We shall be able to communicate with one another instantly, irrespective of distance. Not only this, but through television and telephony we shall see and hear one another as perfectly as though we were face to face, despite intervening distances of thousands of miles; and the instruments through which we shall be able to do his will be amazingly simple compared with our present telephone. A man will be able to carry one in his vest pocket.

"We shall be able to witness and hear events—the inauguration of a President, the playing of a world series game, the havoc of an earthquake or the terror of a battle—just as though we were present.

"When the wireless transmission of power is made commercial, transport and transmission will be revolutionized. Already motion pictures have been transmitted by wireless over a short distance. Later the distance will be illimitable, and by later I mean only a few years hence. Pictures are transmitted over wires—they were telegraphed successfully through the point system thirty years ago. When wireless transmission of power becomes general, these methods will be as crude as is the steam locomotive compared with the electric train.

Woman — free and regal

All railroads will be electrified, and if there are enough museums to hold them the steam locomotives will be grotesque antiques for our immediate posterity.

"Perhaps the most valuable application of wireless energy will be the propulsion of flying machines, which will carry no fuel and will be free from any limitations of the present airplanes and dirigibles. We shall ride from New York to Europe in a few hours. International boundaries will be largely obliterated and a great step will be made toward the unification and harmonious existence of the various races inhabiting the globe. Wireless will not only make possible the supply of energy to region, however inaccessible, but it will be effective politically by harmonizing international interests; it will create understanding instead of differences.

"Modern systems of power transmission will become antiquated. Compact relay stations one half or one quarter the size of our modern power plants will be the basis of operation—in the air and under the sea, for water will effect small loss in conveying energy by wireless."

Mr. Tesla foresees great changes in our daily life. "Present wireless receiving apparatus," says he, "will be scrapped for much simpler machines; static and all forms of interference will be eliminated, so that innumerable transmitters and receivers may be operated without interference. It is more than probable that the household's daily newspaper will be printed 'wirelessly' in the home during the night. Domestic management—the problems of heat, light and household mechanics—will be freed from all labor through beneficent wireless power.

"I foresee the development of the flying machine exceeding that of the automobile, and I expect Mr. Ford to make large contributions toward this progress. The problem of parking automobiles and furnishing separate roads for commercial and pleasure traffic will be solved. Belted parking towers will arise in our large cities, and the roads will be multiplied through sheer necessity, or finally rendered unnecessary when civilization exchanges wheels for wings.

The world's internal reservoirs of heat, indicated by frequent volcanic eruptions, will be tapped for industrial purposes. In an article I wrote twenty years ago I defined a process for continuously converting to human use part of the heat received from the sun by the atmosphere. Experts have jumped to the conclusion that I am attempting to realize a perpetual-motion scheme. But my process has been carefully worked out. It is rational."

Mr. Tesla regards the emergence of woman as one of the most profound portents for the future.

"It is clear to any trained observer," he says, "and even to the sociologically untrained, that a new attitude toward sex discrimination has come over the world through the centuries, receiving an abrupt stimulus just before and after the World War.

"This struggle of the human female toward sex equality will end in a new sex order, with the female as superior. The modern woman, who anticipates in merely superficial phenomena the advancement of her sex, is but a surface symptom of something deeper and more potent fermenting in the bosom of the race.

"It is not in the shallow physical imitation of men that women will assert first their equality and later their superiority, but in the awakening of the intellect of women.

"Through countless generations, from the very beginning, the social subservience of women resulted naturally in the partial atrophy or at least the hereditary suspension of mental qualities which we now know the female sex to be endowed with no less than men.

The queen is the center of life

"But the female mind has demonstrated a capacity for all the mental acquirements and achievements of men, and as generations ensue that capacity will be expanded; the average woman will be as well educated as the average man, and then better educated, for the dormant faculties of her brain will be stimulated to an activity that will be all the more intense and powerful because of centuries of repose. Woman will ignore precedent and startle civilization with their progress.

"The acquisition of new fields of endeavor by women, their gradual usurpation of leadership, will dull and finally dissipate feminine sensibilities, will choke the maternal instinct, so that marriage and motherhood may become abhorrent and human civilization draw closer and closer to the perfect civilization of the bee."

The significance of this lies in the principle dominating the economy of the bee — the most highly organized and intelligently coordinated system of any form of non-rational animal life — the all-governing supremacy of the instinct for immortality which makes divinity out of motherhood.

The center of all bee life is the queen. She dominates the hive, not through hereditary right, for any egg may be hatched into a reigning queen, but because she is the womb of this insect race.

We can only sit and wonder

There are the vast, desexualized armies of workers whose sole aim and happiness in life is hard work. It is the perfection of communism, of socialized, cooperative life wherein all things, including the young, are the property and concern of all.

Then there are the virgin bees, the princess bees, the females which are selected from the eggs of the queen when they are hatched and preserved in case an unfruitful queen should bring disappointment to the hive. And there are the male bees, few in number, unclean of habit, tolerated only because they are necessary to mate with the queen.

When the time is ripe for the queen to take her nuptial flight the male bees are drilled and regimented. The queen passes the drones which guard the gate of the hive, and the male bees follow her in rustling array. Strongest of all the inhabitants of the hive, more powerful than any of her subjects, the queen launches into the air, spiraling upward and upward, the male bees following. Some of the pursuers weaken and fail, drop out of the nuptial chase, but the queen wings higher and higher until a point is reached in the far ether where but one of the male bees remains. By the inflexible law of natural selection he is the strongest, and he mates with the queen. At the moment of marriage his body splits asunder and he perishes.

The queen returns to the hive, impregnated, carrying with her tens of thousands of eggs—a future city of bees, and then begins the cycle of reproduction, the concentration of the teeming life of the hive in unceasing work for the birth of a new generation.

Imagination falters at the prospect of human analogy to this mysterious and superbly dedicated civilization of the bee; but when we consider how the human instinct for race perpetuation dominates life in its normal and exaggerated and perverse manifestations, there is ironic justice in the possibility that this instinct, with the continuing intellectual advance of women, may be finally expressed after the manner of the bee, though it will take centuries to break down the habits and customs of peoples that bar the way to such a simply and scientifically ordered civilization.

We have seen a beginning of this in the United States. In Wisconsin the sterilization of confirmed criminals and pre-marriage examination of males is required by law, while the doctrine of eugenics is now boldly preached where a few decades ago its advocacy was a statutory offense.

Old men have dreamed dreams and young men have seen visions from the beginning of time. We of today can only sit and wonder when a scientist has his say.

January 30, 1926

WORLD SYSTEM OF WIRELESS TRANSMISSION OF ENERGY

—Telegraph and Telegraph Age, October 16, 1927

The transmission of power without wires is not a theory or a mere possibility, as it appears to most people, but a fact demonstrated by me in experiments which have extended for years. Nor did the idea present itself to me all of a sudden, but was the result of a very slow and gradual development and a logical consequence of my investigations which were earnestly undertaken in 1893 when I gave the world the first outline of my system of broadcasting wireless energy for all purposes. In several demonstrative lectures before scientific societies during the preceding three years, I showed that it was not necessary to use two wires in transmitting electrical energy, but that one only might be employed equally well. My experiments with currents of high frequencies were the first ever performed in public and elicited the keenest interest on account of the possibilities they opened up and striking character of the phenomena. Few of the experts familiar with the up-to-date appliances will appreciate the difficulty of my task with the elementary devices I had then at command, as accurate adjustments for resonance had to be made in every experiment.

The transmission of energy through a single conductor without return having been found practicable it occurred to me that possibly even that one wire might be dispensed with and the earth used to convey the energy from the transmitter to the receiver.

High frequency dynamo and tesla coil

Manifestly, currents such as were ordinarily employed in the arts and industries were unsuitable and I had to devise special generators and transformers for furnishing impulses of the requisite quality. First, I perfected high frequency dynamos which were of two types, one with a direct current field excitation and the other in which the magnet was energized by alternating currents of different phase, producing a rotating magnetic field. Both of these have found employment in connection with my broadcasting wireless system. In the first machine I exhibited an efficiency of 90% was attained, but it was necessary to run it in hydrogen or rarefied air to minimize the otherwise prohibitive windage loss and deafening noise.

In order to overcome the inherent limitations of such machines I next concentrated my efforts on the perfection of a peculiar transformer consisting of several tuned circuits in inductive relation which received the primary energy from oscillatory discharges of condensers. This apparatus, originally identified with my name and considered by the leading scientific men my best achievement, is now used in every wireless transmitter and receiver throughout the world. It has enabled me to obtain currents of any desired frequency, electromotive force and volume, and to produce a great variety of electrical, chemical, thermal, light and other effects, Roentgen, cathodic and other rays of transcending intensities. I have employed it in my investigations of the constitution of matter and radioactivity, published from 1896 to 1898 in the Electrical Review in which it was demonstrated, prior to the discovery of Radium by Mme. Sklodowska and Pierre Curie, that radio-activity is a common property of matter and that such bodies emit small particles of various sizes and great velocities, a view which was received with incredulity but finally recognized as true. It has been put to innumerable uses and proved in the hands of others a veritable lamp of Aladdin. As I think of my earliest coils, which were nothing more than scientific toys, the subsequent development appears to me like a dream.

The "Magnifying Transmitter" and earth resonance

While I was perfectly convinced, from the outset, that success would be ultimately achieved, it was not until by slow improvement I evolved the so-called "Magnifying Transmitter" that I obtained convincing evidence of the feasibility of wireless power transmission on a vast scale for all industrial purposes.

The chief discovery, which satisfied me thoroughly as to the practicability of my plan, was made in 1899 at Colorado Springs, where I carried on tests with a generator of fifteen hundred kilowatt capacity and ascertained that under certain conditions the current was capable of passing across the entire globe and returning from the antipodes to its origin with undiminished strength. It was a result so unbelievable that the revelation at first almost stunned me. I saw in a flash that by properly organized apparatus at sending and receiving stations, power virtually in unlimited amounts could be conveyed through the earth at any distance, limited only by the physical dimensions of the globe, with an efficiency as high as 99.5%.

The mode of propagation of the currents from my transmitter through the terrestrial globe is most extraordinary considering the spread of the electrification of the surface. The wave starts with a theoretically infinite speed, slowing down first very quickly and afterward at a lesser rate until the distance is about six thousand miles, when it proceeds with the speed of light. From there on it again increases in speed, slowly at first, and then more rapidly, reaching the antipode with approximately infinite velocity. The law of motion can be expressed by stating that the waves on the terrestrial surface sweep in equal intervals of time over equal area, but it must be understood that the current penetrates deep into the earth and the effects produced on the receivers are the same as if the whole flow was confined to the earth's axis joining the transmitter with the antipode. The mean surface speed is thus about 471,200 kilometers per second — 57% greater than that of the so-called Hertz waves — which should propagate with the velocity of light if they exist. The same constant was found by the noted American astronomer, Capt. J.T.T. See, in his mathematical investigations, for the smallest particles of the ether which he fittingly designates as "etherons." But while in the light of his theory this speed is a physical reality, the spread of the currents at the terrestrial surface is much like the passage of the moon's shadow over the globe.

It will be difficult for most people engaged in practical pursuits to measure or even to form an adequate conception of the intensity of inspiration and force I derive from that part of my work which has passed into history. I have every reason to consider myself one of the most fortunate men, for I experience incessantly a feeling of inexpressible satisfaction that my alternating system is universally employed in the transmission and distribution of heat, light and power and that also my wireless system, in all its essential features, is used throughout the world for conveying intelligence. But my pioneer efforts in this later field are still grossly misunderstood.

Short wave broadcasting and "beam" transmission

Nothing illustrates this better than the recent demonstrations of a number of experts with very short waves, which have created the impression that power will be eventually transmitted by such means. In reality, experiments of this kind are the very denial of the possibility of economic transmission of energy. I have investigated this special subject experimentally during a great number of years, using sometimes waves as short as one millimeter, and have found even these unsuitable for such a purpose, not to say that their production is inseparable from great waste.

In order to secure good results by this method it would be necessary to employ radiations of a wave length incomparably smaller than the dimensions of the reflector, as radiant heat, light, infra-red or ultra-violet rays.

Notwithstanding my repeated explanations experts do not seem to realize that no concentration of energy such as I attain in my wireless power system can or will ever be achieved through the instrumentality of reflectors, for in transmitting energy in this manner the receiver can collect only an amount proportionate to the area exposed to the rays, while in my system it draws the energy from an immense reservoir in ever so much greater quantity. Similar considerations apply to directional transmission by short reflected waves or "beams." If we could produce economically electric vibrations of a frequency approximating that of radiant heat waves, efficient reflectors without appreciable dispersion, and prevent absorption—then such a mode of transmitting energy might become of great importance. But attempts to accomplish this purpose with relatively very low frequencies are sure to prove futile. More than twenty-five years ago my efforts to transmit large amounts of power through the atmosphere resulted in the development of an invention of great promise, which has since been called the "Death Ray," and attributed to Dr. Grindell Mathews, an ingenious and skillful English electrician. The underlying idea was to render the air conducting by suitable ionizing radiations and to convey high tension currents along the path of the rays. Experiments, conducted on a large scale, showed that with pressures of many millions of volts virtually unlimited quantities of energy can be projected to a small distance, as a few hundred feet, which might be satisfactory if the process were more economical and the apparatus less expensive. Since that time I have made important improvements and discovered a new principle which can be successfully applied without difficulty for various purposes in peace and war.

If I have understood the reports correctly, in the "beam transmission" with waves of a few meters length, an oscillating circuit, consisting of a straight vertical conductor, is placed in the focal line of a parabolic surface on which are placed many secondary straight wires parallel to the primary conductor. Now, this disposition is entirely faulty and to all evidence inefficient, as the secondary system does not operate in the manner of a parabolic reflector but merely produces a confused echo. The correct arrangement would require primary and secondary conductors situated in two vertical parallel planes separated by a distance equal to one quarter of the wave length. But even in this best form such a transmitter can only be of doubtful practical value. The two wave trains behind the reflecting, or rather echoing, system do not completely neutralize and there is considerable lateral dissipation. The energy of the primary system diminishes with the square of the distance, and this being also true of the secondary, the useful waves from the latter will suffer diminution of energy in proportion to the fourth power of their length. This means that only a very short wave can be used which, moreover, is unchangeable and difficult to regulate. One must be shortsighted not to perceive that better results will be obtained if the capital is all invested in a single directional system of proper design for the power available increases much more rapidly than the cost of the plant. Assuming even that the beam arrangement works with ideal perfection, it must still be inferior, since the requisite radiant energy is producible at smaller expense with the single system, which has the further advantages that it can be adapted to any wavelength, is equally effective in two directions and therefore of greater earning capacity. So palpably unsound is this scheme that I am at a loss to understand how it could pass the scrutiny of competent experts such as Dr. W. L. Austin and John Stone Stone.

The "World System"

Since I began the construction of the first power plant in 1899 I have expressed myself repeatedly in regard to it and the plans I had previously formed through the medium of the Electrical Review, Electrical World, Electrical Experimenter, Science and Invention and other periodicals, notably the Century Magazine of June, 1900, to which I contributed a lengthy article on the "Problem of Increasing Human Energy"; but certain facts must still be told. In the first place the fundamental difference between the broadcasting system as now practiced and the one I expect to inaugurate is that at present the transmitter emits energy in all directions, while in the system I have devised only force is conveyed to all points of the earth, the energy itself traveling in definite paths determined beforehand. Perhaps the most wonderful feature is that the energy travels chiefly along an orthodromic line, that is, the shortest distance between two points at the surface of the globe, and reaches the receiver without the slightest dispersion, so that an incomparably greater amount is collected than is possible by radiations. I have thus provided a perfect means for transmitting power in any desired direction far more economically and without any such qualitative and quantitative limitations as the use of reflectors would necessarily involve.

Another distinction is that my system is based entirely on resonance, while in present practice reliance is placed chiefly on amplification by auxiliary devices generally consisting of various forms of vacuum tubes which have been brought to remarkable perfection. The foundation to their use was laid by Sir William Crookes, who discovered in 1876 that a highly heated conductor emits electrified particles. In 1882 a young French electrician, Vissiere by name, observed that a current issues from the filament of an incandescent lamp and made careful measurements with specially prepared bulbs, some of which I had opportunity to witness in Ivry-sur-Seine, a suburb of Paris, at that time. But these phenomena found no application in the art until in 1892 I produced a vacuum tube detector superior in sensitiveness to any other form of which I have knowledge. Amazing progress has been since then achieved, but the employment of the modem vacuum detectors and amplifiers is an impediment to advance in the right direction and most of the troubles experienced in broadcasting are due to this cause. Until quite recently the transmitted waves were lacking in uniformity of length, rendering accurate attunement impossible. This defect has been in a measure remedied by control through quartz crystals, and now, for the first time, it is practicable to carry out important refinements for bettering the service.

The electro-mechanical process of producing isochronous oscillations is one of my earliest inventions and I have applied it in many ways with great success. Its application to the operation of existing plants secures important advantages, but in spite of this and other improvements a change in the present apparatus and method of broadcasting is becoming daily more imperative and for this reason I am anxious to resume the introduction of my "World System" with novel transmitters of great effectiveness and receivers of elementary simplicity. In my apparatus the isochronism is so perfect and attunement sharp to such a degree that in the transmission of speech, pictures or similar operations, the frequency or wavelength is varied only through a minute range which need not be more than one hundredth of 1% if desired. Statics and all other interferences are completely eliminated and the service is unaffected by weather, seasonal or diurnal changes of any kind. The system lends itself particularly to World Wireless Telephony and Telegraphy as the current from the transmitter can be kept virtually constant and the control effected by a simple microphone without the elaborate means now employed. Any reasonable number of simultaneous and non-interfering messages is practicable and a speed of many thousands of words per minute can be attained in telegraphic transmission. The same principles are also applicable to operation through wires and cables. In 1903 I proposed to the Western Union and Postal Telegraph companies such multiplex transmission for their lines but received no encouragement mainly because the volume of business did not call for a great increase of working capacity. At a later date my improvements were introduced as "Wired Wireless." A quite inappropriate name inasmuch as the waves radiated from the wire are completely lost and of no effect on the receiver.

My plans for a power plant have been developed to the point of application, but I am still unable to say when I shall begin active work. There are no such difficulties in the way as confronted me from the outset, for at that time I was alone; now many are convinced that my undertaking is rational and practical. Needless to say that I am using every effort to give to the world my best and most important work as soon as possible and free of all blemish and flaw. I have in view a number of places which seem well suited for the purpose, but my warmest wish is to transmit power from Niagara Falls, where the first triumph with my alternating system was achieved.

One of the most important uses of wireless energy will be undoubtedly for the propulsion of flying machines to which power can be readily supplied without ground connection, for although the flow of the currents is confined to the earth an electro magnetic field is created in the atmosphere surrounding it. If conductors or circuits accurately attuned and properly positioned are carried by the plane, energy is drawn into these circuits much the same as a fluid will pass through a hole created in the container. With an industrial plant of great capacity sufficient power can be derived in this manner to propel any kind of aerial machine. This I have always considered as the best and permanent solution of the problems of flight. No fuel of any kind will be required as the propulsion will be accomplished by light electric motors operated at great speed. Nevertheless, anticipating slow progress, I am developing a novel type of flying machine which seems to be well suited for meeting the present necessity of a safe, small and compact "aerial fliver" capable of rising and descending vertically.

Television, as conceived by me in 1893, will be another valuable and timely application. At that time I advanced the idea that the formation of a clear mental image of external objects is accompanied by a reflex action on the retina, making it possible to read thoughts and even to project the images conceived on a screen and render them visible to an audience. This would be of inestimable consequence on all human relations but the idea can not be realized until some way is found to lay bare the retina. Continued reflections on this subject led me to evolve apparatus for transmitting instantaneously true vision without any moving devices, and in 1900 I had already solved three of the problems which confronted me, namely: to individualize and isolate a very great number of channels or "nerves"; to convey to the receiving apparatus energy in sufficient amount, and, to make the vision of the moving images independent of distance. Eventually also I hope to overcome the shortcomings of the selenium cell by a different device.

I am most interested, however, in the perfection of broadcasting which is now carried on with unfit apparatus and on a commercially defective plan. The transmitters have to be greatly improved and the receivers simplified and in the distribution of wireless energy for all purposes the precedent established by the telegraph, telephone and power companies must be followed, for while the means are different the service is of the same character. Technical invention is akin to architecture and the experts must in time come to the same conclusions I have reached long ago. Sooner or later my power system will have to be adopted in its entirety and so far as I am concerned it is as good as done. If I were ever assailed by doubt of ultimate success I would dismiss it by remembering the words of that great philosopher, Lord Kelvin, who after witnessing some of my experiments said to me with tears in his eyes: "I am sure you will do it."

October 16, 1927

NIKOLA TESLA TELLS OF NEW RADIO THEORIES: AN INTERVIEW WITH NIKOLA TESLA

—*New York Herald Tribune, September 22, 1929*

Does Not Believe in Hertz Waves and Heaviside Layer, Interview Discloses The model of a "Tesla Coil" which will be featured in the historic exhibit of the radio show reawakens interest in its inventor.

It is not generally appreciated that this curious apparatus, often associated with pretty or spectacular demonstrations of high voltage electricity, is really a fundamental part of modern radio. For all the tuning apparatus and circuits in every transmitting and receiving set are simply variations of Tesla coils and Tesla coil circuits.

It was for this invention, and other inventions and principles concerned with tuning, heterodyning, and the generation of continuous waves, which were made at least several years before the very first experiments of Marconi, that many of our most reputable engineers have conceded to Nikola Tesla the title of "Father of Radio".

Mr. Tesla, still actively working, was interviewed last week to get his ideas regarding the prospects of the radio of 1930, and beyond. As a prophet, however, he balked. He had repeated time and again his visions for the future. As far back as 1900, he had contemplated a world-wireless system which included broadcasting, picture transmission, international time service, and in addition television and the distribution of electrical power. Part of this early prophecy has been realized—what remained, still stood as his prediction..

Disputes Hertz waves

What, then, about power transmission by radio? Laurence M. Cockaday, the technical editor of this radio section, had expressed the opinion several weeks ago that, with present apparatus at least, it was hardly feasible. Mr. Tesla agreed to discuss the point at length. As a result, he made public for the first time one of the most extraordinary conclusions—that Hertz waves do not exist! If his theory is true, there may be found in it more adequate explanations of "dead spots", fading, reflection and a dozen other problems that have always puzzled the profession.

The inventor began by referring to Cockaday's article:

"I have read the article, and I quite agree with the opinion expressed—that wireless power transmission is impractical with present apparatus. This conclusion will be naturally reached by any one who recognizes the nature of the agent by which the impulses are transmitted in present wireless practice.

"When Dr. Heinrich Hertz undertook his experiments from 1887 to 1889 his object was to demonstrate a theory postulating a medium filling all space, called the ether, which was structureless, of inconceivable tenuity and yet solid and possessed of rigidity incomparably greater than that of the hardest steel. He obtained certain results and the whole world acclaimed them as an experimental verification of that cherished theory. But in reality what he observed tended to prove just its fallacy.

"I had maintained for many years before that such a medium as supposed could not exist, and that we must rather accept the view that all space is filled with a gaseous substance. On repeating the Hertz experiments with much improved and very powerful apparatus, I satisfied myself that what he had observed was nothing else but effects of longitudinal waves in a gaseous medium, that is to say, waves, propagated by alternate compression and expansion. He had observed waves in the ether much of the nature of sound waves in the air.

"Up to 1896, however, I did not succeed in obtaining a positive experimental proof of the existence of such a medium. But in that year I brought out a new form of vacuum tube capable of being charged to any desired potential, and operated it with effective pressures of about 4,000,000 volts. I produced cathodic and other rays of transcending intensity. The effects, according to my view, were due to minute particles of matter carrying enormous electrical charges, which, for want of a better name, I designated as matter not further decomposable. Subsequently those particles were called electrons.

"One of the first striking observations made with my tubes was that a purplish glow for several feet around the end of the tube was formed, and I readily ascertained that it was due to the escape of the charges of the particles as soon as they passed out into the air; for it was only in a nearly perfect vacuum that these charges could be confined to them. The coronal discharge proved that there must be a medium besides air in the space, composed of particles immeasurably smaller than those of air, as otherwise such a discharge would not be possible. On further investigation I found that this gas was so light that a volume equal to that of the earth would weigh only about one-twentieth of a pound.

"The velocity of any sound wave depends on a certain ratio between elasticity and density, and for this ether or universal gas the ratio is 800,000,000,000 times greater than for air. This means that the velocity of the sound waves propagated through the ether is about 300,000 times greater than that of the sound waves in air, which travel at approximately 1,085 feet a second. Consequently the speed in ether is 900,000 x 1,085 feet, or 186,000 miles, and that is the speed of light.

"As the waves of this kind are all the more penetrative the shorter they are, I have for years urged the wireless experts to use such waves in order to get good results, but it took a long time before they settled upon this practice.

"Although the world is still skeptical as to the feasibility of my undertaking, I note that some advanced experts, at least, share my views, and I hope that before long wireless power transmission will be as common as transmission by wires."

According to Mr. Tesla, the present broadcasting station does not propagate Hertzian waves, as has always been supposed, but acts more like an "ether whistle"—transmitting waves through the ether similar to the waves transmitted by an ordinary whistle through the air. He also expressed his disbelief in the Heaviside layer, and claimed that the reflection of waves back toward the earth was due to the change of medium encountered at the vacuous boundary of the atmosphere.

At Colorado Springs, about thirty years ago, this scientist had a Tesla coil seventy-five feet in diameter which produced voltages above 12,000,000, and sparks over 100 feet long. Electrical flashes were created which were the nearest approach to lightning that man has ever made. During his experiments there, of over a year, Tesla claims that he transmitted a considerable amount of electrical current to the other side of the earth. It was upon these, and later experiments that he bases his present prediction.

September 22, 1929

OUR FUTURE MOTIVE POWER

—Everyday Science and Mechanics, December 1931

The material as well as intellectual progress of Man is becoming ever more dependent on the natural forces and energies he is putting to his service. While not exactly a true measure of well being and enlightenment, the amount of power used is a reliable indication of the degree of safety, comfort and convenience, without which the human race would be subject to increasing suffering and want and civilization might perish.

Virtually all of our energies are derived from the sun, and the greatest triumph we have achieved in the utilization of its undying fire is the harnessing of waterfalls. The hydro-electric process, now universally employed, enables us to obtain as much as 85% of the solar energy with machines of elementary simplicity which, by resorting to the latest improvements in the technical arts, might be made capable of enduring for centuries. These advantages are entirely exceptional, very serious handicaps and great, unavoidable losses confronting us in all other transformations of the forces of nature. It is, therefore, desirable in the interest of the world as a whole, that this precious resource should be exploited to the limit. Judging from the average height of the water discharged annually from the clouds, and the mean fall over the aggregate land surface, the total terrestrial water power may be theoretically estimated at ten billions of horse power. Of course, only a part of that is suited for practical development and relatively little is actually utilized—twenty-five percent, perhaps, in the most advanced countries, less in others, and there are some in which not even the ground has been broken. Great waterfalls exist in many inaccessible regions of the globe and new ones are being discovered, all of which will be eventually harnessed when the wireless transmission of energy is commercialized. There is foundation for hope, however, that our present limitations in the amount of available power may be removed in the future. Three-quarters of the earth's surface are covered by the oceans and the rainfall over all this vast area is useless for our purpose. Much thought has been given to artificial production of rain, but none of the means proposed offers the slightest chance of success. Besides, so far only the precipitation in a limited region was contemplated, leaving the total quantity of moisture for the *?*untired land unchanged except as modified through the natural tendency of the oceans to divert more and more water from the continents. The real and important problem for us to solve is not to bring about precipitation in any chosen locality, but to reverse this natural process, draw the vapors from the seas and thereby increase, at will, the rainfall on the land. Can this be done?

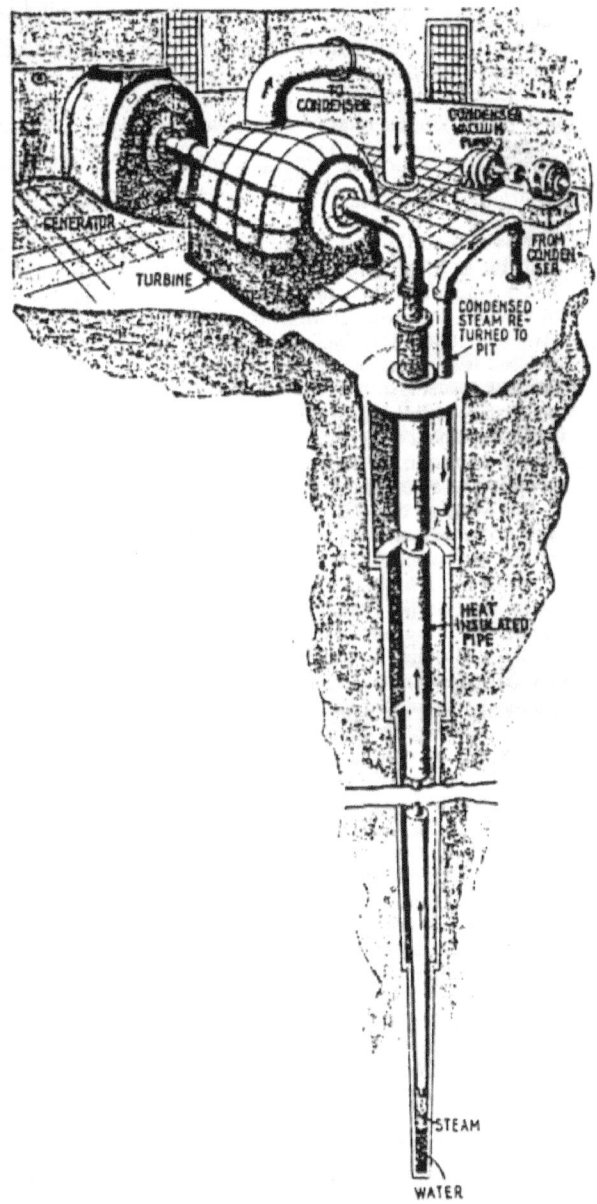

The arrangement of one of the great terrestrial power plants of the future. Water is circulated to the bottom of the shaft, returning as steam to drive the turbine, and then returned to liquid form in the condenser, in an unending cycle.

The sun raises the water to a height where it remains in a state of delicate suspension until a disturbance, of relatively insignificant energy, causes condensation at a place where the balance is most easily disturbed. The action, once started, spreads like a conflagration for a vacuum is formed and the air rushing in, being cooled by expansion, enhances further condensation in the surrounding masses of the cloud. All life on the globe is absolutely dependent on this gigantic trigger mechanism of nature and my extended observations have shown that the complex effects of lightning are, in most cases, the chief controlling agents. This theory, formulated by me in 1892, was borne out in some later experiments I made with artificial lightning bolts over 100 feet long, according to which it appears possible, by great power plants suitably distributed and operated at the proper times, to draw unlimited quantities of water from the oceans to the continents. The machines being driven by the waterfalls, all the work would be performed by the sun, while we would have merely to release the trigger. In this manner we might obtain sufficient energy from falling water to provide for all our necessities. More than this, we could create new lakes and rivers, induce a luxuriant flora and fauna and convert even the arid sands of deserts into rich, fertile soil.

But the full realization of this idea is very remote. The hard fact is that unless new resources are opened up, energy derived from fuel will remain our chief reliance. The thermo-dynamic process is wasteful and barbarous, especially when burning coal, the mining of which, despite of modern improvements, still involves dangers to the unfortunates who are condemned to toil deep in the bowels of the earth. Oil and natural gas are immensely superior in this and other respects and their use is rapidly extending. It is quite evident, though, that this squandering cannot go on indefinitely, for geological investigations prove our fuel stores to be limited. So great has been the drain on them of late years that the specter of exhaustion is looming up threateningly in the distance, and everywhere the minds of engineers and inventors are bent upon increasing the efficiency of known methods and discovering new sources of power.

Nature has provided an abundant supply of energy in various forms which might be utilized if proper means and ways can be devised. The sun's rays falling upon the earth's surface represent a quantity of energy so enormous that but a small part of it could meet all our demands. By normal incidence the rate is mechanically equivalent to about 95 foot pounds per square foot per second, or nearly 7300 horse power per acre of ground. In the equatorial regions the mean annual rate is approximately 2326 and in our latitudes 1737 horse power for the same area. By using the heat to generate steam and operating a turbine under high vacuum probably 200 horse power per acre could be obtained as net useful power in these parts. This would be very satisfactory were it not for the cost of the apparatus which is greatly increased by the necessity of employing a storage plant sufficient to carry the load almost three-quarters of the time.

The ambitious scheme proposed here draws power from the depths of the sea, utilizing the warmth of one layer, brought into contact with the cold of another, to operate great power plants. Its practicability as well as the theory of its operation, is analyzed in this remarkable article.

The energy of light rays, constituting about 10% of the total radiation, might be captured by a cold and highly efficient process in photo-electric cells which may become, on this account, of practical importance in the future.

Some progress in this direction has already been achieved. But for the time being it appears from a careful estimate, that solar power derived from radiant heat and light, even in the tropics, offers small opportunities for practical exploitation. The existing handicaps will be largely removed when the wireless method of power transmission comes into use. Many plants situated in hot zones, could then be operatively connected in a great super-power system to supply energy, at a constant rate, to all points of the globe.

The sun emits, however, a peculiar radiation of great energy which I discovered in 1899. Two years previous I had been engaged in an investigation of radio-activity which led me to the conclusion that the phenomena observed were not due to molecular forces residing in the substances themselves, but were caused by a cosmic ray of extraordinary penetrativeness. That it emanated from the sun was an obvious inference, for although many heavenly bodies are undoubtedly possessed of a similar property, the total radiation which the earth receives from all the suns and stars of the universe is only a little more than one-quarter of 1% of that it gets from our luminary. Hence, to look for the cosmic ray elsewhere is much like *chercher le midi dans les environs de quatorze heures*. My theory was strikingly confirmed when I found that the sun does, indeed, emit a ray marvelous in the inconceivable minuteness of its particles and transcending speed of their motion, vastly exceeding that of light. This ray, by impinging against the cosmic dust generates a secondary radiation, relatively very feeble but fairly penetrative, the intensity of which is, of course, almost the same in all directions. German scientists who investigated it in 1901 assumed that it came from the stars and since that time the fantastic idea has been advanced that it has its origin in new matter constantly created in interstellar space!! We may be sure that there is no place in the universe where such a flagrant violation of natural laws, as the flowing of water uphill, is possible. Perhaps, some time in the future when our means of investigation will be immeasurably improved, we may find ways of capturing this force and utilizing it for the attainment of results beyond our present imagining.

The tides are often considered as a source of motive power and not a few engineers have expressed themselves favorably in regard to their use. But as a matter of fact, the energy is, in most places, insignificant, the harnessing of the waterfall over an acre of ground yielding but little more than one horse power. Only in exceptional locations can the power of the tides be profitably developed.

It has been the dream of many an inventor to utilize the energy of ocean waves, which is considerable. But although numerous schemes have been advanced and much ingenuity shown in devising the mechanical means, nothing of commercial value has so far resulted and the prospects are very poor on account of technical difficulties and the erratic character of this power source.

The force of the wind can be much more easily put to our service and has been in practical use since times immemorial. It is invaluable in ship propulsion and the windmill must be seriously regarded as a power generator. If the cost of this commodity should greatly increase we will be likely to see the countries dotted with these time-honored contrivances. Unfortunately, the value of all this resources is very much reduced by periodic and casual variations, and we are driven to search for a source of constant twenty-four hour power comparable to that of a waterfall. Thus we are led to consider terrestrial heat as a possible fountain of unvarying energy supply.

Terrestrial energy

It is noteworthy that already in 1852 Lord Kelvin called attention to natural heat as a source of power available to Man :. But, contrary of his habit of going to the bottom of every subject of his investigations, he contended himself with the mere suggestion. Later, when the laws of thermo-dynamics became well understood, the prospects of utilizing temperature differences in the ocean, solid earth or the atmosphere, have been often examined. It is well-known that there exists, in tropical seas, a difference of $50°F$ between the surface water and that three miles below. The temperature of the former, being subject to variations, averages $82°F$, while that of the latter is normally at least, at $32°F$, or nearly so, as the result of the slowly influx of the ice-cold polar stream. In solid land these relations are reversed, the temperature increasing about $1°F$ for every 64 feet of descent. Very great differences are also known to exist in the atmosphere, the temperature diminishing with the distance above the earth's surface according to a complex function.

But while all this was of common knowledge for at least 75 years and the utilization of the heat of the earth for power purposes a subject of speculation, no decided attempt to this end seems to have been made until an American engineer, whose name I have been unable to ascertain, proposed to operate engines by steam generated in high vacuum from the warm surface water and condensed by the cold water pumped from a great depth. A fully and carefully worked out plan of this kind, supported by figures and estimates, was submitted by him to prominent capitalists and business men of New York about 50 years ago. He not only contemplated the production and distribution of power for general use but intended even to propel boats by energy derived in the same manner, using, preferably, ether as working fluid. On account of his death, or for other reasons, the project was not carried into practice.

Of this I learned much later when I interested in my alternating system Alfred S. Brown, a well known technical expert, called upon to examine the merits of my investigations, and C. F. Peck, a distinguished lawyer, who organized a company for their commercial introduction. These men were among the first approached by the engineer and considered his plan rational in principle, but the pipe lines, pumps, engines, boilers and condensers involved too great an outlay and, besides, a profitable disposal of the power was difficult and uncertain. My discovery of the rotating magnetic field brought about a change in the situation and in their attitude. They thought that if the energy could be economically transmitted to distant places by my system and the cost of the ocean plant substantially reduced, this inexhaustible source might be successfully exploited. Mr. Peck had influential connections, among them John C. Moore, the founder of the banking house bearing his name. With the exception of the late J. P. Morgan, who towered above all the Wall Street people like Samson over the Philistines, Moore was probably the strongest personality. I was given to understand that if I could evolve a plan satisfactory to Mr. Brown and other engineers, all the capital required for an enterprise on a very large scale, as contemplated by them, would be promptly furnished. No encouragement from my associates was needed for determining me to undertake the task, as the idea appeared, at first, wonderfully promising and attractive although there was nothing about it fundamentally new.

Undoubtedly, the essential conditions required to operate a steam or other thermo-dynamic engine could be fulfilled, a considerable temperature difference being available at all times. No proof had to be furnished that heat would flow from a higher to a lower level and could be transformed into mechanical work. Nor was it necessary to show that the surface water, although much below its normal boiling point of 212° F, can be readily converted into steam by subjecting it to a vacuum which causes ebullition at any temperature however low. It is of common knowledge that, due to this same effect, beans cannot be cooked or eggs hard-boiled on high mountains. Also, for a like reason turbines have been wrecked in steam power plants with the boilers completely shut off, the slightly warm water in the system of connecting pipes being evaporated under a high vacuum inadvertently applied. This behavior of water, or liquids in general, was long before beautifully exemplified in the classical device called — cryophoros — consisting of two communicating and exhausted bulbs partially filled with liquid, which is evaporated in one and condensed in the other. It was invented by W. H. Wollaston, a great

English scientific man and investigator (1766-1828), who first commercialized platinum and was credited by some to have anticipated Faraday in the discovery of electromagnetic rotation. The original instrument brought out at the beginning of the nineteenth century had one of the bulbs packaged in ice with the result of freezing water in the other. Conformably to the views of that time it was thought that the cold of the ice was carried to the water and so the Greek name, meaning—cold-carrier—, was given to the device. But now we know that the process is of opposite character, the freezing being brought about by the transport of the latent heat of evaporation from the warm to the cold bulb. One would naturally infer that the operation would cease as soon as the water is frozen at the surface, but curiously enough the ice itself continues to yield steam and it is only because of this that all of the water is solidified. We may imagine how puzzling this phenomenon appeared more than one century ago!

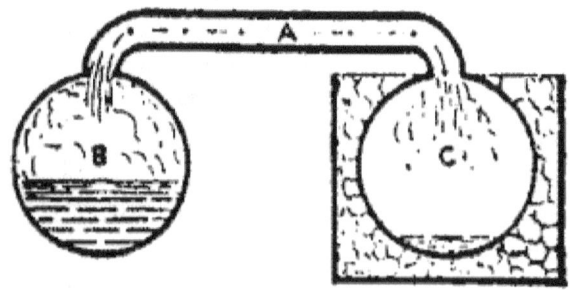

The cryophorus is well-known as a scientific toy, exemplifying also the principle of refrigerating machinery.

The ocean plant proposed by the engineer was nothing else but Wollaston's device of huge proportions, adapted for continuous operation and having an engine interposed between the two communicating vessels. In estimating its thermodynamic performance the first results I arrived at through the medium of pad and pencil fairly bewildered me. To illustrate by an example, suppose that equal quantities, say, one-half pound of the warm and of the cold water, respectively, at 82° and 32° Fahrenheit, are mixed or put in thermal equilibrium otherwise. The first will then give up to the second 12.5 heat units, mechanically equivalent to 9725 foot pounds—the same energy which would be developed in the fall of one pound from so great an altitude as 9725 feet. The dream of my life had been to harness Niagara, but here was a fall sixty times higher and of unlimited volume. To raise the cold water to the surface from any depth whatever, required but a trifling effort and as other losses also seemed negligible I concluded that if only a small portion of this hypothetical fall could be utilized, one of the greatest problems confronting humanity would be solved for all times to come.

I knew that it was to good to be true, nevertheless I followed this ignis fatuus for years until, little by little, through close reasoning, calculation and experiment, I got the true bearings in the swamp of my ignorance and doubt. Then this scheme of harnessing the ocean revealed itself to my mind as one of the crudest imaginable. Just to transport a little heat, water has to be pumped and disposed of in quantities so enormous that a large installation of this type would present new problems in engineering. Contrary to the opinion I had previously formed, this involved the expenditure of a great amount of energy. Then I realized that the gases contained in the water can be only partially extracted and have to be continuously removed from the condenser to prevent the rise of back pressure which might reduce the speed and eventually stop the engine.

Furthermore, due to certain conditions, the deep sea water must enter the pipe warmer than it should conformably to soundings, so that the full temperature difference cannot be obtained, and I discovered other peculiar causes which, after some time, might seriously interfere with the proper functioning of the mechanism. The steam, raised directly from the surface water, is of the poorest quality, mere mist under small pressure, and its consumption per horsepower hour was likely to be twenty times greater than in modem plants. In hydro-electric stations, as before stated, 85% of the energy of falling water may be captured, while in this case hardly more than two-tenths of 1% of the theoretical fall can be utilized. Worst of all, the size and cost of the equipment is utterly out of proportion to the greatest possible returns. These and other limitations and difficulties forced themselves upon me in studying the plans as first submitted.

The introduction of my alternating system started a scramble for the most valuable water power sites and no attempt was made to harness the ocean. But my interest was aroused to such a degree that I continued the work and made a number of improvements which are thought to possess some merit. Satisfied that pipes supported by floats or hung in submarine abysses were impracticable, I proposed a sloping tunnel, lined with heat insulating cement, affording a smooth and unbroken passage for the deep sea water. I found ways of simplifying and cheapening the apparatus and making it more effective by reducing the moisture of the steam and otherwise, and these advances may eventually prove of practical value.

To conduce to a more ready understanding of the evolution of the ocean power plant from the cryophoros and of the nature of some of my improvements, reference may be made to the drawings in which Fig. 1 represents the original device of Wollaston, comprising two highly exhausted vessels B and C, respectively, the boiler and condenser, connected through a channel A. The first named vessel being partly filled with water or other liquid and the second packed in a freezing mixture, the vacuum causes the water when slightly warmed to boil furiously and the well known effect is observed. As the steam generated in the boiler rushes into the condenser with great speed, it is capable of producing a considerable mechanical effort.

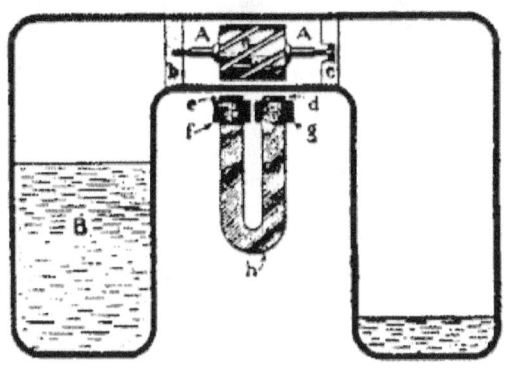

Fig. 2: Plan of a system whereby the transfer of vapor between two vessels at different temperatures drives the armature of electrical generator.

Fig. 2 illustrates how the thermo-dynamic transformation of energy may be effected to obtain useful external work. This particular arrangement is chosen in order to dispense with the necessity of a connection to the outside which would call for the employment of a vacuum pump. A steel armature a, of a diameter nearly equal to that of the channel A, connecting vessels B and C, and shaped like a fan, is supported in virtually frictionless bearings b and c, of which the latter may be designed for taking up the thrust. Surrounding the armature, or turbine rotor, and in close proximity to the same, are soft-iron projections as d and e, wound with coils f and 9 and forming part of a permanent magnet h. The rapid rotation of the armature results in a periodic shifting of magnetic lines from one to the other set of projections, this inducing in the coils currents which may be utilized.

The next step is to adapt the device for continuous operation. This may be done in two ways: by supplying the evaporating and condensing water directly to the vessels B and C, or by merely transmitting and abstracting heat through their walls, in which case the working fluid is entirely separated and circulated in a closed circuit.

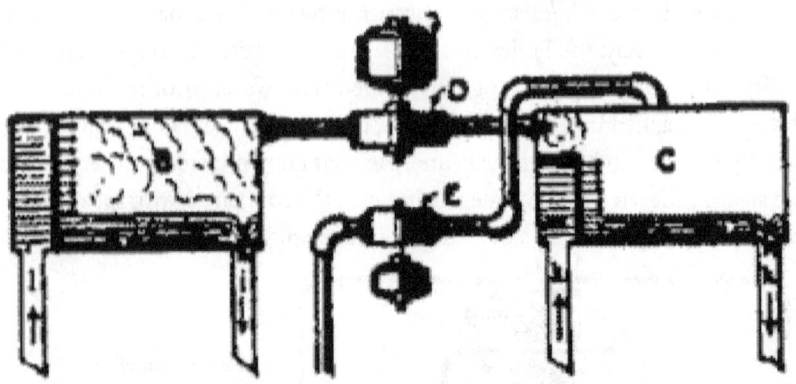

Fig. 3: A more complete sketch of the thermodynamic system, in which the necessary degree of vacuum is produced by the suction pump E.

The first plan is diagrammatically shown in Fig. 3. The vessels B and C are cylinders joined by a turbine D through suitable pipe connections. A suction pump E, constructed for producing a very high vacuum, is attached to the condenser C and may be driven by the turbine through a gear or, as indicated, by an induction motor energized with alternating currents from the dynamo F coupled to the turbine. The water being under atmospheric pressure would flow into the evacuated vessels at too great a speed occasioning corresponding losses, and for this reason it is necessary to supply and drain it through balancing barometric columns ii and kk of proper height thereby insuring the desired circulation, the direction of which is indicated by arrows. Since the latent heat absorbed in evaporation and set free in condensation is very great, an immense quantity of water must be circulated through the vessels in order to prevent changes of temperature sufficient to seriously reduce the performance of the apparatus. In addition to the devices shown, separators must be employed for extracting gases from the water before its entrance into the boiler and condenser. These cannot be of the effective centrifugal type as they would entail too great a loss of energy. The only kind practicable is that used from the earliest beginnings of modern hydraulic, the action of which is based on a slow reversal of direction of flow and accomplishes only partial *degasification*. It should be noted that the gases, by rapid expansion and attendant cooling, impair greatly the quality of the steam and also, more or less, the vacuum in the vessels. One of my improvements is to supply the water in the form of jets, as represented, which furnish the necessary evaporating and condensing surface while at the same time carrying away gases which would be liberated if the water were admitted as usual.

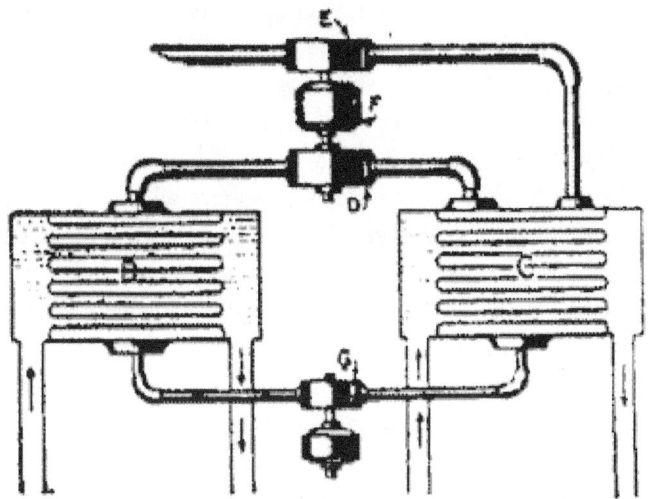

Fig. 4: Here the water, or other fluid operating the turbine D is kept in a closed system, circulating through condensers immersed in water of different temperatures.

A careful study of the scheme illustrated in Fig. 3 has satisfied me that it is, for a number of reasons, disadvantageous and less practicable than that shown in Fig. 4. In this instance the vessels B and C are surface condensers of ordinary design but of very great active areas in view of the excessive steam consumption and small differences of temperature supplied. They may be of the same size, for although the passage of heat from the hot to the cold water takes place through the steam, the law of mixtures is obeyed, the maximum transfer occurring when the quantities of both are equal. Were it not for that the performance might be appreciable improved by supplying the hot water, which has to pass only through short pipes, in greater quantity. The vessels are connected through a turbine D coupled to a generator F, as before, and besides the suction pump E a deep well pump G is employed to force the condensate into the boiler. The water should be sweet and thoroughly degasified yielding steam of good quality and greatly reducing the work of the pumps, and both boiler and condenser should be completely immersed in the circulating media to minimize the heat losses. The important practical advantages of this plan are that any suitable working fluids and units of very great capacity may be used.

Technical experts who may examine the merits of the ocean power scheme will be apt to dismiss lightly the loss of energy involved in the propulsion of the hot and cold water which in reality, may be very serious on account of the lift above the mean ocean level. The outlets are unavoidably very large and if their centers are from three to four feet above the mean level to insure normal functioning at high tide the pumping losses will be considerable. Furthermore, the water is subjected to repeated changes in direction and velocity and suffers a frictional loss of head, especially in the long conduit, all of which may be equivalent to an additional life of a few feet, making the total, say, 7 feet, conservatively estimated.

Now, in the Gulf of Mexico or in Cuban waters, where my associates intended to build plants, the temperature difference between the hot and cold water will be hardly more than 36° F as an annual average, and with the poor steam obtainable the circulation may be as much as 12 Ibs. of each per horse power per second. Consequently, the mechanical work may be estimated at 108 foot pounds per second and this figure must be almost doubled because the overall efficiency of induction motor-driven pump units, which have to be employed, is not much above fifty percent, as a rule. Since one horse power is a rate of 550 foot pounds per second this means a loss of about 40 percent. Besides, the operation of the degasifiers, vacuum and deep well pumps, will consume energy which has to be supplied from the turbo-generator and taken at nearly twice its value for the reason pointed out. All these losses may be reduced in various ways but not to a very great extent, and the example clearly shows the desirability of doing away with them. This argument is applicable, even with greater force, to the cost of the pumping outfits of which I will endeavor to convey an idea by assuming that a 30,000 horse power plant is installed, requiring not less than 300,000 pounds of hot and of cold water per second, which means, approximately, 4,700 cubic feet of each. As a velocity of 3 feet per second should not be exceeded, two pumps of meeting the requirements would have intake and outlet openings of 1800 square feet, with the usual allowances. Evidently such monstrous machines could not be used, for one reason, not to mention others, that the lift would be very great and the loss incurred prohibitive. This brings to light a bad feature of the scheme illustrated in Fig. 3, namely, that it is impracticable to have recourse to very large units and thereby secure the customary advantages. A great number of small units must be of necessity used and it follows that the larger the plant the poorer it will be. Instead of the two pumps each with openings of 1800 square feet, at least one hundred motor-driven pumps with orifices of 36 square feet and a corresponding number of boilers and condensers with enormous inlet and outlet pipes would have to be employed, and at a staggering cost.

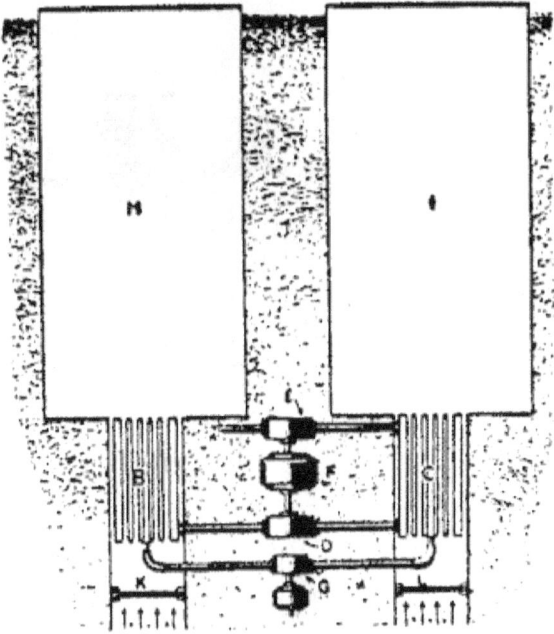

Fig. 5: The basins H and I are filled and emptied by the tide, saving much of the energy otherwise expended in pumping.

These and other similar considerations have prompted me to devise the plan schematically shown in Fig. 5 in which I do away entirely with the water pumps by relying wholly on ebb and tide to bring about the required circulation of the heating and cooling media, thus simplifying the plant and obviating great losses and expenditures. The installation comprises two very large basins lined with heat insulating cement designated by H and I and provided with suitable supports for heat-insulating roofs or covers the function of which is to minimize losses by radiation and influx of heat, respectively, from the hot and to the cold water. Each of the basins has a controllable opening, respectively K and L situated close to the bottom, where also the boiler B and condenser C, are located. The latter are connected through a turbine D coupled to a generator F, constituting a unit of large capacity. As in the case before described, a suction pump E and a deep well pump G are provided, driven by induction motors energized from the generator. All this machinery is placed on a common foundation, as indicated. The basins are filled at high tide and the outflow during the period of ebb is controlled so as to secure the best results. Although the power plant is subject to periodic variations, the plant can be operated satisfactorily without the employment of batteries or other means of storage and thus the cost of this commodity may be greatly reduced.

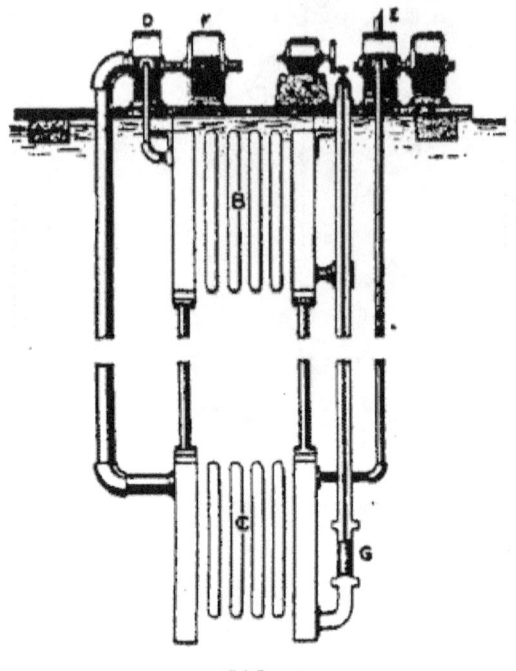

FIG. 6

Fig. 6: A floating thermo-electric power plant, in which the condenser C is suspended beneath the boiler B and the condensate circulates vertically.

Another way of deriving power from the temperature differences in the ocean without the use of water pumps is illustrated in Fig. 6. The apparatus comprises the same essential parts which have been already described, namely, a tubular boiler B and like condenser C connected through a turbine D driving a generator F, a high vacuum pump E and a small reciprocating deep-well pump G for lifting the condensate from the condenser into the boiler. The latter is supported in the warm surface water by a floating structure carrying all the machinery , while the former is suspended at a suitable depth in the cold water. Both of these parts are arranged with the tubes in vertical position insuring a good circulation of the heating and cooling media. This arrangement is very simple and effective but the raising of the condensate by pump G consumes considerable work. I have designed wireless power plants on this plan with practical objects in view and they may perhaps find valuable uses in the future.

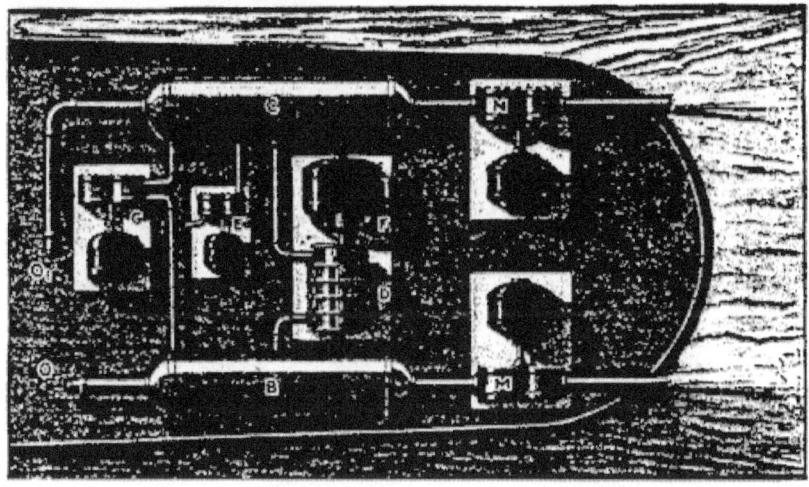

Fig. 7: Design of a vessel to be propelled by energy derived from temperature differences in the water. The symbols designating the operating mechanism are explained in the text.

Fig. 7 represents a partial view of a boat with apparatus for propelling it solely by the heat energy abstracted from the water. I was not informed just how the American engineer intended to propel his vessel and the scheme illustrated is my own. Two rotary pumps M and N are employed to force the warm and cold water, respectively, through the tubes of the boiler B and condenser c. This apparatus is placed slightly below the waterline for minimizing the losses involved in the circulation of the heating and cooling media. The pumps are supposed to be driven by induction motors, as illustrated, and are connected to the discharge pipes and other parts in such a way that the water cannot enter the hold of the vessel. The boiler intake O is near the oceans surface while that of the condenser is at the requisite depth; a streamline shaped duct O 1, open in front, being employed for the purpose. As the temperature of the water diminishes very rapidly through a limited distance from the surface of the ocean, sufficient energy can be abstracted from the water using a duct of fifty feet length to propel the boat by the streams escaping through the discharge pipes. No other means of propulsion is necessary and even the steering can be accomplished by suitably regulating the volume of the two streams discharged astern as shown. The turbine D, generator F, high vacuum pump E, deepwell pump G and other parts serve the same purposes as before. Some stored energy must be provided to start the vacuum pump and thereby initiate the operation of the apparatus.

The ocean power plan of the illustration seems very inviting when considering that the energy obtained is proportionate to the quantity of the water pumped and, therefore, virtually unlimited. But it must be remembered that the true merit of such a scheme can only be measured by the returns. We have still greater and more readily available resources which are unused because they are unprofitable. On closer investigation many discouraging facts are unearthed. The deep sea water is, normally, at a low temperature but at any time a warm current of water may be produced and render the plant useless. It has been observed that there may be differences of 35° F or even more in the temperature of water at the same depth. Just as convection currents occur in the air so they may also be produced in the ocean and this is an ever present menace to an undertaking of this kind. It also appears the actually obtainable temperature difference must be always appreciably smaller than might be inferred from soundings. The raising of the denser water below a certain level into the less dense above it involves a performance of work which must be done by the pump but is not lost as the volume of the water discharged at the top of the deep-sea duct is correspondingly increased. This does not hold true of the water above the level and, consequently, the flow into the intake of the duct takes place preponderatingly from above, that is, in the direction of the downward convection current. Owing to this, warm water from above enters the intake thus reducing the temperature difference. Another curious fact cannot be ignored. The sea is densely populated with organisms which are subject to changes caused by age. As they grow older the life stream deposits more and more solid matter, they become specifically heavier and sink gradually until finally at great depths life is extinct. If this floating matter could be removed by the pump at a constant rate, as the water, it would give relatively little trouble. But as water is removed the concentration of this matter continuously increases and may become so great as to interfere seriously with the operation of the plant. Impairment of performance, if not interruption, might also be brought about by rust and deposits in the pipes, loosening of joints and other mishaps and for this reason I consider a tunnel as the practicable means for transporting the cold water.

 I have studied this plan of power production from all angles and have devised apparatus for bringing down all losses to what I might call the irreducible minimum and still I find the performance too small to enable successful competition with the present methods.

 The utilization of temperature differences in the solid earth presents several important advantages. It would make it unnecessary to go to the tropics where power is of smaller value. Indeed, the colder the climate the better. A shaft could be sunk in the midst of a densely populated district and a great saving effected in the cost of distribution. The shaft would be costly, of course, but the apparatus cheap, simple and efficient. The first drawing illustrates its essential parts comprising a boiler at a great depth, a condenser, cooled by river or other water available, on the ground, a turbine coupled to a generator, and a motor-driven high vacuum pump. The steam or vapor generated in the boiler is conveyed to the turbine and condenser through an insulated central pipe while another smaller pipe, likewise provided with an *adiathermanous* covering serves to feed the condensate into the boiler by gravity. All that is necessary to open up unlimited resources of power throughout the world is to find some economic and speedy way of sinking deep shafts.

* * * * *

Whether we shall have to rely on power derived from terrestrial heat must be left to the future. If we should exhaust our present resources without opening up new ones, this possibility might arise. Undoubtedly, our stores of coal and oil will be eventually used up and there is not enough water power to supply our needs. The idea of obtaining motive energy from atoms or change of elements is unscientific and illusionary and cannot be condemned too emphatically. The same is true of the scheme of harnessing the energy supposed to be liberated at such temperatures as 40,000,000° C (Centigrade) recently suggested. The fundamental fallacy in all these proposals is that it takes more energy to disintegrate than can be usefully recovered even in an ideal process.

Glaringly fallacious theories are responsible for such chimerical hopes. Probably the worst of these is the electron theory. Of the four or five atomic structures which have been suggested not a single one is possible. Not more than one in a thousand men of science knows that an electron—whatever it be—can only exist in the perfect vacuum of intermolecular and interstellar spaces or highly exhausted tubes and that the nucleus stripped of electrons, is devoid of energy.

It was clear to me many years ago that a new and better source of power had to be discovered to meet the ever increasing demands of mankind. In a lecture delivered before the *American Institute of Electrical Engineers* at Columbia University May 20, 1891, I said: "We are whirling through endless space with inconceivable speed, all around us everything is spinning, everything is moving, everywhere is energy. There must be some way of availing ourselves of this energy more directly. Then, with the light obtained from the medium, with the power derived from it, with every form of energy obtained without effort, from the store forever inexhaustible, humanity will advance with giant strides."

I have thought and worked with this object in view unremittingly and am glad to say that I have sufficient theoretical and experimental evidence to fill me with hope, not to say confidence, that my efforts of years will be rewarded and that we shall have at our disposal a new source of power, superior even to the hydro-electric, which may be obtained by means of simple apparatus everywhere and in almost constant and unlimited amount.

*****

Excerpt from "Sea Power Plant Designed by Tesla," "... while the condensed water flows by gravity through another pipe reaching to depth at which the temperature of the ground exceeds that of the condensate. By circulating the steam in great volume through the turbine and condenser I am able to maintain a considerable temperature difference between the ground and the interior of the shaft, so that a very great quantity of heat flows into the same continually; to be transformed into mechanical work. The only requisite is a sufficient volume of cooling water.

"By this method it is practicable to supply all the power which a small community may require from a shaft of moderate depth, certainly less than a mile. And for isolated dwellings a few hundred feet depth would be ample, particularly if such a fluid such as ether is employed for running the turbine."

New York Times, Nov. 8, 1931.
December 1931

TESLA COSMIC RAY MOTOR MAY TRANSMIT POWER ROUND EARTH

—Brooklyn Eagle. July 10, 1932

Famed Scientist, on Eve of 76th Birthday, Says He Has Succeeded in Harnessing 'Penetrating Rays' to Operate Small Motive Device, by John J. A. O'Neill, Science Editor of the Eagle.

"I have harnessed the cosmic rays and caused them to operate a motive device," declared Nikola Tesla, famous scientist, in an interview last evening on the eve of his 76th birthday.

Tesla, who all his life has worked in seclusion and struggles to avoid publicity with all the vigor with which movie stars court it, permits a handful of "science writers" to violate the rules as a sort of birthday party.

It is very much of an ordeal to the tall, straight, meticulously attired gentleman whose inventions have been epoch-making and who is unable to understand why the public should be interested in him.

"Cosmic ray investigation is a subject that is very close to me. I was the first to discover these rays and I naturally feel toward them as I would toward my own flesh and blood," said Dr. Tesla.

His statement is borne out by reference to clippings of interviews with him more than a quarter of a century ago in which he discussed "penetrating rays" and to which not much attention was given as no one was able to comprehend the nature of them as he discussed them.

"I have advanced a theory of the cosmic rays and at every step of my investigations I have found it completely justified." said Dr. Tesla.

Not much power yet

Dr. Tesla stated that the amount of power he was able to develop in the device was insignificant.

I asked him if its power output was of the same magnitude as that of Crookes' radiometer, the device with four vanes in a glass tube that are rotated by sunlight, and which is often seen in jewelers' windows. He stated that the power output was many thousand times that of a Crookes' radiometer.

"The attractive features of the Cosmic rays is their constancy. They shower down on us throughout the whole 24 hours, and if a plant is developed to use their power it will not require devices for storing energy as would be necessary with devices using wind, tide or sunlight."

Exceed velocity of light

"All of my investigations seem to point to the conclusion that they are small particles, each carrying so small a charge that we are justified in calling them neutrons. They move with great velocity, exceeding that of light.

"More than 25 years ago I began my efforts to harness the cosmic rays and I can now state that I have succeeded in operating a motive device by means of them."

I was able to prevail upon Dr. Tesla to give me some idea of the principle upon which his cosmic ray motor works.

"I will tell you in the most general way," he said. "The cosmic ray ionizes the air, setting free many charges — ions and electrons. These charges are captured in a condenser which is made to discharge through the circuit of the motor."

Hopes to build large motor

"I have hopes of building my motor on a large scale, but circumstances have not been favorable to carrying out my plan."

I asked Dr. Tesla if his plan for transmission of power between planets involved the use of cosmic rays, and he stated that the two projects have no connection whatever. He stated that he has continued his experimental work in the laboratory on the interplanetary power transmission project and is certain of its feasibility.

I also asked him if he is still at work on the project which he inaugurated in the '90's of transmitting power wirelessly anywhere on earth. He is at work on it, he said, and it could be put into operation.

Cited two principles

He at that time announced two principles which could be used in this project. In one the ionizing of the upper air would make it as good a conductor of electricity as a metal.

In the other the power would be transmitted by creating "standing waves" in the earth by charging the earth with a giant electrical oscillator that would make the earth vibrate electrically in the same way a bell vibrates mechanically when it is struck with a hammer.

"I do not use the plan involving the conductivity of the upper strata of the air," he said, "but I use the conductivity of the earth itself, and in this I need no wires to send electrical energy to any part of the globe."

July 10, 1932

PIONEER RADIO ENGINEER GIVES VIEWS ON POWER

—New York Herald Tribune, September 11, 1932

Tesla Says Wireless Waves Are Not Electromagnetic, But Sound In Nature Holds Space Not Curved—Predicts Power Transmission to Other Planets

The assumption of the Maxwellian ether was thought necessary to explain the propagation of light by transverse vibrations, which can only occur in a solid. So fascinating was this theory that even at present it has many supporters, despite the manifest impossibility of a medium, perfectly mobile and tenuous to a degree inconceivable, and yet extremely rigid, like steel. As a result some illusionary ideas have been formed and various phenomena erroneously interpreted. The so-called Hertz waves are still considered a reality proving that light is electrical in its nature, and also that the ether is capable of transmitting transverse vibrations of frequencies however low. This view has become untenable since I showed that the universal medium is a gaseous body in which only longitudinal pulses can be propagated, involving alternating compressions and expansions similar to those produced by sound waves in the air. Thus, a wireless transmitter does not emit Hertz waves which are a myth, but sound waves in the ether, behaving in every respect like those in the air, except that, owing to the great elastic force and extremely small density of the medium, their speed is that of light.

Suggested short waves early

Since waves of this kind are all the more penetrating, the shorter they are, I have urged the experts engaged in the commercial application of the wireless art to employ very short waves, but for a long time my suggestions were not heeded. Eventually, though, this was done, and gradually the wavelengths were reduced to but a few meters. Invariably it was found that these waves, just as those in the air, follow the curvature of the earth and bend around obstacles, a peculiarity exhibited to a much lesser degree by transverse vibrations in a solid. Recently, however, ultrashort waves have been experimented with and the fact that they also have the same property was hailed as a great discovery, offering the stupendous promise to make wireless transmission infinitely simpler and cheaper.

It is of interest to know what wireless experts have expected, knowing that waves a few meters long are transmitted clear to the antipodes. Is there any reason that they would behave radically different when their length is reduced to about half of one meter?

Waves go around world

As the general knowledge of this subject seems very limited, I may state, that even waves only one or two millimeters long, which I produced thirty-three years ago, provided that they carry sufficient energy, can be transmitted around the globe. This is not so much due to refraction and reflection as to the properties of a gaseous medium and certain peculiar action, which I shall explain some time in the future. At present it may be sufficient to call attention to an important fact in this connection, namely, that this bending of the beam projected from reflector does not affect in the least its behavior in other respects. As regards deflection in a horizontal plane, it acts just as though it were straight. To be explicit the horizontal deviations are comparatively slight. In a proposed ultrashort wave transmission, the vertical bending, far from being an advantage, is a serious drawback, as it increased greatly the liability of disturbances by obstacles at the earth's surface. The downward deflection always occurs, irrespective of wavelength, and also if the beam is thrown upward at an angle to the horizontal, and this tendency is, according to my finding, all the more pronounced the bigger the planet. On a body as large as the sun, it would be impossible to project a disturbance of this kind to any considerable distance except along the surface.

It might be inferred that I am alluding to the curvature of space supposed to exist according to the teachings of relativity, but nothing could be further from my mind. I hold that space cannot be curved, for the simple reason that it can have no properties. It might as well be said that God has properties. He has not, but only attributes and these are of our own making. Of properties we can only speak when dealing with matter filling the space. To say that in the presence of large bodies space becomes curved, is equivalent to stating that something can act upon nothing. I, for one, refuse to subscribe to such a view.

Need radio channels

The chief object of employing very short waves is to provide an increased number of channels required to satisfy the ever-growing demand for wireless appliances. But this is only because the transmitting and receiving apparatus, as generally employed, is ill-conceived and not well adapted for selection. The transmitter generates several systems of waves, all of which, except one, are useless. As a consequence, only an infinitesimal amount of energy reaches the receiver and dependence is placed on extreme amplification, which can be easily affected by the use of the so-called three-electrode tubes. This invention has been credited to others, but as a matter of fact, it was brought out by me in 1892, the principle being described and illustrated in my lecture before the Franklin Institute and National Electric Light Association. In my original device I put around the incandescent filament a conducting member, which I called a "sieve." This device is connected to a wire leading outside of the bulb and serves to modify the stream of particles projected from the filament according to the charge imparted to it. In this manner a new kind of detector, rectifier and amplifier was provided. Many forms of tubes on this principle were constructed by me and various interesting effects obtained by their means shown to visitors in my laboratory from 1893 to 1899, when I undertook the erection of an experimental world-system wireless plant at Colorado Springs.

During the last thirty-two years these tubes have been made veritable marvels of mechanical perfection, but while helpful in many ways they have drawn the experts away from the simpler and much superior arrangement, which I attempted to introduce in 1901. My plans involved the use of a highly effective and efficient transmitter conveying to any receiver at whatever distance, a relatively large amount of energy. The receiver is itself a device of elementary simplicity partaking of the characteristics of the ear, except that it is immensely more sensitive. In such a system resonant amplification is the only one necessary and the selectivity is so great that any desired number of separate channels can be provided without going to waves shorter than a few meters.

For this reason, and because of other shortcomings, I do not attach much importance to the employment of waves, which are now being experimented with. Besides, I am contemplating the practical use of another principle, which I have discovered and which is almost unlimited in the number of channels and in the energy three-electrode tubes. This invention has been credited to others, but as a matter of fact, it was brought out by me in 1892, the principle being transmitted. It should enable us to obtain many important results heretofore considered impossible. With the knowledge of the facts before me, I do not think it hazardous to predict that we will be enabled to illuminate the whole sky at night and that eventually we will flash power in virtually unlimited amounts to planets. It would not surprise me at all if an experiment to transmit thousands of horsepower to the moon by this new method were made in a few years from now.

September 11, 1932

THE ETERNAL SOURCE OF ENERGY OF THE UNIVERSE, ORIGIN AND INTENSITY OF COSMIC RAYS[1]

—October 13, 1932, New York

A little over one century ago many astronomers, including Laplace still thought that the system of heavenly bodies was unalterable and that they would perform their motions in the same manner through an eternity. But the gradual perfection of instruments and refinement of methods of investigation, achieved since that time, has led to the recognition that there is a continuous change going on in the celestial regions subjecting all bodies to ever varying influence. Where this change is leading to, and what is to be its final phase, have become questions of supreme scientific interest. In a communication to the Royal Society of Edinburgh dated April 19, 1852 and the Philosophical Magazine of October of the same year, Lord Kelvin drew attention to the general tendency in nature towards dissipation of mechanical energy, a fact borne out in daily observation of thermo-dynamic and dynamo-thermic processes and one of ominous significance. It meant that the driving force of the universe was steadily decreasing and that ultimately all of its motive energy will be exhausted none remaining available for mechanical work. In the macro-cosmos, with its countless conception, this process might require billion of years for its consummation; but in the infinitesimal worlds of the micro-cosmos it must have been quickly completed. Such being the case then, according to an experimental findings and deductions of positive science, any material substance (cooled down to the absolute zero of temperature) should be devoid of an internal movement and energy, so to speak, dead.

This idea of the great philosopher, who later honored me with his friendship, had a fascinating effect on my mind and in meditating over it I was struck by the thought that if there is energy within the substance it can only come from without. This truth was so manifest to me that I expressed it in the following axiom: "There is no energy in matter except that absorbed from the medium." Lord Kelvin gave us a picture of a dying universe, of a clockwork wound up and running down, inevitably doomed to come to a full stop in the far, far off future. It was a gloomy view incompatible with artistic, scientific and mechanical sense. I asked myself again and again, was there not some force winding up the clock as it runs down? The axiom I had formulated gave me a clue. If all energy is supplied to matter from without then this all important function must be performed by the medium. Yes—but how?

I pondered over this oldest and greatest of all riddles of physical science a long time in vain, despairingly remind of the words of the poet:

> *Where, boundless nature, can I hold you fast?*
> *And where you breasts? Wells that sustain*
> *All life—the heaven and the earth are nursed.*
> Goethe. Faust

What I strove for seemed unattainable, but a kind fate favored me and a few inspired experiments lifted the veil. It was a revelation wonderful and incredible explaining many mysteries of nature and disclosing as in a lightening flash the illusionary character of some modem theories incidentally also bearing out the universal truth of the above axiom.

When radio-active rays were discovered their investigators believed them to be due to liberation of atomic energy in the form of waves. This being impossible in the light of the preceding I concluded that they were produced by some external disturbance and composed of electrified particles. My theory was not seriously taken although it appeared simple and plausible. Suppose that bullets are fired against a wall. Where a missile strikes the material is crushed and spatters in all directions radial from the place of impact In this example it is perfectly clear that the energy of the flying pieces can only be derived from that of the bullets. But in manifestation of radio-activity no such proof could be advanced and it was, therefore, of the first importance to demonstrate experimentally the existence of this miraculous disturbance in the medium. I was rewarded in these efforts with quick success largely because of the efficient method I adopted which consisted in deriving from a great mass of air, ionized by the disturbance, a current, storing its energy in a condenser and discharging the same through an indicating device. This plan did away with the limitations and incertitude of the electroscope first employed and was described by me in articles and patents from 1900 to 1905. It was logical to expect, judging from the behavior of known radiations, that the chief source of the new rays would be the sun, but this supposition was contradicted by observations and theoretical considerations which disclosed some surprising facts in this connection.

Light and heat rays are absorbed in their passage through a medium in a certain proportion to its density. The ether, although the most tenuous of all substances, is no exception to this rule. Its density has been first estimated by Lord Kelvin and conformably to his finding a column of one square centimeter cross section and of a length such that light, traveling at a rate of three hundred thousands kilometers per second, would require one year to traverse it, should weigh 4.8 grams. This is just about the weigh of a prism of ordinary glass of the same cross section and two centimeters length which, therefore, may be assumed as the equivalent of the ether column in absorption. A column of the ether one thousand times longer would thus absorb as much light as twenty meters of glass. However, there are suns at distances of many thousands of light years and it is evident that virtually no light from them can reach the earth. But if these suns emit rays immensely more penetrative than those of light they will be slightly dimmed and so the aggregate amount of radiations pouring upon the earth from all sides will be overwhelmingly greater than that supplied to it by our luminary. If light and heat rays would be as penetrative as the cosmic, so fierce would be the perpetual glare and so scorching the heat that life on this and other planets could not exist.

Rays in every respect similar to the cosmic are produced by my vacuum tubes when operated at pressures of ten millions of volts or more, but even if it were not confirmed by experiment, the theory I advanced in 1897 would afford the simplest and most probable explanation of the phenomena. Is not the universe with its infinite and impenetrable boundary a perfect vacuum tube of dimensions and power inconceivable? Are not its fiery suns electrodes at temperatures far beyond any we can apply in the puny and crude contrivances of our making? Is it not a fact that the suns and stars are under immense electrical pressures transcending any that man can ever produce and is this not equally true of the vacuum in celestial space? Finally, can there be any doubt that cosmic dust and meteoric matter present an infinitude of targets acting as reflectors and transformers of energy? If under ideal working conditions, and with apparatus on a scale beyond the grasp of the human mind, rays of surpassing intensity and penetrative power would not be generated, then, indeed, nature has made an unique exception to its laws.

It has been suggested that the cosmic rays are electrons or that they are the result of creation of new matter in the interstellar deserts. These views are too fantastic to be even for a moment seriously considered. They are natural outcroppings of this age of deep but unrational thinking, of impossible theories, the latest of which might, perhaps, deal with the curvature of time. What this world of ours would be if time were curved:

As there exists considerable doubt in regard to the manner in which the intensity of the cosmic rays varies with altitude the following simple formula derived from my early experimental data may be welcome to those who are interested in the subject.

```
I = (W+P) / (W+p)
```

In this expression W is the weight in kilograms of a column of lead of one square centimeter cross section and one hundred and eighty centimeters length, P the normal pressure of the atmosphere at sea level in kilograms per square centimeter, p the atmospheric pressure at the altitude under consideration and in like measure and I the intensity of the radiation in terms of that at sea level which is taken as unit. Substituting the actual values for W and P, respectively 1.9809 and 1.0133 kilograms, the formula reduces to

```
I = 2.99421 / (1.9809 + p)
```

Obviously, at sea level p = P hence the intensity is equal to 1, this being the unit of measurement. On the other hand, at the extreme limit of the atmosphere p = 0 and the intensity I = 1.5115.

The maximum increase with height is, consequently, a little over fifty-one percent. This formula, based on my finding that the absorption is proportionate to the density of the medium whatever it be, is fairly accurate. Other investigators might find different values for W but they will undoubtedly observe the same character of dependence, namely, that the intensity increases proportionately to the height for a few kilometers and then at a gradually lessening rate.

(1) Tesliana, special edition, Belgrade, Yugoslavia, 1995, pp. 56-59.

TESLA 'HARNESSES' COSMIC ENERGY

—Philadelphia Public Ledger, November 2, 1933

Inventor Announces Discovery of Power to Displace Fuel in Driving Machinery— Calls Sun Main Source

A principle by which power for driving the machinery of the world may be derived from the cosmic energy which operates the universe, has been discovered by Nikola Tesla, noted physicist and inventor of scientific devices, he announced today.

This principle, which taps a source of power described as "everywhere present in unlimited quantities" and which may be transmitted by wire or wireless from central plants to any part of the globe, will eliminate the need of coal, oil, gas or any other of the common fuels, he said.

Dr. Tesla in a statement today at his hotel indicated the time was not far distant when the principle would be ready for practical commercial development.

Asked whether the sudden introduction of his principle would upset the present economic system, Dr. Tesla replied, "It is badly upset already." He added that now as never before was the time ripe for the development of new resources. While in its present form the theory calls for the development of the energy in central plants requiring vast machinery, Dr. Tesla said he might be able to work out a plan for its use by individuals.

The central source of cosmic energy for the earth is the sun, Dr. Tesla said, but "night will not interrupt the flow of the new power supply."

November 2, 1933

TESLA INVENTS PEACE RAY

—*New York Sun. July 10, 1934*

Tesla Describes His Beam of Destructive Energy

Invention of a "beam of matter moving at high velocity" which would act as a "beam of destructive energy" was announced today by Dr. Nikola Tesla, the inventor, in his annual birthday interview. Dr. Tesla is 78, and for the past several years has made his anniversary the occasion for announcement of scientific discoveries.

The beam, as described by the inventor to rather bewildered reporters, would be projected on land from power houses set 200 miles or so apart and would provide an impenetrable wall for a country in time of war. Anything with which the ray came in contact would be destroyed, the inventor indicated. Planes would fall, armies would be wiped out and even the smallest country might so insure "security" against which nothing could avail.

Dr. Tesla announced that he plans to suggest his method at Geneva as an insurance of peace.

July 10, 1934

TESLA ON POWER DEVELOPMENT AND FUTURE MARVELS

—*New York World Telegram, July 24, 1934*

I am a reader of your excellent paper and frequently preserve excerpts of interest to me for future reference.

One of these is an article by William Engle, in your issue of June 29, 1934, dealing with hydro-electric development in which the author characterizes my recent announcement of a new inexhaustible source of power as "nebulous."

A preliminary information is necessarily incomplete, but I always make sure that it is based on demonstrated fact and accurate as far as it goes. My illustrious namesake, Copernicus, used to go twenty times over his scientific statements before giving them out; nevertheless, compared with the attention I bestow upon my own, he might have been considered a careless man.

The author of the article gives an eloquent account of water power development, recalling vividly to my mind the almost miraculous way in which success with my alternating system was achieved. As I review the past, I realize how fortunate it was that at the time when, after years of fruitless talking to deaf ears, I finally managed tb be heard by a few, there was a man in the electrical industry towering above all others, like Samson over the Philistines. A genius of the first degree, inventive ability and mastery of business, a man truly great, of phenomenal powers—George Westinghouse. He espoused my cause and undertook to wage a war against overwhelming odds.

The alternating current was completely discredited, decried as deadly and of no commercial value. Edison thought that the wires might be used for hanging laundry to dry. Steinmetz had a very poor opinion of my induction motor. The old interests were powerful and resolved to fight any encroachment on their business by all means fair or foul. But Westinghouse was not dismayed and threw all his energy and resources into the battle of the century. More than once he came near to being snuffed out, but finally he routed his opponents and put the new industry on a firm foundation. It was a monumental achievement unparalleled in the history of technical development. The service he rendered to the world is beyond estimate.

But it took another human dynamo, a genius of a different kind—Samuel Insull—to enlarge on the work of Westinghouse and apply the system on a colossal scale.

Insull concentrated his efforts on cheapening the production, transmission and distribution of power. He recognized early the economic advantages of large units and prevailed upon the manufacturers to supply him with huge turbo-generators, regardless of cost. He introduced other improvements raising the efficiency and range of central stations and finally realized, practically and successfully, the Super Power System which I had barely suggested in 1893. The results he obtained were such as to astonish engineers, and his bold example was quickly followed here as well as in other countries, saving immense sums of money to the consumers.

At present the work of Westinghouse and Insull is carried further in every corner of the globe, providing new resources, transforming cities and communities and contributing to the safety, comfort and convenience of hundreds of millions. Let us thank the stars that these great pioneers lived in our time, as otherwise we might have had to wait a century for the benefits we now enjoy.

* * * * *

Another item of interest to me is your flattering editorial of July 12, 1934, with a fly in the ointment since you state that examination of performance does not in recent cases fulfill my prophecy. Perhaps not, but on the whole I have been extraordinarily successful. You would be surprised to know how many of my discoveries and inventions are in extensive use. To give an illustration, I may refer to my wireless system of transmission of energy which is looked upon by many as a pipe dream.

These uninformed people should be told that "wireless" is not a single invention but an art involving the use of many of them, and of them I have contributed the fundamental and most essential, and they are universally employed. There is as yet no pressing necessity for wireless transmission of power in industrial amounts, but as soon as it arises the system will be applied and with perfect success.

Still another item which has interested me is a report from Washington in the World-Telegram of July 13, 1934, to the effect that scientists doubt the death ray effect. I am quite in agreement with these doubters and probably more pessimistic in this respect than anybody else for I speak from long experience.

Rays of the requisite energy can not be produced, and then, again, their intensity diminishes with the square of the distance. Not so the agent I employ, which will enable us to transmit to a distant point billions of times more energy than is possible by any kind of ray.

We are all fallible, but as I examine the subject in the light of my present theoretical and experimental knowledge I am filled with deep conviction that I am giving to the world something far beyond the wildest dreams of inventors of all time.

New York.
July 24, 1934

DR. TESLA VISIONS THE END OF AIRCRAFT IN WAR

—*Every Week Magazine, Oct. 21, 1934*

"America Enters War!" "United States Joins Allies!" "Congress Declares War!" The newsboys were screaming the headlines through the rainy April night. Men and women stood on corners, talking, talking, talking

The drift of the days went on. Troop trains pulled out of the stations, from Centreville, Mississippi, up to Bangor, Maine. The drums blew. The ships sailed and the casualty lists came back. One by one the gold stars replaced the white

And 1917 drifted into 1918.

Dr. Nikola Tesla was in his laboratory trying hard to solve a problem of ages. Once in a while he raised his head to listen. Then he turned back to his experiments. He was going to end war!

The noted inventor, 78 years old now, already had 700 inventions to his credit. This was to be his greatest.

Years marched on. The fanfare and the drums were done. The dead were buried. The living came home.

Now, 15 years after the war has ended, Tesla, one of the greatest inventors of all time, has announced that his invention to end all wars, by a perfect means of defense which any nation can employ, is ready. Soon, he says, he will take it to Geneva to present it to the Peace Conference.

Whether it is a dream or reality may soon be known. He claims to have created a new agent, silent and invisible, which kills without trace and yet pierces the thickest armor. It is a beam of death and destruction formed of minute particles of matter carrying such tremendous energy that they could bring down a fleet of 10,000 attacking planes and wipe out an army of millions at a distance of 250 miles.

"The invention," said Dr. Tesla, "will make war impossible for it will surround any country using this means with an impenetrable, invisible wall of protection. Plants for the generating of this beam will be erected along the coasts and near cities. One plant will afford perfect safety within an area of 40.000 square miles.

"The beam will be effective at any distance at which the object to be destroyed can be perceived through a telescope. Every country will have to adopt this invention, for without it a nation will be helpless.

"The beam, intended chiefly for defense, will be projected from an electric power plant, ready to be put in action at the first sign of danger. The cost of operation will be insignificant, as the plant is chiefly intended for use in emergency. But to make the investment profitable in times of peace it may be commercially employed for a number of purposes."

Dr. Tesla wishes it to be understood that the means he has perfected has nothing in common with the so-called "death ray."

"It is impossible to develop such a ray. I worked on that idea for many years." he says, "before my ignorance was dispelled and I became convinced that it could not be realized. This new beam of mine consists of minute bullets moving at a terrific speed, and any amount of power desired can be transmitted by them. The whole plant is just a gun, but one which is incomparably superior to the present."

The picture of the protected world, in which men will devote their time in pursuits of peace, is a strangely fascinating one.

Imagine the map of the world, every country surrounded by great plants which will offer absolute protection to the nation itself and instant death to any intruders. Only ships flying white flags of peace can sail into a foreign harbor.

The power plants, resembling forts placed at strategic distances along a country's border, will be on guard. As they are immovable, they will constitute essentially means for defense, and by making invasion impossible will greatly advance the cause of peace.

If, occasionally, nations decide that they must have war just for the thrill of a throbbing drum and a singing bugle, it can be staged on the sea, Dr. Tesla says. Navy supremacy will banish aircraft.

"The airplane will cease to be used as a means of offense," the great inventor explains. "It will be used entirely for peace, as it should he. An airplane, through the very nature of its construction, can not carry with it a generating plant for the beam. If it comes in contact with a country which is protected, it has no chance.

"The battleships will ride to sea safe from air raids, for they will be equipped with smaller plants for generating a beam of sufficient power to destroy any attacking airplane. But they will not be permitted to come near the shore Of a protected country and attack it with any chance of success.

"The nation which has the best equipped battleships, however, will gain the supremacy of the seas. Submarines will be obsolete, for the methods of detecting them will be perfected to such a degree that there will be no longer any advantage in submerging. When a submarine is located the beams will function under water, though not quite so effectively as in air."

Four new inventions of Dr. Tesla are involved in the creation of the beam.

"Briefly, the first comprises a method and apparatus for producing rays and other manifestations of energy in free air, eliminating the high vacuum heretofore indispensable." he explains.

"The second one is the process for producing electrical force of immense power.

"The third method amplifies the process, and the fourth produces a tremendous electrical repelling force."

In times of peace such a plant can be used to transmit power in any amount up to its full capacity and to any place on the earth visible through a telescope, according to its inventor. Voltages never before attained, of 50,000.000 volts or more, will have to be applied.

The man who is responsible for so many discoveries and improvements has devoted his entire life to his scientific pursuits. Tall, lean, reserved, his path goes between the two small laboratories and the various manufacturing plants with which he has contact.

Born in Yugoslavia, Tesla comes from a race of inventors.

"On my mother's side, for three generations, almost all members of the families were inventors," he says. "My mother was Georgianna Mandic, who was noted as an inventor of household appliances. One of the things which she perfected was her own weaving machine.

"Her family can be traced back to the seventh century, in the historical records. My grandfather was an officer in Napoleon's army."

Tesla began to invent at the age of six. As he grew up his interest focused in the laboratory.

"I sleep about one and one-half hours a night." the inventor says. "I think that is enough for any man. When I was young I needed more sleep. But age doesn't require so much. There are so many things to do I do not want to spend time sleeping needlessly. In my family all were poor sleepers. Time spent in sleep is lost time, we always felt."

Tesla, busy with his 700 inventions, never had time for marriage. He never had a girl in his young days. He never had a romance. There was no leisure for them.

His diet is simple. He lives chiefly on vegetables, cereals and milk. The menu includes onions, spinach, celery, carrots, lettuce, with potatoes occasionally. Whites of eggs and milk complete the diet. There is no meat on his vegetable plate. He never smokes or tastes tea, coffee, alcoholic beverages or any other stimulant.

While he is perfecting the beam which will defend nations from attach, the inventor is playing with other ideas. He goes from one to the other, he says, as this or that gains paramount interest or some new clue is suggested.

"But what is giving me more fun than anything I have done for a long, long time," Dr. Tesla explains, "is an electric bath which I hope to have ready for general use very soon.

"It doesn't require much room. There is a platform on which the person stands. He turns on the current. Instantly all foreign material such as dust, dandruff, scales on the skin and microbes is thrown off from the body. The nerves, too, are exhilarated and strengthened. The 'bath' is excellent for medical as well as for cleaning purposes."

However, the war picture gives the master inventor more satisfaction than the minor inventions. He is rejoicing because his instrument of death will save millions of lives and inestimable property.

His only regret is that there may be another war before the discoveries he has made have been placed before the Disarmament Conference at Geneva, and generally adopted by the nations of the world.

"The next war, and I am afraid that there will be one before long," he says, "will be fought in the air. But if the beam is adopted war in the air will cease.

"Whatever battles there are thereafter will be confined to the sea. But no nation will dare to attack another nation when every country is armed. There will be a general feeling of safety throughout the world."

Oct. 21, 1934

THE NEW ART OF PROJECTING CONCENTRATED NON-DISPERSIVE ENERGY THROUGH NATURAL MEDIA

—Circa May 16, 1935.

Briefly Exposed by Nikola Tesla

The advances described are the result of my research carried on for many years with the chief object of transmitting electrical energy to great distances. The first important practical realization of these efforts was the alternating current power system now in universal use. I then turned my attention to wireless transmission and was fortunate enough to achieve similar success in this fruitful field, my discoveries and inventions being employed throughout the world. In the course of this work, I mastered the technique of high potentials sufficiently for enabling me to construct and operate, in 1899, a wireless transmitter developing up to twenty million volts. Some time before I contemplated the possibility of transmitting such high tension currents over a narrow beam of radiant energy ionizing the air and rendering it, in measure, conductive. After preliminary laboratory experiments, I made tests on a large scale with the transmitter referred to and a beam of ultra-violet rays of great energy in an attempt to conduct the current to the high rarefied strata of the air and thus create an auroral such as might be utilized for illumination, especially of oceans at night. I found that there was some virtue in the principal but the results did not justify the hope of important practical applications although, some years later, several inventors claimed to have produced a "death ray" in this manner. While the published reports to this effect were entirely unfounded, I believe that with the new transmitter to be built, this and many other wonders will be achieved. Much time was devoted by me to the transmission of radiant energy, in various forms, by reflectors and I perfected means for increasing enormously the intensity of the effects, but was baffled in all my efforts to materially reduce dispersion and became fully convinced that this handicap could only be overcome by conveying the power through the medium of small particles projected, at prodigious velocity, from the transmitter. Electro-static repulsion was the only means to this end and apparatus of stupendous force would have to be developed, but granted that sufficient speed and energy could be realized with a single row of minute bodies then there would be no dispersion whatever even at great distance. Since the cross section of the carriers might be reduced to almost microscopic dimensions an immense concentration of energy, irrespective of distance, could be attained.

When I undertook to carry out this plan in practice, the difficulties seemed insurmountable. In the first place, a closed vacuum tube could not be employed as no window could withstand the force of the impact. This made it absolutely necessary to project the particles in free air which meant that each could hold only an insignificant charge. Thus, no matter how high the potential of the terminal, the force of repulsion would be necessarily too small for the purpose contemplated. . . . But by the application of my discoveries and inventions it is possible to increase the force of repulsion more than a million times and what was heretofore impossible, is rendered easy of accomplishment. The successful carrying out of the plan involves a number of more or less important improvements but the principal among these include the following:

1. A new form of high vacuum tube open to the atmosphere.
2. Provisions for imparting to a minute particle an extremely high charge.
3. A new terminal of relatively small dimensions and enormous potential.
4. An electro-static generator on a new principle and of very great power.

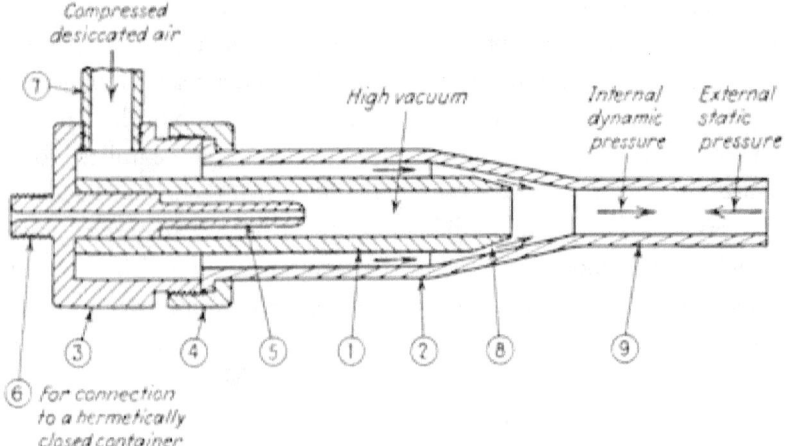

Fig. 1: Illustrating open vacuum tube

These devices and methods of operation will be explained by reference to the attached drawings in which Fig. 1 and Fig. 2 represent forms of the new open tube.

In Fig. 1, the device consists of an inner cylindrical conduit 1, cemented to a metallic socket 3, and an outer conduit 2, which is tightly screwed to the socket by a nut 4, and has on the open side a taper with a cylindrical end 9, of the same inside diameter as conduit 1. The socket 3, is bored out to provide a large chamber around the inner conduit and carries a pipe 7, through which thoroughly desiccated air or other gas under suitable pressure is supplied. The open end of the inner and the tapering part of the outer conduit are ground to form an expanding nozzle 8, through which the air escapes into the atmosphere thereby creating a high vacuum in the inner conduit. The socket 3 has a small central hole and is provided with an inside extension 5, and a threaded outside projection 6, the latter serving for connection to a container supplying automatically suitable particles or material for same while the former fulfill the purpose of charging them as they emerge from the hole. The conduit 1 and 2, may be made of fused quartz, *pyrex glass* or other refractory material and it is obviously desirable that all the parts of the apparatus have small and nearly equal coefficients of thermal expansion especially when the working medium, which might also be superheated steam, is at an elevated temperature.

It will be observed that in this tube I do away with the solid wall or window indispensable in all types heretofore employed, producing the high vacuum required and preventing the inrush of the air by a gaseous jet of high velocity. Evidently, to secure this result, the dynamic pressure of the jet must be at least equal to the external static pressure.

Expressed in symbols:

```
V2w/2g = P
```

Assuming equality:
```
V = √2g P/w
```

in which equation V is the speed of the jet at its entrance to channel 8 in meters, g the acceleration of gravity likewise in meters, P the external pressure in kilograms per square meter and w the normal weight of the air in kilograms per cubic meter. Now:

```
g = 9.81 meters
P = 10332.9 kilograms
w = 1.2929 kilograms
```

These values give
```
V = 396 meters
```

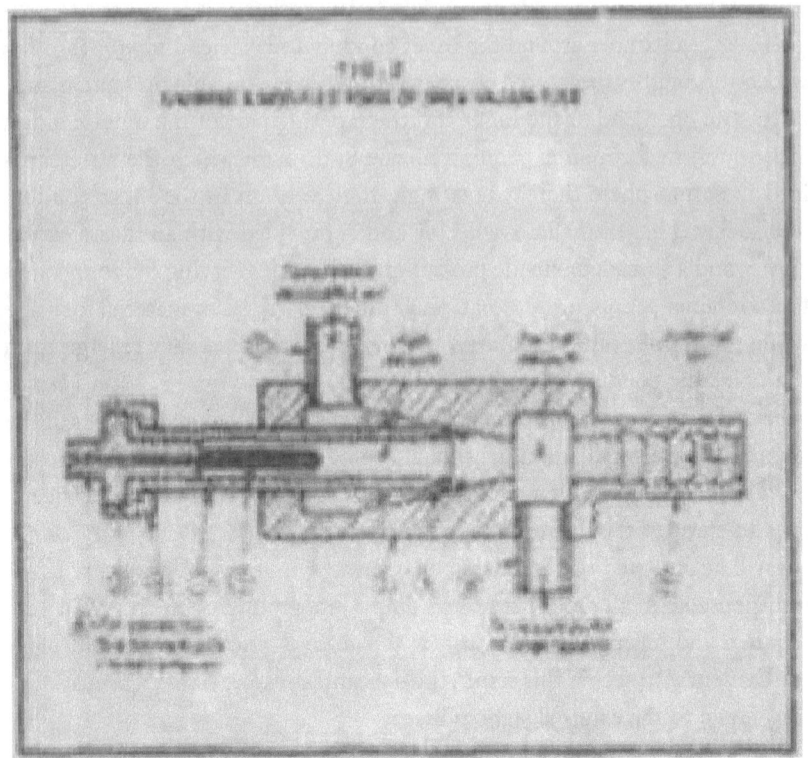

Fig. 2: Showing a modified form of open vacuum tube

Some allowance should be made for the frictional loss in the nozzle and the outlet channel and also for the deflection of the jet. For most purposes, the velocity need not be much greater, but as the degree of rarefaction depends on the square of V, it is desirable to obtain as high a value as practicable. Usually, vacuums obtained by a mercury vapor pump are considered very high. In those, the velocity is only 280 meters per second but the vapor is 6.9 times heavier than air. Therefore to get the same vacuum in the air jet, its speed should be $280 \times [6.9]^{1/2} = 735$ meters. With a working medium at high temperature and pressure, both within practicable limits, this value can be attained and even exceeded. Thus, a gaseous jet of very high velocity affords a means for closing the end of the tube, more perfectly than any window that can be made while at the same time permitting and facilitating the exit of the particles. Referring to Fig. 2., it shows schematically a modified form of my tube intended for various scientific and practical uses when it may be preferable or necessary not to discharge the jet through the open end. The construction of the device will be easily understood in view of the foregoing description like parts being similarly designated. A cylindrical conduit 1, is provided as in fig. 1, but the outer cone is replaced by a block 2, of lava or other insulating material shaped as indicated and firmly cemented to the conduit 1, which is hermetically joined by a nut 4, to a metallic plug 3, having a central hole, and extensions 3 and 6, serving the purpose stated above. The working fluid, as compressed desiccated air, is supplied by means of a pipe 7, to a large annular space around conduit 1, and escapes through an expanding nozzle 8, formed by the tapering part of the block and the end of the conduit, into a chamber connected by a pipe 10, to a vacuum pump of large capacity—not shown on the drawing—for carrying off not only the air issuing from the nozzle but also that rushing in from the outside through the open end 9. In order to minimize the volume of the latter, I avail myself of an invention of mine known as the "valvular conduit" by providing the wall of the open end 9, with recesses as indicated giving rise to whirls and eddies which use up some of the energy of the stream and reduce its velocity. In this way, a pressure of about 100 millimeters of mercury can be readily maintained in the chamber increasing greatly the expansion ration of the air and its speed through the nozzle.

It is hardly necessary to remark that my open vacuum tubes require mechanical power for operation which may range from 10 to 20 H.P., but this drawback is insignificant when considering the important advantages they offer and I anticipate that they will be extensively employed.

It remains to be explained how such a tube is utilized for imparting to a particle to be projected a very great charge. Imagine that the small spherical body be placed in a nearly perfect vacuum and electrically connected to the large sphere forming the high potential terminal of the transmitter. By virtue of the connection, the small sphere will then be at the potential of the large one no matter what its distance from the same but the quantity of electricity stored on the small sphere will vary greatly with the distance and be proportionate to the difference of its potential and that of the adjacent medium. If the small sphere is very close to the large one, this difference will be insignificant and so to the charge; but if the small sphere is at a great distance from the large one where the potential imparted by the same to the medium approximates zero, the quantity of electricity stored on the small one will be relatively enormous and equal to Qr/R. To illustrate, if $r = 1/100$ e.s. and $R = 1000$ e.s. and $Q = 108$ e.s. units, as before assumed, then $Q = 1000$ e.s. units which is a hundred thousand times more than previously obtainable. At a distance $2R$ from the center of the terminal, at which the difference between the potential of the small sphere and the adjacent medium will be half of the total, or 15,000,000 volts, Q will be 500 e.s. units and from theoretical considerations, it appears that the best results will be secured if the particle is charged in high vacuum at that distance. It can be accomplished all the more easily the smaller the radius of the terminal and this is one of the reasons why my improvement, illustrated in Fig. 3, is of great practical importance.

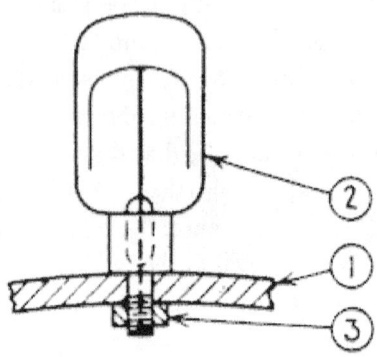

Enlarged View of One
of the Attachments

FIG. 3: New terminal for exceedingly high potentials consisting of spherical frame attachments [see fig. 4 for terminal.]

Diagram indicating distribution of charges

As will appear from the inspection of the drawing, the spherical frame of the terminal is equipped with devices, one of which is shown in the enlarged view below and comprises a bulb 2, of glass or other insulating material and an electrode of thin sheet suitable rounded. The latter is joined by a supporting wire to a metallic socket adapted for fastening to the frame 1, by means of nut 3. The bulb is exhausted to the very highest vacuum obtainable and the electrode can be charged to an immense density. Thus, it is made possible to raise the potential of the terminal to any value desired, so to speak, without limit, and the usual losses are avoided. I am confident that as much as one hundred million volts will be reached with such a transmitter providing a tool on inestimable value for practical purposes as well as scientific research.

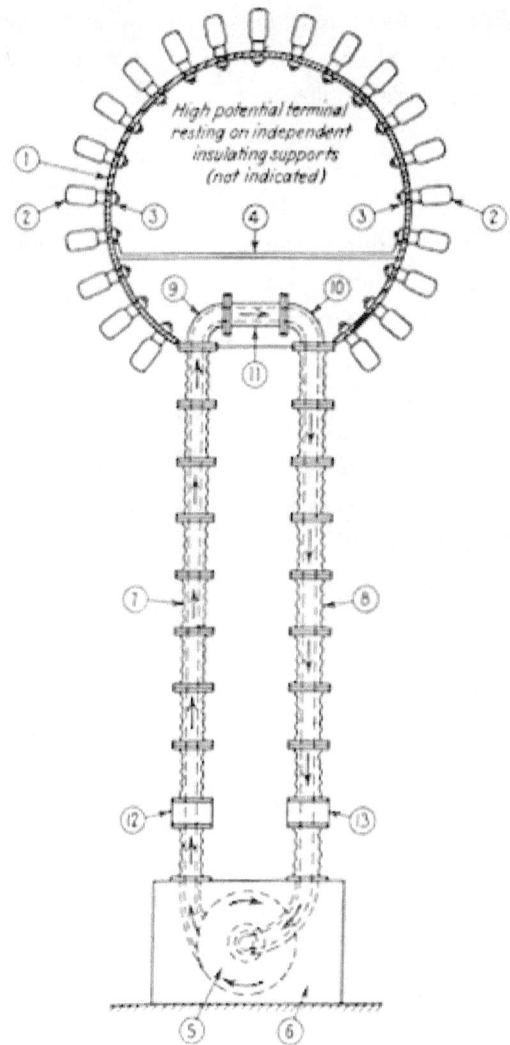

Fig. 4: Schematic illustration of new high potential generator

Perhaps the most important of these inventions is the new high potential electro-static generator, schematically represented in Fig. 4, which is provided with my improved terminal consisting of a spherical metallic frame 1, with attachments 2, adapted to be fastened to the former by nuts 3, as above described. The terminal has a platform 4, in the interior of the frame intended for supporting machinery, instruments and observers, and is carried to a suitable elevation on insulating columns omitted from the drawing for the sake of simplicity. To energize the terminal, air under pressure is driven at high speed through a hermetically closed channel comprising a turbo compressor 5, with intake and outlet connections, conduits 7 and 8, special fittings 9 and 10, and a short pipe 11. The conduits 7 and 8, are preferably composed of pieces of glazed porcelain bolted tightly together, the joints being made airtight by suitable packing and are corrugated on the outside to minimize electrical leakage. The fittings 9 and 10 and pipe 11, may also be of the same kind of material. The air before entering and after leaving the compressor, as well as all apparatus within the airtight enclosure 6, is effectively cooled and maintained at a constant temperature by means as ordinarily employed which was not thought necessary to illustrate. The operation of the machine will be understood most readily by likening the moving column of air to a running belt. When the air, leaving the compressor, reaches the device 12, containing discharge points electrified by a direct current of high tension, it is ionized and the charge imparted to it is carried upward to the special fitting 9, where it is drawn off by sucking points and charges the terminal. On the return to the compressor the air passes through special fitting 10, where it receives electricity of the opposite sign conveying it to the device 13, and from there to the ground. These actions are repeated with great rapidity. The generator can be made self-exciting by suitable connections. For several reasons, I estimate that a machine as described will have an output of many times greater than a belt generator of the same size and, besides, it has several other important construction and operative advantages.

To give an approximate estimate of performance, reference is made to diagram in Fig. 5, representing a spherical terminal and an open vacuum tube for projecting particles. Suppose that d be the distance from the center o at which a particle of radius r = 1/100 c.m. is charged in vacuum to the potential of the terminal, as before explained, and that D is the distance from center O at which the particle leaves the vacuous space, then, in passing through the distance D-d it will be accelerated to a velocity

$$V_1 = \sqrt{2Qq}\,(D-d)\,/\,md\,D \text{ centimeters per second}$$

In its transit from distance D to a very much greater distance an additional velocity of

$$V_2 = \sqrt{2Qq'}\,/\,m\,D \text{ centimeters per second}$$

q' being, theoretically, smaller than q. But I have found that although the particle in contact with air is neutralized rapidly yet, on account of its small surface, magnitude of the charge and prodigious speed, a very great distance is traversed without material reduction of the charge so that, without appreciable error, q' may be considered equal to q. Thus, the total velocity attained will be

$$V = V_1 + V_2 = \sqrt{2Qq}(D-d) / mD + \sqrt{2Qq'} / mD \text{ centimeters per second}$$

in which expression Q and q are in e.s. units, D and d in centimeters and m the mass of the particle in grams. But the calculation may be simplified, for if the charge is virtually constant through a great distance, the velocity finally attained will be

$$V = \sqrt{2Qq} / md \text{ centimeters per second}$$

Assume now that the terminal is equivalent to a sphere of radius $R = 250$ centimeters which heretofore could only be charged to a potential of 100 x 250 = 25,000 e.s. units or 7,500,000 volts but, by taking advantage of my improvements, can be readily charged to 2×10^5 e.s. units or 6×10^7 volts in which case the quantity of electricity stored will be $Q = 2 \times 10^5 \times 250 = 5 \times 10^7$ e.s. units. If, for best effect, the particle is charged in vacuum at a distance $d = 2R = 500$ centimeters where the difference between its potential and adjacent medium is 3×10^7 volts or 10^5 e.s. units, then $q/r = 10^5$ and $q = 10^5 = 1000$ e.s. units. The particle will have a volume of $4\pi/3 \times 10^6$ cubic centimeters and if it be tungsten, it will weigh about $4\pi \times 18/3 \times 10^6 = 7686/10^{11}$ gram. Substituting these values

$$V = \sqrt{2 \times 5 \times 10^7 \times 1000 \times 10^{11} / 1000 \times 7686 \times 500} = 1{,}613{,}000 \text{ centimeters or } 16{,}130 \text{ meters per second.}$$

This finding may be checked by using the relation between the joule's equivalent and the kinetic energy. Here the joules are $3 \times 10^7 \times 1000 / 3 \times 10^9 = 10$ and approximately equal to 10^6 gram-centimeters. Consequently,

$$mV^2 / 2 = 10^6$$
$$V^2 = 2 \times 10^6 \times 10^{11} / 7686 \text{ and}$$
$$V = 1{,}613{,}000 \text{ centimeters or } 16{,}130 \text{ meters}$$

as found above by my formula which is always applicable while the latter rule is not.

Since a joule is equivalent to about 10,000 gram-centimeters, the kinetic energy is equal to 105 gram-centimeters or 1 kilogram-meter.

In order to determine the probable trajectory the air resistance encountered by the particle has to be estimated from practical data and theoretical consideration. Very extensive ballistic tests by French experts have established conclusively that up to a velocity of 400 meters per second, the resistance increases as the square of the speed but from there on, to the highest velocities attained, the increase is directly proportional to the speed. On the other hand, it has been found in tests with rifles that an ordinary bullet, 8 millimeters in diameter and three times as long, fired at 400 meters per second, encounters a mean resistance of about 0.02 kilogram and from these facts, it can be inferred that the average resistance of the particle at the maximum speed V might be of the order of 1/64,000 of a kilogram and if so, the trajectory should be approximately 64,000 meters or 64 kilometers. Obviously, resistance data cannot be accurate, but as the mechanical effects can be increased many times, there should be no difficulty in securing the practically required range with a transmitter as described. In all probability, when the technique is perfected, results will be obtained which are thought impossible at present. Such a particle, notwithstanding its minute volume of 1/250,000 cubic centimeter, would be very destructive. It would pierce the usual protecting covering of aeroplanes, put machinery out of commission and ignite fuel and explosives. To combatants, it would be deadly at any distance well within its full range. Projected almost simultaneously in great numbers, the particles would produce intense heating effects. In action, against aeroplanes, the range would be very much greater on account of the smaller density of the air. Evidently, the smaller the particles, the greater will be their speed. For instance, if r = 1/10,000 centimeter, a velocity of 160,000 meters per second will be attained. An enormous increase in speed and range would be secured with particles of a diameter smaller than 800 times the molecular diameter.

It is important to devise a thoroughly practical and simple means for supplying particles and I have invented two which seem to meet this requirement. One is to feed tungsten or other wire from a spool in a closed container joined hermetically to the projector, the rotation of the spool being under control of the operator. Using wire 2/100 centimeters in diameter, twenty cubic centimeters of the same would provide material for 5,000,000 particles. The other device consists of a closed container fixed to the projector and filled with mercury which can be expanded by external and controllable application of heat and forced, under great pressure, through a minute hole in the extreme end of the extension reaching to the distance d as before illustrated and explained. The droplet torn off and projected would have the hardness of steel owing to the great capillary pressure. If mercury can be used for the purpose, this means is ideally simple and cheap.

Reprinted from Nikola Tesla's Teleforce & Telegeodynamics Proposals, Tesla Presents series Part 4, Leland I. Anderson, Editor

Correspondence

Hotel New Yorker
New York, N. Y.
April 7, 1934
S. W. Kintner, Esq.
Westinghouse Electric & Mfg. Co.
Pittsburgh, Pa.

My dear Mr. Kintner;

I was glad that you did not put the matter before Mr. Merrick for I found after careful thought and figuring that it would take much more money to carry out my proposal which I made to you on the spur of the moment stimulated by the pleasure of our meeting and your warm response. The Westinghouse people made a friendly gesture and I wanted to meet them in the same spirit by giving them the first opportunity on discoveries which I honestly believe to be more important than any of recorded in the history of invention

I have groped for years trying to find some solution of felt most pressing problem of humanity—that of insuring peace and, little by little, I have been led to the ideal means to this end, for they will afford perfect protection to every country without providing a new component for attack. The International Peace Conference will insist on its immediate and universal adoption, for as long as the countries are imperfectly protected invasions are sure to occur.

I note your suggestion but am at a loss to see how to carry it out. Rest assured though, that I shall always hold your people in high regard and if I ever find it in my power to advance their interest I shall spare no effort.

The skepticism of your expert was expected. He is probably under the sway of the modern illusionary ideas and the abler he is the more apt he is to be in error. But I have demonstrated all the principals involved and am going ahead with perfect confidence which all the experts in the world could not shake.

Yours very truly,
Nikola Tesla

* * * * *

Hotel New Yorker
November 29, 1934
J.P. Morgan Esq.
23 Wall Street
New York

Dear Mr. Morgan:

I have made recent discoveries of inestimable value which are referred in the marked passage of the clipping enclosed. Their practical application should yield an immense fortune.

The flying machine has completely demoralized the world, so much that in some cities, as London and Paris, people are in moral fear from aerial bombing. The new means I have perfected afford absolute protection against this and other forms of attack.

You know how your father assisted me in the development of my wireless system. He did not get any returns but I am convinced that if he were living he would be gratified by the knowledge that my inventions are universally applied. I still gratefully remember your own support although the war deprived me of the success I had achieved. I not only lost everything in those two undertakings but was for years compelled to pay off all sorts of unfair claims. It was only a little while ago that I managed to settle the last one and terminate the tormenting nightmare.

These new discoveries, which I have carried out experimentally on a limited scale, have created a profound impression. One of the most pressing problems seems to be the protection of London and I am writing to some influential friends in England hoping that my plan will be adopted without delay. The Russians are very anxious to render their borders safe against Japanese invasion and I have made them a proposal which is being seriously considered. I have many admirers there especially on account of the introduction of my alternating system to an extent unprecedented. Some years ago Lenin made me twice in succession very tempting offers to come to Russia but I could not tear myself from my laboratory work.

Words cannot express how much I am aching for the same facilities which I then had at my disposal and for the opportunity of squaring my account with your father's estate and yourself. I am no longer a dreamer but a practical man of great experience gained in long and bitter trials. If I had now twenty five thousand dollars to secure my property and make convincing demonstrations I could acquire in a short time colossal wealth. Would you be willing to advance me this sum if I pledged to you these inventions?

Mr. Morgan you are still able to help an undying cause but how long will you be in this privileged position? We are in the clutches of a political party which caters openly and brazenly to the mob and believes that by pouring out billions of public money, still unequalled, it can remain in power indefinitely. The democratic principles are forsaken and individual liberty and incentives are made a joke. The "New Deal" is a perpetual motion scheme which can never work but is given a semblance of operativeness by unceasing supply of the peoples capital. Most of the measures adopted are a bid for votes and some are destructive to established industries and decidedly socialistic. The next step might be the distribution of wealth by excessive taxing if not conscription.

With best wishes and respectful regards believe me as ever yours most faithfully

N. Tesla

*****

WESTERN UNION TELEGRAM
from Nikola Tesla
to Sava Kosanovic
New York, N.Y.
March 1, 1941

I thank Dr. Macek and you for the happy news. It is important that you know the following: [In] eight years I developed a new title using 50 of my patents of which one third are not applied. In the system there are no electrons. Energy goes into the same direction without any distribution [dissipation] and the same on all sides of distance. It contains neutrons. [In] the air [its size] is equal to a diameter of hydrogen. It can destroy the largest ships afloat. There is unlimited distance of travel. The same is for airplanes.

One will need nine stations: for Serbia; three for Croatia and two for Slovenia and everyone needs 200 KW which can defend our dear homeland against any type of attack.

The contents of one bomb can be exploded in the air. I add that in the station one must have a small generator or battery of 30 volts for activation.

Express my deepest respect to Dr. Macek and accept the warmest greetings and thanks.

Your uncle, Nikola Tesla

*****

WESTERN UNION TELEGRAM
from Nikola Tesla
to Sava Kosanovic
New York, N.Y.
March 4, 1941

As though I am poor with words. I still didn't explain it enough what would be necessary to increase up to twelve stations: eight in Croatia, each of the same construction like at Wardenclyffe and only 20 meters high — a ball five meters in diameter — the station would be using diesel oil for energy with mechanical action — my air turbines, steam powered, electrically or other manners of transforming into alternating electrical current with sixty billion volts pressure without danger. I am waiting for Governor Subasic to select one station on top of Mt. Lovcen. There will not be any light, electrical energy will deliver particles through space with the speed of 118,837,370,000 centimeters per second. This is 394,579 the speed of light. As I said about airplanes it can be used for tanks, trucks, automobiles and various machines in factories, with hydroelectrical wheels and unlimited other machines. The particles can be larger than that of the diameter of an Hydrogen atom with metals of all kinds of materials and sent to all distances and good results in war and bring about peace. Particles are practical with neutrons, because, they are 3,723 times lighter than electricity or electrons that cannot penetrate space for great distances. In my attempts with an effective 20 million volts, electrons carried 40 times more electricity than normally and penetrated two meters in depth and terrible damage in a moment each. I have to finish because that I give you a fresh view.

Warm Greetings, I remain your uncle, Nikola

A MACHINE TO END WAR

A Famous Inventor, Picturing Life 100 Years from Now, Reveals an Astounding Scientific Venture Which He Believes Will Change the Course of History.
Liberty, February 1937, by Nikola Tesla as told to George Sylvester Viereck

Tesla. "It seems," he says, "that I have always been ahead of my time."

Editor's Note: Nikola Tesla, now in his seventy-eighth year, has been called the father of radio, television, power transmission, the induction motor, and the robot, and the discoverer of the cosmic ray. Recently he has announced a heretofore unknown source of energy present everywhere in unlimited amounts, and he is now working upon a device which he believes will make war impracticable.

Tesla and Edison have often been represented as rivals. They were rivals, to a certain extent, in the battle between the alternating and direct current in which Tesla championed the former. He won; the great power plants at Niagara Falls and elsewhere are founded on the Tesla system. Otherwise the two men were merely opposites. Edison had a genius for practical inventions immediately applicable. Tesla, whose inventions were far ahead of the time, aroused antagonisms which delayed the fruition of his ideas for years.

However, great physicists like Kelvin and Crookes spoke of his inventions as marvelous. "Tesla," said Professor A. E. Kennelly of Harvard University when the Edison medal was presented to the inventor, "set wheels going round all over the world... What he showed was a revelation to science and art unto ail time."

"Were we," remarks B. A. Behrend, distinguished author and engineer," to seize and to eliminate the results of Mr. Tesla's work, the wheels of industry would cease to turn, our electric cars and trains would stop, our towns would be dark, our mills would be dead and idle."

Forecasting is perilous. No man can look very far into the future. Progress and invention evolve in directions other than those anticipated. Such has been my experience, although I may flatter myself that many of the developments which I forecast have been verified by events in the first third of the twentieth century.

It seems that I have always been ahead of my time. I had to wait nineteen years before Niagara was harnessed by my system, fifteen years before the basic inventions for wireless which I gave to the world in 1893 were applied universally. I announced the cosmic ray and my theory of radio activity in 1896. One of my most important discoveries—terrestrial resonance—which is the foundation of wireless power transmission and which I announced in 1899, is not understood even today. Nearly two years after I had flashed an electric current around the globe, Edison, Steinmetz, Marconi, and others declared that it would not be possible to transmit even signals by wireless across the Atlantic. Having anticipated so many important developments, it is not without assurance that I attempt to predict what life is likely to be in the twenty-first century.

Life is and will ever remain an equation incapable of solution, but it contains certain known factors. We may definitely say that it is a movement even if we do not fully understand its nature. Movement implies a body which is being moved and a force which propels it against resistance. Man, in the large, is a mass urged on by a force. Hence the general laws governing movement in the realm of mechanics are applicable to humanity.

There are three ways by which the energy which determines human progress can be increased: First, we may increase the mass. This, in the case of humanity, would mean the improvement of living conditions, health, eugenics, etc. Second, we may reduce the frictional forces which impede progress, such as ignorance, insanity, and religious fanaticism. Third, we may multiply the energy of the human mass by enchaining the forces of the universe, like those of the sun, the ocean, the winds and tides.

The first method increases food and well-being. The second tends to bring peace. The third enhances our ability to work and to achieve. There can be no progress that is not constantly directed toward increasing well-being, peace, and achievement. Here the mechanistic conception of life is one with the teachings of Buddha and the Sermon on the Mount.

While I am not a believer in the orthodox sense, I commend religion, first, because every individual should have some ideal—religious, artistic, scientific, or humanitarian—to give significance to his life. Second, because all the great religions contain wise prescriptions relating to the conduct of life, which hold good now as they did when they were promulgated.

There is no conflict between the ideal of religion and the ideal of science, but science is opposed to theological dogmas because science is founded on fact. To me, the universe is simply a great machine which never came into being and never will end. The human being is no exception to the natural order. Man, like the universe, is a machine. Nothing enters our minds or determines our actions which is not directly or indirectly a response to stimuli beating upon our sense organs from without. Owing to the similarity of our construction and the sameness of our environment, we respond in like manner to similar stimuli, and from the concordance of our reactions, understanding is barn. In the course of ages, mechanisms of infinite complexity are developed, but what we call "soul " or "spirit," is nothing more than the sum of the functionings of the body. When this functioning ceases, the "soul" or the "spirit" ceases likewise.

I expressed these ideas long before the behaviorists, led by Pavlov in Russia and by Watson in the United States, proclaimed their new psychology. This apparently mechanistic conception is not antagonistic to an ethical conception of life. The acceptance by mankind at large of these tenets will not destroy religious ideals. Today Buddhism and Christianity are the greatest religions both in number of disciples and in importance. I believe that the essence of both will he the religion of the human race in the twenty-first century.

The year 2100 will see eugenics universally established. In past ages, the law governing the survival of the fittest roughly weeded out the less desirable strains. Then man's new sense of pity began to interfere with the ruthless workings of nature. As a result, we continue to keep alive and to breed the unfit. The only method compatible with our notions of civilization and the race is to prevent the breeding of the unfit by sterilization and the deliberate guidance of the mating instinct, Several European countries and a number of states of the American Union sterilize the criminal and the insane. This is not sufficient. The trend of opinion among eugenists is that we must make marriage more difficult. Certainly no one who is not a desirable parent should be permitted to produce progeny. A century from now it will no more occur to a normal person to mate with a person eugenically unfit than to marry a habitual criminal.

Hygiene, physical culture will be recognized branches of education and government. The Secretary of Hygiene or Physical Culture will he far more important in the cabinet of the President of the United States who holds office in the year 2035 than the Secretary of War. The pollution of our beaches such as exists today around New York City will seem as unthinkable to our children and grandchildren as life without plumbing seems to us. Our water supply will he far more carefully supervised, and only a lunatic will drink unsterilized water.

More people die or grow sick from polluted water than from coffee, tea, tobacco, and other stimulants. I myself eschew all stimulants. I also practically abstain from meat. I am convinced that within a century coffee, tea, and tobacco will be no longer in vogue. Alcohol, however, will still be used. It is not a stimulant but a veritable elixir of life. The abolition of stimulants will not come about forcibly. It will simply be no longer fashionable to poison the system with harmful ingredients. Bernarr Macfadden has shown how it is possible to provide palatable food based upon natural products such as milk, honey, and wheat. I believe that the food which is served today in his penny restaurants will be the basis of epicurean meals in the smartest banquet halls of the twenty-first century.

There will be enough wheat and wheat products to feed the entire world, including the teeming millions of China and India, now chronically on the verge of starvation. The earth is bountiful, and where her bounty fails, nitrogen drawn from the air will refertilize her womb. I developed a process for this purpose in 1900. It was perfected fourteen years later under the stress of war by German chemists.

Long before the next century dawns, systematic reforestation and the scientific management of natural resources will have made an end of all devastating droughts, forest fires, and floods. The universal utilization of water power and its long-distance transmission will supply every household with cheap power and will dispense with the necessity of burning fuel. The struggle for existence being lessened, there should be development along ideal rather than material lines.

Today the most civilized countries of the world spend a maximum of their income on war and a minimum on education. The twenty-first century will reverse this order. It will be more glorious to fight against ignorance than to die on the field of battle. The discovery of a new scientific truth will be more important than the squabbles of diplomats. Even the newspapers of our own day are beginning to treat scientific discoveries and the creation of fresh philosophical concepts as news. The newspapers of the twenty-first century will give a mere "stick" in the back pages to accounts of crime or political controversies, but will headline on the front pages the proclamation of a new scientific hypothesis.

"It will be possible to destroy anything approaching within 200 miles. My invention will provide a wall of power," declares Tesla.

Progress along such lines will be impossible while nations persist in the savage practice of killing each other off. I inherited from my father, an erudite man who labored hard for peace, an ineradicable hatred of war. Like other inventors, I believed at one time that war could he stopped by making it more destructive. But I found that I was mistaken. I underestimated man's combative instinct, which it will take more than a century to breed out. We cannot abolish war by outlawing it. We cannot end it by disarming the strong. War can be stopped, not by making the strong weak but by making every nation, weak or strong, able to defend itself.

Hitherto all devices that could be used for defense could also be utilized to serve for aggression. This nullified the value of the improvement for purposes of peace. But I was fortunate enough to evolve a new idea and to perfect means which can be used chiefly for defense. If it is adopted, it will revolutionize the relations between nations. It will make any country, large or small, impregnable against armies, airplanes, and other means for attack. My invention requires a large plant, but once it is established it will he possible to destroy anything, men or machines, approaching within a radius of 200 miles. It will, so to speak, provide a wall of power offering an insuperable obstacle against any effective aggression.

If no country can be attacked successfully, there can be no purpose in war. My discovery ends the menace of airplanes or submarines, but it insures the supremacy of the battleship, because battleships may be provided with some of the required equipment. There might still be war at sea, but no warship could successfully attack the shore line, as the coast equipment will be superior to the armament of any battleship.

I want to state explicitly that this invention of mine does not contemplate the use of any so-called "death rays." Rays are not applicable because they cannot be produced in requisite quantities and diminish rapidly in intensity with distance. All the energy of New York City (approximately two million horsepower) transformed into rays and projected twenty miles, could not kill a human being, because, according to a well known law of physics, it would disperse to such an extent as to be ineffectual.

My apparatus projects particles which may be relatively large or of microscopic dimensions, enabling us to convey to a small area at a great distance trillions of times more energy than is possible with rays of any kind. Many thousands of horsepower can thus be transmitted by a stream thinner than a hair, so that nothing can resist. This wonderful feature will make it possible, among other things, to achieve undreamed-of results in television, for there will be almost no limit to the intensity of illumination, the size of the picture, or distance of projection.

I do not say that there may not be several destructive wars before the world accepts my gift. I may not live to see its acceptance. But I am convinced that a century from now every nation will render itself immune from attack by my device or by a device based upon a similar principle.

At present we suffer from the derangement of our civilization because we have not yet completely adjusted ourselves to the machine age. The solution of our problems does not lie in destroying but in mastering the machine.

Innumerable activities still performed by human hands today will be performed by automatons. At this very moment scientists working in the laboratories of American universities are attempting to create what has been described as a "thinking machine." I anticipated this development.

I actually constructed "robots." Today the robot is an accepted fact, but the principle has not been pushed far enough. In the twenty-first century the robot will take the place which slave labor occupied in ancient civilization. There is no reason at all why most of this should not come to pass in less than a century, treeing mankind to pursue its higher aspirations.

And unless mankind's attention is too violently diverted by external wars and internal revolutions, there is no reason why the electric millennium should not begin in a few decades.

February 1937

TESLA PREDICTS SHIPS POWERED BY SHORE BEAM

New York Herald Tribune, June 5, 1935. Scoffs at Normandy's "Speed". Sees Success for His Plan to Use Stratosphere Ray Would Light Sea at Night Says French Liner's System Copied His in U. S. Boats.

Dr. Nikola Tesla, scientist and seer whose discoveries in the fields of polyphase electrical current and wireless place him in the front rank of modern inventors, refused yesterday to be awed by the record speed achievement of the French liner Normandy in crossing the Atlantic in 4 days 11 hours 42 minutes and predicted that enormous ships would cross the ocean at far greater speeds by means of a high-tension current projected from power plants on shore to vessels at sea through the upper reaches of the atmosphere.

In his room at the Hotel New Yorker, dressed in a blue bathrobe, blue socks and red slippers, Dr. Tesla expounded the principles of his fabulous method of power transmission — a method which he has been developing at irregular intervals from as far back as 1897. The virtues of stratosphere transmission, he said, lay not only in its potential increase of a vessel's speed but also in its power to eliminate the dangers of nocturnal navigation.

In short, high-tension currents of electricity passing through the stratosphere would light the sky and to a degree turn night into day. With power plants stationed at intermediate posts such as upon the Azores and Bermuda, vessels could cross the Atlantic, propelled and safeguarded at the same time by electricity generated ashore. There would no longer be danger of boiler explosions nor hazards of collisions at sea. Even on moonless, cloudy nights, there still would gleam overhead the faint rays of surging electrical currents, so strong that pilots would be able to distinguish objects miles away.

Normandy uses U.S. cruiser system

Dr. Tesla, a tall, slender man with straight silvery hair, lean features and bright blue eyes that belie his seventy-eight years, prefaced his prophecies by pointing out that the Normandy — s system of power generation and application was not new — but one which had been adopted long ago in some of the United States cruisers. The principle is one of his own inventions.

"The Normandy," he said, "employs an 'electric drive' in which turbines drive generators and generators supply the current to independent motors. In this case the turbines are driven by steam, the generators are of the three-phase type and the motors are of the induction type.

"In many respects the machinery installed on the United States cruisers by former Secretary Josephus Daniels is more remarkable than that on the Normandy on account of the limitations of available space. Moreover, while the Normandy develops only 160,000 horsepower, the cruisers each develop 185,000 horsepower. These cruisers employ the most remarkable engine plants in the world, and I believe that this drive would not have been employed on the Normandy had it not been for the pioneering work done in the United States.

"In view of the adoption on such a large scale of these inventions of mine, it is interesting to recall that I was violently attacked only a few years ago by a professor of marine engineering at Columbia, who claimed the electrical drive was not feasible and that it was folly to undertake it.

"However splendid the machinery on the Normandy might be, the time is not distant when we will have much simpler and better means of propulsion."

Cites his force beam as one way

Here Dr. Tesla recalled the possibilities of his force beam of particles which he announced last year as a potential defensive weapon of great value. One of its aspects is a death ray capable of destroying airplanes and armies. Another is a means of power transmission which could be used to relay immense voltages of power over distances limited only by the curvature of the earth.

The difficulties inherent in using this method as a means of propulsion for oceangoing ships, however, were seen by Dr. Tesla to lie in the necessity of vast outlays of capital and concerted harmonious endeavor by the chief nations of the world. The latter, he said, would be impossible to achieve at the present time. A third difficulty would be the task of keeping a ship at sea constantly in touch with a threadlike beam of particles from ashore.

Dr. Tesla, therefore, suggested that his other scheme, of stratosphere transmission of electricity, would be a far more feasible means of marine propulsion. The principles of the two plans are entirely distinct. The force beam is a thin barrage of tiny particles discharged at tremendous velocities from a kind of electrical gun. The other invention, which he has not hitherto discussed publicly, is of transmitting high tension currents through the upper air, and receiving them by means of a vertical ionizing beam which would be a sort of invisible electrode. He discussed this yesterday:

Started new idea in 1897

"There is a method of conveying great power to ships at sea which would be able to propel them across oceans at high speed. This method I conceived between 1897 and 1899, and in Colorado Springs in 1899. I made experiments along this line on a large scale.

"The principle is this: A ray of great ionizing power is used to give to the atmosphere great powers of conduction. A high tension current of 10,000,000 to 12,000,000 volts is then passed along this ray to the upper strata of the air, which strata can be broken down very readily and will conduct electricity very well.

"A ship would have to have equipment for producing a similar ionizing ray. The current which has passed through the stratosphere will strike this ray, travel down it and pass into the engines which propel the ship.

Pet scheme to light ocean

"I will confess that I was disappointed when I first made tests along this line on a large scale. They did not yield practical results. At the time I used about 8,000,000 to 12,000,000 volts of electricity. As a source of ionizing rays I employed a powerful arc reflected up into the sky. At the time I was trying only to connect a high tension current and the upper strata of the air, because my pet scheme for years has been to light the ocean at night.

"However, since 1902 I have made many improvements in my method which I know now will assure success. A power plant upon the Azores, for instance, could send a current up into the stratosphere and illuminate the sky sufficiently for pilots to discern objects upon the ocean at a safe distance."

Dr. Tesla said that he was working constantly every day to perfect his force beam, his method of stratosphere transmission of power, and a number of other inventions the nature of which he was not ready to disclose. When it was called to his attention that he was working pretty hard for a man who would be seventy-nine years old next month, he replied:

"Why, I'm young I never think of my age. Really, you know, I'm just a youngster."

June 5, 1935

TESLA TRIES TO PREVENT WORLD WAR TWO

—*The unpublished Chapter 34 of Prodigal Genius, by John J. O'Neill*

When Tesla was talking as a scientist he was opposed to wars on moral, economic and all practical and theoretical grounds. But like most scientists, when he stopped thinking as a scientist, and let his emotions rule his thoughts, he found exceptions in which he felt some wars and situations were justifiable. As a scientist he was unwilling to have the discoveries of scientists applied to the purposes of war makers, but when in the emotional phase of his nature took the ruling position he was then willing to apply his genius to devising measures that would prevent wars by supplying protective devices.

This attitude is exemplified in the following statement which he prepared in the twenties but did not publish:

"At present many of the ablest minds are trying to devise expedients for preventing a repetition of the awful conflict which is only theoretically ended and the duration and main issues of which I correctly predicted in an article printed in the Sun of December 20, 1914. The League is not a remedy but, on the contrary, in the opinion of a number of competent men, may bring about results just the opposite. It is particularly regrettable that a punitive policy was adopted in framing the terms of peace because a few years hence it will be possible for nations to fight without armies, ships or guns, by weapons far more terrible, to the destructive action and range of which there is virtually no limit. Any city at any distance whatsoever from the enemy can be destroyed by him and no power on earth can stop him from doing so. If we want to avert an impending calamity and a state of things which may transform this globe into an inferno, we should push the development of flying machines and wireless transmission of energy without an instant's delay and with all the power and resources of the nation."

Tesla saw preventative possibilities in his new invention which embodied "death ray" characteristics and which was made several years after the foregoing statement was written. He saw it providing a curtain of protection which any country, no matter how small, could use as a protection against invasion. While he might offer it as a defensive weapon, however, there would be nothing to stop military men from using it as a weapon of offense.[1]

While I did not know the nature of Tesla's plan I was convinced that it did embody many discoveries that would be of commercial value, and these were the angles he should seek to develop. I felt that if he could be induced to develop some minor phase of his work that would have immediate commercial use he could derive an income from it which would enable him to proceed with his more elaborate plans. To this end I sought to gain some insight into his thoughts, that would enable me to get a practical plan into operation. This was no secret to Tesla and he successfully parried every thrust I made.

The clearest conception I got, and that was largely from scattered remarks, and by making deductions from them, concerned a possible manner in which one phase of his curtain of protection might operate. This was a "war" angle and as such it did not interest me, but since it involved "lightning balls," or "fireballs," I was very curious. Fireballs had always fascinated me, and I had read everything I could lay my hands on about them.

A fireball is a strange phenomenon associated with lightning. Some of the energy of the lightning stroke appears to become locked into a ball shaped structure which may be of any size from a couple of inches to a foot in diameter. It looks like a perfect sphere, brightly incandescent and floats like a bubble, being easily carried by air currents. They may last for a short time, from a fraction of a second to many seconds. In this interval, during which they stay fairly close to the ground, they may come close to many objects without damaging them or being damaged by them. Suddenly, for no known reason, the ball explodes doing as much damage as a bomb, if close to structures, and no damage if in the open.

The fireball looked to me like a gigantically enlarged model of the tiny electron, one of the building blocks of matter, which acts as if it were just a spherical area of space in which an amount of energy was crystallized to give it structure. I felt that if it were possible to discover how a large amount of energy was stored in this fairy bubble structure of a fireball a new insight might be gained into the structure of the electron and other fundamental particles of matter. Also this method of storing energy could be applied to a thousand useful purposes.

When I approached Tesla with pleas along this line to develop this possible phase of his discovery he would evade direct reply by indulging in a, not always, tolerant lecture on my gullibility in believing theories about the complex structure of the atom. While he had in earlier years discussed some of his experiences with fireballs in his laboratory at Colorado Springs and explained his theory of their formation, he would not in later years permit himself to be drawn into a discussion of them as a possible part of his system. This, of course, made me suspicious that the clue was "hot" but I could be completely wrong in my conclusions. Tesla was very quick in detecting my technique when I sought to narrow the field down by trying to get him to deny statements when he was adamant to direct questions.

Tesla became familiar with the destructive characteristics of fireballs in his experiments at Colorado Springs in 1899. He produced them quite by accident and saw them, more than once, explode and shatter his tall mast and also destroy apparatus within his laboratory. The destructive action accompanying the disintegration of a fireball, he declared, takes place with inconceivable violence.

He studied the process by which they were produced, not because he wanted to produce them but in order to eliminate the conditions in which they were created. It is not pleasant, he related, to have fireballs explode in your vicinity for they will destroy anything they come in contact with.

It will be necessary to reconstruct his statements from very fragmentary notes and a long distance memory.

Parasitic oscillations, or circuits, within the main circuit were a source of danger from this cause. Points of resistance in the main circuits could result in minor oscillating circuits between terminals or between two points of resistance and these minor circuits would have a very much higher period of oscillation than the main circuit and could be set into oscillation by the main current of lower frequency.

Even when the principle oscillating circuit was adjusted for the greatest efficiency of operation by the diminution of all sources of losses the fireballs continued to occur but these were due to stray high frequency charges from random earth currents.

From these experiences it became apparent that the fireballs resulted from the interaction of two frequencies, a stray higher frequency wave imposed on the lower frequency free oscillations of the main circuit.

As the free oscillation of the main circuit builds up from the zero point to the quarter wave length node it passes through various rates of change. In a current of shorter wavelength the rates of change will be steeper. When the two currents react on each other the resultant complex will contain a wave in which there is an extremely steep rate of change, and for the briefest instant currents may move at a tremendous rate, at the rate of millions of horsepower.

This condition acts as a trigger which may cause the total energy of the powerful longer wave to be discharged in an infinitesimally small interval of time and at a proportionally tremendously great rate of energy movement which cannot confine itself to the metal circuit and is released into surrounding space with inconceivable violence.

It is but a step, from learning how a high frequency current can explosively discharge a lower frequency current, to using the principle to design a system in which these explosions can be produced by intent. The following process appears a possible one but no evidence is available that it is the one Tesla evolved: An oscillator, such as he used to send power wirelessly around the earth at Colorado Springs, is set in operation at a frequency to which a given warship is resonant. The complex structure of a ship would provide a great number of spots in which electrical oscillations will be set up of a much higher frequency than those coursing through the ship as a whole. These parasite currents will react on the main current causing the production of fireballs which by their explosions will destroy the ship, even more effectively than the explosion of the magazine which would also take place. A second oscillator may be used to transmit the shorter wavelength current.

Somewhat later I learned the reason for Tesla's reticence to discuss the details. This came shortly after Stanley Baldwin replaced Neville Chamberlin as Prime Minister of Great Britain.

Tesla revealed that he had carried on negotiations with Prime Minister Chamberlin for the sale of his ray system to Great Britain for $30,000,000 on the basis of his presentation that the device would provide complete protection for the British Isles against any enemy approaching by sea or air, and would provide an offensive weapon to which there was no defense. He was convinced, he declared, of the sincerity of Mr. Chamberlin and his intent to adopt the device as it would have prevented the outbreak of the then threatening war, and would have made possible the continuation—under the duress which this weapon would have made possible—of the working agreement involving France, Germany and Britain to maintain the status quo in Europe. When Chamberlin failed, at the Munich conference, to retain this state of European equilibrium it was necessary to get rid of Chamberlin and install a new Prime Minister who could make the effort to shift one corner of the triangle from Germany to Russia. Baldwin found no virtue in Tesla's plan and peremptorily ended negotiations.

Tesla was greatly disappointed by the collapse of his negotiations with the British Government. With it there collapsed his hopes of providing a demonstration of his most recent, and, what he considered, his most important discoveries. He did not, however, dwell on the subject; beyond the single conversation he did not mention the matter again. He did not get another chance to finance the demonstration of these discoveries.

During the period in which the negotiations were being carried on, Tesla declared, efforts had been made to steal the invention. His room had been entered and his papers examined but the thieves, or spies, left empty handed. There was no danger, he said, that his invention could be stolen for he had at no time committed any part of it to paper. He could trust his memory to preserve every fine detail of his investigations. This was true, he said, of all of his later major discoveries.

The nature of his system makes little difference now; he has gone and has taken it with him. Perhaps, if there is any communication from beyond the veil that separates this life from whatever exists hereafter, Tesla may look down upon earth's struggling mortals and find some way of dropping a hint concerning what he accomplished; but, if the situation is such that this cannot take place, then we must await until the human race produces another Tesla.

(1) The "death ray," also known as "teleforce," was Tesla's particle-beam defense weapon. See Nikola Tesla's Teleforce & Telegeodynamics Proposals for details.

MECHANICAL THERAPY

—Undated

In order to convey a clear idea of the significance and revolutionary character of this discovery it is indispensable to make a brief statement regarding Electrical Therapy.

Fifty years ago, while investigating high frequency currents developed by me at that time, I observed that they produced certain physiological effects offering new and great possibilities in medical treatment. My first announcement spread like fire and experiments were undertaken by a host of experts here and in other countries. When a famous French physician, Dr. D'Arsonval, declared that he had made the same discovery, a heated controversy relative to priority was started. The French, eager to honor their countryman, made him a member of the Academy, ignoring entirely my earlier publication. Resolved to take steps for vindicating my claim, I went to Paris, where I met Dr. D'Arsonval. His personal charm disarmed me completely and I abandoned my intention, content to rest on the record. It shows that my disclosure antedated his and also that he used my apparatus in his demonstrations. The final judgement is left to posterity.

Since the beginning the growth of the new art and industry has been phenomenal, some manufacturers turning out daily hundreds of sets. Many millions are now in use throughout the world. The currents furnished by them have proved an ideal tonic for the human nerve system. They promote heart action and digestion, induce healthful sleep, rid the skin of destructive exudations and cure colds and fever by the warmth they create. They vivify atrophied or paralyzed parts of the body, allay all kinds of suffering and save annually thousands of lives. Leaders in the profession have assured me that I have done more for humanity by this medical treatment than by all my other discoveries and inventions. Be that as it may, I feel certain that the Mechanical Therapy, which I am about to give to the world, will be of incomparably greater benefit. Its discovery was made accidentally under the following circumstances.

I had installed at the laboratory, 25 South Fifth Avenue, one of my mechanical oscillators with the object of using it in the exact determination of various physical constants. The machine was bolted in vertical position to a platform supported on elastic cushions and, when operated by compressed air, performed minute oscillations absolutely isochronous, that is to say, consuming rigorously equal intervals of time. So perfect was its functioning in this respect that clocks driven by it indicated the hour with astronomical precision. One day, as I was making some observations, I stepped on the platform and the vibrations imparted to it by the machine were transmitted to my body. The sensation experienced was as strange as agreeable, and I asked my assistants to try. They did so and were mystified and pleased like myself. But a few minutes later some of us, who had stayed longer on the platform, felt an unspeakable and pressing necessity which had to be promptly satisfied, and then the stupendous truth dawned upon me. Evidently, these isochronous rapid oscillations stimulated powerfully the peristaltic movements which propel the food-stuffs through the alimentary channels. A means was thus provided whereby their contents can be perfectly regulated and controlled at will, and without the use of drugs, specific remedies or internal applications whatever.

When I began to practice with my assistants Mechanical Therapy we used to finish our meals quickly and rush back to the laboratory. We suffered from dyspepsia and various stomach troubles, biliousness, constipation, flatulence and other disturbances, all natural results of such irregular habit. But only after a week of application, during which I improved the technique and my assistants learned how to take the treatment to their best advantage, all these forms of sickness disappeared as by enchantment and for nearly four years, while the machine was in use, we were all in excellent health. I cured a number of people, among them my great friend Mark Twain whose books saved my life. He came to the laboratory in the worst shape suffering from a variety of distressing and dangerous ailments but in less than two months he regained his old vigor and ability of enjoying life to the fullest extent. Shortly after, a great calamity befell me: my laboratory was destroyed by fire. Nothing was insured and the loss of priceless apparatus and records gave me a terrific shock from which I did not recover for several years. The enforced discontinuance of Mechanical Therapy also caused me deep regret. I had evolved a wonderful remedy for ills of inestimable value to mankind and invented apparatus offering unbounded commercial possibilities but when I came to consider practical introduction I realized that it was entirely unsuitable. It was big, heavy and noisy, called for a continuous supply of oil, part of which was discharged into the room as fine spray; it consumed considerable power and required a number of objectionable accessories. During the succeeding years I made great improvements and finally evolved a design which leaves nothing to be desired. The machine will be very small and light, operate noiselessly without any lubricant, consume a trifling amount of energy and will be, to my knowledge, the most beautiful device ever put on the market. The intention is to exhibit it in action at the occasion of my annual reception in honor of the Press which has been, unfortunately, delayed this year, and I anticipate that it will elicit great interest and receive great publicity. Unless I am grossly mistaken it will be introduced very extensively and, eventually, there will be one in every household.

The practical application of Mechanical Therapy through my oscillators will profoundly affect human life. By insuring perfect regularity of evacuations the body will function better in every respect and life will become ever so much safer and more enjoyable. One of the most important results will be the great reduction — amounting possibly to seventy-five percent — in the number of heart failures, which are mostly caused by some acute upset of the digestive process and normal operation of the stomach. Another vital improvement will be derived from the quickened removal of toxic excretions of organs affected by disease. It is reasonable to expect that through this and other healthful actions ulcers and similar internal lesions or abscesses will be cured and relief might be obtained even in cases of a cancer or other malignant growth. Skilled physicians and surgeons will be able to perform veritable miracles with such oscillations. They stimulate strongly the liver, spleen, kidneys, bladder and other organs and by these desirable actions they must contribute not a little to well being. Persons suffering from anemia of any form will be especially helped by the treatment. But the greatest benefit will be derived from it by women who will be able to reduce without the usual tantalizing abstinence, privation, sacrifice of time and money and torture they have to endure. They will improve much in appearance, acquire clear eyes and complexions and it may be safely predicted that long continued treatment will bring forth feminine beauty never seen before. It is not to be forgotten that the elimination of countless drugs, patent medicines and specific remedies of all kinds taken internally, by which millions of people doom themselves to an early grave, will be of untold good to humanity.

www.ingramcontent.com/pod-product-compliance
Lightning Source LLC
Chambersburg PA
CBHW020937230426
43666CB00005B/64